The Rare Earths in Modern Science and Technology

Volume 3

The Rare Earths in Modern Science and Technology

Volume 3

Edited by

Gregory J. McCarthy
North Dakota State University, Fargo, North Dakota

Herbert B. Silber
University of Texas at San Antonio, San Antonio, Texas

and

James J. Rhyne
National Bureau of Standards, Washington, D.C.

Associate Editor
Faye M. Kalina
North Dakota State University, Fargo, North Dakota

Editorial Assistants
Linda R. Haugrud and Joyce L. Mortensen
North Dakota State University, Fargo, North Dakota

PLENUM PRESS · NEW YORK AND LONDON

Library of Congress Cataloging in Publication Data

Rare Earth Research Conference.
  The rare earths in modern science and technology.

  Vol. 3–   edited by G. J. McCarthy, H. B. Silber, and J. J. Rhyne.
  Includes bibliographical references and indexes.
  1. Earths, Rare—Congresses. I. McCarthy, Gregory J. II. Rhyne, J. J. III. Silber, Herbert B.
QD172.R2R27 1978   546′.4   78-5365
                                                                           AACR1
ISBN-13: 978-1-4613-3408-8   e-ISBN-13: 978-1-4613-3406-4

DOI: 10.1007/978-1-4613-3406-4

Proceedings of the 15th Rare Earths Research Conference, held
June 15-18, 1981, at the University of Missouri at Rolla

Softcover reprint of the hardcover 1st edition 1982

A Division of Plenum Publishing Corporation
233 Spring Street, New York, N.Y. 10013

FOREWORD

The Fifteenth Rare Earth Research Conference was held
June 15-18, 1981 on the Rolla campus of the University of Missouri.
The conference was hosted by the Graduate Center for Materials
Research, the College of Arts and Science, and the School of Mines
and Metallurgy.

It was expected that the conference would provide a forum for
critical examination and review of the current and important trends
in rare earth science and technology. To this end, over 170 papers
were presented in both oral and poster sessions by researchers
representing some nineteen countries. The program committee was
particularly gratified to see the diversity of effort being devoted
to rare earth research by different disciplines all over the world.
The collection of refereed papers in this volume attests to the
fact that the objectives of the program committee were indeed
realized.

A high point of the meeting was the presentation of the
Frank H. Spedding Award to a most distinguished colleague, Professor
Georg Busch, Eidgenossische Technische Hochschule, Zurich. Prof-
essor W. Edward Wallace, University of Pittsburgh, recipient of
the first Frank H. Spedding Award made the presentation to
Professor Busch who then gave the Plenary Address.

Oral sessions began with keynote addresses delivered by:
B.D. Sykes, University of Alberta; J.E. Greedan, McMaster Univer-
sity; M. Tecotzky, U.S.R. Chemicals; J.R. Jackman, Reactive Metals
and Alloys; J. Owens, Harshaw Chemical; W.T. Carnall, Argonne
National Laboratory; M.B. Maples, University of California-San
Diego; R. Lemaire, Neel Laboratory, Grenoble, France; W.J. Evans,
University of Chicago; E. Kaldis, ETH, Zurich, Switzerland;
R.G. Barnes, Iowa State University; and P. Muntz, Universitat
Konstanz/FRG.

The success of any scientific meeting depends largely upon
the voluntary efforts of many dedicated people. I first wish to

acknowledge with much appreciation the advice and assistance given me upon assuming my duties as president of the board and conference chairman by the past conference chairman and secretary, Professor John Gruber and Faye Kalina.

The success of the program is testimony to the talents of Professor Gregory Choppin and the excellent job done by his program committee and session chairpersons.

Program Committee

G. Barlow	R. Haire	H. Silber
E. Greedan	J. Kaczmarec	S. Taher
K. Gschneidner	H. Marks	A. Tauber

Session Chairpersons

J.C. Achard	F.L. Carter
J.L. Atwood	W.T. Carnall
E. Banks	J. Chrysochoos
B.J. Beaudry	J. Deportes
B.A. Bilal	T. Donahue
J-C. Bunzli	L. Eyring
O.N. Carlson	J. Gruber
P. Caro	K. Gschneidner
A. Percheron-Guegan	O. Serra
R.G. Haire	A. Dean Sherry
J.M. Haschke	S.M.A. Taher
J. Kaczmarec	W.E. Wallace
H. Kirchmayr	G. Gorller-Walrand
S.K. Malik	W.C. Weimer
T.J. Marks	W. Yelon
L. Niinisto	

I appreciate the many hours of work contributed by those who served on the following committees:

Local Committee	Selection Committee for the Frank H. Spedding Award Recipient
Adrian Daane, Chairperson	
Harlan Anderson	W.E. Wallace, Chairperson
Gordon Lewis	Joseph Cannon
Thomas O'Keefe	Paul Caro
D. Vincent Roach	J.B. Gruber
Donald Sparlin	Fred Rothwarf
Manfred Wuttig	
William Yelon	

Arlene James, Chairperson
Jean Daane
Eunice French

We are all deeply indebted to the co-editors of the Conference Proceedings who gave many hours to the organization and refereeing of the papers in bringing you Volume 3 of "The Rare Earths in Modern Science and Technology".

We are also very appreciative of the monetary support provided to the Conference by the following donors:

National Science Foundation, Washington, D.C., U.S.A.
Petroleum Research Fund (American Chemical Society),
     Washington, D.C., U.S.A.
Modern Metals, Union/Molycorp, Los Angeles, California, U.S.A.
Raytheon Corp., Lexington, Massachusetts, U.S.A.
Reactive Metals and Alloys Corp., West Pittsburgh,
     Pennsylvania, U.S.A.
Rhone-Poulenc Industries (Chimie Fine Division), Paris,
     Cedex 08, France
Ronson Metals Corp., Newark, New Jersey, U.S.A.
University of Missouri-Rolla, Rolla, Missouri, U.S.A.
North Dakota State University, Fargo, North Dakota, U.S.A.

I must acknowledge, especially, the invaluable help and counsel of Norma Fleming, Conference Coordinator, and her staff, and of Professor Adrian Daane, Chairman of the Local Committee.

I thank very much our secretaries: Faye Kalina for assisting the co-editors with the manuscripts and proceedings, Janet Thompson and Nina Haas, Materials Research, UMR, for assisting with all phases of the Conference activities; Jane Bunting and Audrey Thompson, Extension Office, UMR, for handling mailings and registration; and Gislaine Meneroud, Neel Laboratory, Grenoble, for assisting with registration.

Lastly, I wish to thank all the conference participants for their kindness and congeniality. I look forward to seeing all of you again at the Sixteenth Rare Earth Research Conference at Florida State University in April 1983.

William J. James
General Conference Chairman
University of Missouri-Rolla
November 1981

PREFACE

A coherent picture of research progress and new developments involving the rare earths can be difficult to develop due to the wide dispersal of relevant papers throughout the physics, chemistry and materials literature. We have once again taken advantage of the international gathering of scientists for a Rare Earth Research Conference to present under one cover a comprehensive update of the rare earths in modern science and technology. Authors presenting papers in Rolla were invited to submit papers or notes for this volume. All submissions were refereed. We have included three types of contributions: longer invited review papers, shorter research reports and one or two page notes. The first two are meant to have this volume as their sole publication outlet while the notes are typically from authors who have submitted, or are planning, a full publication elsewhere.

Among the 120 contributions in this volume the reader will find many papers in two of the areas where rare earths have been utilized most heavily over the last two decades, luminescent and magnetic materials. Professor Busch's Spedding Award Address describes the pioneering work of his group at the ETH, Zurich, on such materials and provides us with historical insight into research on the rare earths when their complex electronic structure, spectra and interactions in solids were just beginning to be understood. The role of rare earths in steelmaking is also reviewed. This volume also has strong components in the physical and structural chemistry of rare earth compounds and in the innovative applications of rare earths in bioinorganic, organometallic and coordination chemistry. We also welcome our first contributions from scientists in the People's Republic of China.

H.B. Silber, J.J. Rhyne and I would especially like to acknowledge the many referees whose efforts contributed so much to the quality of this volume.

Gregory J. McCarthy
Fargo, North Dakota
October, 1981

Presentation of the Frank H. Spedding Award
for Outstanding Contributions to the Science and Technology
of the Rare Earths to
Professor Georg Busch

GEORG BUSCH

Our Spedding Award recipient has had a scientific career of almost incredible depth and range. Prof. Busch's early interests lay with the phenomenon of ferroelectricity. In that, he had interests in common with the late Prof. Berndt Matthias, whose origin was also in the Eidgenössische Technische Hochschule Zurich.

Prof. Busch became interested in the 1940's in semiconductor physics and in the early 1950's he began a search for a ferromagnetic semiconductor. In time he realized that his great quest led

inexorably to the rare earths. (My feeling is that when you need
a certain material property you can nearly always find it among
the rare earth systems.) In his studies of rare earth-containing
III-V semiconductors, such as GdN and CeP, and II-VI semiconductors,
such as EuS and TmS, he found his ferromagnetic superconductor.

In his work on the rare earth chalcogenides and pnictides he
wrote a new chapter in the solid state sciences. Recognizing the
importance of chemical control, Prof. Busch established a Solid
State Chemistry Group within his Solid State Physics Institute and
they produced and studied in great depth single crystals of the
rare earth pnictides and chalcogenides which had been characterized
in the most exhaustive detail.

His work of the 60's and 70's involved synthesis and studies
of the structural, thermodynamic, magnetic and optical properties
of this large class of inorganic compounds. Prof. Busch's work
involved classical studies such as determination of specific heats
and bulk magnetization but also involved more recently introduced
techniques such as ESR, XPS and spin-polarized photoemission. The
latter is, I believe, a first in Prof. Busch's laboratory. If
you wish to know all about his rare earth chalcogenide and rare
earth pnictide work, then you have the job of reading about 110
papers which he has published on these materials.

Recently his attention has turned to intermetallic hydrides --
hydrogen storage systems -- structures of complex hydrides and
surface features. In the early 1960's our group in Pittsburgh
began work on rare earth nitrides. We soon encountered the
stunningly brilliant work of Busch and Vogt on rare earth nitrides.
And we decided upon other things to do. In our work on hydrides
we recognize Busch's group at Zurich as a force to be reckoned
with. Or to put it in less pejorative language, the Zurich group
under Busch is a world center of excellence in hydride research.
And from this excellence I am convinced all the world will profit.

It is a very great pleasure for me to present Prof. G. Busch,
the 1982 Spedding Award recipient.

                                    W. E. Wallace
                                    University of Pittsburgh
                                    June, 1981

CONTENTS

SPECTROSCOPY

PHASE EQUILIBRIA AND THERMODYNAMICS

STRUCTURAL AND SOLID STATE CHEMISTRY

CONTENTS

MAGNETIC PROPERTIES

CONTENTS

HYDRIDES

CONTENTS

NEW APPLICATIONS

RARE EARTH REMINISCENCES

G. Busch

Eidgenössische Technische Hochschule Zürich, Laboratorium für Festkörperphysik, CH 8093, Zürich, Switzerland

Mr. Chairman, Ladies and Gentlemen, Dear Professor Wallace:

To be selected as the 1981 recipient of the Frank H. Spedding Award came as a real surprise to me. Needless to say that I feel greatly honored myself by your appreciation, but also my former colleagues and collaborators at the Solid State Physics Laboratory at ETH were delighted to hear the news. It gives me great pleasure to accept the award and I would like to thank the members of the selection committee most sincerely and especially Professor Wallace for his extremely kind laudation. What I am going to tell you will be a kind of a potpourri of old facts, personal reminiscences and some new results.

Let me start with the Austrian chemist Auer-von Welsbach (1). He not only succeeded in separating praseodymium and neodymium for the first time nearly a hundred years ago but he also revolutionized illumination at the end of last century by his invention of the so-called Auer mantle. It consists of a cotton tissue soaked in a solution of thorium - and cerium-nitrate. When heated by a gasflame it emits a very bright greenish light. Its spectrum is shown in Fig. 1. It is not the spectrum of a black body. As can be seen there is practically no radiation in the near infrared. We still had this kind of illumination in our home more than 60 years ago, and here begin my personal reminiscences.

Paul Scherrer was a young professor of physics at ETH in the early twenties. He was famous for his brilliant demonstration lectures he gave every Thursday afternoon and I went there as a boy. In fall 1925 he lectured on the structure of the atom--without Schrödinger's equation of course! But he showed and explained beautiful

1

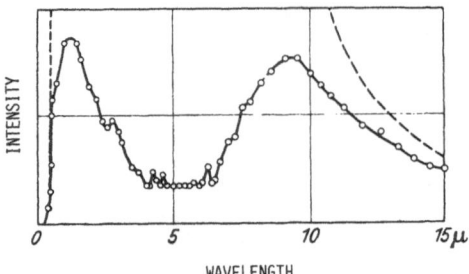

Figure 1. Emission spectrum of the Auer-mantle at
           1800 K compared to black body radiation
           at the same temperature (broken line).

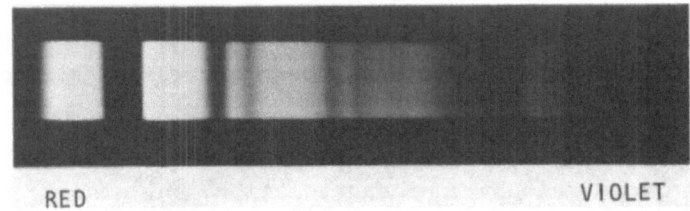

Figure 2. Absorption spectrum of didymium-glass in
           the visible range at room temperature.

physics experiments, and among these was the absorption spectrum of a
cube of didymium-glass in the visible range. The wonderful spectrum
shown in Fig. 2 deeply impressed me and I should never forget it. It
remained in my memory as a kind of a color symphony. I also studied a
book on inorganic chemistry published in 1923 by Fritz Ephraim (2).
I think that it was the most systématic and illuminating book on inorganic chemistry I ever came across and I was fascinated by the short
chapter on rare earths. I learned that with a simple spectroscope one
is able to detect the absorption lines of rare earths, in minerals for
instance, even in reflected light.

     Years later I learned that the characteristic absorption bands
become extremely sharp lines at sufficiently low temperatures. This
was observed for the first time by Jean Becquerel (3) in 1907. He
investigated minerals like xenotime, parisite, bastnaesite and others
at low temperatures at the Kamerlingh Onnes Laboratory in Leiden. He
also observed the Zeeman splitting of several lines in an external
magnetic field. It was not trivial at all that he only found Zeeman
doublets and triplets, i.e. a normal splitting. In a fundamental
paper published in 1929 (4) he also gave an explanation of his observations which was essentially correct: due to the electric crystal

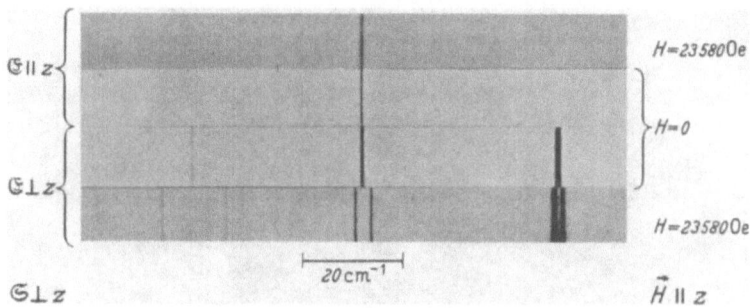

Figure 3.   ZEEMAN-splitting in $Er(C_2H_5SO_4)_3 \cdot 9H_2O$ (7).

field the degeneracy of the complex spectral terms of the rare earth
ions is lifted by the Stark-effect. This was before <u>Bethe</u> (5) and
<u>Kramers</u> (6) had developed their group theoretical representation of
crystal field symmetry! Early results by <u>Hellwege</u> (7) et al. are
shown in Fig. 3. Similar observations were made by <u>Freed</u> and <u>Spedding</u>
(8) in 1931.

I knew about the sharp lines of rare earth ions in crystals as I
finished my thesis on KH_2PO_4 and isomorphous substances in 1938. Ferroelectricity was still a curiosity and not at all understood. I
believed in some kind of ordering of hydrogen-bridges, but nobody knew
of anything better. To get some more information I had the idea to
dope my crystals with rare earths as probes for the internal electric
field. Below the Curie temperature I expected a spontaneous Stark-
splitting as a consequence of the spontaneous electric polarization.
Because of unfavorable working conditions in Zürich and the lack of
money to buy rare earths I had to abandon the idea and to my knowledge
the experiment has never been done or the results were not conclusive.
(It might be interesting to look at the new perovskite-type materials
like $BaPrO_3$ and others, which will be presented by <u>Morss and Mensi</u> (9)
at this conference.) So I left the rare earth field--forever I
thought. This was just before World War II.

And then Frank <u>Spedding</u> and his colleagues, by their unique and
outstanding achievements in the field, made rare earth metals and compounds available to chemists and physicists in the whole world. A
renaissance of spectroscopy began, mostly due to the general availability of helium temperatures. Fundamental work was done by <u>Hellwege</u>
(10), <u>Dieke</u> (11), and <u>Gruber</u> and others which allowed the experimental
proof of <u>Bethe</u> and <u>Kramers</u> theories. A wonderful experiment--in my
opinion--was done in 1966 by <u>Randazzo</u> (12). In 1962 <u>Wolf</u> (13) et al.
found that $GdCl_3$ became ferromagnetic below 2.2 K. Due to the Weiss
internal field a spontaneous Zeeman splitting of some of the sharp
absorption lines was to be expected. That is exactly what Randazzo
observed in $GdCl_3$ crystals doped with Er^{3+}. His results are shown in

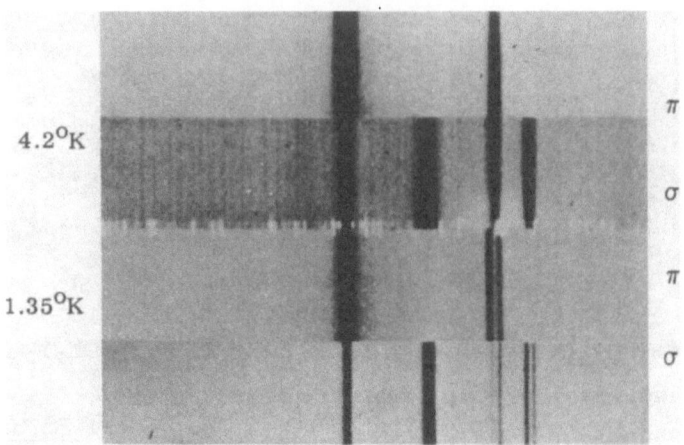

Figure 4.   Spontaneous ZEEMAN-splitting of Er^{3+} lines in
            $GdCl_3$ above and below the Curie temperature (12).

Fig. 4. Today exchange-splitting is a more fashionable way to describe the same phenomenon. Exchange splitting was described by
Sievers and Tinkham (14) on Yb-Fe-Garnet in 1961 and by Hellwege (15)
et al. on Dy-Al-Garnet in 1960 and they were able to estimate the
strength of the internal Weiss field. Also the metamagnetic transition of $DyAlO_3$ manifests itself in a splitting and a shift of selected
Dy-lines as was observed by Schuchert (16) in 1968. In summary, the
line spectra of rare earth solids have been extensively studied and
are quite well understood today.

    It took me nearly twenty years to come back to rare earths. In
1952, I attended the spring meeting of the American Physical Society
in Columbus, Ohio. On this occasion a dramatic disagreement took
place between John Slater and Clarence Zener about the role of conduction electrons in ferromagnetic exchange interactions. Apparently
the questions could not be answered by theoretical arguments. Why not
try an experiment instead? So the idea occurred to me to search for a
ferromagnetic semiconductor giving the possibility to change the concentration of conduction electrons by doping or by changing the temperature. I sketched a hypothetical curve of the spontaneous magnetization as a function of temperature for a semiconductor with a specific doping. The original of my product of imagination is shown in
Fig. 5. If conduction electrons are necessary to induce ferromagnetic
coupling of adjacent magnetic moments, there ought to be a critical
concentration where the ferromagnetic (not ferri-, nor antiferromagnetic) interaction is switched on. Therefore, besides the normal
Curie point, T_c, one should find another lower lying critical temperature, T_x. Below T_x, antiferromagnetic order or no order at all might
exist. To my knowledge no such thing has been found yet, and quite

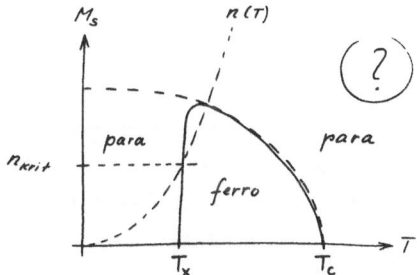

Ferromagnetischer Halbleiter

Figure 5. Hypothetical temperature variation of the
spontaneous magnetization of a ferromagnetic
semiconductor.

recently--not earlier fortunately--I heard about some theoretical
arguments, that this cannot be.

In 1953, my former student Fritz Hulliger (17) began his theses
and his search for a ferromagnetic semiconductor with high hopes. He
prepared several hundred compounds containing practically all the 3d-
transition elements. He found a lot of new metals, semimetals and
semiconductors with interesting magnetic properties, but not what I
hoped him to find. We felt that the behavior of the mostly non-local-
ized 3d-electrons and their contributions to magnetism, electrical
conductivity and chemical bonding was too complex to be fully understood. We hoped that in rare earth compounds, due to the localization
of the 4f-electrons the situation might be easier to overlook. This
was sheer speculation, of course, and experts tried to convince me
that non-metallic ferromagnets would never exist in nature at all.
All the same we started to work on rare earth compounds and we began
with gadolinium--for financial reasons. This was a mistake. In 1960
came the blow!

Tsubokawa (18) found $CrBr_3$ to become ferromagnetic below 4.3 K
in 1960. In 1961, Matthias (19) et al. discovered EuO with a Curie
temperature of about 69 K and, as already mentioned, the third ferromagnetic insulator became known in 1962. No doubt, EuO and also EuS
go ferromagnetic at low temperatures and are poor conductors at room
temperature and above. But would they stay insulators also below the
Curie temperature? Could not, by some unknown magnetic interaction,
the band structure be changed and an overlap of consecutive energy
bands occur? EuS and EuSe are transparent in the visible and show
beautiful red colors by absorption. Why not measure the optical absorption down to low temperatures? I told Peter Wachter, then a post-
doctorate at our laboratory, to do the experiment. What he found is
shown in Fig. 6. The energy gap does persist indeed when the temperature is lowered below the Curie temperature but there is a pronounced

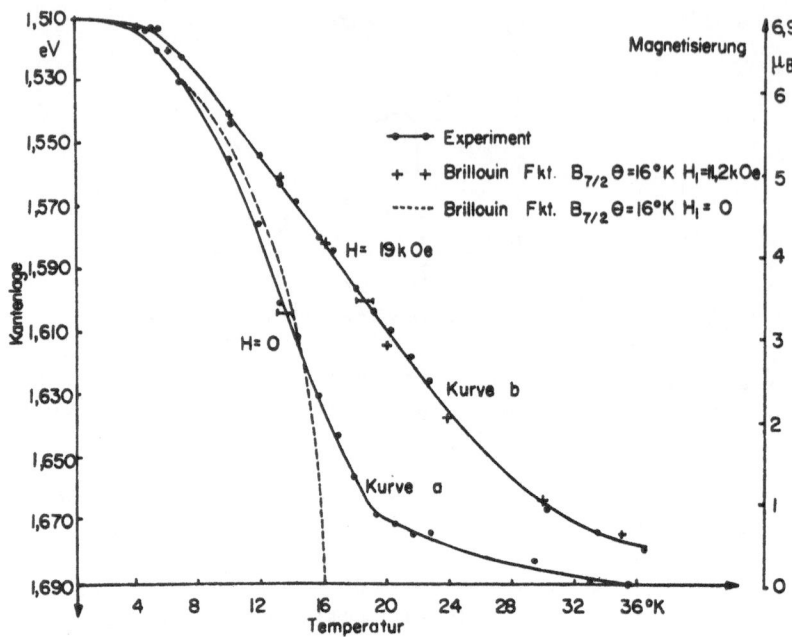

Figure 6.   Energy gap of EuS as a function of temperature.

red-shift (20) of the absorption edge which seems to be typical for
all kinds of ferromagnetic insulators or semiconductors. Apparently
the red-shift is strongly correlated to the variation of the spontaneous and the induced magnetization of the specimen and is due to a
splitting of the conduction band. Various mechanisms to explain this
new magneto-optic effect have been proposed in the past. Quite recently Nolting and Oleś (21) have published a more elaborate theory,
showing that the splitting of the conduction band is much more complex
than so far assumed, and sets in already above the Curie temperature.
The experimental facts are very well explained indeed.

     When the energy gap of the europium chalcogenides changes so
drastically with magnetization, what about the work-function of these
materials? To measure it by photo-electric methods seemed to be the
most appropriate way to answer this question. With this in mind an
extended research project on photoemission was started at our laboratory. One of the most recent results obtained by Felix Meier (22) on
Gd-doped EuO is shown in Fig. 7. As can be seen also the photothreshold is reduced substantially when the substance becomes ferromagnetic. A schematic interpretation of the phenomenon is given in
Fig. 8. The diagram on the left side shows the top most energy levels
of EuS above the Curie temperature. There is no magnetic order either
of the 4f or of the conduction electrons. Due to the strong doping,
the Fermi-level lies above the bottom of the conduction band

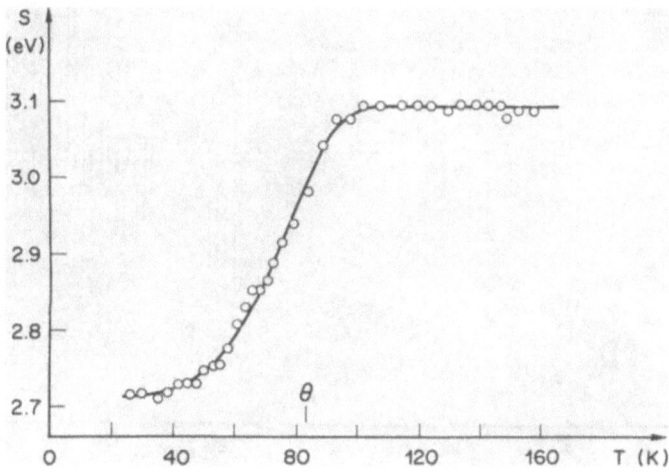

Figure 7.   Photothreshold S of EuO + 4.3% Gd as a function of temperature (22).

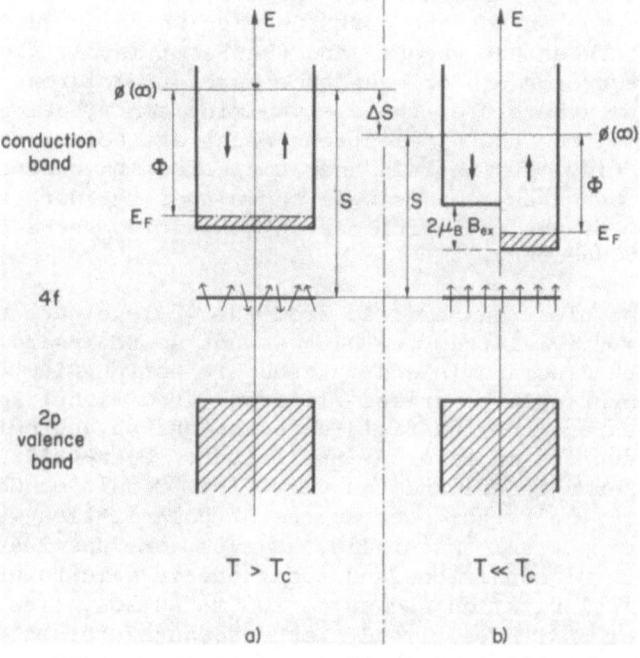

Figure 8.   Schematic energy diagrams of EuO above and below Curie temperature. Red shift and change of pho threshold.

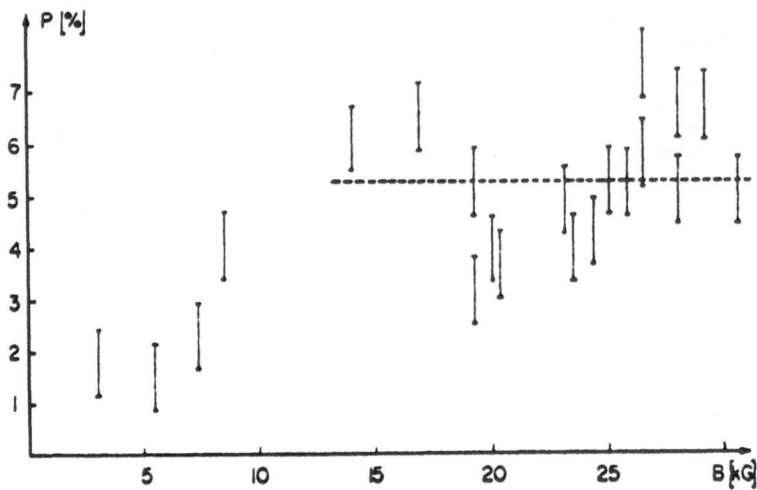

Figure 9.   Polarization of photoelectrons from
            ferromagnetic gadolinium.

corresponding to a strong degree of degeneracy. The right side of
Fig. 8 shows the situation below the Curie point. What one measures
by photoemission is not the work function which by definition is the
difference between the vacuum- and the Fermi level. In spite of the
doping there are orders of magnitude more 4f electrons than conduction
electrons and, therefore, the overwhelming contribution to the photo-
current stems from the 4f electrons which are polarized below T_c. At
the present time, I am afraid that a clearcut theoretical interpretation cannot be given. It is well known that the work function of a
solid is one of the most problematic quantities to be calculated with
satisfactory accuracy.

     That the electrons emitted from the 4f level are spin-polarized
looks nearly trivial today. This was not so ten years ago! In 1969
Hans Cristoph Siegmann joined my group in Zürich with the idea to look
for emission of spin-polarized electrons from magnetized solids. We
considered experiments on field-emission and on photoemission and
luckily we decided in favor of the latter. Spin-polarized photoelectrons were found indeed and the very first result on Gd is shown in
Fig. 9 (23). The rather poor degree of polarization was somewhat dis-
appointing of course. Meanwhile, however, one has learned to improve
the experimental conditions and today one is able to produce beams of
nearly 100% polarization and up to 100 mA at 50%. The high energy
physicists at SLAC have already taken advantage of this fact. Apart
from that, spin-polarized photoelectrons prove to be a powerful tool
to study the surface of solids and various surface reactions.

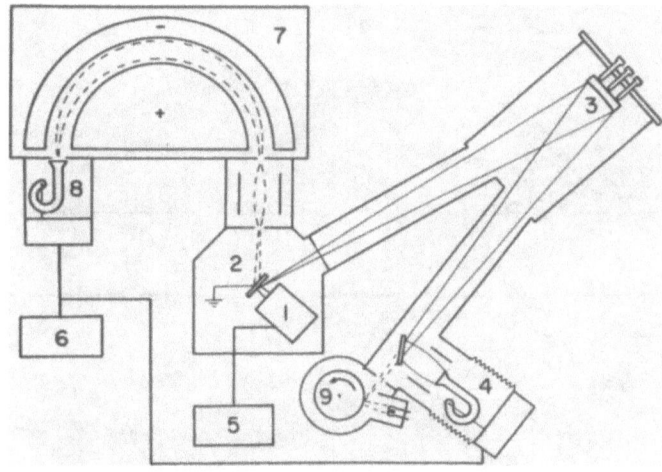

Figure 10.   Combined XPS-BIS apparatus. 1-electron gun;
2-sample; 3-x-ray monochromator; 4-x-ray photon
detector; 5-electron gun power supply; 6-counting
electronics and data outputs; 7-photoelectron
energy analyzer; 8-photoelectron detector; and
9-x-ray tube.

Some fifteen years ago we started to study photoemission in the
vacuum ultraviolet and the x-ray region aiming at information on the
density of states of solids and possibly of liquids. The method
called ESCA was invented by Siegbahn (24) at the University of Uppsala
in the early 1950's and has subsequently led to many important applications in solid state physics. The spectrometer developed by Yves
Baer is shown schematically in Fig. 10. The specimen is irradiated by
a powerful beam of monochromatized x-rays and the energy distribution
of the emitted electrons is measured by means of an electrostatic
analyzer. The whole system can be pumped down to better than 10^{-11}
Torr and the resolving power is approximately 0.3 eV. Recently, a
system to measure Bremsstrahl-Isochromate-Spectra (BIS) was added to
the apparatus which now permits investigations not only of occupied
but also unoccupied energy levels of a solid. As an example results
on La, Eu and Gd are shown in Fig. 11 (25). The two nearly identical
peaks of Gd below and above the Fermi level correspond to the seven
occupied and the seven unoccupied 4f states consistent with our understanding of energy levels of rare earths.

In my opinion, the magnetic, magneto-optical and optical phenomena of rare earth compounds have been extensively studied in the last
twenty years and are quite well understood. This is not the case with
respect to transport properties such as electrical and thermal conductivity, thermoelectric power and galvanomagnetic effects. I am not

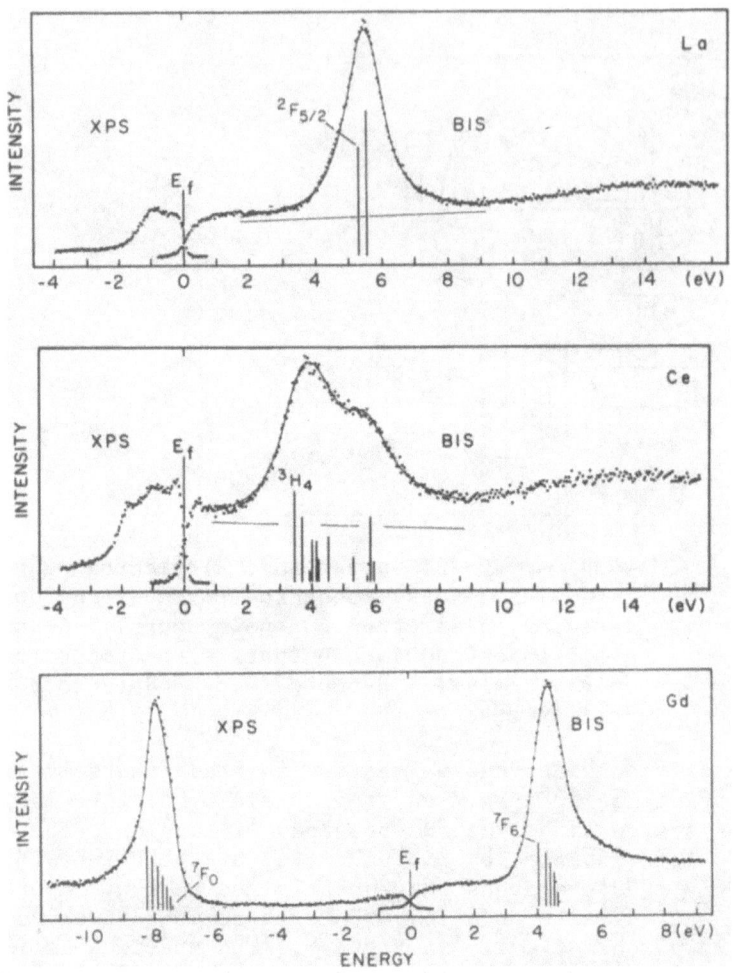

Figure 11. XPS and BIS spectra of La, Ce and Gd (25).

very familiar with metals and semimetals and I think it would be use-
less to talk about this field in the presence of experts as Professors
Wallace and Gschneidner. What I have in mind are the simple binary
compounds with sodium chloride type structures, such as the chalcogenides and the pnictides, GdP or GdS. The phosphide looks like silicon
and the sulfide shows a beautiful golden color. Are these substances
metals, semimetals or semiconductors and what about the oxides, sulfides, selenides and tellurides and the nitrides, phosphides, arsenides and antimonides of all the rest of the rare earth family? Some
of them show metallic or semimetallic properties. However, although
they are metallic, a certain polarity due to the partly ionic character of the chemical bonding should exist. Therefore, the term "polar

metal" seems appropriate to me. I have no idea whether this makes
sense or whether the polarity is completely shielded by conduction
electrons. It would be interesting to know more about the charge-
density distribution within their crystal lattices. The rare earth
compounds also have opened up the vast and very fashionable field of
valence instabilities.

A lot of careful work has still to be done to clarify many hang-
ing questions. One should realize, that when working on compounds, we
are exposed to the full drama of semiconductor physics and chemistry.
Crystal imperfections, impurities including gases and non-stoichiometry are known to influence transport properties very drastically or
may play an even dominating role. Whoever has been active in the
field of semiconductor physics knows about the--today--amusing discussions and controversies fifty years ago about whether silicon is a
metal or a semiconductor.

Only the most advanced and careful synthetical and analytical
methods are good enough to solve the many open questions. A very
close cooperation between solid state physicists and solid state
chemists seems to me indispensable--not only in rare earth research.

Admittedly, this is a very personal view of rare earth research
and I must apologize for not having mentioned many highlights in the
field as for instance Lawson and Tangs (26) observation of a 12%
volume contraction of cerium in 1949 and the interpretation in terms
of a valence change from 4 to 3 by Zachariasen, Pauling and Gschneidner, (27), the spiral spin structures of the heavy rare earths found
by the Oak Ridge Neutron Diffraction group (28), the beautiful neutron
diffraction work on the Gd^{160} isotope by Moon, Koehler, Cable and
Child (29) in 1972 and the fundamental investigation by Libowitz (30)
and Wallace (31) on rare earth hydrides and the remarkable work on
rare earth-cobalt alloys by Wallace, Nesbitt and Wernik (32).

Finally I would like to thank my former colleagues and collaborators of the Laboratorium für Festkörperphysik at ETH Zürich:
Y. Baer, W. Baltensperger, M. Campagna, P. Cotti, R. Hauger, F. Hulliger, P. Junod, E. Kaldis, F. Lévy, B. Magyar, J. Muheim, P. Munz,
B. Natterer, H.R. Neukomm, K. Sattler, P. Schwob, H.C. Siegmann,
P. Streit, W. Stutius, R. Verreault, H. Vetsch, O. Vogt, P. Wachter
and S. Yuan, many students and all the members of the administrative
and technical staff at the institute. Without their continuous and
enthusiastic help and loyal cooperation during 25 years I never would
have been recipient of the Frank H. Spedding Award.

REFERENCES

1. C. Auer-von Welsbach; Ueber das Gasglühlicht. Wochenschrift d.
    Nieder-Oesterreichischen Gewerbevereins 1886.

2. F. Ephraim, Anorganische Chemie, Verlag Theodor Steinkopff, Dresden u. Leipzig, 1923.
3. J. Becquerel, Le Radium 4, 328 (1907). J. Becquerel and H. Kamerling Onnes; Le Radium 5, 227 (1908).
4. J. Becquerel, Z. Phys. 58, 205 (1929).
5. H.A. Bethe, Ann. Phys. 3, 133 (1929).
6. H.A. Kramers, Proc. Acad. Sci. Amsterdam 32, 1176 (1929).
7. K.H. Hellwege, S. Hüfner and H.G. Kahle, Z. Physik 160, 149 (1960).
8. S. Freed and F.H. Spedding, Phys. Rev. 35, 1408 (1930), 38, 670 (1931); F.H. Spedding, Phys. Rev. 38, 2080 (1931).
9. L.R. Morss and N. Mensi, this volume.
10. K.H. Hellwege and Coworkers, Numerous publications predominantly in Z. Physik beginning 1939.
11. G.H. Dieke, Spectra and Energy Levels of Rare Earth Ions in Crystals. M.H. Crosswhite and Hannah Crosswhite (Eds.), Interscience Publishers (1968).
12. D.J. Randazzo, Ph.D. Thesis, John Hopkins Univ., 1966; Phys. Letters 28a, 269 (1968); J. Chem. Phys. 49, 1808 (1968).
13. W.P. Wolf, M.J.M. Leask, B. Mangum and A.F.G. Wyatt; J. Phys. Soc. Japan 17, Supp. B1, 487 (1962).
14. A.J. Sievers and M. Tinkham, Phys. Rev. 124, 321 (1961); 129, 1995 (1963).
15. K.H. Hellwege, S. Hüfner, M. Schinkmann and A. Schmidt, Phys. Letters 12, 109 (1964).
16. H. Schuchert, Dissertation, Techn. Hochschule Darmstadt (1968).
17. F. Hulliger, Helv. Phys. Acta 32, 615 (1959).
18. J. Tsubokawa, J. Phys. Soc. Japan 15, 1664 (1960).
19. B.T. Matthias, R.M. Bozorth and J.H. van Vleck, Phys. Rev. Letters 7, 160 (1961).
20. G. Busch, P. Junod and P. Wachter, Phys. Letters 12, 11 (1964).
21. W. Nolting and A.M. Oleś, Phys. Rev. B 23, 4122 (1981).
22. F. Meier, J. de Physique; Colloque Suppl. 6, 41, C 5-9 (1980).
23. G. Busch, M. Campagna, P. Cotti and H.Ch. Siegmann, Phys. Rev. Letters 22, 597 (1969).
24. K. Siegbahn, C. Nordling, A. Fahlman, R. Nordberg, K. Hamrin, J. Hedman, G. Johansson, T. Bergmark, S.E. Karlsson, I. Lindgren and B. Lindberg; ESCA Atomic, Molecular and Solid State Structure Studied by means of Electron Spectroscopy. Nova Acta Regiae Societatis Scientiarium Upsaliensis, Uppsala 1967.
25. J. Lang, Y. Baer and P.A. Cox, J. Phys. F 11, 121 (1981).
26. A.W. Lawson and T.Y. Tang, Phys. Rev. 76, 301 (1949).
27. K.A. Gschneidner, Valence Instabilities, P.D. Parks (Ed.), Plenum Press, New York (1977).
28. W.C. Koehler, J.W. Cable, E.O. Wallace, M.K. Wilkinson, J. App. Phys. 33, Suppl. 1124 (1962).
29. R.M. Moon, W.C. Koehler, J.W. Cable and R. Child, Phys. Rev. B 5, 997 (1972).
30. G.G. Libowitz, Berichte der Bunsengesellschaft 76, 837 (1972).

31. W.E. Wallace, cf. Hydrogen in Metals I + II; G. Alefeld and J. Völkl, Topics in Applied Physics Vol. 28, Springer-Verlag 1978.
32. H.R. Kirchmayer and C.A. Poldy, in Handbook of the Physics and Chemistry of Rare Earths, K.A. Gschneidner, Jr. and L. Eyring (Eds.), North Holland Publishing Company (1979).

MAGNETIC RESONANCE STUDIES OF METAL CATION TRANSPORT ACROSS

BIOLOGICAL MEMBRANES: USE OF PARAMAGNETIC LANTHANIDE IONS

James A. Balschi, Vincent P. Cirillo[†], William J. leNoble,
Martin M. Pike, Everett C. Schreiber, Jr.,
Sanford R. Simon[†], and Charles S. Springer , Jr.*

Depts. of Chemistry and [†]Biochemistry
State University of New York at Stony Brook (SUNY)
Stony Brook, New York 11794

Paramagnetic lanthanide ions ($*Ln^{+3}$) have proved to be very
useful in effecting magnetic resonance studies of the transport of
metal cations across biological membranes; a very important process
(1) usually facilitated by integral membrane proteins (2). We will
discuss two different types of study: 1) the transport of $*Ln^{+3}$ ions
as surrogates for physiological metal cations (Na^+, K^+, Mg^{+2} and
Ca^{+2}) 2) the use of anionic complexes of $*Ln^{+3}$ ions to study the
transport of physiological cations directly by nuclear magnetic
resonance (NMR).

The first type of study has employed small unilamellar vesicles
(SUV, 100-500 Å, OD) which are prepared (by sonication of aqueous
phospholipid dispersions) with diamagnetic metal cations, Ln^{+3},
present both inside and outside the vesicles. After vesicle formation is complete $*Ln^{+3}$ are introduced only outside the vesicles.
Although, in principle, any paramagnetic cation will work, the
paramagnetic lanthanide cations offer the optimal magnetic properties. One can adjust the concentrations such that the
osmolarities are the same on each side of the membrane. This method
has manifestations in both NMR and electron paramagnetic resonance
(EPR). The NMR approach has already been analyzed in detail (3).

We have recently developed an EPR analog of this technique and
give some initial results here. If one observes the EPR spectrum
of a mixture of SUV, formed from egg lecithin (EL), and the
nitroxide free radical tempone (T, O=⟨N⟩-O), one sees the expected
hyperfine triplet resonance. The T is thought to distribute itself
between the aqueous phase and the lipid head-group surface region
of the bilayer membrane (4). Thus, its resonances are monitors
for these regions of the solution. If the solution is then made

ca. 60mM in Gd^{+3} (EL, ca. 70 mM; T, ca. 2 mM), the spectrum exhibits two isochronous (same g values) triplets; one with broad hyperfine components (peak-to-peak line-width ca. 4 gauss) and the other with sharp components (ca.0.7 gauss).

We assign the broad triplet to those T molecules outside the vesicles where they are subject to relaxation by the paramagnetic Gd^{+3} ion (^8S ground state). The components of this triplet exhibit essentially identical linewidths to those we observe for a solution containing 60 mM Gd^{+3} and 2 mM T with no EL. (In titration experiments with such solutions, we find that the observed line-width continuously increases with the molar ratio of Gd^{+3} to T. This indicates that the Gd^{+3} T interaction is fast on the EPR time-scale.) Thus, the broad tiplet is the total resonance of T in 4 different states: free and Gd^{+3}-perturbed in the outside solution, free and Gd^{+3}-perturbed on the outside vesicle surfaces (where Gd^{+3} also binds (3)). Membrane surface-bound T and aqueous T are in slow exchange on the EPR timescale, but the resonance of these states overlap so much that they are only distinguishable when the totally deuterated molecule is used (4).

We assign the sharp triplet to those T molecules on the insides of the vesicles, which are impermeable to Gd^{+3}. The components of this triplet have essentially identical linewidths to those of T dissolved in pure water. Although T is added only after the SUV are formed, it enters the vesicles very quickly on the laboratory timescale. However, the fact that broad and sharp resonances are distinct indicates that vesicle permeation by T is slow on the EPR timescale. Thus our assignment would have the sharp triplet as the total resonance of T free and bound to the inside surfaces of the vesicles.

For species taken up by SUV, the molar ratio of inside species to total species ranges from ca 0.02 (at this EL concentration), when the species is not bound to the membrane, to ca. 0.33 when the species is completely bound to the membrane surface (3). We have performed a rudimentary hand resolution and integration of the downfield peak into its broad and sharp components. We find that the ratio of the area of the sharp component to the total area of the downfield peak is 0.19, well within the above range. This con-firms that some of the inside T is bound to the membrane. If the mixture is subjected to resonication, the sharp components dis-appear as resonication progresses. Resonication is known to make SUV permeable to $*Ln^{+3}$ ions by an "all or nothing" process (3). Thus, the Gd^{+3} taken into the vesicles relaxes the inside tempone resonances.

When the anionic ionophore lasalocid-A (X-537A) is introduced into the solution (as the Li^+ salt, X^-), the spectrum becomes time-dependent. The sharp components are observed to gradually disappear. The X^- is known to transport $*Ln^{+3}$ ions into SUV by an "intermediate slow leakage" process (3). The increasing relaxation of the inside T resonance serves to report the transport of Gd^{+3} into the vesicles.

We made a crude estimate of the height of the sharp downfield component for each of the spectra as time progress. Assuming the linewidth to be unchanged, the height is proportional to the peak area. A plot of the natural logarithm of the ratio of the height at any time to the extrapolated height at zero time versus time yielded a reasonable straight line. The slope corresponded to a first-order rate constant for transport of 17×10^{-3} min^{-1} (ambient temperature, pH=4, 0.5 mM X^-).

An advantage of the EPR version of this technique is that we can easily obtain spectra as a function of pressure (5). Thus, an analogous experiment was repeated in our pressure cell (5) at the same temperature (ambient) but at 272 atm. The transport process was found to be much retarded. Analogous treatment of this data results in a rate constant of 3.7×10^{-3} min^{-1}.

Assuming a constant value for the volume of activation, ΔV^*, we can use the relationship $(\Delta(lnk)/\Delta P)_T = \Delta V^*/(RT)$ to estimate ΔV^* from the above results. Doing this, we obtain the strikingly large value of +140 cc/mol. Of course, this value must be verified by using computer-resolved and integrated peak areas and by obtaining the rate constants over a range of pressures (we have been able to obtain spectra at pressures above 1000 atm. on our apparatus). The temperature will be controlled and the ionic strength adjusted so as to avoid the build up of diffusion potentials (3).

The volume of activation represents the net volume change occurring in all the steps up to and including the rate-determining step of the transport process. Its interpretation will prove interesting. A major contribution to ΔV^* might be expected to be the partial or complete dehydration of the Gd^{+3} aquo ion prior to chelation by two or more X^- molecules (6). This is thought to be a step common to all ionophore mediated transport. Such a step usually leads to a volume increase because coordinated water molecules are compressed due to electrostriction. However, the value of ΔV^* given above is sufficiently large that contributions from the X^- and phospholipid molecules are almost certainly also indicated.

As a diagnostic and straightforward as the above method is (in both its NMR and EPR manifestations), it has two major drawbacks with regard to studies of biological transport. First, since the NMR version employs the resonances of the lipid molecules themselves, it is limited to use with small vesicles because the corresponding

resonances of larger vesicles, organelles, cells, or organs are
much too broad for this high-resultion technique. Second, the
technique does involve the study of surrogate cations and, although
many species have been found to transport these (including ionophores,
detergents, and proteins (3)) biological systems are notoriously
selective. One can be sure that many membrane transporting proteins
are quite selective for a particular physiological metal cation.
For these reasons, we have been working to develop a new form of the
NMR technique, capable of directly monitoring the transport of
physiological cations in larger vesicles, organelles, living cells,
and possibly intact organs in vivo.

NMR spectroscopists have long known the properties of the magnetic isotopes of the physiological cations Na^+, K^+, Mg^{+2} and Ca^{+2}
and that direct NMR detection of all of these cations is possible
(7). Except for $^{23}Na^+$, however, these are very difficult to study
because of isotopic rarity, small magnetic moment, or both. Recent
high-field developments have increased to feasible levels the
sensitivities of NMR spectrometers to enriched samples. Since the
essence of the NMR method described above is the difference in resonance frequencies for species on the two sides of the membrane, we
thought it incumbent to develop a family of shift reagents, anionic
dysprosium(III) complexes, for these cationic nuclei. We have demonstrated the effectiveness of our shift reagents with natural
abundance $^{23}Na^+$ and $^7Li^+$ NMR experiments on a Varian XL-100.

For example, we prepared the bis nitrilotriacetate (NTA^{-3})
complex of Dy^{+3} in situ by the following reaction (TEA is

$$Dy(OH)_3 + 2H_3NTA + 3TEA \rightarrow 3HTEA^+ + Dy(NTA)_2^{-3} + 3H_2O \qquad (1)$$

$(N(CH_2CH_2OH)_3)$. This complex causes a continuous upfield isotropic
hyperfine shift of the $^{23}Na^+$ resonance as the molar ratio (ρ) of
$Dy(NTA)_2^{-3}/Na^+$ is increased (ca. 4 ppm at a ρ of ca. 0.6,
$[Na^+]$ = ca. 100 mM). This indicates that there is a labile (on the
NMR timescale) interaction between Na^+ and $Dy(NTA)_2^{-3}$ (reaction 2),

$$Na^+ + Dy(NTA)_2^{-3} \rightleftarrows Na^{+\cdot}Dy(NTA)_2^{-3} \qquad (2)$$

probably involving ion-pairing (8,9). If the solution is made
increasingly concentrated in the diamagnetic lanthanide Lu^{+3}, the
hyperfine shift decreases; the peak moves back towards its unshifted
position. The ligand exchange reactions (3) undoubtedly render the
Dy^{+3} ion

$$Lu^{+3} + Dy(NTA)_2^{-3} \rightleftarrows Lu(NTA)_i^{+3(1-i)} + Dy(NTA)_{2-i}^{+3(i-1)}; \quad i = 1,2 \qquad (3)$$

into forms ($Dy(NTA)$, $i = 1$, and Dy^{+3}, $i = 2$) which do not act as
shift reagents for $^{23}Na^+$. Some shift-reagent induced broadening of

the shifted Na^+ peak is observed but is small compared to the magnitude of the shift. The shifted line appears to retain a Lorentzian shape, indicating that the spectrum remains in the "fast motional narrowing" condition (10) (^{23}Na, I = 3/2, is a quadrupolar nucleus).

If we prepare large unilamellar vesicles (LUV), 44mM in EL, by a chromatographic detergent-removal technique reported to yield vesicles with an average diameter of ca. 1000 $\overset{o}{A}$ (11), and do so with a 15:1 Na^+ concentration gradient (inside over outside), we observe a strong, narrow $^{23}Na^+$ resonance. If we then make the outside aqueous space ca. 20 mM in $(HTEA)_3Dy(NTA)_2$, the single resonance is separated clearly into a doublet, the smaller component (38%) being shifted upfield by a ca. 80 Hz, the larger component remaining unshifted (even ca. 16 hours after the gradient was created). This is exactly the shift and relative areas expected if the upfield resonance is assigned to the outside sodium and the downfield peak to those Na^+ ions trapped inside the vesicles. The shift reagent anion is impermeant to the LUV.

The situation is ideal for sensitive monitoring of Na^+ transport. If we add the known ionophore, valinomycin, Na^+ efflux is observed (the downfield resonance loses intensity at the expense of the upfield resonance); reaching equilibrium with an exponential time course whose time constant depends on the valinomycin/lipid molar ratio. After the system has reached equilibrium, we can add the more efficient ionophore X^-. The remaining peak representing the inner Na^+ appears to broaden and coalesce with the outer resonance as if the ionophore was catalyzing the equilibrium transport to the point where it becomes fast on the NMR timescale. We have observed X^- to do this in inverted micelle systems (6). Such equilibrium transport can be quantitated by total lineshape analysis (TLA). Thus, this method will be very useful in testing the leakiness of various LUV preparations and in studying passive and active transport in membrane protein reconsituted vesicles.

Finally, we have recently demonstrated another potentially very important application of our shift reagents: i.e. the study of metal cation transport in living cells and unicellular organisms. Live baker's yeast (Saccaromyces cerevisiae) cells were incubated in 0.2 M Na_3citrate long enough for the cells to become sufficiently loaded with Na^+. Then, the cells were centrifuged, separated from the supernatant, washed once with water, and resuspended (40%) in D_2O. The $Dy(NTA)_2^{3-}$ shift reagent was introduced to the outside solution. The first spectrum was obtained ca. 11 minutes after the cells were resuspended in D_2O. It shows two resonances in exactly the position expected for inside and outside Na^+ ions. Thus, some sodium ions have already been transported out. The lines are noticeably broader than those observed in the LUV experiments

(above) but they can be clearly distinguished. Subsequent spectra
clearly show Na$^+$ ions undergoing transport out of the cells. The
peak intensity ratio follows a reasonably exponential time course
with a first order rate constant of ca. 6.6 x 10^{-3} min^{-1}. We have
also observed an analogous splitting of the ^{7}Li$^+$ resonance in a
mutant strain of yeast.

ACKNOWLEDGEMENT

We would like to thank the National Science Foundation (Grants
#PCM 78-07918 and #PCM 80-03739) for support of this work.

REFERENCES

1. E. Racker, Acc. Chem. Res., 12: 338 (1979).

2. A.S. Hobbs, and R.W. Albers, Ann. Rev. Biphys. Bioeng., 9:
    259 (1980).

3. D.Z. Ting, P.S. Hagan, S.I. Chan, J.D. Doll, and C.S. Springer,
    Biophys. Jour., 34: 189 (1981).

4. C.F. Polnazek, S. Schreier, K.W. Butler, and I.C.P. Smith,
    Jour. Amer. Chem. Soc., 100: 8233 (1978).

5. W.J. leNoble, and P. Staub, Jour. Organomet. Chem., 156:
    25 (1978).

6. S.T. Chen, C.S. Springer, submitted for publication.

7. S.Forsen, "NMR Studies of Ion Binding to Biological
    Macromolecules", in Proc. Eur. Conf. on NMR of Macromolecules;
    E. by F. Conti; Levici; Sassari, Sardenia; (1979), p. 243.

8. G.A. Elgavish, and J. Reuben, Jour. Amer. Chem. Soc., 99:
    1762 (1977).

9. H. Degani, and G.A. Elgavish, FEBS Lett, 90: 357 (1978).

10. M.M. Civan, and M. Shporer, Chapt. 1 in "Biological Magnetic
    Resonance", Volume 1, Ed. by L.J. Berliner, and J. Reuben,
    Plenum, New York (1978).

11. H.G. Enoch, and P. Strittmatter, Proc. Nat. Acad. Sci., 76:
    145 (1979).

INTER-LANTHANIDE ION ENERGY TRANSFER DISTANCE MEASUREMENTS IN BIOLOGICAL SYSTEMS

W.DeW. Horrocks, Jr., M.-J. Rhee, A.P. Snyder,
T. Choosri, and V.K. Arkle

Department of Chemistry, The Pennsylvania State University, University Park, PA 16802

We have recently reported (1) the utility of laser induced luminescence excitation spectroscopy of the $^5D_0 \leftarrow \, ^7F_0$ transition of Eu(III) for the detection and characterization of individual, distinct metal ion sites in calcium-binding proteins. Owing to the nondegenerate nature of the two states involved, the observation of more than one excitation signal implies the existence of two or more nonequivalent binding sites. In systems where energy donor ions (e.g., Eu(III), Tb(III)) and energy acceptor ions (e.g., Ho(III), Nd(III)) are bound at different nearby sites, Förster-type non-radiative energy transfer occurs. This energy transfer is monitored by observing the decrease in the lifetime, τ, of the excited state of the donor caused by the presence of an energy acceptor ion and provides a means of estimating the distance between the two binding sites (2).

Before such measurements are carried out, Eu(III) binding is characterized by excitation spectroscopy to determine the number and type of binding sites present. We have carried out such experiments on thermolysin, parvalbumin, and calmodulin, among others. In the first two cases lanthanide ion, Ln(III), binding has been studied by crystallographic techniques (3-5). Our method allows direct solid-state-solution-state comparisons. In the case of thermolysin the solid-state and solution-state spectra correspond exactly and confirm the crystallographic finding that Ln(III) ion binding may occur at Ca site S(1) alone or at Ca sites S(1), S(3), and S(4), depending on conditions. The case of parvalbumin is more complicated, but extremely interesting. We have shown (6) that as Eu(III) ions are added to a solution of parvalbumin at pH 6.5, they bind simultaneously to the primary CD and EF sites and to a third, more solvent accessible site as well. Furthermore, we found that bind-

21

ing at the third site can be eliminated by reducing the pH to 4.0.
Our results are in apparent contradiction of the crystallographic
results (4,5) which show no evidence for a third site and that only
the EF site is occupied when crystals are grown with less than one
equivalent of Eu(III). In order to resolve this matter we grew
crystals of parvalbumin from a solution containing 0.5 equivalents
of Eu(III), the solution excitation spectrum of which shows occupancy of all three sites. The excitation spectrum of the resulting
crystals shows only a single peak characteristic of occupancy of
the EF site, confirming the crystallographic results and demonstrating an important difference between metal ion binding to parvalbumin
in the solution and crystalline states. The excitation spectrum of
Eu(III) ions bound to calmodulin is less well resolved, but our
results are consistent with initial occupancy of calcium sites I and
II followed by binding at sites III and IV as was reported for
Tb(III) (7,8).

With the metal ion binding thus characterized we have carried
out inter-Ln(III) ion energy transfer measurements between sites
S(1) and S(4) of thermolysin (9), the CD and EF sites of parvalbumin (7), and sites I and II of calmodulin. With the exception
of the Tb(III) → Nd(III) pair, distances obtained in thermolysin
and parvalbumin are in good agreement with the fortuitously identical
x-ray structural distances of 11.8 Å found by crystallographic studies of these proteins (10,11). In the case of calmodulin, preliminary results yield a distance of 11.2 Å, extremely close to the
parvalbumin results and consistent with the postulated homology of
the two proteins (12).

1. W. DeW. Horrocks, Jr. and D.R. Sudnick, Science 206:1194 (1979).
2. W.DeW. Horrocks, Jr., M.-J. Rhee, A.P. Snyder, and D.R. Sudnick,
    J. Am. Chem. Soc. 102:3650 (1980).
3. B.W. Matthews and L.H. Weaver, Biochemistry 13:1719 (1974).
4. P. C. Moews and R. H. Kretsinger, J. Mol. Biol. 91:229 (1975).
5. J. Sowadsky, G. Cornick, and R.H. Kretsinger, J. Mol. Biol. 124;
    123 (1978).
6. M.-J. Rhee, D.R. Sudnick, V.K. Arkle, and W.DeW. Horrocks, Jr.,
    Biochemistry, 20:3328 (1981).
7. M.-C. Kilhoffer, J.G. Demaille, and D. Gerard, FEBS Lett. 116;
    269 (1980).
8. M.-C. Kilhoffer, D. Gerard, and J.G. Demaille, FEBS Lett. 120:
    99 (1980).
9. A.P. Snyder, D.R. Sudnick, V.K. Arkle, and W.DeW. Horrocks, Jr.,
    Biochemistry 20:3334 (1981).
10. R.H. Kretsinger and C.E. Nockholds, J. Biol. Chem. 248:3313
    (1973).

11. B.W. Matthews, L.H. Weaver, and W.R. Kester, J. Biol. Chem. 249:8030 (1974).
12. C.B. Klee, T.H. Crouch, and P.G. Richman, Annu. Rev. Biochem. 49:489 (1980).
13. This research was supported by NIH grant GM23599.

GADOLINIUM AS AN EPR AND NMR PROBE OF CA^{2+} SITES IN BIOLOGICAL SYSTEMS

Eileen M. Stephens* and Charles M. Grisham[†]

Department of Chemistry, University of Virginia

Charlottesville, Virginia 22901

In spite of an increasing awareness of the importance of calcium in biological systems (1), Ca^{2+} has received little attention compared to most of the biologically relevant transition metals. Ca^{2+} is not easily studied by physical methods: the absence of unpaired electrons makes it unsuitable for electron paramagnetic resonance (EPR) studies; its NMR signal is extremely weak, and its electronic transitions cannot be studied by conventional optical spectroscopic methods. Biochemists as well as inorganic chemists have used the trivalent lanthanides as analogues for Ca^{2+}. The ability of the lanthanides to replace Ca^{2+} in, for example, Ca^{2+} binding proteins is well established (2). Several investigators, including Horrocks (3) and Martin and Richardson (4) have utilized lanthanide ions, including terbium (Tb^{3+}) and europium (Eu^{3+}), as luminescence probes of Ca^{2+} binding sites in biological systems. Gd^{3+}, with seven unpaired electrons, has been used extensively as a paramagnetic probe in NMR studies (5,6) and in single crystal EPR (7,8). Until recently (9,10), however, its usefulness as a EPR probe in biological systems has been overlooked or discounted (11). We have conducted an extensive study of the EPR properties of Gd^{3+} and have found that EPR spectra of solutions and glasses can be obtained which are highly sensitive to the symmetry, ligands and coordination geometry of Gd^{3+}.

---

*Fellow of the Muscular Dystrophy Association of America, 1980–81.
[†]Research Career Development Awardee of the National Institutes of Health (AM00613).

Gd^{3+} has an $^8S_{7/2}$ ground state and two naturally occurring isotopes with nuclear spin: ^{155}Gd and ^{157}Gd (I=3/2 for both). The hyperfine coupling constant of about 5 gauss is resolvable only in single crystal studies. Thus the spin Hamiltonian for EPR spectra of Gd^{3+} in solution and/or in the powder state is:

$$\mathcal{H} = \beta\hat{S}\cdot\hat{\hat{g}}\cdot\hat{H} + \hat{S}\cdot\hat{\hat{D}}\cdot\hat{S}$$

where the first term is the electronic Zeeman interaction term and the second is the dipolar interaction term, which includes the zero-field splitting effects, spin-orbit coupling and inter-actions between neighboring spin systems. To a first order approximation, the energies of S-state ions are not affected by zero-field and spin-orbit interactions, but, as will become obvious below, higher order analyses including zero field and spin-orbit effects are essential to the interpretation of Gd^{3+} EPR spectra. The interactions of neighboring spin systems, how-ever, will not be considered here. Thus the dipolar term can be written as

$$\hat{S}\cdot\hat{\hat{D}}\cdot\hat{S} = \mathcal{H}_v + \mathcal{H}_{LS}$$

where $\mathcal{H}_v$ is the zero field term and $\mathcal{H}_{LS}$ is the spin-orbit coupling term. $\mathcal{H}_v$ can be written as

$$\mathcal{H}_v = \sum_{k,q} B_q^{(k)}\cdot U_q^{(k)}(S)$$

where the $B_q^{(k)}$ are the crystal field coefficients, $U_q^{(k)}(S)$ are the intraconfigurational unit tensor operators defined in terms of the total spin operator S, and k = 2, 4, 6. In the Independent Systems Crystal Field (ISCF) model, Faulkner and Richardson (12) have refined $\mathcal{H}_v$ as shown:

$$\mathcal{H}_v = \sum_{k,q} [B_q^{(k)}(chg) + B_q^{(k)}(pol)]U_q^{(k)}(S)$$

where $B_q^{(k)}(chg) = Gd^{3+}$ multipole-ligand net charge interaction term, and $B_q^{(k)}(pol) = Gd^{3+}$ multipole-ligand dipole interaction term.

The X-band EPR spectrum of $GdCl_3$ in aqueous solution at room temperature consists of a single transition of linewidth 530 G at g = 2. However, as shown in Figure 1A, the Gd^{3+} EPR spectrum in H_2O/glycerol glass at 77°K includes a number of weak transitions in the 1700-2400 gauss range. At 9.0 GHz, these transitions are only 1-2% of the intensity of the g = 2 signal.

At progressively lower microwave frequencies the low field transitions become relatively more intense. At 1.98 GHz (Figure 1D) the g = 2 transition (indicated with the arrow) is dwarfed by the low field transitions.

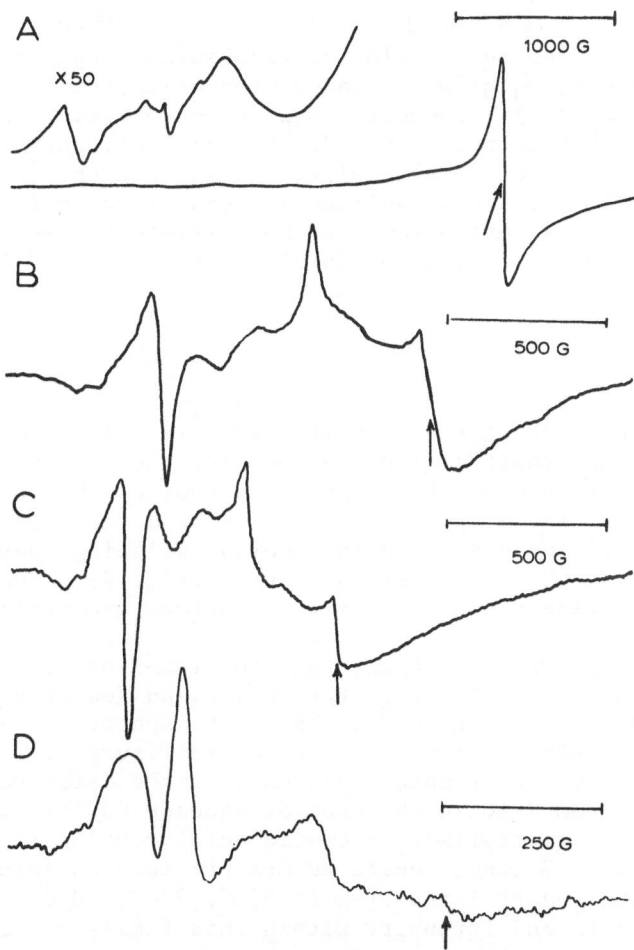

Figure 1.   EPR spectra of GdCl$_3$ in 1:1 water:glycerol at 77°K, (A) at 9.10 GHz, scan range 0 to 4000 gauss, (B) 3.94 GHz, scan range 0 to 2000 gauss, (C) 2.89 GHz, scan range 0 to 2000 gauss, (D) 1.98 GHz, scan range 0 to 1000 gauss. Arrows indicate position of "g = 2" transition at (A) 3300 gauss, (B) 1370 gauss, (C) 1040 gauss and (D) 725 gauss.

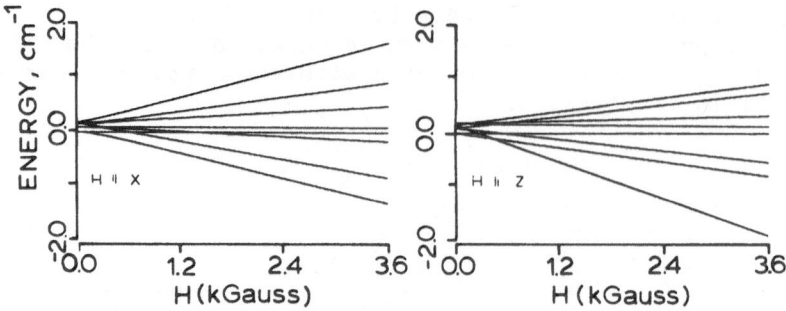

Figure 2. Zeeman energy levels for gadolinium ethylsulfate,
           $Gd(C_2H_5SO_4)_3 \cdot 6H_2O)$, calculated using the ISCF theory
           (see text). Symmetry, D_{3h}; electrostatic radial func-
           tions from reference 18; crystal field parameters chosen
           to give zero field splittings in reference 13. Both
           gadolinium ethyl sulfate and gadolinium chloride (Figure
           3A and 4) are assumed to be completely dissociated in
           aqueous solution and should give "free-ion" spectra.

     The enhancement of the transitions at low field can be attri-
buted to the fact that, at low frequencies, the magnitude of the
microwave energy quantum, $h\nu$, approaches that of the crystal field
splitting (13). The energy levels "cross" and the wave functions
mix as the field increases and the Zeeman splitting approaches
and then exceeds the zero-field splitting (Fig. 2). These
phenomena give rise to multiple fine structure transitions.

     It is not surprising, then, that the low-field transitions
appear to be very sensitive to the nature and geometry of the
ligands coordinated to the Gd^{3+}. The 9 GHz spectra of aqueous
Gd^{3+} and two complexes of Gd^{3+} are shown in Figure 3. In these
complexes, the number of water molecules in the first coordination
sphere varies from 9-10 in the case of aqueous Gd^{3+} to zero in the
case of the MIDA (methylimidodiactetic acid) complex (14). The
EDTA complex with 3 inner sphere waters provides an intermediate
case (15). The spectral features at 560G, 720G, 1035G and 1150G
vary in linewidth and intensity within this family of complexes,
and the region from 1350G to 1500G, while less well defined, is
also sensitive to the nature of the complex. Small changes in
the nature and number of ligands can produce large changes in this
region of the spectrum (16).

     Clearly, there exists a substantial amount of information in
these spectra. Whether this spectral information can be correlated with molecular structure depends greatly on whether these

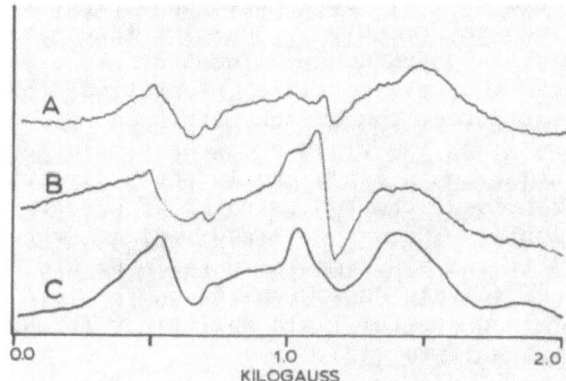

Figure 3.   Low field experimental EPR spectra at 9.06 GHz.
(A) 10 mM $GdCl_3$, pH 7.0, (B) 16 mM $GdCl_3$, 20 mM
NaEDTA, pH 7.0, (C) 5 mM $GdCl_3$, 20 mM MIDA, pH 9.7.
All samples in 1:1 water:glycerol at 77°K.

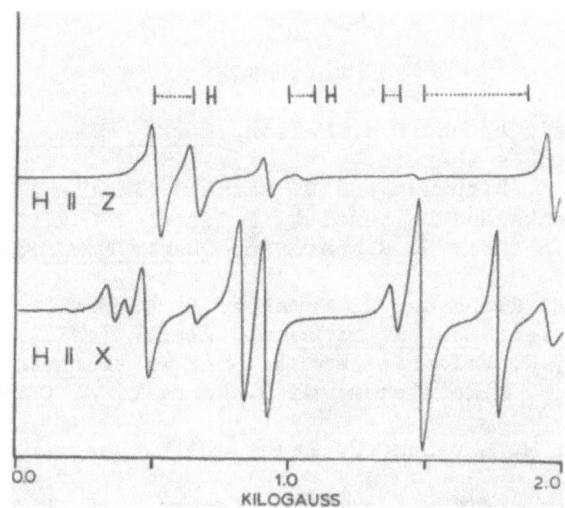

Figure 4.   Low field portion of simulated single crystal spectra
of gadolinium ethyl sulfate. Dotted lines show width
and position of low field transitions in spectra of
aqueous $GdCl_3$. See legend for Figure 2.

spectra can be simulated from sets of assumed parameters derived
from model structures. We have undertaken such a simulation and
have based our calculations on the formalism of Faulkner and
Richardson (17). The spectra generated by these simulations are

sensitive to crystal field parameters and orientation of the
crystal axes, and the relative transition intensities are sensitive to the microwave frequency. These single crystal spectra
must be averaged over all possible orientations to obtain powder-
type spectra for direct comparison with spectra such as Figure 3.
Shown in Figure 4 are low field areas of simulated spectra of
aqueous Gd^{3+} oriented in the y and in the z directions. These
orientations represent the two extremes of effects for an axially
symmetric molecule. The dashed lines indicate the field positions
of transitions in the experimental data in Figure 3. The crystal
field parameters in this case were chosen to yield the experimentally determined crystal field splittings for hydrated
gadolinium ethyl sulfate (13).

If unique correlations can be made between experimental and
simulated Gd^{3+} EPR spectra, it will provide the biological
chemist with a powerful structural probe. Gd^{3+} EPR should be
particularly useful with Ca^{2+}-activated membrane proteins, which
heretofore have uniformly resisted the efforts of the crystallo-
graphers.

REFERENCES

1. R.H. Kretsinger and D.J. Nelson, Coord. Chem. Rev., 18:29 (1976)
   and references therein.
2. J. Gomez, E. Birnbaum and D. Darnall, Biochem., 13:3745 (1974).
3. W.D. Horrocks and D. Sudnick, Science, 206:1194 (1976).
4. R.B. Martin and F.S. Richardson, Quart. Rev. Biochem., 12:181
   (1979).
5. R. Dwek, R. Richards, K. Moralee, E. Nieboer, R.J.P. Williams,
   and A. Xavier, Eur. J. Biochem., 21:204 (1971).
6. G. Cottam, K. Valentine and A. Sherry, Biochem., 13:3532 (1974).
7. S. Misra, P. Mikolajczak and S. Korczak, J. Chem. Phys., 74:922
   (1981).
8. M. Rappaz, M. Abraham, J. Ramey and L. Boatner, Phys. Rev. B, 23:
   1012 (1981).
9. E. Stephens and C.M. Grisham, Biochem., 18:4876 (1979).
10. G. Reed, Polymer Preprints, 20(2):203 (1979).
11. E. Geraldes and R. Williams, J. Chem. Soc. (Dalton), 1721 (1977).
12. T.R. Faulkner and F.S. Richardson, Molecular Phys., 39:75 (1980).
13. B.G. Wybourne, Phys. Rev., 148:317 (1966).
14. S. Salama and F.S. Richardson, J. Phys. Chem., 84:512 (1980).
15. J.L. Hoard, B. Lee and M.D. Lind, J. Am. Chem. Soc., 87:1612
    (1965).
16. E.M. Stephens and C.M. Grisham, manuscript in preparation.
17. T.R. Faulkner, J.P. Morley and F.S. Richardson, Molecular Phys.,
    40:1481 (1980).
18. W.T. Carnall, P.R. Fields and K. Rajnak, J. Chem. Phys., 49:4443
    (1968).

OPTICAL ACTIVITY OF MIXED-LIGAND COMPLEXES OF Tb(III) WITH

PYRIDINE-2,6-DICARBOXYLIC ACID AND HYDROXYCARBOXYLIC ACIDS

Harry G. Brittain[*]

Seton Hall University, Department of Chemistry
South Orange, N.J. 07079

ABSTRACT

The interaction between Tb(III) and several α-hydroxy-carboxylic acids has been studied by means of circularly polarized luminescence (CPL) spectroscopy. Complexes having the general formula Tb(DPA)(L) in ratios of 1:2:1 and 1:1:1 were studied (where DPA signifies pyridine-2,6-dicarboxylate) were studied in an attempt to use CPL spectroscopy as a probe of the solution chemistry existing between the Tb/DPA complexes and the chiral hydroxycarboxylic acids. The chiral L ligands used were L-hydroxy-isocaproic acid, L-arginnic acid, L-hydroxyglutaric acid, and D-isocitric acid. All of the ligands were found to bind Tb(III) in a bidentate fashion, except for isocitrate which appeared to be terdentate. Equilibrium quotients for the interaction of the Tb/DPA complexes were calculated from the CPL data.

INTRODUCTION

The lanthanide complexes of hydroxycarboxylic acids have been of interest since the early observation that these ligands could be used in the ion exchange separation of the rare-earth cations, and many workers have studied the solution phase equilibria of these complexes (1). On the other hand, little attention has been paid to chiroptical studies of the lanthanide complexes, in spite of the degree of information that can be gleaned from such studies and the commercial availability of the resolved ligands. In our laboratory, we have shown that CPL spectroscopy is the best

* Teacher-Scholar of the Camille and Henry Dreyfus Foundation

chiroptical method for the study of chiral lanthanide complexes, since the emission bands do not overlap and the technique is very sensitive. In earlier studies, we have examined the solution phase chemistry of Tb/DPA mixed-ligand complexes which contained L-malic acid (2) and L-lactic, L-mandelic, and L-phenyllactic acids (3), and have been able to deduce information regarding the mode of metal/ligand bonding in these complexes. In the present report, we extend our studies to include four new hydroxycarboxylic acid ligands and as a result have been able to develop new details of the solution phase bonding.

EXPERIMENTAL

The procedures required to obtain the Tb(DPA)(L) and Tb(DPA)$_2$(L) mixed-ligand complexes have been described previously in great detail (2,3), as has been the CPL spectrometer (4). L- α -hydroxy-isocaproic (HIC), L-arginnic (ARG), L-hydroxyglutaric (HGT), and D-isocitric (ICT) acids were used as received from Sigma, with

HIC, R = $-CH_2-CH(CH_3)-CH_3$

ARG, R = $-CH_2-CH_2-CH_2-C(NH)-NH_2$

$$R \longrightarrow CH \longrightarrow COOH$$
$$|$$
$$OH$$

HGT, R = $-CH_2-CH_2-COOH$

ICT, R = $-CH(COOH)-CH_2-COOH$

RESULTS AND DISCUSSION

The binding of a chiral hydroxycarboxylic acid ligand by either Tb(DPA) or Tb(DPA)$_2$ results in the generation of strong CPL in the Tb(III) emission bands. This CPL is a reliable indicator of the absolute configuration of the chiral ligand, and we have found that the $^5D_4 \rightarrow {}^7F_5$ Tb(III) emission band (4-5) yields the maximum amount of information (2-4). In one previous study (5), we showed that the CPL associated with Tb/DPA complexes containing monodentate chiral ligands was considerably weaker than the CPL of complexes known to bind in a bidentate manner, and the CPL lineshape was also quite different (3). We then concluded that the hydroxycarboxylic acids bind to the Tb(III) ion in a bidentate fashion, even in acid solution where the hydroxyl group cannot be ionized (3).

In the present work, we find that the CPL of mixed-ligand complexes which contain HIC, ARG, or HGT is identical to the line-shapes reported before (3), and that this CPL is invariant from pH 3.0 to 7.0. It is therefore possible to conclude that these ligands also bind in the same bidentate manner. Such a conclusion is not surprising for the HIC and ARG ligands, but HGT contains an additional carboxylic acid group which might become involved in bonding.

No change in CPL lineshape was observed with the HGT complexes as
the pH was raised from 3.0 to 10.0 (as did happen with the malic
acid ligand (2) in a previous work), and we thus conclude that the
chelate ring that would be formed if HGT bound one Tb(III) between
the two carboxyl groups is too large to be stable. A similar conclusion was reached when considering the CPL of mixed-ligand complexes containing glutamic acid (4).

The ICT ligand represents a more difficult situation to describe,
since it contains two asymmetric atoms (as the <u>threo</u>-isomer, with
the absolute configuration at the hydroxycarboxylic acid carbon
being opposite to that of HIC, ARG, or HGT). While the CPL associated with the 4-5 Tb(III) emission is very similar to that seen
for the other ligands (aside from being opposite in sign due to the
aforementioned difference in absolute configuration), the CPL of
the other Tb(III) emission bands is sufficently different in character as to indicate that a different mode of bonding must exist
(6). Careful comparison to our earlier works (2-4) reveals that
none of the observed CPL patterns exactly matches the ICT spectra.
More detailed analysis (including a study of the CPL spectra associated with three other Tb(III) emission bands) clearly shows that
simple bidentate attachment of the ICT ligand at the hydroxycarboxylic acid functionality is insufficient to account for the observed
spectra (6). We conclude that the ICT ligand binds to the Tb(III)
ion in a terdentate manner using two carboxyl groups and the un-
ionized hydroxyl group. Unlike the Tb(DPA) complexes of L-malic
acid, no CPL sign inversion is noted at high pH and therefore no
conformational change of the Tb(DPA)(ICT) complex accompanies
other ligand proton ionizations.

One can place the CPL results on a quantitative scale by use of
the Luminescence dissymmetry factor, which is defined simply as the
ratio of the CPL to the total luminescence (7). The dissymmetry
factor thus is a dimensionless quantity, which can be related to the
rotational strength of the transition (7). For vicinal or conformational effects, the chirality experienced by the Tb(III) ion should
be additive for each chiral ligand bound (8). Values for the dissymmetry factors of the Tb(DPA) complexes are essentially the same
as for the Tb(DPA)$_2$ complexes, therefore indicating that only one of
the hydroxycarboxylic acid ligands used in this study can be bound
in the mixed-ligand complex. On the other hand, the dissymmetry
factors for the ICT complexes reveal that two ICT ligands can be
bound by the Tb(III) complex.

With stoichiometric information in hand, it is an easy matter
to calculate equilibrium quotients for the addition of the chiral
ligands to the Tb/DPA complexes. The CPL intensity is directly
related to the amount of mixed-ligand complex that is formed, and
with knowledge of starting concentrations one can calculate the
equilibrium quotients of Table I (also shown are the dissymmetry

Table I. Limiting Dissymmetry Factors and Equilibrium Quotients
          at pH 5 of the Tb/DPA-Hydroxycarboxylate Complexes

| Ligand | Tb(DPA)(L) | | Tb(DPA)$_2$(L) | |
	g_{lum} x 10^2	K_1	g_{lum} x 10^2	K_1
HIC	-9.5	175	-9.7	81
ARG	-6.2	184	-5.9	55
HGT	-3.5	110	-3.4	30
ICT	3.0	380	1.4	165

factors of the fully formed complexes. One may immediately see that
the formation constants of the Tb(DPA) complexes are considerably
reduced from the corresponding constants of Tb(III) with the same
ligands. While K_1 normally ranges from 300 to 800 (1), we find
that for the Tb(DPA) mixed-ligand complexes K_1 ranges from 110 to
184, and for the Tb(DPA)$_2$ complexes K_1 ranges from 30 to 81. We
feel that these trends are a combination of increased steric crowding at the Tb (III) caused by the presence of bulky DPA ligands combined with a decreased effective positive charge resulting from the
binding of negative DPA ligands. The equilibrium quotients associated with the ICT ligand are quite a bit larger, and we feel that
this feature is a consequence of the terdentate bonding associated
with this ligand. In fact, ICT is observed to form a bis-complex
with Tb(DPA), and we have calculated that K_2 equals 50 for this
complex.

REFERENCES

1) A. E. Martell and R. M. Smith, Critical Stability Constants,
    Vol. 3, Plenum Press, New York, pp 24-60 (1977).
2) H. G. Brittain, Inorg. Chem., 19:2136 (1980).
3) H. G. Brittain, Inorg. Chem., 20:959 (1981).
4) H. G. Brittain, J. Am. Chem., Soc., 102:3693 (1980).
5) H. G. Brittain, Inorg. Chem., Acta., 53:L7 (1981).
6) H. G. Brittain, Inorg. Chem., in the press.
7) F. S. Richardson and J. P. Riehl, Chem. Rev., 77:773 (1977).
8) P. E. Schipper, J. Am. Chem. Soc., 100:1433 (1978).

TERBIUM LUMINESCENCE AS A PROBE OF LANTHANIDE COORDINATION IN SOLUTION

Frederick S. Richardson

Department of Chemistry, University of Virginia

Charlottesville, Virginia 22901

Similarities between the coordination chemistries of Ca^{2+} and trivalent lanthanide ions (Ln^{3+}) make the latter very attractive as "substitutional" spectroscopic probes of the Ca^{2+} binding properties of biomolecular systems. Unlike Ca^{2+}, the Ln^{3+} ions possess magnetic and optical spectroscopic properties which are readily accessible to experimental measurement and sensitive to ligand coordination properties. In aqueous solution media, terbium (Tb^{3+}) luminescence provides an especially sensitive probe for investigating Tb^{3+}/biomolecule interactions and coordination. Of special interest are (a) the induction of circular intensity differentials in the terbium luminescence by chiral ligands (commonly referred to as circularly polarized luminescence or CPL), and (b) the modulation of terbium luminescence intensities and lifetimes effected by specific kinds of Tb^{3+}-ligand interactions. An example of the latter is the sensitization and enhancement of terbium luminescence by a ligand possessing a near-UV chromophore which can function as an energy donor to the Tb^{3+} ion in a nonradiative energy transfer process. Measurements of these properties can be used to obtain binding constants and to characterize the nature of the lanthanide binding sites in biomolecular systems. Most studies reported to date have focused on Tb^{3+}-amino acid complexes and the binding of Tb^{3+} to protein molecules. Here we report luminescence studies on Tb^{3+} binding to small carbohydrates (monosaccharides and disaccharides), a series of ribonucleosides, and an ionophore with antibiotic activity.

CPL/emission studies were carried out on a series of Tb^{3+}-sugar and Tb^{3+}-sugar acid systems in DMSO and in aqueous solution. Strong emission optical activity (CPL) was observed for each of the sugar acid systems in DMSO and in aqueous solution at pH>4.5, indicating significant Tb^{3+}-sugar acid complexation. However, for the Tb^{3+}-sugar systems strong CPL was generated only in those cases where the sugar molecule could exist in a conformational form having three -OH ligating groups in an axial-equatorial-axial sequence (as, for example, in D-ribose, D-talose, and D-tagatose). In these cases the sugar molecule can chelate to the Tb^{3+} ion in a terdentate fashion via three weakly coordinating (but favorably positioned) hydroxyl ligating groups. Only very weak CPL was observed in those cases where the sugar molecule could not achieve an axial-equatorial-axial sequence of -OH groups.

CPL/emission and terbium luminescence enhancement studies were carried out on a seris of Tb^{3+}-ribonucleoside and Tb^{3+}-2'-deoxy-ribonucleoside systems in aqueous solution at neutral pH. In these studies, $[Tb^{3+}]$ = 7.5 mM and the $[Tb^{3+}]$:[nucleoside] ratios were either 1:1, 1:2, or 1:5. The terbium luminescence enhancement studies were carried out by using near-UV excitation of the base moieties in the nucleoside ligands. It was found that strong CPL is always accompanied by a large terbium luminescence intensity enhancement and vice versa. This observation implies that Tb^{3+}-base coordination is essential to strong Tb^{3+}-nucleoside complexation. Strong CPL and terbium luminescence enhancement were found only for those systems in which the ribosyl moiety of the nucleo-side could present two adjacent -OH groups as coordination donors, and the base moiety could present a carbonyl donor group. In these cases, the nucleoside can function as a terdentate ligand for Tb^{3+}. Uridine, inosine, and cytidine were found to generate the strongest CPL and terbium luminescence enhancement.

Antibiotic lasalocid A (X537A) is a known ionophore for both monovalent and divalent cations. It is also known to bind Ln^{3+} ions. Its structure includes a salicylic acid moiety at one end and an additional five oxygen atoms throughout the remainder of its structure. Six of the eight oxygen atoms in this molecule can serve simultaneously as metal ion ligators, and the molecule contains nine asymmetric carbon atoms in its skeleton. We have used the CPL/emission and terbium luminescence enhancement techniques to probe the structural details of Tb^{3+}-lasalocid A complexes in methanol and methanol-water solutions. In the terbium luminescence enhancement studies, the salicylic acid moiety was found to be an especially efficient energy donor for Tb^{3+} when it was bound.

The studies reported here show terbium luminescence enhancement and CPL/emission to be very sensitive probes of Ln^{3+}-ligand complex formation and structure.

ACKNOWLEDGMENTS

This work was supported by the National Science Foundation (NSF Grant CHE80-04209).

A ^{160}Tb(III) PROBE OF THE CALCIUM BINDING SITES OF MUSCLE CALCIUM BINDING PARVALBUMIN

Federico Gonzalez-Fernandez and Donald J. Nelson
Department of Chemistry
Jeppson Laboratory
Clark University
Worcester, Massachusetts   01610

ABSTRACT

A flow dialysis apparatus has been constructed which is suitable for the study of proteins which bind metal ions tightly (e.g., $pK_d > 6$). In the present experiments, muscle calcium binding parvalbumin was labelled with radioactive terbium-160, and dialyzed against buffer containing various competing metal cations. The rate of terbium-160 exchange was followed with a NaI(Tl) γ-ray scintillation detector. The following order of decreasing rates of terbium-160 exchange were observed: Tb(III)>Yb(III)>Gd(III)>La(III)> Cd(II)>Ca(II)>Mg(II)>Ba(II)>Sr(II)>Ni(II)>Co(II)>Zn(II).

INTRODUCTION

Muscle calcium binding parvalbumins (MCBP) are low molecular weight (mol. wt. approximately 12,000) acidic proteins present in chordates (1). They, or related cytosolic calcium binding proteins, may also be present in invertebrates as well (2). The crystal structure of one isotype from carp, MCBP-3, has been determined and refined to 1.9 Å resolution (3). The structure consists of six α-helical regions, designated helices A through F. A calcium ion is bound in the loop between helices C and D. A second calcium is bound in the EF loop. In each binding site, the ligands can be assigned to the vertices of an octahedron. All ligands are oxygen atoms of side chain groups except at one vertex where a peptide oxygen coordinates the calcium ion. All ligands of the CD calcium come from the protein; one ligand at the EF site is H_2O. While the CD and EF sites have similar affinities for calcium ion ($pK_d \approx 8.3$), Sowadski et al. (4) have shown, by x-ray difference Fourier

37

techniques, that trivalent terbium ion could replace the EF calcium without replacing the CD calcium, in the 2.8 M ammonium sulfate solution required to grow and maintain crystals. The preferential replacement of the EF calcium ion with initial additions of terbium ion is also supported by the carbon-13 NMR solution studies on lanthanide-substituted parvalbumin by Nelson et al. (5).

A number of dialysis techniques have been used to study metal ion binding to proteins with varying degrees of success (6). Most of the conventional methods are generally applicable only to systems with relatively low stability constants (i.e., $pK_d<5$). We have modified the flow-dialysis apparatus of Colowick and Womack (7) to make it especially suitable for the study of tight binding metallo-proteins.

METHODS

The Protein Solution

Carp muscle calcium binding parvalbumins (MCBP) were isolated from the white muscle of the common mirror carp (Cyprinus carpio) according to the procedure of Pechere et al. (8). The three major isotypes, MCBP-2, MCBP-3 and MCBP-5, were separated on a DEAE-52 cellulose ion-exchange column. Isotype three was used in all experiments except where otherwise indicated.

Prior to each set of exchange experiments, a fresh 10 ml stock of labelled protein solution was prepared. All solutions were pre-pared in piperazine buffer (10 mM piperazine, 20 mM KCl, pH=6.5). The concentration of parvalbumin was 0.02 mM except for those experiments presented in Figure 5 where it was 0.1 mM. The protein was labelled by introducing into the stock solution 200 μl of a Tb(III)-160 solution, prepared by adding 4.0 ml of distilled water to 0.1 ml of 1 microcurie, carrier-free $^{160}TbCl_3$ (New England Nuclear, Cambridge, Mass. U.S.A.). Following overnight incubation, the protein-terbium mixture was exhaustively dialyzed against piperazine buffer at 5°C to remove any unbound $^{160}Tb(III)$. The amount of terbium added was negligible compared to the MCBP-3 con-centration (i.e., ($^{160}Tb(III)/(MCBP-3)<<0.1$), to favor uptake of the lanthanide ion by the solvent-exposed EF site (5).

The Flow Dialysis System

The flow dialysis apparatus is shown schematically in Figure 1. The dialysis cell, which was milled from two blocks of solid lucite, was composed of a buffer half-cell and a protein half-cell. The buffer half-cell contained a 5 cm^3 cavity and was equipped with inlet and outlet ports. The protein half-cell contained a 1 cm^3 cavity. The two half-cells were bolted together, separated only by

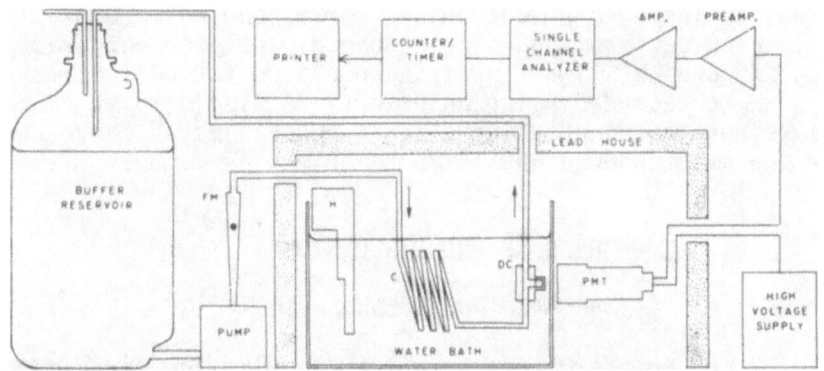

Fig. 1. The flow dialysis system. The protein sample labelled
with terbium-160 was placed in the 1 cm³ chamber of the dialysis
cell (DC). Buffer usually containing a competing metal ion was
pumped through the buffer side of the dialysis cell at a rate of
200 ml/min. The "washout" of terbium-160 was followed with a NaI
(Tl) scintillation detector, optically coupled to a photomultiplier
tube (PMT), (FM=flow meter; H=heater; C=coil of tubing).

a piece of dialysis membrane. The area of diffusion across the
membrane was 1 cm². Buffer, usually containing a competitive metal
ion (e.g., Ca(II), Tb(III), etc.) was pumped through the buffer
half-cell at 200 ml/min with a peristaltic pump.

The dialysis apparatus was operated in a lead house to reduce
background radiation. The cell was mounted approximately one cm
from a 3 in. x 3 in. cylindrical Harshaw NaI(Tl) scintillation crystal detector, optically coupled to a photomultiplier tube. A bias
voltage of ca. 1.5 kV was applied to the photomultiplier tube using
an Ortec 456 power supply. Since the external reservoir was located
outside the lead house, only ¹⁶⁰Tb(III) bound to protein was
detected.

The Counting Apparatus

The output of the photomultiplier tube was amplified using an
Ortec No. 113 preamplifier and then further amplified and shaped
with an Elscint CAV-3 linear amplifier. In order to minimize the
effect of background radiation and to effectively integrate the
¹⁶⁰Tb(III) 0.879 MeV γ-ray peak, a single-channel analyzer (Ortec

420-A) was employed to select pulses corresponding to 0.879 ± 0.50 MeV. Logic pulses from the single channel analyzer were routed to an Ortec 776 counter/timer. Data accumulated for dwell times of 2000 sec. were recorded using an Ortec 777A line printer. Exchange half-lives were obtained using a least-squares decay curve analysis program implemented on a PDP 11/70 computer.

RESULTS AND DISCUSSION

The Alkaline Earth Series

Fig. 2 presents exchange data for the alkaline earth metal ions Ca(II), Mg(II), Sr(II) and Ba(II). As expected, calcium ion displaces ^{160}Tb(III) from the solvent-exposed EF site more readily than any of the other alkaline earth metal ions tested. The smaller Mg(II) ion is found to displace bound terbium more readily than the larger alkaline earth metals Sr(II) and Ba(II).

To understand the exchange half-life pattern for the alkaline earth series presented in Fig. 2, metal ion charge density as well as ionic radius must be considered. A cation must first shed its outer sphere of coordinating water molecules before it can interact with the ligands in the metal ion binding site of the protein. Among the alkaline earth metal ions tested, Mg(II) has the most difficulty in shedding solvent water because of its high charge density. Eigen (9) determined that the rate of water substitution is more than three orders of magnitude slower for Mg(II) (10^5 sec^{-1}) than for Ca(II) (5×10^8 sec^{-1}). The comparatively long ^{160}Tb(III) exchange half-life for Mg(II) can be attributed in part to this solvent effect. In contrast, Sr(II) and Ba(II), despite their low charge density, are simply too large to be accommodated as well as Ca(II) in the parvalbumin binding site.

The Lanthanides: La(III), Gd(III), Yb(III), Tb(III) and Er(III).

Terbium-160 exchange half-lives are more than an order of magnitude shorter for the lanthanide metal ions than the alkaline earth metals. This result is presented in Fig. 3A where calcium ion is compared with three members of the lanthanide series. The half-life for the exchange of ^{160}Tb(III) by non-radioactive Tb(III), which is not shown in Fig. 3A, is 10 hours (correlation coefficient= 0.9993). The Ca(II) data in Fig. 2 are the same as that in Fig. 3A except expanded vertically by a factor of 4.7.

The comparatively short exchange half-lives observed for all the lanthanides can be attributed to their higher ionic charge (+3) and the fact that their ionic radii are very similar to that of divalent calcium. At 1 mM levels, the exchange half-lives for the various lanthanide ions tested are not significantly different

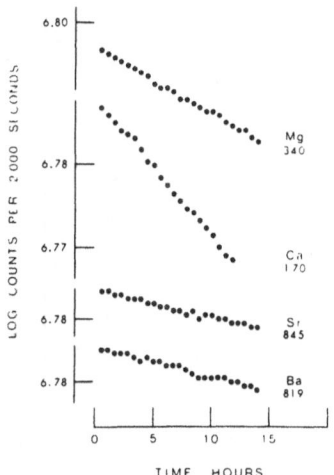

Fig. 2. Time-activity spectra for the exchange of MCBP-3 bound
terbium-160 with competing Mg(II), Ca(II), Sr(II) and Ba(II) in
the washout buffer. Each point corresponds to 2000 sec. of data
acquisition. The concentration of the competing metal was 1.0 mM.
Protein concentration was 0.02 mM. The numbers beneath the atomic
symbols are the half-lives for [160]Tb exchange in hours.

(i.e., half-lives≈10 hrs.). To obtain greater time resolution
needed to discriminate amongst the various lanthanides, lanthanide
ion concentration in the washout buffer was lowered from 1.0 mM to
0.1 mM (Fig. 3B). Lowering the lanthanide ion concentration by an
order of magnitude increased the exchange half-lives 20 to 80%.
The half-life for La(III) is significantly longer than those of the
other lanthanides tested. This is explained in part by the fact
that the ionic radius of La(III) (0.115 nm) is significantly larger
than the ionic radius of Ca(II) (0.099 nm) and the ionic radii of
the other lanthanides examined (Gd(III), 0.102 nm; Tb(III), 0.100 nm
and Yb(III), 0.094 nm) (6). Tb(III), the lanthanide ion most
commonly used as a probe of calcium binding sites in biological
systems, exhibits the shortest exchange half-life.

The Transition Metals: Co(II), Ni(II), Zn(II) and Cd(II)

The fourth row transition metals Co(II), Ni(II) and Zn(II) have
a longer [160]Tb(III) exchange half-life than the alkaline earth
metals (Fig. 4). The average half-life for [160]Tb(III) exchange is
almost an order of magnitude longer for these metals than for
Ca(II). Co(II), Ni(II) and Zn(II) are not well accommodated by
either the geometry of the binding site or the ligand types at the
parvalbumin binding site. Co(II), Ni(II) and Zn(II) do have a high

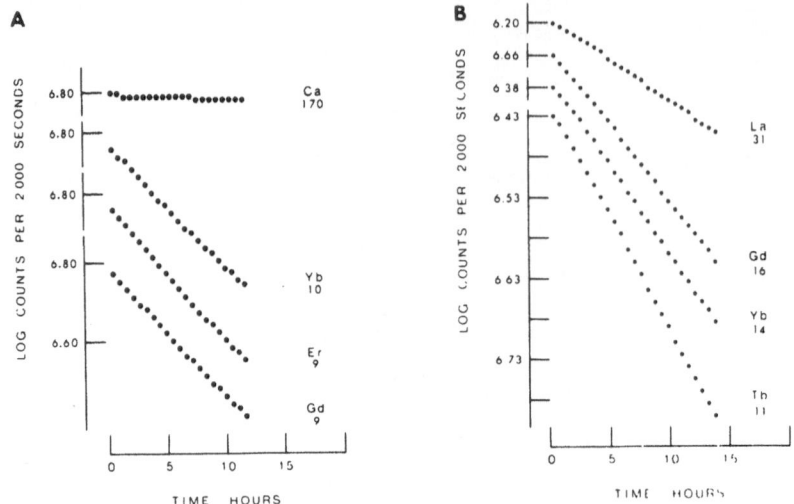

Fig. 3. Time-activity spectra for the exchange of MCBP-3 bound terbium-160 with competing Ca(II) and various members of the lanthanide series. Each point corresponds to 2000 sec. of data acquisition. Protein concentration was 0.02 mM. The numbers beneath the atomic symbols are the half-lives for ^{160}Tb(III) exchange in hours. A: Concentration of competing metal was 1.0 mM B: Concentration of competing metal was 0.1 mM.

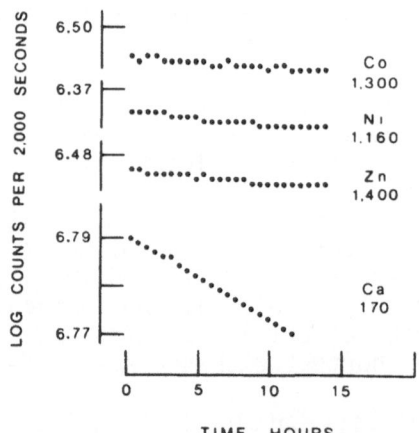

Fig. 4. Comparison of fourth row transition metal ions Co(II), Ni(II), and Zn(II) with Ca(II). The half-lives for ^{160}Tb(III) exchange are shown beneath the atomic symbol for the competing metal ion. Protein concentration was 0.02 mM. Concentration of competing metal ion was 1.0 mM.

affinity for metallo-enzymes such as carbonic anhydrase and carboxy-
peptidase; however, the coordination geometry is tetrahedral, rather
than octahedral, and the preferred ligand for these metals is
nitrogen, rather than oxygen. Six-coordinate complexes of Co(II),
Ni(II) and Zn(II) do exist, but the preferred geometry is trigonal
prismatic, not octahedral. (The reduced preference for the octahedron can be explained by the angular overlap metal (10).)

Metal ion exchange experiments employing Cd(II), a fifth row
transition metal ion in Group IIB, yielded particularly interesting
and biologically relevant results. The similar ionic radii of
Cd(II) and Ca(II) (0.092 nm and 0.099 nm, respectively) suggests
that Cd(II) might bind readily to the protein, provided there are
no coordination geometry or ligand type constraints. Fig. 5
presents 160Tb(III) exchange data for Cd(II) and Ca(II). (The
Ca(II)-induced exchange half-lives presented in Figs. 2 and 5 are
not directly comparable, as the protein concentration (MCBP-5)
employed in the experiments represented in Fig. 5 are five times
higher than that associated with Fig. 2).

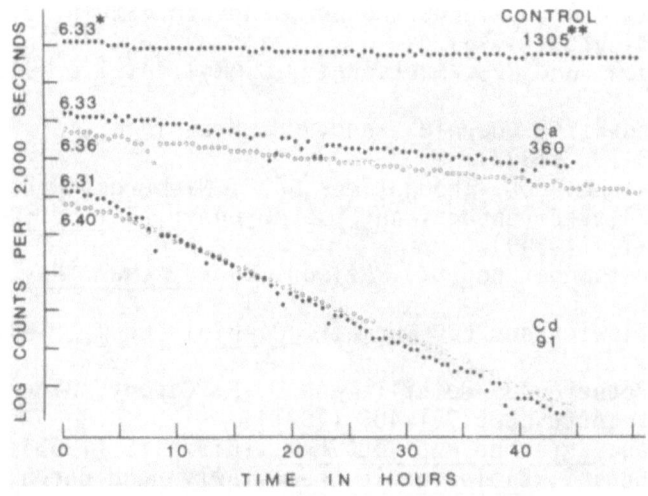

Fig. 5. Comparison of Ca(II)- and Cd(II)-induced exchange of
160Tb(III) from parvalbumin. Each division in the vertical scale
equals 0.025 log units. In the control experiment, no competing
metal ion was present in the washout buffer. Open circles are
repeat experiments. Protein concentration was 0.1 mM. Concentration of competing metal ion was 3 mM. *=Initial log counts per
2000 sec. **=Average half-life in hrs. Error (one-half the
difference between half-lives): Ca, t1/2 = 360 ± 24 hrs.;t
Cd, t1/2 = 91 ± 5 hrs.

The Cd(II) exchange half-life is shorter by a factor of 4 than that of Ca(II), and by implication (see Fig. 4) well over an order-of-magnitude shorter than that of the fourth row transition metals examined (i.e., Co(II), Ni(II) and Zn(II)). It is interesting to find a dramatic difference between the two group IIB metal ions, Zn(II) and Cd(II). This difference can be attributed to the coordination preference of these two metals. According to Cotton and Wilkinson (11), "there is no ligand field stabilization effect in Zn(II) and Cd(II) ions because of their completed d shells, (and therefore) their stereochemistry is determined solely by consideration of size, electrostatic forces, and covalent bonding forces. The effect of size is to make Cd(II) more likely than Zn(II) to assume a coordination number of six." The reproducible finding that the Cd(II) exchange half-life is substantially shorter than the half-life for the naturally occurring metal ion, Ca(II), has clear toxicological relevance to the problem of "Cd(II)-poisoning" of biologically essential calcium systems, and should be investigated further.

REFERENCES

1. R.H. Kretsinger, C.R.C. Crit. Rev. Biochem, 8:119-174 (1980).
2. L. Wnuk, J.A. Cox, L.G. Kohler, and E.A. Stein, J. Biol. Chem., 254:5284-5289 (1979).
3. P.C. Moews and R.H. Kretsinger, J. Mol. Biol., 91:201-228 (1975).
4. J. Sowadski, G. Cornick, and R.H. Kretsinger, J. Mol. Biol., 124:123-132 (1978).
5. D.J. Nelson, A.D. Theoharides, A.C. Nieburgs, R.K. Murray, F. Gonzalez-Fernandez, and D.S. Brenner, Int. J. Quantum Chem., 16:159-174 (1979).
6. R.H. Kretsinger and D.J. Nelson, Coord. Chem. Rev., 18:29-124 (1976).
7. S.P. Colowick and F.C. Womack, J. Biol. Chem., 244:744-777 (1969).
8. J.-F. Pechere, J. Demaille and J.-P. Capony, Biochim. et. Biophys. Acta, 236:391-408 (1971).
9. M. Eigen, Pure and Applied Chem., 6:97-115 (1963).
10. E.-I. Ochiai, Bioinorganic Chem., Allyn and Bacon, Boston, pp. 365-370 (1977).
11. F.A. Cotton and G. Wilkinson, Adv. Inorganic Chem., John Wiley and Sons, New York, p. 504 (1972).

THE LANTHANIDES AS STRUCTURAL PROBES IN PEPTIDES

Robert E. Lenkinski, and Richard L. Stephens*

University of Guelph, Guelph,
Ontario, Canada N1G 2W1
*University of Alabama in Birmingham
Birmingham, AL 35294

We have been interested in developing a methodology based on the analysis of paramagnetic lanthanide ion perturbations observed in the NMR parameters of the peptide. One of the apparent limitations of this approach is the fact that in aqueous solution the peptide must contain a suitable metal ion binding site. Recently we have been able to show, however, that a relatively minor chemical modification of a neurohypophyseal hormone (oxytocin) produced a metal ion binding derivative (1). In the present communication we will discuss the results of NMR experiments performed on two peptides as an illustration of our approach. These peptides are angiotensin II, (Asp-Arg-Val-Tyr-Ile-His-Pro-Phe) a hormone which contains a metal ion binding site and [Glu^4]-oxytocin
(oxytocin ≡ Cys-Tyr-Ile-Gln-Asn-Cys-Pro-Leu-Gly-NH$_2$).

THEORETICAL BACKGROUND

The line broadenings (transverse relaxation rate enhancements) of a given resonance in the presence of a paramagnetic ion can be expressed as

$$1/T_{2p} = \pi(\Delta\nu_{1/2}{}^M - \Delta\nu_{1/2}{}^O) \tag{1}$$

where $\Delta\nu_{1/2}{}^M$ and $\Delta\nu_{1/2}{}^O$ are the linewidths at half-height (in hertz) of the resonance with and without the paramagnetic metal ion, respectively, and $1/T_{2p}$ is the paramagnetic contribution to the transverse relaxation rate in reciprocal seconds. In the absence of any outer-sphere relaxation and when $1/T_{2M} \geq \Delta\omega_M$ (where ω_M is the chemical-shift difference between the free and bound states, and T_{2M} is the relaxation rate of the resonance in its complexed state), the concentration dependence of $1/T_{2p}$ is given by (2,3)

$$1/T_{2p} = P_M/(T_{2M} + \tau_m) \tag{2}$$

where P_M is the fraction of ligand bound to the metal ion and τ_m is the mean residence time of the ligand in its complexed state. When the relaxation enhancement arises solely through the dipolar interaction, the following equation holds (4,5):

$$1/T_{2M} = \frac{\gamma_I^2 \beta^2 S(S+1)g^2}{15r^6} \left[4\tau_c + \frac{3\tau_c}{1+\omega_I^2\tau_c^2} + \frac{13\tau_c}{1+\omega_S^2\tau_c^2} \right] \tag{3}$$

ω_I and ω_S are the nuclear and electronic Larmor frequencies related to the resonance frequency by $\omega_I = 2\pi\nu$ and $\omega_S = 660\ \omega_I$, γ_I is the magnetogyric ratio of the nucleus being observed, β is the Bohr magneton, S is the total electron spin, and r is the distance between the nucleus and the paramagnetic ion. The correlation time τ_c in Eq. (3) is defined as

$$1/\tau_c = 1/\tau_s + 1/\tau_r + 1/\tau_m \tag{4}$$

where τ_r is the rotational correlation time of the complex, τ_s is the electron spin relaxation time of the paramagnetic electron, and τ_m is the mean residence time of the ligand in the metal complex.

LANTHANIDE BINDING STUDIES OF ANGIOTENSIN II

The effects of the successive additions of Eu^{3+} are shown in Figure 1. A scatchard plot of representative europium induced shifts is linear with an intercept on the x-axis near unity indicating the formation of a 1:1 metal hormone complex. A value for the dissociation constant of $6.5 \times 10^{-3} M^{-1}$ (pH-6.5) was obtained by non-linear least squares fit of the Eu^{3+} induced shifts. It is important to realize that the value of this dissociation constant should be very similar in the Gd^{3+} hormone complex. (See Reuben (6) for other examples).

The effects of La^{3+} on the lowfield region of the angiotensin II spectrum are shown in Fig. 2. The minor shifts of the His^6 C2-H and the Phe^8 NH are believed to be due, at least in part, by a slight variation in pH between the two samples. Otherwise, there are no major changes in either chemical shift or the NH-C$^\alpha$H coupling constants. This lack of any significant perturbation on the angiotensin II NMR spectrum in the presence of La^{3+} indicates that the backbone conformation of the hormone is essentially the same in the free peptide and the metal complex. Therefore, we suggest that conclusions reached about the topology of the metal-peptide complexes can be expected to be applicable to the uncomplexed hormone with no more than minor refinements.

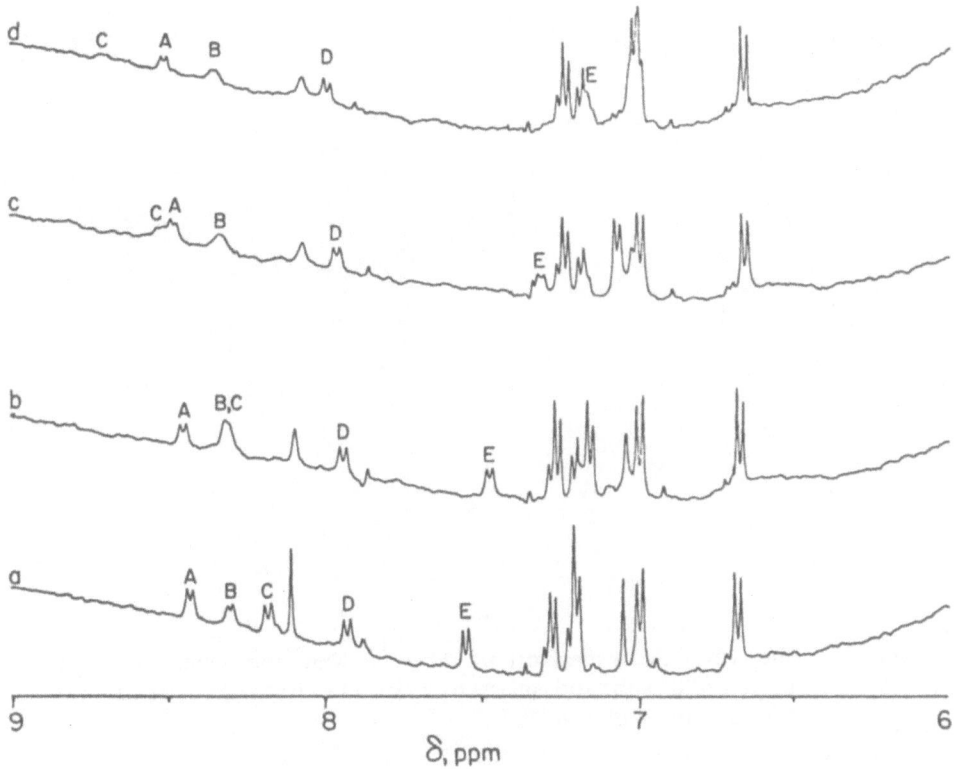

Figure 1. The low-field region of the ^{1}H NMR spectrum of angiotensin II at 400 MHz and pH = 6.5 in the presence of a) no EuCl$_3$, b) 3.10 mM EuCl$_3$, c) 12.4 mM EuCl$_3$ and d) 24.8 mM EuCl$_3$. The five NH resonances observed are labeled A-E. For complete assignments see Figure 2.

The effects of Gd^{3+} on the peptide NH resonances of angiotensin II were monitored in H$_2$O. The addition of Gd^{3+} results in differential broadenings of the five observed NH resonances (note that at pH's above 4.0 the NH of Arg^2 becomes too broad to observe). The variation in the linewidths of these resonances with increasing Gd^{3+} concentrations were analyzed in terms of equations (2) and (3) to give the NH – Gd^{3+} distances shown in Table 1. These calculations were performed following an approach which we have outlined in detail in a previous paper (7).

LANTHANIDE BINDING STUDIES OF [Glu^4] OXYTOCIN

The effects of the presence of La^{3+} ions on the downfield

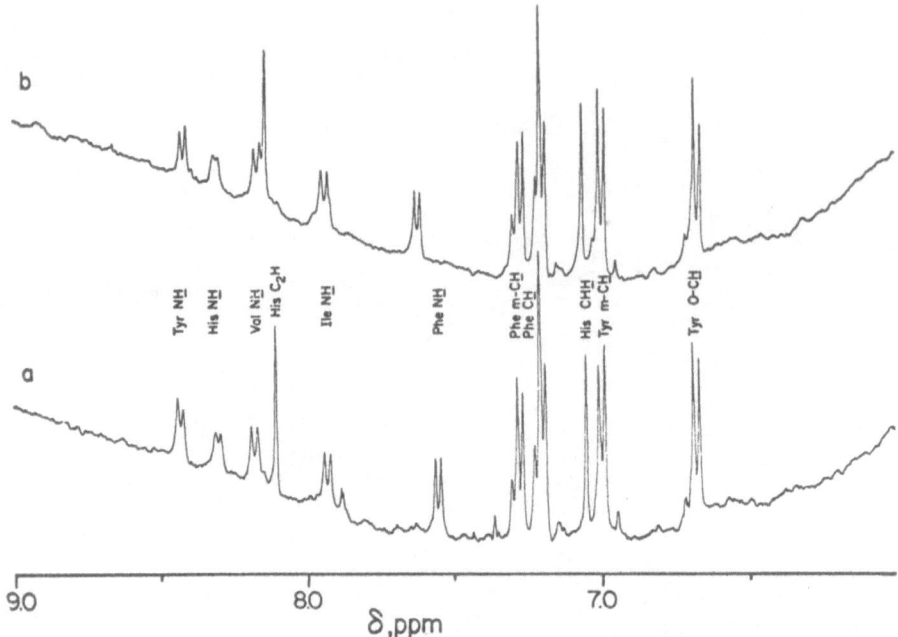

Figure 2. The low-field region of the ^{1}H NMR spectrum of angiotensin
II at 400 MHz and pH = 6.5 in the presence of a) no La^{3+}
and b) 34.0 mM La^{3+}.

Table 1. Average metal-nuclear distances calculated for
the Gd^{3+} - angiotensin II complex.

Atom	$1/T_{2M}$(Hz)[a]	$\hat{R}$(Å)[b]	R_{min}(Å)[c]	R_{max}(Å)[c]
Val^3 NH	2621.3 ± 695.6	6.6	6.3	6.9
Tyr^4 NH	588.5 ± 153.6	8.4	8.1	8.8
Ile^5 NH	106.2 ± 94.1	11.2	10.1	16.1
His^6 NH	459.6 ± 50.5	8.8	8.6	8.9
Phe^8 NH	2130.7 ± 313.4	6.8	6.6	7.0
His^6 C2-H	1389.2 ± 350.7	7.3	7.1	7.7

[a]Determined from the slopes of plots of $1/T_{2p}$ vs. P_M. Errors represent 90% confidence intervals of the slopes of $1/T_{2p}$ vs. P_M.
[b]Estimated metal-proton distances.
[c]Estimated 90% confidence limits of $\hat{R}$ calculated from the confidence limits of $1/T_{2M}$.

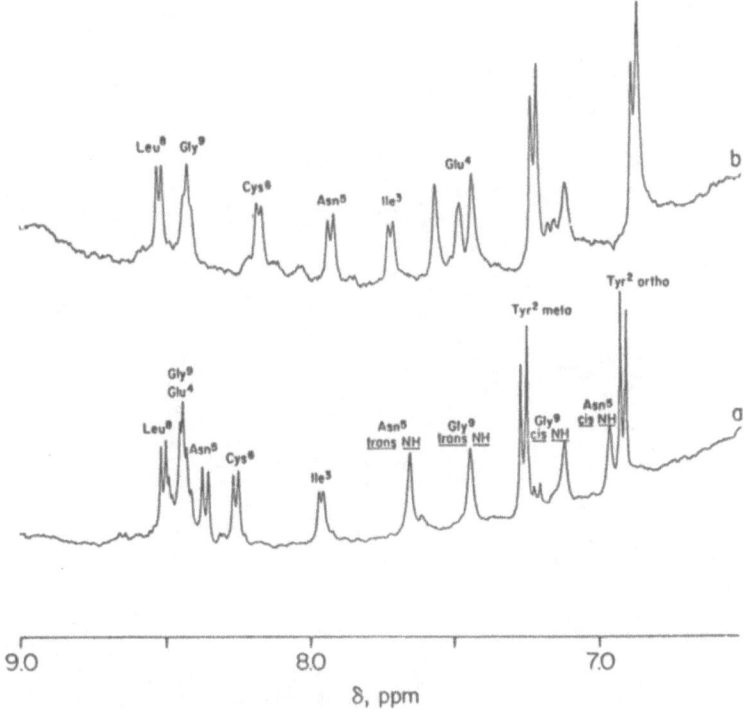

Figure 3. The downfield region of the ^{1}H n.m.r. spectrum of [Glu^4] oxytocin in H$_2$O, 9 mM, pH 6.1 (a) with no La^{3+} present and (b) in the presence of 43 mM La^{3+}. Each spectrum is the sum of 128 scans.

portion of the ^{1}H NMR spectrum of [Glu^4] oxytocin are shown in Fig. 3. The presence of La^{3+} ions results in large perturbations in the chemical shifts of the peptide NH's of Glu^4 and Asn^5; smaller perturbations in the chemical shifts of peptide NH's of Ile^3 and Cys^6 and the cis NH of the primary amide of Asn^5. Note that these shifts are accompanied by little or no change in the coupling constants of the peptide NH's indicating that the conformation(s) of the peptide backbone of this molecule is probably not significantly altered on the binding of metal ions. The aforementioned chemical shift perturbations could result from electrostatic effects not associated with conformational changes.

The effects of the addition of Yb^{3+} on the downfield portion of the ^{1}H NMR spectrum of [Glu^4] oxytocin are shown in Fig. 4. Note that the chemical shift perturbations are concentration dependent indicating that the molecule is in fast exchange between its free and complexed state on the ^{1}H NMR chemical shift time scale. Scatchard analysis of the shift data indicated that the complex has a 1:1 stoichiometry. The limiting Yb^{3+} shifts obtained from our analysis are given in Table 2. We have also measured the Gd^{3+} induced

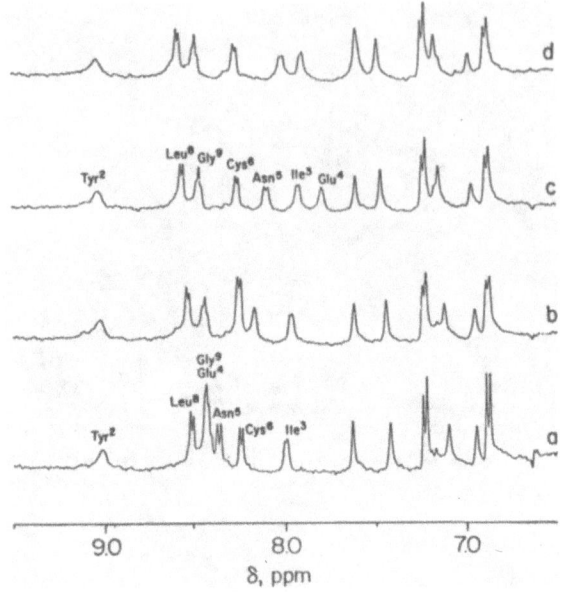

Figure 4.   The downfield region of the ^{1}H NMR spectrum of [Glu[4]]
            oxytocin in H_2O, 9 mM, pH = 5.8 (a) no Yb^{3+}, (b) 2.82 mM
            Yb^{3+}, (c) 5.22 mM Yb^{3+}, and (d) 10.7 mM Yb^{3+}. Each spectra
            is the sum of 128 scans.

linebroadening in the downfield portion of the proton spectrum of
[Glu[4]] oxytocin. These results were analyzed in a manner similar to
that described in a previous report (6). The distances obtained from
this analysis are listed in Table 2.

    The closest peptide NH is clearly that of the Glu[4] residue indi-
cating that the sidechain carboxyl forms at least part of the metal
binding site. The peptide NH of Asn[5] is also relatively close to the
metal ion as well. Note that the NH's of the Asn[5] sidechain amide are
also relatively near to the metal ion binding site with the cis NH
being somewhat closer than the trans NH. An examination of a model
for the Asn[5] residue indicates that the calculated distances are consistent with the participation of the sidechain carbonyl of Asn[5] as
well as the carboxyl group of Glu[4] in metal binding. Note that the
meta protons of the tyrosine ring are farther removed from the metal
binding site than the corresponding ortho protons, indicating that on
the average there is some preferred orientation of this sidechain
with respect to the metal ion.

Table 2. Gd^{3+}-induced relaxation rate enhancements, average metal-nuclear distance and Yb^{3+}-induced limiting shifts observed in the downfield region of the ^1H NMR spectrum of $[Glu^4]$ oxytocin in H_2O.

Resonance	$1/T_{2M}$,[a] Hz	$<r>_{AV}$, Å	Yb^{3+} Limiting Shifts[e] Hz
m-Tyr^2 CH	5.4×10^3	7.0 ± 0.5	f
o-Tyr^2 CH	1.4×10^4	5.5 ± 0.5	f
Tyr^2 NH	b	>8	f
Ile^3 NH	c	–	−80
Glu^4 NH	d	<4.5	−728
Asn^5 NH	3.9×10^4	4.8 ± 0.5	−320
Asn^5 cis NH	5.4×10^4	4.7 ± 0.5	30
Asn^5 trans NH	1.2×10^4	6.3 ± 0.4	−25
Cys^6 NH	9.5×10^3	6.2 ± 0.5	16
Leu^8 NH	b	>8	49
Gly^9 NH	b	>8	51
Gly^9 cis NH	b	>8	61
Gly^9 trans NH	b	>8	54

[a] Obtained from the slope of plots of $1/T_{2p}$ vs P_M.
[b] Not significantly broadened at highest Gd^{3+} concentration used.
[c] Could not be measured because of overlap.
[d] Broadened beyond detection at lowest Gd^{3+} concentration employed.
[e] A positive sign indicates a downfield shift.
[f] Shift was too small to measure.

REFERENCES

1. R. Walter, C.W. Smith, K.P. Sarathy, R.P. Pillai, N.R. Krishna, R.E. Lenkinski, J.D. Glickson and V.J. Hruby, Int. J. Peptides and Proteins, in press (1981).
2. T.J. Swift and R.E. Connick, J. Chem. Phys., 37:307 (1962).
3. Z. Luz and S. Meiboom, J. Chem. Phys., 40:2686 (1964).
4. I. Solomon, Phys. Rev., 99:559 (1955).
5. N. Bloembergen, J. Chem. Phys., 27:572 (1957).
6. J. Reuben, in Handbook on the Physics and Chemistry of the Rare Earths, K.A. Gschneidner and L. Eyring (eds.), North Holland, New York, pp. 515-552 (1979).
7. R.E. Lenkinski, J.L. Dallas and J.D. Glickson, J. Amer. Chem. Soc., 101:3071 (1979).

MULTINUCLEAR NMR STUDY OF THREE AQUEOUS LANTHANIDE SHIFT REAGENTS:

COMPLEXES WITH EDTA AND TWO MACROCYCLIC LIGANDS

Charles C. Bryden, Charles N. Reilley, Kenan Laboratories
of Chemistry, University of North Carolina, Chapel Hill,
N.C. 27514; Jean F. Desreux, Analytical and Radiochemistry, University of Liege, Sart Tilman, B-4000
Liege, Belgium

INTRODUCTION

Lanthanide aquo ions have frequently been used as shift
reagents for the determination of the structure of water soluble
molecules; more recently, lanthanide EDTA complexes have been used
for the same purpose (1,2). The applicability and utility of
aqueous shift reagents could be further increased if problems such
as limited pH range, presence of higher than 1:1 complexes, contact
shifts, and non-axial symmetry could be overcome. With these
considerations in mind, we have investigated the shift reagent
properties of the lanthanide complexes of two ligands, DOTA and NOTA,
and we have compared them to the LnEDTA shift reagents.

DOTA   NOTA   EDTA

The complexes with DOTA and NOTA are axially symmetric. Moreover,
the LnDOTA complexes are eight-coordinate, and are unlikely to form
higher than 1:1 adducts with substrate molecules because of steric
crowding (7). Potentiometric titrations show that LnDOTA complexes
do not coordinate OH^- even at pH 12. These features suggest that
LnNOTA and LnDOTA complexes could be superior to LnEDTA as aqueous
shift reagents. To determine the relative utility of the three
types of complexes as aqueous shift reagents with respect to pos-
itively charged, neutral, and negatively charged substrates, we

53

measured the ^{23}Na shifts of Na^+, the ^{2}H and ^{17}O shifts of deuterated water, and the ^{35}Cl shifts of Cl^- in 0.1M solutions.

EXPERIMENTAL

The ligand DOTA was prepared as described in (7), and the ligand NOTA was prepared as the trisodium salt according to (8). All spectra were obtained on a Varian XL-100/12 NMR spectrometer modified for multinuclear observation as described in (9).

DISCUSSION

The paramagnetic shift is a combination of a contact (through-bonds) shift and a dipolar (through-space) shift, expressed by the equation $\delta = F<S_z> + GC^D$ where the measured shift in ppm is δ, the first term on the right hand side represents the contact shift as the product of a ligand dependent parameter F and a lanthanide ion dependent parameter $<S_z>$, and the second term on the right hand side represents the dipolar shift as the product of a ligand dependent parameter G and a lanthanide ion dependent parameter C^D (3). The lanthanide dependent parameters $<S_z>$ and C^D have been calculated for all the lanthanides (4-6) and the separation between the contact and the dipolar contributions was performed as described elsewhere (3,9). When the shift reagent is axially symmetric (at least a C_3 axis), G is proportional to the crystal field coefficient $<r^2>A_2^0$, and a geometric term $(3\cos^2\theta - 1)r^3$, where r and θ are the coordinates of the observed nucleus. It is the dipolar shift that is of interest for the determination of the structures of substrate molecules.

The NMR shifts induced by the DOTA, EDTA and NOTA complexes and by the aquo lanthanides (4,10) are presented in Figs. 1 and 2. The solid line in each plot represents the dependence of the pseudo-contact parameter C_D or the contact parameter $<S_z>$ scaled to the observed shifts.

The most outstanding aspect of the results in Figs. 1 and 2 is that all of the shifts in DOTA solutions follow the theoretical dipolar shifts. In solutions of the other three complexes, the water ^{17}O shifts follow the theoretical contact shifts. It is likely that the very different shifting patterns of the complexes originate from their structures in solution. NMR studies (7) have indicated that the DOTA complexes are sterically very crowded, the encapsulated metal ions being completely surrounded by the tetra-azacycle and by the four acetate groups. Furthermore, it can be assumed that these substrates are located near to the carboxylate groups of DOTA in the vicinity of the C_4 axis since the dipolar shifts of all the nuclei investigated exhibit the same sign. In contrast, the NOTA, EDTA, and aquo complexes have more open, less

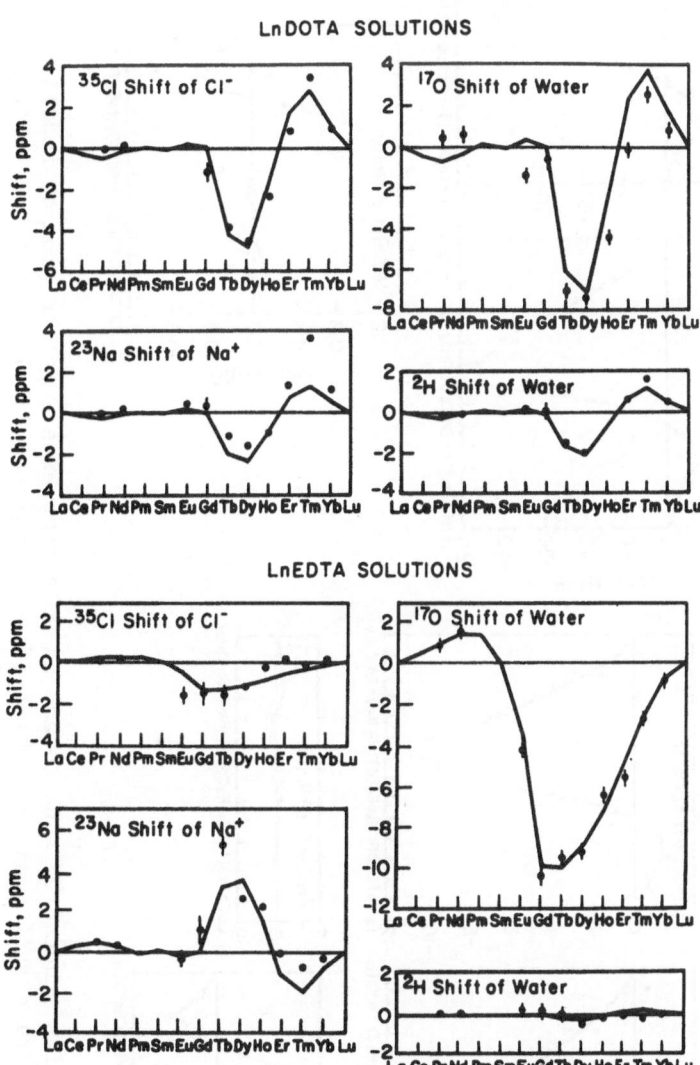

Figure 1.   Measured vs. theoretical contact or dipolar NMR shifts
            induced by Ln DOTA and EDTA.

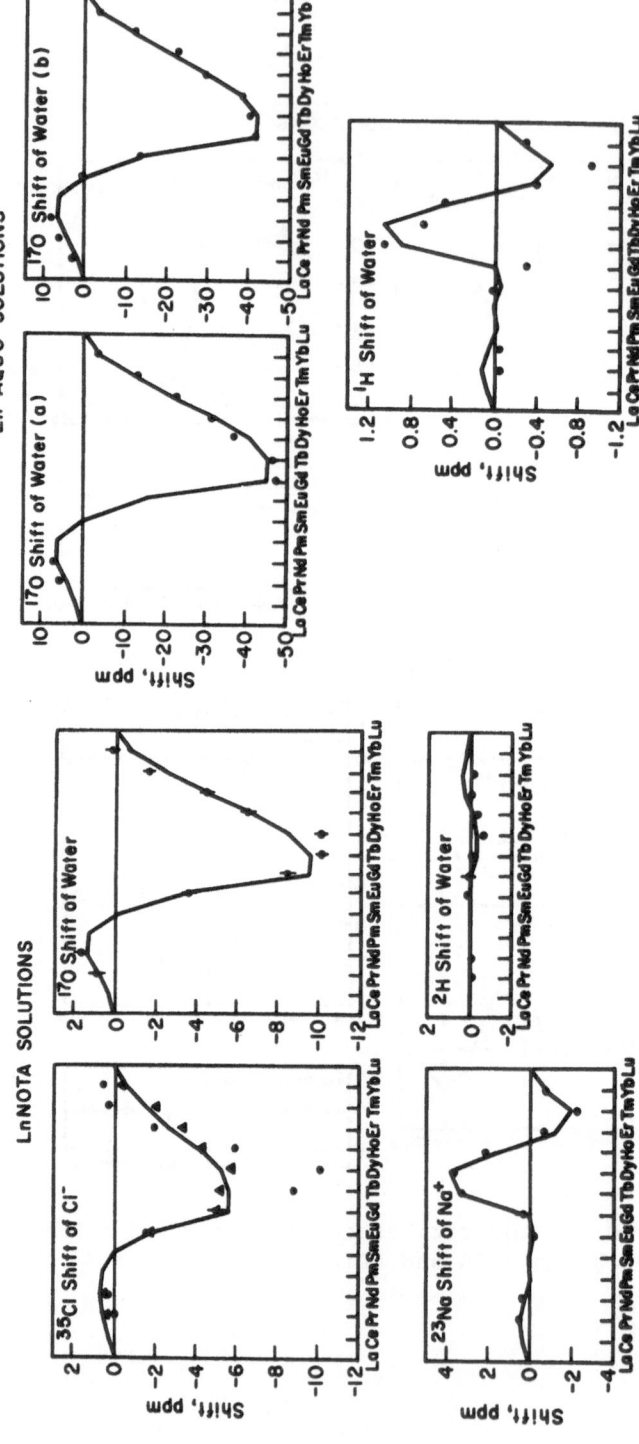

Figure 2. NMR shifts induced by aquo Ln and Ln NOTA complexes. In the plot of LnNOTA ^{35}Cl shifts, • = observed shift, and ▲ = shift after calculated dipolar shift was subtracted. The Ln aquo ^{1}H shifts, and the Ln aquo ^{17}O shifts labeled (a) are from (10), the ^{17}O shifts labeled (b) are from (4). All Ln aquo shifts were extrapolated to 0.1 M.

crowded structures. A water molecule or a chloride ion can thus be directly coordinated to the lanthanide ion in these complexes in keeping with the large contact shifts observed.

As the contact shifts contain no structural information, the best shift reagents of those studied clearly are the Ln DOTA complexes.

ACKNOWLEDGMENTS

We gratefully acknowledge research support from the National Science Foundation and the F.R.N.S. of Belgium. J.F.D. is a Chercheur Qualifié at this institution.

REFERENCES

1. C.M. Dobson, R.J.P. Williams, A.V. Xavier, J. Chem. Soc., Dalton Trans., 1762 (1974).
2. G.A. Elgavish, J. Reuben, J. Am. Chem. Soc., 98:4755 (1976).
3. C.N. Reilley, B.W. Good, J.F. Desreux, Anal. Chem., 47:2110 (1975).
4. W.B. Lewis, J.A. Jackson, J.F. Lemons, H. Taube, J. Chem. Phys., 36:694 (1962).
5. R.M. Golding, M.P. Halton, Aust. J. Chem., 25:2577 (1972).
6. B. Bleaney, J. Magn. Reson., 8:91 (1972).
7. J.F. Desreux, Inorg. Chem., 19:1319 (1980).
8. D.S. Everhart, R.F. Evilia, unpublished work, University of New Orleans.
9. B.W. Good, Ph.D. Thesis, University of North Carolina, 1978.
10. J. Reuben, D. Fiat, J. Chem. Phys., 51:4909 (1969).

STRUCTURES OF DITHIOPHOSPHINATE COMPLEXES OF THE LANTHANIDES IN THE SOLID AND SOLUTION BY X-RAY DIFFRACTION AND PARAMAGNETIC NMR

A. Alan Pinkerton

Institut de Chimie Minérale et Analytique
Université de Lausanne, Place du Château 3
CH-1005 Lausanne, Switzerland

The determination of the structures of lanthanide complexes in solution poses certain problems. Even the determination of the point symmetry of the lanthanide ion from spectroscopic data is not always unambiguous as the crystal field splittings are small and may not always be resolved at normal temperatures. Paramagnetic NMR is a powerful tool for obtaining geometric information in solution. However, the total observed isotropic shifts are the sum of a dipolar term and a scalar term, and only the dipolar contribution is structure dependent (1). In the past this has limited the experiment to the measurement of proton shifts which are predominantly dipolar in origin. We have since developed a method for separating the dipolar and scalar contributions to the total observed shifts (2) and thus may study any nucleus in a complex that has a non-zero spin. A bonus in this procedure is that we also obtain the value of the hyperfine coupling of the observed nucleus with the unpaired electron spin at the same time, thus obtaining a measure of the covalency of the metal-ligand bond.

We have been studying dithiophosphinate ($R_2PS_2^-$) complexes of the lanthanides for several years (3) and have now measured the ^{31}P and ^1H NMR of several series of compounds of the general formula $Ln(S_2PR_2)_4^-$ (R = alkyl or alkoxy) in CD_2Cl_2 solution. The results have been analysed to obtain the phosphorus hyperfine coupling constants and relative dipolar shifts. Typically, we observe a discontinuity in these values at holmium which we attribute to a structural change. This is manifested by a reduction in the hyperfine coupling constant by a factor of ca. 2, and in the relative dipolar shifts by an order of magnitude.

Extrapolating from the solid to solution is always dangerous, and especially so for easily deformable lanthanide complexes. However, the nature of the observed structural change may be ascertained by looking at the solid state structures of these compounds determined by X-ray crystallography. In the solid state no analogous structural break is observed. In all the crystal structures determined (six so far) for these eight co-ordinate species, the co-ordination geometry is dodecahedral (mmmm isomer) and no evidence for the alternative square antiprism has been found. Close examination of these structures shows that the transformation of a dodecahedron into a square antiprism requires no change in the phosphorus position, but only a small twist around the ligand two-fold axis. However, in making this small change in geometry, the direction of the main symmetry axis ($\bar{4}$ for the dodecahedron and $\bar{8}$ for the square antiprism) changes by 90°.

The magnitude of the phosphorus dipolar shift is proportional to $(3\cos^2\theta - 1)r^{-3}$ where r is the distance of the phosphorus nucleus from the metal ion, and θ is the angle between the vector r and the principal axis of the magnetic susceptibility tensor (assumed to be colinear with the main symmetry axis). Clearly changing the co-ordination geometry by a simple rotation of the ligands will not change r, but the new direction of the main symmetry axis will change θ. On the basis of the crystal structures that we have determined, we obtain values of θ of ca. 68° for the dodecahedron, whereas θ for the corresponding square antiprism would be ca. 53°, close to the magic angle of 54.7° where the geometric term goes to zero. Thus, the phosphorus dipolar shift should be large for a dodecahedral structure but tend to zero for a square antiprism.

We thus conclude that the change in structure indicated by the NMR data is from a dodecahedron for the light members of the series to a square antiprism for the heavy members. This is in contrast to the situation observed in the solid which demonstrates the danger in making simple extrapolations.

REFERENCES

1) R.E. Sievers, Nuclear Magnetic Resonance Shift Reagents, Academic Press, New York, (1973).

2) A.A. Pinkerton and W.L. Earl, JCS Dalton, 267 (1978).

3) A.A. Pinkerton and D. Schwarzenbach, JCS Dalton, 1300 (1980), and references therein.

RECENT ADVANCES IN THE LOW VALENT APPROACH TO f-ELEMENT ORGANO-METALLIC CHEMISTRY

William J. Evans

Department of Chemistry, University of Chicago

5735 S. Ellis Ave., Chicago, Illinois 60637

ABSTRACT

The low valent organometallic chemistry of the lanthanide elements has been investigated by cocondensing lanthanide metal vapor with neutral unsaturated hydrocarbons at -196°C. Preparative-scale reactions of this type provide a variety of new classes of organolanthanide complexes as well as insight into the lanthanide metal unsaturated hydrocarbon interaction. The utility of this low valent approach in preparing divalent and trivalent organolanthanide hydrides and soluble divalent organosamarium complexes is also discussed.

INTRODUCTION

The organometallic chemistry of the f elements traditionally has been limited in two respects: it has involved primarily a single oxidation state for the metal and a narrow range of organic ligands. For the lanthanides, the +3 oxidation state is the predominant valence state, whereas for the two most extensively studied actinides, uranium and thorium, the +4 oxidation state is most common. In these oxidation states, the f element organometallics are largely ionic in character, and consequently, stabilization is effected by optimization of electrostatic interactions and by steric inhibition of decomposition pathways using bulky ligands. The organic ligands which best satisfy these requirements are large, stable, organic anions. Pre-eminent among these are the cyclopentadienide and cyclooctatetraenide ions, $C_5H_5^-$ and $C_8H_8^{2-}$, respectively. Cyclopentadienyl derivatives constitute the most common class of organo-f-element complexes and much of organo-f-element chemistry involves their ionic metathesis chemistry and their reaction chemistry with small molecules.

We have examined a different approach to f element chemistry in the past few years (1-7). Investigating specifically the lanthanide elements, we have studied their low valent chemistry in efforts to demonstrate that a more extensive chemistry is available to these elements than is found in their traditional, ionic, trivalent complexes and to provide a basis for seeking unusual catalytic transformations which involve f elements. Although the lanthanides are sometimes regarded as trivalent versions of the alkali and alkaline earth metals and at other times are compared to transition metals, in fact, the lanthanides are unique. Their radial and electronic properties, ionization potentials and electron affinities clearly distinguish them from other metals. Associated with this unique combination of physical properties should be some unusual chemistry. Therefore, we sought to determine if unusual chemical properties could be identified and exploited in the lower oxidation states of these metals.

One experimental approach which has been useful in the study of low valent lanthanide chemistry is the metal vapor technique (8,9). In this method, an ingot of lanthanide metal is vaporized in a high vacuum reactor by resistively heating a tungsten boat or basket containing the lanthanide. The metal vapor is cocondensed with a potential ligand on the walls of the reactor which are cooled to -196°C. This technique provides a reaction system in which the initial interaction is that between a zero valent lanthanide and the reactant. We have used this approach extensively to study the reactivity of zero valent lanthanides with neutral unsaturated hydrocarbons.

SYNTHETIC RESULTS

Cocondensation of lanthanide metal vapor with unsaturated hydrocarbons allows the synthesis of a wide variety of new classes of organolanthanide complexes on a preparative scale (1, 3,4,7). Yields vary depending on the metal/ligand combination, but as much as two to three grams of isolated product can be obtained in some reaction systems. Empirical formulas of some representative reaction products (determined by complete elemental analysis) are shown in eqn. (1)-(5).

$$Ln + H_2C=CH-CH=CH_2 \rightarrow Ln(C_4H_6)_3 \qquad Nd,Sm,Er \qquad (1)$$

$$Ln + H_2C=C(CH_3)-C(CH_3)=CH_2 \rightarrow Ln(C_6H_{10})_2 \qquad La,Nd,Sm,Er \qquad (2)$$

$$Ln + CH_3CH=CH_2 \rightarrow Ln(C_3H_6)_3 \qquad Er \qquad (3)$$

$$Ln + CH_3CH_2C\equiv CCH_2CH_3 \rightarrow Ln_2(C_6H_{10})_3 \qquad Nd,Er \qquad (4)$$

$$Ln + CH_3CH_2C\equiv CCH_2CH_3 \rightarrow Ln(C_6H_{10}) \qquad Sm,Yb \qquad (5)$$

These complexes differ from traditional organolanthanides in several respects. First of all, the observed stoichiometries have low ligand to metal ratios. For example, the ytterbium and samarium 3-hexyne products have formal ligand to metal ratios of one, whereas most organolanthanides are commonly nine or ten coordinate (cf. $[(CH_3C_5H_4)_3Nd]_4$ (10)). Second, the stoichiometries vary in an unusual manner depending on ligand and metal. For example, 2,3-dimethyl substitution of the butadiene ligand changes the ligand to metal ratio from three (eqn. 1) to two (eqn. 2). In the 3-hexyne system, changing from neodymium and erbium to ytterbium and samarium changes the ratio from 1.5 (eqn. 4) to one (eqn. 5). This latter result conforms to a general, observed trend that ytterbium and frequently samarium display reactivity different from the other lanthanides in these metal vapor syntheses. For example, ytterbium reacts with butadienes to form a polymeric product rather than isolable complexes as in eqns. (1) and (2).

The physical properties of these complexes also differ from those of traditional trivalent organolanthanides. The optical spectra of these complexes are unusual in that they display strong charge transfer bands in the near infrared-visible region instead of the usual sharp and weak 4f-4f transitions. Room temperature magnetic moments are often outside the range of values previously reported for organolanthanide compounds.

The solution behavior is also unusual. These complexes are highly associated in solution and display an unusual molecular weight dependence on solvent. For example, isopiestic and cryoscopic molecular weight measurements indicate that the 3-hexyne product, $[ErC_9H_{15}]_n$, is dimeric in arenes, i.e. n=2 $(Er_2(C_6H_{10})_3)$, but in concentrated solution or in tetrahydrofuran (THF), it is highly associated with n > 10. This is just opposite the trend found for traditional organolanthanides which are more highly associated in arenes than in THF. For example, $(C_5H_5)_2LnCl$ and $(C_5H_5)_2LnCH_3$ are dimeric in benzene, but exist as monomeric THF adducts in THF (11, 12). Since the metal vaporization products oligomerize rather than crystallize in concentrated solution, these species have not yet been structurally characterized by X-ray diffraction.

The most important property of these metal vapor complexes is that they have the capacity to catalytically activate molecular hydrogen in homogeneous solution (3,4). Catalytic hydrogenation of alkynes and alkenes is effected by these species at room temperature under one atm. of hydrogen often with high stereospecificity (3-hexyne $\rightarrow$ > 96% cis-3-hexene (3)). Hence, the initial exploratory synthetic studies conducted as part of this low valent metal vaporization approach to lanthanide chemistry not only demonstrate that a wider variety of organic ligands can be used to generate organo-

lanthanide complexes (i.e. one is not limited to common stable organic anions), but these studies also demonstrate that the lanthanides have the capacity to function in homogeneous catalytic reactions involving small molecule transfer.

Once the general reactivity of the elemental lanthanide metals
with neutral unsaturated hydrocarbons was established, subsequent
experiments were directed to defining more precisely the mode of
metal hydrocarbon interaction. The origin of these studies can
best be understood by considering the several ways in which a lanthanide metal (or any metal atom (9)) can interact with an unsaturated hydrocarbon. Three structural possibilities for the
erbium 3-hexyne product, $[ErC_9H_{15}]_2$, are shown below ($R = CH_3CH_2$).

B.
$$CH_3-CH-C{\equiv}C-CH-CH_3$$
$$H-Er \qquad Er-H$$
$$RC{\equiv}CCH \qquad HCC{\equiv}CR$$
$$CH_3 \qquad CH_3$$

B'.
$$CH_3-C=C=C-CH_2CH_3$$
$$H-Er \quad Er-H$$
$$CH_3CH=C=CR \quad RC=C=CHCH_3$$

Most simply, an alkyne can add to the metal to form a π complex as
in A. This structure has an alternate cyclometallopropene form, A',
which may be more reasonable for these complexes considering the
electropositive nature of the lanthanides. Alternatively, the metal
may not interact with the unsaturated bond at all, but instead insert into C-H bonds by oxidative addition as in B. This type of reaction could occur in many ways. Structure B is drawn so that the

metal is fully oxidized to the common +3 oxidation state and there are two types of carbon–carbon multiple bonds as is consistent with the observed absorptions at 1870 and 1790 cm^{-1} in the infrared spectrum. The formation of such a structure would require a multi-step reaction, since the metal undergoes a three electron oxidation via two electron oxidative addition steps. Such two electron oxidative addition reactions might be expected to be more facile for europium, ytterbium and samarium, which have more stable divalent states. A third possibility for the reaction of a lanthanide metal with an alkyne is radical formation (structure C), as has been postulated for the Al/HC≡CH system based on matrix epr spectroscopy (13). If the monovalent erbium in the initially formed radical oxidatively inserts into a C–H linkage and the resulting Er–H adds to excess 3-hexyne in the matrix, the radical forming reaction can be rationalized to form a product with the exact formula determined experimentally (in contrast to structure B which is low in hydrogen).

These structural possibilities have been investigated in several ways: syntheses of modified crystalline derivatives have been attempted, syntheses of model complexes containing some of the ligands postulated in these structures have been examined, and additional metal vapor reactions designed to test these reaction pathways have been carried out. The synthesis of a crystalline cyclopentadienyl derivative of a metal vapor product was attempted by adding methylcyclopentadiene to the 3-hexyne product obtainable in highest yield, namely YbC_6H_{10}. Although a crystalline product was isolated from this reaction, it unfortunately did not retain any of the original alkyne ligand: the product was characterized by X-ray diffraction as the divalent $[(\mu-CH_3C_5H_4)(CH_3C_5H_4)Yb(THF))]_n$ (5). Hence, the only information gained about the metal vapor product was that it was capable of reducing methylcyclopentadiene. Even though this investigation failed to provide information on the alkyne, it did represent the first reported structural determination of a divalent organolanthanide complex. This complex displayed an interesting polymeric structure in which the monomeric units are linked into extended linear chains by one bridging methylcyclopentadienide ligand per monomer. Both sides of the bridging ligand are coordinated in a pentahapto fashion to ytterbium atoms. Each divalent ytterbium is formally ten coordinate due to a terminal ring, a THF and the bridging rings above and below the metal. This complex is of further interest vis-a-vis the metal vapor products in that it is a low valent lanthanide complex which obviously has a strong tendency to oligomerize. In THF solution, however, isopiestic molecular weight measurements indicate that this complex is a monomer. In contrast, the 3-hexyne metal vapor products are highly oligomerized in THF solution.

A second attempt at derivatization of the metal vapor products was the formation of an oxygen bonded derivative by CO insertion. A reaction occurs when YbC_6H_{10} is treated with CO, but the product

is insoluble in THF (14). In contrast, when CO reacts with the
cyclopentadienyl alkyl, $(C_5H_5)_2Lu(t-C_4H_9)(THF)$, a two step reaction
occurs which forms soluble products at both stages (15). Initially,
one molecule of CO inserts into the Lu-alkyl bond to form a dihapto-
acyl complex characterized by infrared (ν_{CO} = 1490) and ^1H and ^{13}C
NMR spectroscopy of the ^{12}CO and ^{13}CO products (eqn. 6, R = $t-C_4H_9$).
A second molecule of CO then reacts to form a dimer containing an

$$(C_5H_5)_2LuR(THF) + CO \rightarrow (C_5H_5)_2Lu-\overset{\overset{O}{\|}}{C}R \leftrightarrow (C_5H_5)_2Lu \overset{O}{\underset{}{\nwarrow}} :CR \tag{6}$$

enedione diolate moiety which coordinates to the lutetium via 6-
membered rings (eqn. 7). The complex was characterized by analyti-

$$(C_5H_5)_2Lu \overset{O}{\nwarrow} :CR \xrightarrow{CO} (C_5H_5)_2Lu \tag{7}$$

cal and spectroscopic methods and by X-ray crystallography. The
observed chemistry has strong parallels in early transition metal
(16) and actinide chemistry (17). Having established how trivalent
lanthanides interact with CO and having determined the necessary
spectral parameters, further studies of the reaction of CO with
lanthanide metal vapor products can now be carried out.

The structural possibilities which were proposed for the metal
vapor reaction products were also investigated by attempting the
independent synthesis of model complexes which contain the postulated
ligands. Specifically, the synthesis of lanthanide complexes containing allenyl and hydride ligands (as proposed in structure B) was
examined. Although lithium salts of the allenyl anions $C_6H_9^-$ and
$C_8H_8^{2-}$ can be readily prepared from 3-hexyne by reaction with $t-$
C_4H_9Li, they react with $ErCl_3$ to form mixtures containing mono- and
dianions. Since neither of these anions are particularly bulky,
none of the several trivalent erbium allenyl complexes possible,
$Er(C_6H_8)(C_6H_9)$, $Er(C_6H_9)_2Cl$, $LiEr(C_6H_8)(C_6H_9)Cl$, etc., are particularly favored. Consequently, a mixture of complexes results which is
difficult to characterize and compare with the metal vapor products.

Independent synthesis of an organolanthanide complex containing
a hydride has also been difficult. Initially, the synthesis of a
cyclopentadienyl ytterbium hydride was attempted via hydrogenolysis
(eqn. 8). Decomposition of the green toluene insoluble product with

$$[(CH_3C_5H_4)_2YbCH_3]_2 + 2 H_2 \xrightarrow{\text{toluene}} 2 (CH_3C_5H_4)_2YbH + 2 CH_4 \quad (8)$$

D_2O and CH_3I formed the expected HD and CH_4, respectively, indicating the presence of hydride, but the yields were lower than
anticipated (5). Further characterization indicated that the green
product was a mixture containing $(CH_3C_5H_4)_2Yb$, suggesting that the
initially formed trivalent ytterbium hydride was unstable with respect to the divalent complex in this system.

To avoid such an Yb^{3+}-H to Yb^{2+} decomposition and to allow NMR
characterization of the product, the synthesis of cyclopentadienyl
lutetium hydrides was attempted (18). Hydrogenolysis of
$(C_5H_5)_2Lu(t-C_4H_9)(THF)$ in THF failed to give a reaction over a six
hour period at room temperature. On the other hand, in toluene,
the hydrogenolysis reaction was rapid, eliminating the expected 2-
methylpropane in good yield. Unfortunately, the primary product in
this case had very little solubility in THF. Although this material
appeared to contain hydride ligands based on chemical decomposition
studies, and although the soluble product appeared to be the desired
$(C_5H_5)_2LuH(THF)$, the yield was low. Alternative routes to this hydride were investigated including thermal and photochemical β-hydro-
gen elimination reactions (eqn. 9). These reactions similarly gave

$$(C_5H_5)_2Lu(t-C_4H_9)(THF) \xrightarrow[\text{or } \Delta]{h\nu} (C_5H_5)_2LuH(THF) + H_2C=C(CH_3)_2 \quad (9)$$

low yields of soluble hydride products.

The structural possibilities described above for the metal
vapor products, A-C, were investigated not only by the solution
methods just described, but also by metal vapor methods. Since
the reactions leading to structures B and C required that the substituent groups attached to the unsaturated bond contain hydrogen,
it was of interest to investigate metal vapor reactions involving
smaller unsaturated hydrocarbons which lacked these potentially reactive sites. Accordingly, the reactions of ethene and 1,2-propa-
diene with lanthanide vapor (Sm, Er and Yb) were studied (7). Propene reactions were also investigated for comparison. Unfortunately, with the exception of the erbium propene reaction product (eqn.
6), all the products of these small ligand reactions were insoluble.
Nevertheless, hydrolysis of the reaction products indicated that a
variety of reaction pathways were traversed by the lanthanide metals
in these systems: oxidative addition of the metal to C-H bonds,
two electron reduction of carbon-carbon multiple bonds (cf. structure A'), cleavage of carbon-carbon multiple bonds, and homologa-
tion, oligomerization and dehydrogenation of the organic ligand.
The first two reactions listed are the most prevalent, and quantitative studies of the ethene reactions indicated that these reactions are more facile for ytterbium and samarium than for erbium.

Since each of these two reactions involves a formal two electron
oxidation of the metal, this result is consistent with the fact that
Yb^{2+} and Sm^{2+} are more accessible and stable than Er^{2+}. This differentiation between the zero valent chemistry of ytterbium and
samarium, on one hand, and erbium, on the other, was noted earlier.

The facile formation of lanthanide hydrides via oxidative addition in the above reactions with small hydrocarbons (cf. the difficulties of solution syntheses of hydrides) suggested that the metal
vapor approach would be an excellent synthetic route to lanthanide
hydrides if sufficiently large organic substrates could be used to
disperse the energy of the metal vapor, minimize the number of side
reactions, and provide soluble products. This goal was successfully
realized using terminal alkynes as substrates (7). Vaporization of
ytterbium into a 1-hexyne matrix forms purple products which are
formulated as $[HYb_2(C \equiv CC_4H_9)_3]_n$ based on elemental, chemical and
spectroscopic analyses. Although the reaction sequence presumably
involves initial oxidative addition of a C-H bond to ytterbium to
form $HYbC \equiv CC_4H_9$, the observed product stoichiometry indicates the
overall reaction sequence is more complex. For example, the coordinatively unsaturated $HYbC \equiv CR$ ($R = n-C_4H_9$) could increase the
steric bulk around the metal center by reaction with additional 1-
hexyne to form $Yb(C \equiv CR)_2$ (with elimination of hydrogen). This intermediate could then add to $HYbC \equiv CR$ to form the observed product
(eqn. 11). Since alkynide ligands have a strong tendency to form

$$HYbC \equiv CR + Yb(C \equiv CR)_2 \rightarrow HYb \quad YbC \equiv CR \rightarrow \quad Yb \quad Yb \quad Yb \quad Yb \qquad (10)$$

bridged, bimetallic lanthanide complexes (19), a high molecular
weight bridged structure would not be unexpected in this case. In
contrast to the difficulty encountered in synthesizing soluble lanthanide hydrides by the solution methods described above, soluble
ytterbium hydrides are easily obtained in up to one gram quantities
by this metal vapor approach. Decomposition of an Yb^{3+}-H is not a
problem here since the metal already is in the lower +2 oxidation
state. Analogous 1-hexyne reactions with erbium and samarium form
THF soluble trivalent hydrides of formula $[HLn(C \equiv CC_4H_9)_2]_n$. These
results are consistent with the fact that ytterbium has the most
stable divalent state of these three metals.

Synthesis of lanthanide hydrides from another relatively acidic
organic hydrocarbon, pentamethylcyclopentadiene, has also been in-

vestigated. Vaporization of samarium metal into a hexane solution of C_5Me_5H in a rotary metal vapor reactor allows the synthesis of purple, THF soluble, divalent samarium products. Evidence for both $(C_5Me_5)SmH(THF)_x$ $(+ D_2O \rightarrow HD + \frac{1}{2} D_2)$ and $(C_5Me_5)_2Sm(THF)_2$ has been obtained. The latter complex has been shown to be a monomer by X-ray diffraction (like the pentamethylcyclopentadienyl Yb analogs (20,21)) and is the first divalent organosamarium complex to be crystallographically characterized. Indeed, these complexes are the first <u>soluble</u> divalent organosamarium complexes known. Since Sm^{2+} is the most powerful reducing agent of the common divalent lanthanides (Eu,Yb,Sm), these complexes are expected to display a wide range of reactivity in oxidative reactions.

CONCLUSION

Several general conclusions can be drawn from all of these low valent lanthanide studies. First, organolanthanide chemistry is not limited only to stable organic anions. A wide variety of neutral unsaturated hydrocarbons will interact with the lanthanides and form isolable complexes. These complexes may be extremely useful in identifying new types of organolanthanide reactivity, catalytic activity and bonding capacity. Second, using lanthanide metals as reactants, many reaction pathways become available in organolanthanide chemistry. The results of a specific reaction will depend on the particular metal/ligand combination used and one cannot treat the lanthanides simply as a homologous series of chemically similar metals when considering their low valent chemistry. Third, the reactivity accessible by this low valent approach can provide synthetic routes to complexes not easily obtainable by traditional methods as demonstrated by the syntheses of divalent and trivalent hydrides and the soluble, crystalline, divalent samarium complex. Finally, it is clear that much still needs to be learned about low valent lanthanide chemistry. Further benefits from this low valent approach should be realized as structural and reactivity features are more fully elucidated in the future.

ACKNOWLEDGMENT

I wish to thank the National Science Foundation for support of this work and the Camille and Henry Dreyfus Foundation for a Teacher-Scholar Grant.

REFERENCES

(1) W.J. Evans, S.C. Engerer and A.C. Neville, <u>J. Am. Chem. Soc.</u>, 100: 331 (1978).
(2) W.J. Evans, A.L. Wayda, C.W. Chang and W.M. Cwirla, <u>ibid.</u>, 100: 333 (1978).
(3) W.J. Evans, S.C. Engerer, P.A. Piliero and A.L. Wayda, <u>J. Chem. Soc. Chem. Commun.</u>, 1007 (1979).

(4) W.J. Evans, S.C. Engerer, P.A. Piliero and A.L. Wayda, <u>Funda-mental Research in Homogeneous Catalysis</u>, Vol. 3, M. Tsutsui, ed., Plenum Publishing Corp., New York, 941 (1979).

(5) H.A. Zinnen, J.J. Pluth and W.J. Evans, <u>J. Chem. Soc. Chem. Commun.</u>, 810 (1980).

(6) W.J. Evans, S.C. Engerer and P.A. Piliero, 178th ACS National Meeting, Washington, D.C., September, 1979, INOR 188.

(7) W.J. Evans, K.M. Coleson, S.C. Engerer, A.L. Wayda and H.A. Zinnen, Second Chemical Congress of the North American Continent, Las Vegas, Nevada, August, 1980, INOR 297.

(8) P.S. Skell and M.J. McGlinchey, <u>Angew. Chem., Internat. Ed., Engl.</u>, 14: 195 (1975); P.L. Timms and T.W. Turney, <u>Adv. Organo-met. Chem.</u>, 15: 53 (1977); K.J. Klabunde, <u>Chemistry of Free Atoms and Particles</u>, Academic Press, New York (1980).

(9) J.R. Blackborrow and D. Young, <u>Metal Vapor Synthesis in Organo-metallic Chemistry</u>, Springer Verlag, Berlin, Germany (1979).

(10) J.H. Burns, W.H. Baldwin, and F.H. Fink, <u>Inorg. Chem.</u>, 13: 1916 (1974).

(11) R.E. Maginn, S. Manastyrskyj and M. Dubeck, <u>J. Am. Chem. Soc.</u>, 85: 672 (1963).

(12) J. Holton, M.F. Lappert, D.G.H. Ballard, R. Pearce, J.L. Atwood and W.E. Hunter, <u>J. Chem. Soc. Dalton</u>, 54 (1979).

(13) P.H. Kasai, D. McLeod, Jr. and T. Watanabe, <u>J. Am. Chem. Soc.</u>, 99: 3521 (1977).

(14) W.J. Evans and J.H. Meadows, unpublished results.

(15) W.J. Evans, A.L. Wayda, W.E. Hunter and J.L. Atwood, <u>J. Chem. Soc. Chem. Comm.</u>, in press (1981).

(16) J.M. Manriquez, D.R. McAlister, R.D. Sanner and J.E. Bercaw, <u>J. Am. Chem. Soc.</u>, 100: 2716 (1978).

(17) P.J. Fagan, J.M. Manriquez, T.J. Marks, V.W. Day, S.H. Vollmer and C.S. Day, <u>ibid.</u>, 102: 5393 (1980).

(18) W.J. Evans and A.L. Wayda, I. Bloom and K.M. Coleson, 181st ACS National Meeting, Atlanta, Ga., April, 1981, INOR 219.

(19) J.L. Atwood, W.E. Hunter, A.L. Wayda and W.J. Evans, submitted.

(20) P.L. Watson, <u>J. Chem. Soc. Chem. Comm.</u>, 652 (1980).

(21) T.D. Tilley, R.A. Andersen, B. Spencer, H. Ruben, A. Zalkin and D.H. Templeton, <u>Inorg. Chem.</u>, 19: 2999 (1980).

LASER PHOTOCHEMISTRY OF A URANIUM COMPOUND TAILORED FOR

10μ ABSORPTION: U(OCH3)6

Edward A. Cuellar, Steven S. Miller, Robert C. Teitelbaum, Tobin J. Marks, and Eric Weitz

Department of Chemistry and the Materials Research Center, Northwestern University, Evanston, Illinois 60201

INTRODUCTION

Isotope separation by selective laser-induced multiphoton infrared excitation of polyatomic molecules is now a reality for a variety of elements (1). In light of this progress, it would be challenging from both a scientific and technological standpoint to develop an efficient laser induced multiphoton isotope separation (MIS) process for the element uranium. The key molecular requirements for MIS are existence of an infrared active normal vibrational mode which exhibits a non-zero isotope shift and which absorbs in the spectral region corresponding to the output of an efficient laser system. To date, most isotope separation has involved irradiation of molecular species with a pulsed, discretely tunable CO_2 gas laser having usable spectral output in the 9.2-10.8μ region.

In principle, infrared MIS would appear to be an ideal process for enrichment in ^{235}U of naturally occurring uranium or of uranium tailings produced in gaseous diffusion. However, the seemingly most attractive molecular candidate for the process, UF_6, lacks significant absorption in the CO_2 laser region (2). Thus one is faced with basically two alternatives: develop a laser which operates in the spectral region where UF_6 absorbs (2,3), or tailor a volatile uranium containing compound to possess an isotope-sensitive absorption in the output region of a well developed, reliable laser system such as the CO_2 laser. In this paper we report on our integrated synthetic, spectroscopic, and photochemical efforts involving the latter alternative and one of a number of possible candidates, the prototype molecule, $U(OCH_3)_6$ (4-6).

EXPERIMENTAL

The compounds $U(OCH_3)_6$ and $U(^{18}OCH_3)_6$ were prepared as described elsewhere (5). Vapor pressure measurements were performed with an

MKS-221A capacitance manometer. Infrared spectra were acquired with
either Perkin-Elmer 283 or Nicolet 7199 (FT) infrared spectrophoto-
meters. Raman spectra were obtained with the 6471 Å line of a Spectra Physics krypton ion laser using a Spex 1401 double monochrometer,
photon counting detection, and a spinning sample configuration.

Two lasers were utilized in MIS studies. One was a homebuilt
Rogowski profile CO_2 double discharge TEA laser producing pulses on
the P(38) line of the $00^01 - 10^00$ transition which varied in energy
from 0.9 J to 1.7 J/pulse. Alternatively, a commercial (Lumonics K
202-2) CO_2 laser was used which produces up to 15 J per pulse on the
P branch lines of the 00^01-10^00 transition in CO_2. The output of
this laser was spatially filtered with care, and the maximum obtainable unfocused fluence after filtering was approximately 4 J/cm^2.

With either laser, the sample was irradiated in a 42.0 cm
path length glass cell having KCl or KBr front windows. With the
home-built laser system, a rear reflector was used for double-pass-
ing the beam. With the commerical laser, the beam was allowed to
exit the cell though a KCl or KBr window. Sample containers were
connected to either end of the evacuated irradiation cell. The
unirradiated sample was placed in one compartment and allowed to
pass down the length of the cell during which time it was irradiated
at a rate of 1 Hz. The sample was continuously condensed in a 77°K
container on the opposite end of the cell. The collecting container
consisted of a multifingered manifold so that irradiated samples and
blanks could be collected in separate containers for isotopic analysis without variation of the cell geometry. The quantity of uranium
collected was determined spectrophotometrically. Isotopic analyses
were carried out on Hewlett Packard 5930A or Hewlett Packard 5985
mass spectrometers (4,6).

RESULTS AND DISCUSSION

Uranium hexamethoxide was originally synthesized by Gilman (7)
in low yield via a laborious multistep procedure. Therefore, improved syntheses were developed which can produce large quantities
of uranium hexamethoxide in greater yield and in fewer steps (eqs.
(1),(2)) (5).

$$UF_6 + 2CH_3Si(OCH_3)_3 \longrightarrow U(OCH_3)_6 + 2CH_3SiF_3 \quad (1)$$

$$UF_6 + 6NaOCH_3 \longrightarrow U(OCH_3)_6 + 6NaF \quad (2)$$

Thus, $U(OCH_3)_6$ can now be produced in a single step and in high
yield via reaction of readily available UF_6 with an inexpensive
methoxylating agent. Uranium hexamethoxide is a dark-red, moisture-
sensitive crystalline substance which sublimes readily at 30°C/
10^{-6} mm; its vapor pressure varies from 0.003 torr at 19.0°C to 0.017
torr at 33.0°C. The molecular structure of uranium hexamethoxide

has been determined by single crystal X-ray diffraction (8). The immediate coordination geometry about the uranium atom is essentially octahedral, however the nonlinearity of the U-O-C vectors imparts an actual molecular symmetry of C_i in the solid state (neglecting the hydrogen atoms). It will be seen that vibrational spectroscopic data are in excellent accord with the near octahedral structure.

The vibrational spectrum of $U(OCH_3)_6$ can be assigned by analogy to that of UF_6 (2) and is in accord with an octahedral U-O framework. The infrared spectrum of a Nujol solution of $U(OCH_3)_6$ exhibits a strong band at 464.8 cm^{-1} which upon $^{18}OCH_3$ substitution shifts approximately 14 cm^{-1} to lower frequency. Under idealized O_h symmetry, this transition is assigned to the highest frequency infrared-active U-O (9) stretching mode, the ν_3 (T_{1u}) vibration. The strong band at 1051.5 cm^{-1} shifts upon $^{18}OCH_3$ substitution to 1021.4 cm^{-1}. This band is assigned to a ligand C-O stretching vibration (10). Raman spectra of uranium hexamethoxide reveal two modes in the U-O stretching region (495.5 and 400.6 cm^{-1}) which also shift to lower frequency upon $^{18}OCH_3$ substitution. Neither of these modes appears in the infrared spectra of the molecules, and likewise the infrared-active T_{1u} mode is not observed in the Raman spectra. Thus, the exclusion rule is valid for these pseudooctahedral systems. The two peaks in the low frequency Raman spectrum are assigned (for $U(^{16}OCH_3)_6$) to the ν_1 (A_{1g}) vibration (495.5 cm^{-1}) and to the ν_2 (E_g) vibration at (400.6 cm^{-1}). The other U-O modes (ν_4,ν_5,ν_6) are expected (2) to be largely bending in character, weak, and at lower energies than ν_1,ν_2,ν_3. They have not been observed to date. No significant OCH_3 ligand vibrations are expected in the 400-500 cm^{-1} region (10).

The most intense combination bands in the infrared spectra of UF_6 (2) and other octahedral hexafluorides (11) are the $\nu_1 + \nu_3$ and $\nu_2 + \nu_3$ modes. By analogy, $U(OCH_3)_6$ is expected to exhibit a nonnegligible $\nu_1 + \nu_3$ transition in the 10μ region. Weak absorptions are observed in the 920-970 cm^{-1} region of the $U(OCH_3)_6$ infrared spectrum, and detailed spectral analysis is still in progress. Vibrational spectroscopic data for uranium hexamethoxide fundamental modes are summarized in Table 1.

Table 1. Vibrational spectroscopic data for uranium hexamethoxide[a]

Type of Internal Coordinate Change	Symmetry Type[b]	$U(^{16}OCH_3)_6$ cm^{-1}	$U(^{18}OCH_3)_6$ cm^{-1}
U-O stretch	A_{1g}	495.5	481.0
U-O stretch	T_{1u}	464.8	450.4
U-O stretch	E_g	400.6	386.0
C-O stretch	A_{1g}, T_{1u}, E_g[a]	1051.5	1021.4

[a]All frequencies have an estimated accuracy of ± 1.0 cm^{-1}.
[b]Assuming local O_h symmetry.

Two types of irradiation experiments were conducted with gaseous U(OCH$_3$)$_6$: i) experiments designed to maximize photolysis yield and to elucidate photochemical processes; ii) experiments designed to probe isotopic selectivity. When U(OCH$_3$)$_6$ was passed through the irradiation cell with a line of the 9.6 μ CO$_2$ laser branch in resonance with the C-O stretching fundamental, almost complete dissociation of U(OCH$_3$)$_6$ was achieved, as judged by the absence of uranium-containing products in the collection vessel. During photolysis, the walls of the irradiation cell became coated with a solid tan photoproduct. This compound was identified by infrared spectroscopy (and comparison with an authentic sample) as U(OCH$_3$)$_5$. This compound is known to be considerably less volatile than U(OCH$_3$)$_6$ and the homolytic dissociation shown in eq.(3) is a plausible explanation for its formation.

$$U(OCH_3)_6 + h\nu \longrightarrow U(OCH_3)_5 + CH_3O\cdot \qquad (3)$$

The fate of the methoxy radicals probably involves, among other pathways, hydrogen atom abstraction from other U(OCH$_3$)$_6$ molecules, the known disproportionation reaction to methanol and formaldehyde (12), and possibly methoxy radical abstraction from other U(OCH$_3$)$_6$ molecules. Support for the first two processes is provided by the organic photolysis products: substantial quantities of methanol and formaldehyde are detected by gc/ms.

Attempts at isotope-selective photochemistry involved irradiation with a number of laser lines in the vicinity of the U(OCH$_3$)$_6$ $\nu_1 + \nu_3$ mode, which should exhibit a ^{235}U-^{238}U frequency shift. Irradiation of gaseous U(OCH$_3$)$_6$ with the P(38) line of the 9.6μ CO$_2$ laser branch results in isotopic enrichment of the material collected in the downstream cold finger of the irradiation apparatus. This U(OCH$_3$)$_6$ is enriched in ^{235}U by approximately 2.3% at a fluence level of ~ 4J/cm^2. As a check, these experiments were carried out with both naturally abundant and enriched uranium. Furthermore, preliminary studies of the power dependence of the enrichment offers an explanation for the previously observed variation of enrichment factors on a given laser line (4). The enrichment factor appears to be a strong function of the fluence with both a low fluence and a high fluence threshold for isotope separation (~3J/cm^2 and 5.6J/cm^2, respectively). Since our enrichment factor is actually the result of a convolution of two quantities: the total yield of dissociated molecules and the isotopic selectivity, the fluence threshold is the result of a decrease in either or both of these factors. We postulate that the low fluence threshold is due to a decrease in overall yield (to an instrumentally undetectable enrichment) and that the high fluence threshold is due to a decrease in selectivity, which has also been observed in other systems at high fluences (13). The smaller the isotope shift in a system, the lower in fluence the high fluence threshold would be expected to occur.

The beam profile and the fluence across the beam profile must be highly uniform to insure a reproducible enrichment factor. In early experiments (4), there was difficulty in obtaining uniform beam profiles at the desired fluence level. The commercial TEA laser used for later studies allows a much more uniform beam profile.

The fraction of $U(OCH_3)_6$ actually dissociated in a single pass through the irradiation cell is of great interest in determining the actual efficiency of the isotope separation process. This was determined via condensation of a $U(OCH_3)_6$ sample without the laser operating, then repeating the process for the same period of time with the laser operating (collecting that material in a separate finger of the collection cell), then finally collecting another blank in a separate container. The amount of uranium present in each sample was determined spectrophotometerically. Our results at a fluence level of ~4 J/cm^2 indicate that less than 10% of the $U(OCH_3)_6$ is dissociated in a single pass when irradiation is performed with the P(38) 9.6μ laser wavelength. With an average enrichment factor of 1.023 (4) and less than 10% dissociation, this translates to a selectivity of greater than 1.27 (eq.(4)) (13).

$$\alpha = \ln f / \ln f \ \beta \qquad\qquad (4)$$

Here f is the fraction of material left undissociated and β is the normal isotopic enrichment factor. This result should be regarded as a present lower limit on the selectivity for this system, and a number of approaches to greater enrichment are evident.

CONCLUSIONS

It is clear that significant uranium isotope separation can be achieved by 10μ photolysis of $U(OCH_3)_6$. Promising directions for improvement of yield and selectivity in isotope separation processes cesses utilizing this compound include the use of scavengers, multilaser excitation, and modification of the irradiation geometry. Equally promising is the design and refinement of other volatile uranium compounds for 10μ isotope separation processes. A number of candidates are conceivable, and recent results with uranyl compounds (13) and uranium borodeuteride (14) confirm the attractiveness of such an approach.

ACKNOWLEDGMENTS

We thank the National Science Foundation (T.J.M., CHE8009060 A01; E.W., CHE76-10333 A02) and the Electric Power Research Institute (RP506-7) for support of this research. The use of the Central Facilities of Northwestern University Materials Research Center, (NSF-MRL program grant DMR76-80847) facilitated this work.

REFERENCES

1. (a) V.S. Letokhov and C.B. Moore, "Laser Isotope Separation,"
    in Chemical and Biological Applications of Lasers, Vol. III.,
    C.B. Moore (ed.), Academic Press, New York, pp. 1-165 (1977).
    (b) V.S. Letokhov, "Laser Separation of Isotopes," in Annual
    Review of Physical Chemistry, Vol. 28, B.S. Rabinovitch (ed.),
    Annual Reviews, Inc., Palo Alto, pp. 133-159 (1977).
2. R.J. Jensen and C.P. Robinson, AICHE Symp. Series No. 169,
    73:76 (1977).
3. J.J. Tiee and C. Wittig, Optics Comm. 27:377 (1978).
4. S.S. Miller, D.D. DeFord, T.J. Marks and E. Weitz, J. Am. Chem.
    Soc. 101:1036 (1979).
5. E.A. Cuellar and T.J. Marks, Inorg. Chem. 20:2129 (1981).
6. E.A. Cuellar, S.S. Miller, T.J. Marks, E. Weitz, submitted
    for publication.
7. R.G. Jones, E. Bindschadler, D. Blume, G. Karmas, G.A. Martin,
    Jr., J.R. Thirtle, F.A. Yeoman and H. Gilman, J. Am. Chem.
    Soc., 78:6030 (1956).
8. V.W. Day, C.S. Day, S.S. Miller, T.J. Marks and E. Weitz
    unpublished results.
9. D.C. Bradley, R.C. Mehrotra and D.P. Gaur, Metal Alkoxides,
    Academic Press, London, pp. 116-122 (1978).
10. L.J. Bellamy, The Infrared Spectra of Complex Molecules,
    3rd ed., Chapman and Hall, London, Chapters 5 and 7 (1975).
11. K. Nakamoto, Infrared and Raman Spectra of Inorganic and
    Coordination Compounds, 3rd ed., Wiley-Interscience,
    New York, pp. 151-160 (1978).
12. M.J. Gibian and R.C. Corley, Chem. Rev., 73:441 (1973).
13. D.M. Cox, R.B. Hall, J.A. Horsley, G.M. Kramer,
    P. Rabinowitz, and A. Kaldor, Science, 205:390 (1979).
14. (a) R.T. Paine, R.W. Light and M. Nelson, Spectrochim. Acta,
    35A:213 (1979).
    (b) R.T. Paine, P.R. Schonberg, R.W. Light, W.C. Danen and
    S.M. Freund, J. Inorg. Nucl. Chem., 41:1577 (1979).

NEW STOICHIOMETRIC AND CATALYTIC PENTAMETHYLCYCLOPENTADIENYL

ORGANOACTINIDE CHEMISTRY

Paul J. Fagan, Eric A. Maatta, Afif M. Seyam, and
Tobin J. Marks
Department of Chemistry, Northwestern University
Evanston, IL 60201

ABSTRACT

This paper reviews recent results on the synthesis and
properties of thorium and uranium bis(pentamethylcyclopentadienyl) chlorides, hydrocarbyls, chlorohydrocarbyls, and
hydrides. These are some of the most reactive and chemically
diverse organoactinides prepared to date. Reaction patterns of
the metal-carbon sigma bonds which are discussed include C-H and
H-H activation. Hydrogenolysis has afforded the first molecular
actinide hydrides. The hydrides exhibit appreciable hydridic
character. They also activate molecular hydrogen and olefins.
The olefin activation and hydrogenolysis reactions can be coupled
to effect homogeneous, catalytic olefin hydrogenation. The differences between thorium and uranium chemistry appear largely to
reflect differences in accessible oxidation states and in metal-
ligand bond polarity.

INTRODUCTION

Much of the progress in contemporary actinide organometallic
chemistry has sprung from successes in "tuning" the metal ion
coordination sphere. Thus, complexes of the type $M(\eta^5-C_5H_5)_3R$
(1-4) are thermally stable, but appear to be too coordinatively
congested to allow full reactivity of the most interesting aspect
of the complex: the metal-to-carbon two-electron, two-center
sigma bond. On the other hand, unsaturated homoleptic hydrocarbyls of the type MR_4 (1-4) have such low thermal stability that
it has been impossible to investigate the metal-to-carbon bond

chemistry. The purpose of this article is to briefly review chemistry which has exploited a compromise between the above extremes in coordination sphere saturation and has thus evolved what are probably the most chemically reactive and versatile organoactinides synthesized to date.

SYNTHESIS OF BIS(PENTAMETHYLCYCLOPENTADIENYL) THORIUM AND URANIUM DICHLORIDES, CHLOROHYDROCARBYLS, AND BIS(HYDROCARBYLS)

Two-ring dichloride precursor complexes are readily synthesized from actinide tetrahalides and the pentamethylcyclopentadienyl Grignard reagent (eq.(1)) (5,6). Like all actinide pentamethylcyclopentadienyls, these new complexes are soluble in nonpolar organic solvents and crystallize readily. Mono- or dialkylation with lithium reagents affords a wide variety of chlorohydrocarbyls and bis(hydrocarbyls) in high yield (eqs.(2) and (3)).

$$2[(CH_3)_5C_5]MgX + MCl_4 \xrightarrow{\text{toluene}\atop\text{reflux}} \qquad (1)$$

$$+ 1RLi \longrightarrow \qquad (2)$$

$$+ 2RLi \longrightarrow \qquad (3)$$

M = Th, U
R = hydrocarbyl functionality

The molecular structures are of the C_{2v} pseudotetrahedral, "bent sandwich" type; metrical parameters are unexceptional for tetravalent actinide organometallics. The new $M[(CH_3)_5C_5]_2R_2$ and $M[(CH_3)_5C_5]_2(R)Cl$ compounds display a rich and varied chemistry. In the sections which follow, we focus on those aspects of the chemistry which elucidate the characteristics of the actinide-to-carbon sigma bond and which are particularly relevant to catalysis.

C-H ACTIVATION

The diphenyl complexes metathesize aromatic hydrocarbons; the case of C_6D_6 is illustrated in eq.(4). Thus, the ^1H NMR spectra

$$M[\eta^5-(CH_3)_5C_5]_2(C_6H_5)_2 \quad \underset{C_6H_6}{\overset{C_6D_6}{\rightleftharpoons}} \quad M[\eta^5-(CH_3)_5C_5]_2(C_6D_5)_2 \quad (4)$$

$$M = U, Th$$

of $U[\eta^5-(CH_3)_5C_5]_2(C_6H_5)_2$ solutions freshly prepared in C_6D_6 at room temperature evidence rapid formation of $C_6H_{6-x}D_x$, $x = 0,1,2$ ... and concurrent disappearance of the metal-bound phenyl resonances; no new isotropically shifted resonances are observed. The reaction obeys pseudo-first order kinetics with k = 6.06 ±0.3 x 10^{-2} min^{-1} at 45°C, > 1.0 x 10^{-1} min^{-1} at 70°C. Integration studies indicate essentially quantitative (>95%) conversion of the phenyl protons to benzene protons at the completion of the reaction; the intensity of the $(CH_3)_5C_5$ signal remains unchanged. An ortho hydrogen abstraction process to produce the intermediate benzyne complex is proposed to explain these observations. The highly reactive benzyne reacts with benzene to yield a deuterated diphenyl complex (eq.(5)) or can be trapped with diphenylacetylene (eq.(6)) to yield a metallacycle. That the metallacycle formation indeed involves the interception of a highly reactive intermediate rather than attack on the diphenyl complex is supported by kinetic measurements: the rate of disappearance of the diphenyl complex is unchanged by the addition of diphenylacetylene. From product analysis it can also be determined that the trapping is highly efficient (eq.(5) does not compete effectively with eq.(6)). Interestingly, the rate of benzene metathesis by the organouranium complex is ca. 100 times faster than by $Zr(C_5H_5)_2(aryls)_2$ (7) under the same conditions. Preliminary studies indicate that the thorium bis(neopentyl) complex undergoes an analogous reaction to yield a diphenyl complex and tetramethylsilane.

(5)

(6)

HYDROGEN ACTIVATION: THE SYNTHESIS OF ORGANOACTINIDE HYDRIDES

Hydrogenolysis of the thorium and uranium dialkyls in toluene solution at 25° (1 atm H_2) results in the rapid formation of the first molecular actinide hydrides (eq.(7)).   In experiments with

$$2M[\eta^5-(CH_3)_5C_5]_2R_2 + 4H_2 \xrightarrow[\text{toluene}]{25°} \{M[\eta^5-(CH_3)_5C_5]_2(\mu\text{-}H)H\}_2$$

$$+ 4RH \qquad (7)$$

$$M = Th, U$$

$M[\eta^5-(CH_3)_5C_5]_2(CH_3)_2$ complexes, vacuum line measurements indicated the formation of two equivalents of methane and the absorption of two equivalents of H_2 per equivalent of the actinide dialkyl, as predicted by eq.(7).   If deuterium gas is used, CH_3D is produced quantitatively, and in the case of thorium, $\{Th[\eta^5-(CH_3)_5C_5]_2(\mu\text{-}D)D\}_2$ is the product.   A parallel experiment with the uranium analogue was complicated by rapid exchange of the $\eta^5-(CH_3)_5C_5$ hydrogen atoms with the U-D units.   Interestingly, the analogue to eq.(7) for M = Zr requires several days at room temperature (8).

The new hydrides were characterized by a full complement of physicochemical methods.   Furthermore, the molecular structure of the thorium hydride dimer has been determined by single crystal neutron diffraction (9). Both terminal (two-center, two-electron) and bridging (three-center, two-electron) metal-hydrogen interactions are evident.   The former Th-H distance is 2.03(1) Å, and the latter, 2.29(3) Å; the Th-Th distance of 4.007(8) Å evidences little metal-metal bonding (the corresponding distance in thorium metal is 3.59 Å). ^1H NMR spectroscopy indicates that the bridging and terminal hydrogen atoms are rapidly interconverting down to -90°C in solution (6).

Although the thorium hydride displays high thermal stability that of uranium readily and reversibly loses hydrogen (eq.(8)) to

$$\{[\eta^5-(CH_3)_5C_5]_2UH_2\}_2 \;\rightleftharpoons\; \frac{2}{x}\;\{[\eta^5-(CH_3)_5C_5]_2UH\}_x + H_2 \qquad (8)$$

form a U(III) hydride. The facility of eq.(8) in contrast to the stability of the thorium hydride is understandable in terms of the ready accessibility of the trivalent uranium oxidation state.

REACTIONS OF ORGANOACTINIDE HYDRIDES

The chemistry of organoactinide hydrides evidences marked "hydridic" character. This can be seen in the rapid reactions with ketones (eq.(9)), and alcohols (eq.(10)),

$$\underset{}{\text{Th}} \Big\langle \begin{smallmatrix} H \\ H \end{smallmatrix} + 2CH_3COCH_3 \xrightarrow{\text{toluene}} \text{Th} \Big\langle \begin{smallmatrix} OCH(CH_3)_2 \\ OCH(CH_3)_2 \end{smallmatrix} \qquad (9)$$

$$\underset{}{\text{Th}} \Big\langle \begin{smallmatrix} H \\ H \end{smallmatrix} + 2(CH_3)_3COH \xrightarrow{\text{toluene}} \text{Th} \Big\langle \begin{smallmatrix} OC(CH_3)_3 \\ OC(CH_3)_3 \end{smallmatrix} + 4H_2 \qquad (10)$$

The organoactinide hydrides also activate molecular hydrogen. An example of this remarkably rapid exchange reaction between the metal-bound hydrides of the thorium complex and molecular H_2 is shown in eq.(11). As revealed by the lineshape alterations in

$$\{Th[(CH_3)_5C_5]_2H_2\}_2 \underset{H_2}{\overset{*H_2}{\rightleftharpoons}} \{Th[(CH_3)_5C_5]_2{}^*H_2\}_2 \qquad (11)$$

variable temperature ^1H NMR spectra (0.7 atm H_2, 10 mM hydride), the preexchange lifetime for H_2 nuclei is short on the NMR timescale at 0°C! These results have been verified by magnetization transfer experiments.

The thorium hydride $\{Th[\eta^5-(CH_3)_5C_5]_2(\mu-H)H\}_2$ was also found to activate ethylene. Thus it reacts rapidly and quantitatively with ethylene in toluene solution at room temperature to yield the corresponding diethyl complex (eq.(12)). This compound appears to be indefinitely stable with respect to β-hydride

$$\{Th[\eta^5-(CH_3)_5C_5]_2(\mu-H)H\}_2 + 4 \quad \begin{matrix} H \\ H \end{matrix} C = C \begin{matrix} H \\ H \end{matrix}$$

$$\xrightarrow[25°]{< 10 \text{ min}} \quad 2[\eta^5-(CH_3)_5C_5]_2Th \begin{matrix} C_2H_5 \\ C_2H_5 \end{matrix} \qquad (12)$$

elimination at room temperature both in the solid state and in solution in the dark. Interestingly, however, the heat and/or light produced by the glowbar of an infrared spectrometer induced decomposition. Thus, bubbles of gas were observed in a Nujol mull and strong, broad bands characteristic of bridging hydride ligands, and we tentatively suggest the reaction of eq.(13); we

$$2[\eta^5-(CH_3)_5C_5]_2Th \begin{matrix} C_2H_5 \\ C_2H_5 \end{matrix} \longrightarrow$$

$$[\eta^5-(CH_3)_5C_5]_2Th(Et) \begin{matrix} H \\ H \end{matrix} (Et)Th[\eta^5-(CH_3)_5C_5]_2$$

$$+ \quad 2 \quad \begin{matrix} H \\ H \end{matrix} C = C \begin{matrix} H \\ H \end{matrix} \qquad (13)$$

have previously demonstrated photoinduced β-hydride elimination in the $Th(\eta^5-C_5H_5)_3R$ series (10,11).

CATALYTIC OLEFIN HYDROGENATION

The facile insertion of ethylene into actinide-hydride bonds and the rapid hydrogenolysis of actinide-carbon bonds suggested a possible cycle for olefin hydrogenation. Experiments with the $\{M[(CH_3)_5C_5]_2(\mu-H)H\}_2$ complexes and 1-hexene verified that a sequence of such processes can indeed be made catalytic. Typical initial turnover frequencies for 1-hexene hydrogenation per MH_2 moiety at 25°, 0.9 atm H_2 were 0.5 h^{-1} (M = Th), and 70 h^{-1} (M = U). The organoactinide hydrides can be recovered unchanged when the olefin is exhausted. In a preparative scale reaction, the uranium/1-hexene system functioned for over 800 turnovers with no detectable decay in catalytic efficiency. In regard to other substrates, it was also found that diphenylacetylene could be hydrogenated under the same conditions to yield 1,2-diphenylethane. Based upon the information at hand, a plausible catalytic scenario (invoking either monomer or dimer) can be generated (Fig. 1). For the more reactive uranium

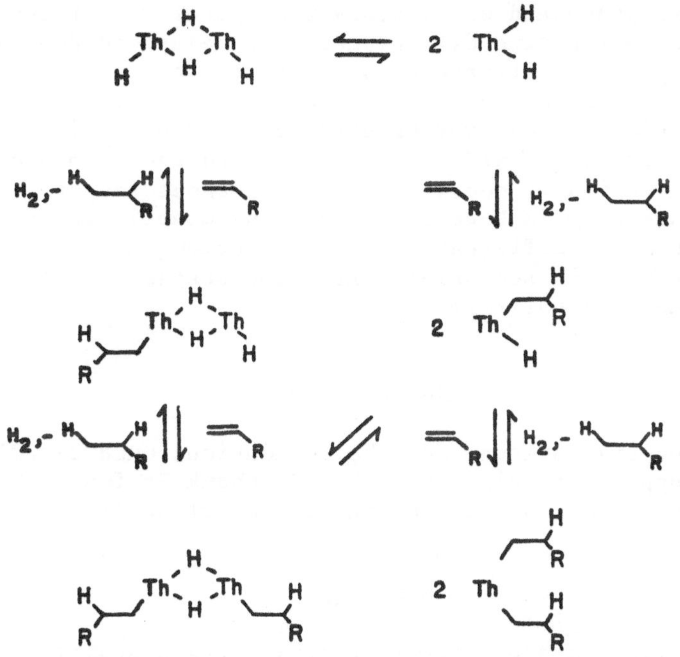

Fig. 1. Proposed mechanism for the catalytic hydrogenation of terminal olefins by $\{Th[(CH_3)_5C_5]_2H_2\}_2$. From reference 6.

hydride, the involvement of the trivalent species $\{U[(CH_3)_5C_5]_2H\}_x$ seems likely, as does the possibility of binuclear oxidative addition/reductive elimination as well as more extensive free radical chemistry.

We have also found that when deposited upon high area metal oxide supports (e.g., dehydroxylated γ-alumina) the $M[\eta^5-(CH_3)_5C_5]_2(CH_3)_2$ complexes are precursors for potent heterogeneous olefin hydrogenation catalysts (12). For example, turnover frequencies for propylene hydrogenation are comparable to those for supported platinum under the same conditions!

CONCLUSIONS

The properties of the bis(pentamethylcyclopentadienyl) actinide hydrocarbyls demonstrate that proper ligational adjustment can lead to thermally stable yet highly reactive actinide hydrocarbyls and hydrides. It is already apparent that such molecules have a rich and varied chemistry. In placing this chemistry in perspective, there appear to be distinct similari-

ties to main group and early transition metal reactivity patterns;
there are also striking differences. In regard to the d-element
Group IVB systems, differences from the actinides in structure
and reactivity appear to reflect the relative availability of
metal oxidation states, the greater polarity of actinide-element
bonds, the larger actinide ionic radii, and the commensurate
increase in available coordination sites about the 5f ion. The
actinide chemistry is however by no means monolithic, and the
observed chemical differences between thorium and uranium arise
from differences in accessible oxidation states as well as in
metal-ligand bond polarity.

ACKNOWLEDGMENTS

We thank the National Science Foundation (CHE8009060) for
generous support of this research. We thank Professor Victor W.
Day for structural information in advance of publication.

REFERENCES

1. T.J. Marks and R.D. Fischer (eds.), Organometallics of the
   f-Elements, Reidel Publishing Co., Dordrecht, Holland (1979).
2. T.J. Marks, Prog. Inorg. Chem., 25:224 (1979).
3. T.J. Marks and R.D. Ernst, "Scandium, Yttrium, and the
   Lanthanides and Actinides," in Comprehensive Organometallic
   Chemistry, G. Wilkinson, F.G.A. Stone and E.W. Abel (eds),
   Pergamon Press, Oxford, in press.
4. P.J. Fagan, J.M. Manriquez, and T.J. Marks, reference 1,
   Chapt. 4.
5. J.M. Manriquez, P.J. Fagan and T.J. Marks, J. Am. Chem. Soc.,
   100:3939 (1978).
6. P.J. Fagan, J.M. Manriquez, E.A. Maatta, A.M. Seyam and T.J.
   Marks, J. Am. Chem. Soc., in press.
7. G. Erker, J. Organometal. Chem., 134:189 (1977).
8. J.M. Manriquez, D.R. McAlister, R.D. Sanner and J.E. Bercaw,
   J. Am. Chem. Soc., 100:2716 (1978).
9. R.W. Broach, A.J. Schultz, J.M. Williams, G.M. Brown, J.M.
   Manriquez, P.J. Fagan and T.J. Marks, Science, 203:172
   (1979).
10. D.G. Kalina, T.J. Marks and W. A. Wachter, J. Am. Chem. Soc.,
    99:3877 (1977).
11. D.G. Kalina, E.A. Mintz and T.J. Marks, submitted for publi-
    cation.
12. R.G. Bowman, R. Nakamura, R.L. Burwell, Jr. and T.J. Marks,
    J. Chem. Soc., Chem. Comm., 257 (1981).

CHOOSING A COORDINATION NUMBER FOR Ln(III) : 1:1, 1:2, 2:1, 3:2
AND 4:3 COMPLEXES WITH CROWN ETHERS

Jean-Claude G. Bünzli , Denis Wessner, Aldo Giorgetti
and Yolande Frésart, Université de Lausanne, Institut
de chimie minérale et analytique, CH-1005 Lausanne
Switzerland

The stoichiometry of the complexes between rare earth salts
and crown ethers depends on the nature of both the macrocycle and
the counteranion, as shown in the Table (1).

The 1:2 sandwich complexes contain only ionic perchlorates;
the formation of 4:3 complexes implies the formation of LnX_6^{3-}
species and such complexes will be isolated only if these species
are stable enough (3). Considering the results reported in the
Table, we may draw the following generalisations (D_i is the ionic
diameter of the Ln(III) ion and D_e is the cavity diameter of the
polyether) : (i) in presence of poorly complexing anions, 1:2 and
1:1 complexes are isolated when D_i/D_e is > 1 and < 1, respectively;
(ii) in presence of more strongly complexing anions, 1:1 complexes
form when D_i/D_e > 1 and complexes with Ln:L ratios ⩾ 1 can be
isolated when D_i/D_e < 1. The results with CF_3COO^- are at variance
with these conclusions and an explanation must await the elucida-
tion of the structure of the 3:2 and 2:1 complexes; the chemical
bonds between 12-crown-4 ether and the rare earth ion is relatively weak and the stoichiometry depends strongly upon the experimental conditions used for the isolation of the complexes.

The following CN could be evidenced, on the basis of IR-
Raman and/or X-ray data : CN = 7, e.g. $PrCl_3 \cdot (12-4)$; CN = 8,
$PrCl_3 \cdot (15-5)$; CN = 9, $Pr(ClO_4)_3 \cdot (12-4) \cdot (15-5)$; CN = 10,
$Eu(NO_3)_3 \cdot (12-4)$ (4); CN = 11, $Eu(NO_3)_3 \cdot (15-5)$ (5); CN = 10 and 12,
$[Nd(NO_3)_3]_4 \cdot (18-6)_3$ (6). Therefore, a judicious choice of the
parameters discussed above allows one to design complexes with a
given CN for the Ln(III) ions.

Table Complexes between lanthanoid salts LnX_3 and unsubstituted crown ethers L we have isolated.

L^a	Ln:L	ClO_4^-	NO_3^-	Cl^-	NCS^-	CF_3COO^-	$CF_3SO_3^-$
12-4	1:2	La-Gd^b					
	1:1		La-Lu	Pr,Eu		La,Ce	Pr
	3:2					Pr,Eu,Er	
	2:1					Nd,Sm	
15-5	1:2	La-Gd					
	1:1		La-Lu	Pr,Eu	Pr		Pr
	4:3		Gd-Lu				
	2:1					La-Eu	
18-6	1:1	Pr,Eu	La-Lu^c	Pr	Pr	Pr	Pr
	4:3		La-Lu	Pr			
21-7	1:1		Er				
	4:3		La-Eu				

^{a} 12-4 means 12-crown-4 ether, etc. ^{b} La-Lu (2)

^{c} Except for Sm and Gd.

REFERENCES

(1) Part 13 of the series 'Complexes of Lanthanoid Salts with Macrocyclic Ligands'.

(2) J.F. Desreux and G. Duyckaerts, Inorganica Chimica Acta, 35:L313 (1979).

(3) J.-C. G. Bünzli and D. Wessner, Helv. Chim. Acta, 64:582 (1981), and references therein.

(4) J.-C. G. Bünzli, B. Klein, D. Wessner and N.W. Alcock, to be published.

(5) J.-C. G. Bünzli, B. Klein, D. Wessner, G. Chapuis and K.J. Schenk, in press.

(6) J.-C. G. Bünzli, B. Klein, D. Wessner, G. Chapuis, K.J. Schenk, G. Bombieri and G. de Paoli, Inorg. Chim. Acta, 54:L43 (1981).

A SYSTEMATIC STUDY OF THE COMPLEXATION OF DI- AND TRI-VALENT LANTHANIDE IONS BY MACROCYCLES OF VARYING SIZE

Jean F. Desreux and Jean Massaux

Analytical and Radiochemistry, University of Liège

Sart Tilman, B-4000 Liège, Belgium

Despite the extensive chemistry reported for the complexes between a host of macrocyclic compounds and the alkali metal ions (1), only a limited number of lanthanide (Ln) complexes have been studied so far. Some unusual features of the complexation of Ln ions in anhydrous propylene carbonate by the eighteen-membered cycle di-tert-butylbenzo-18-crown-6, 4, have recently been reported by the authors (2). This study has now been extended to the macrocycles 1-5 (see Figures 1 and 2 for the structure of the ligands).

EXPERIMENTAL

The stability constants of the macrocyclic complexes were determined in anhydrous propylene carbonate (ionic strength 0.1) by a competitive potentiometric method using Pb(II) or Tl(I) as auxiliary ions. The complexation of Sm(II) and Yb(II) was investigated by cyclic voltammetry and by polarography. Further details are given in ref. 2.

RESULTS AND DISCUSSION

As shown in Fig. 1, the small cycles 12-crown-4, 1, and 15-crown-5, 2, exhibit a similar behavior: the stability of the 1:2 complexes decreases through the Ln series while the formation of the 1:1 complexes is much less dependent on the ionic radius r of the Ln ions. Ligand 1 is too small (radius of cavity 0.57 Å) to accommodate a Ln ion but it easily forms 1:2 complexes which are of the sandwich type (3) and which can be isolated in the solid state (3,4). The decrease in stability of the 1:2 complexes is attributed to the Ln contraction

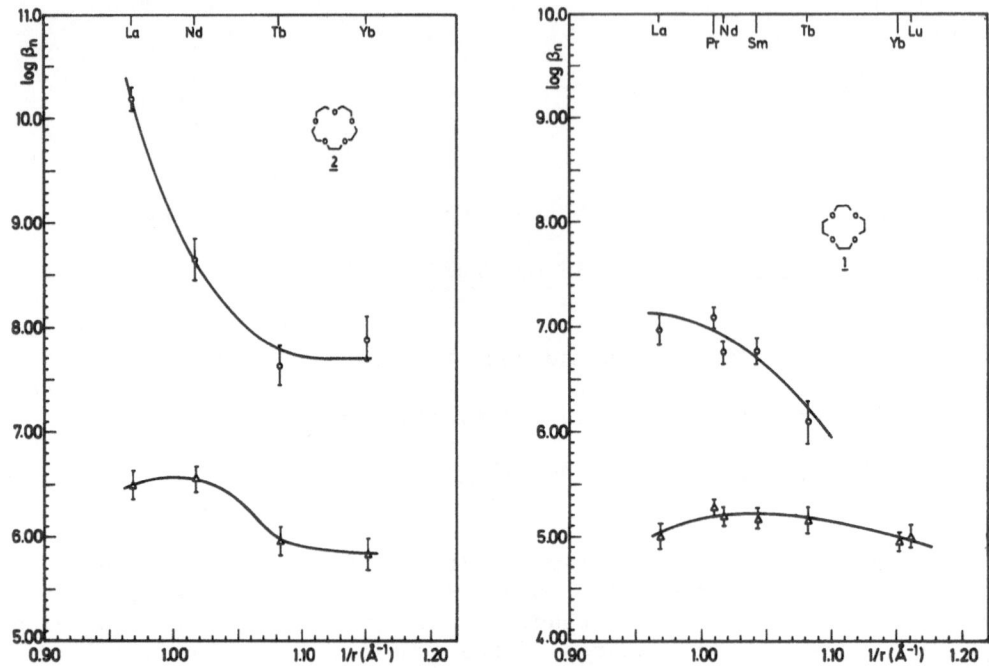

Figure 1.   Stability constants of the 1:1 (lower curves) and 1:2
            (upper curves) Ln complexes with 12-crown-4, 1, and
            15-crown-5, 2, in anhydrous propylene carbonate.

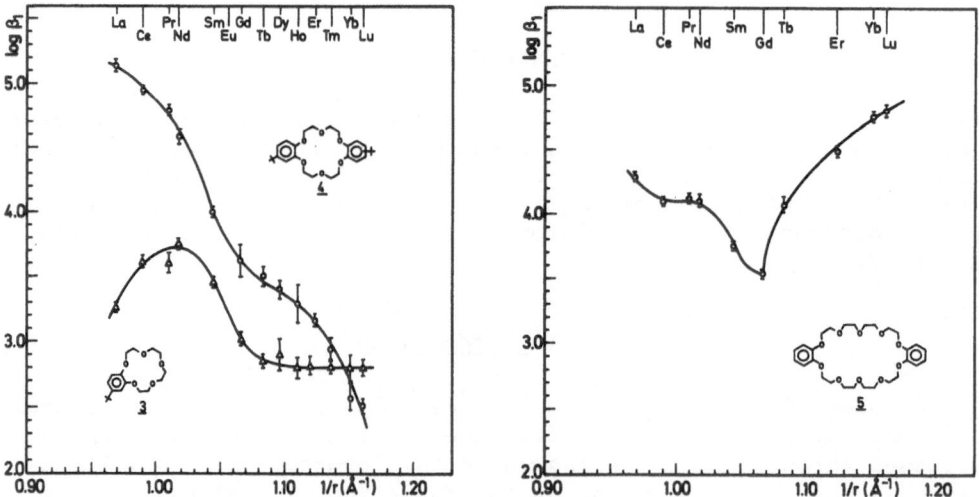

Figure 2.   Stability constants of the 1:1 Ln complexes with di-tert-
            butylbenzo-18-crown-6, 3 (upper curve), tert-butylbenzo-
            15-crown-5, 4 (lower curve) and dibenzo-30-crown-10, 5,
            in anhydrous propylene carbonate.

which leads to an increased solvation and to larger repulsions between the two rings brought closer to each other. The same line of
reasoning can be followed for an interpretation of the complexation
of Ln ions by the somewhat larger macrocycle 2. The complexes are
1.5 to 2 orders of magnitude more stable than in the case of ligand 1,
a difference due to the larger number of coordinating atoms. Also,
2 has a larger internal cavity (radius 0.86-0.92 A) and is better
able to encapsulate the smallest Ln ions.

The presence of an electron acceptor substituent such as an
aromatic ring on a macrocycle has a strong depressing effect on the
complexation as appears from a comparison of the complexing abilities
of cycle 2 (Fig. 1) and of its substituted derivative tert-butylbenzo-
15-crown-5, 3 (Fig. 2). The complexation curve obtained with 3 for
the 1:1 complexes exhibits a maximum, a phenomenon which is not unknown in the field of Ln chemistry (5) but which is unusually large
in the present case. No indication of the formation of 1:2 adducts
was obtained in the experimental conditions $|L| \leqslant 3(|Pb| + |Ln|)$ except
for La(III) ($\log \beta_2 = 5.91 \pm 0.08$).

The binding of the Ln ions by macrocycle 4 has already been
discussed by the authors (2). The stability of the 1:1 complexes
decreases regularly through the Ln series despite the increase in
charge density of the encapsulated ions. Macrocycle 4 is the first
ligand for which such a behavior has been found, the opposite behavior
is observed for all other ligands. Macrocycle 4 is larger than the
Ln ions (radius of cavity 1.26-1.37 Å) and the decrease in stability
is considered to arise from the combined effects of a greater solvation and of a poorer fit between the cavity and the ions when their
size decreases.

The largest of the macrocycles investigated in the present work,
dibenzo-30-crown-10, 5, is a highly flexible ligand which is wrapped
round metal ions such as Na$^+$ or K$^+$ to form a cavity of smaller size
(radius of cavity for KI.5 = 1.45-1.53 Å (6)). The complexation
curve exhibited by 5 is a unique case since an important minimum is
observed at Gd(III). The so-called "Gd break" has been reported for
a large number of ligands which have no structural features in common
but the decrease in stability for Gd(III) is much smaller than in
Fig. 2 which shows a "Gd break" extending over the entire Ln series.
In analogy with the results reported for NaB(C_6H_5)$_4$ in anhydrous
medium (6), it is anticipated that the complexes with the largest Ln
have a "wrap around" structure. Such a structure is becoming less
and less stable when the ionic radius r decreases because of the increased intra-ligand repulsions. After Gd(III), ligand 5 probably
adopts an entirely different more open geometry.

All the crown ethers 1-5 exhibit the unusual property of being
able to stabilize the +2 oxidation state of Sm and Yb either because
the divalent ions are filling better the internal cavity of the large

macrocycles $\underline{4}$ and $\underline{5}$ or because stable 1:2 sandwich complexes are formed with the smaller cycles. For instance, in the case of $\underline{5}$, the stability constants are $\log\beta_1=8.3\pm0.1$ for Sm(II) and $\log\beta_1=7.5\pm0.1$ for Yb(II) while the stability constants of Sm(II).$(\underline{3})_2$ and Yb(II).$(\underline{3})_2$ are $\log\beta_2=10.8\pm0.1$ and 8.4 ± 0.1 respectively. None of the more classical, non-cyclic ligands are able to stabilize the divalent Sm and Yb ions because of their lower charge density.

ACKNOWLEDGMENTS

The authors gratefully acknowledge research support from the Fonds National de la Recherche Scientifique of Belgium. J.F.D. is Chercheur Qualifié at this Institution.

REFERENCES

1. I.M. Kolthoff, Anal. Chem., 51:1R (1979).
2. J. Massaux, J.F. Desreux, C. Delchambre and G. Duyckaerts, Inorg. Chem., 19:1893 (1980).
3. J.F. Desreux and G. Duyckaerts, Inorg. Chim. Acta, 35:L313 (1979)
4. J.-C.G. Bünzli, B. Klein and D. Wessner, Inorg. Chim. Acta, 44:L147 (1980).
5. Y. Hasegawa and G.R. Choppin, Inorg. Chem., 16:2931 (1977).
6. D.G. Parsons, M.R. Truter and J.N. Wingfield, Inorg. Chim. Acta, 14:45 (1975).

OCTAHEDRAL COMPOUNDS OF LANTHANIDES: TRIMETHYLPHOSPHINE OXIDE (tmpo) AS LIGAND

O.A. Serra, Departamento de Quimica FFCLRPUSP Av.
Bandeirantes S/N 14100 Ribeirão Preto SP Brasil
M. Moraes, Institute de Quimica USP, CP 20780 01000
São Paulo SP Brasil (1)

Lanthanide compounds with high symmetry simplify the study of f-f emissions or excitations (2-5). The synthesis and characterization of $Ln(tmpo)_6X_3$; Ln=La,Gd,Eu,Lu and Y; $X=ClO_4^-$ and PF_6^-, and the emission spectra of Eu^{3+} compounds at 77 K and room temperature are presented. The tmpo was prepared as described in the literature (6). The compounds were obtained by mixing ethanolic solutions (previously treated with dimethoxy propane) of the salt and ligand. Elemental analysis for C and H and the complexometric titration of Ln^{3+} agree with the proposed stoichiometry. The compounds are stable to 300°C and are less hygroscopic than tmpo which melts at 137°C. The molar conductivity of $Eu(tmpo)_6(ClO_4)_3$ in nitromethane is 270 ohm^{-1} $mol^{-1}cm^2$, which agrees with 1:3 electrolytes (7). The IR spectra show only one band at 628 cm^{-1} (ν_4) demonstrating that ClO_4^- is not coordinated. The P=O stretching is displaced from 1170 cm^{-1} (tmpo) to 1100 cm^{-1}. Due to the ClO_4^- (ν_3) mode at 1100 cm^{-1} this assignment is possible only in the PF_6 compounds. X-ray diffraction patterns show isomorphism for all the perchlorates.

The 4579 and 5145 Å Argon emissions were used for excitation at low intensity, as shown in Fig. 1 for $Eu(tmpo)_6(ClO_4)_3$. Similar spectra were obtained for the corresponding $La_{.99}Eu_{.01}$ compound with less intense emission lines. The excitation spectra of the Eu^{3+} compounds had only one band for the $^7F_0 \rightarrow {}^5D_1$ transition. The emission spectra show that the degenerate $^5D_0 \rightarrow {}^7F_1$ transition, allowed by magnetic dipole selection rules, is more intense that the electric dipole bands.

These results suggest that the compounds possess octahedral symmetry. A partial energy diagram for $Eu(tmpo)_6(ClO_4)_3$ is presented in Fig. 2. Resolving the secular determinant for $^7F_2$ (no J mixing) in an

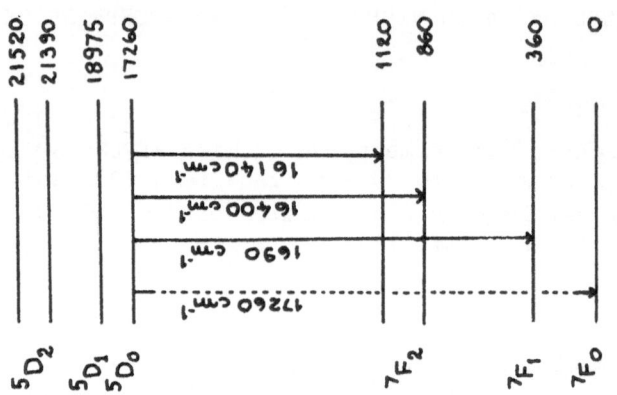

Figure 2. Partial energy level diagram for Eu(tmpo)$_6$ (ClO$_4$)$_3$.

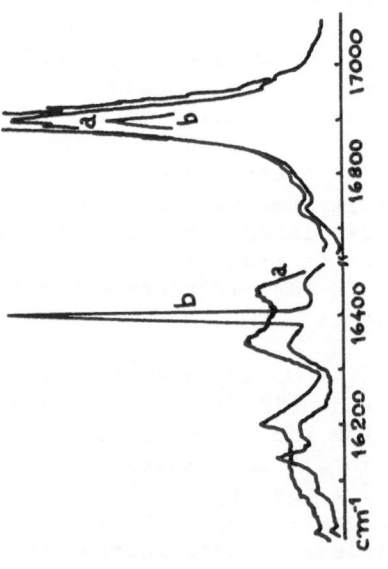

Figure 1. Emission spectra of Eu(tmpo)$_6$ (ClO)$_3$. a) 300 K b) 77 K. Excitation 4579 Å.

O_h crystal field, we obtain $B_0^4 = 1640$ cm^{-1}, since $B_0^4/B_4^4 = \sqrt{14/5}$, the value for B_4^4 is 982 cm^{-1}. Since the value of B_0^4 for $Cs_2NaEuCl_6$ was found (8,9) to be 1384 cm^{-1}, the tmpo is higher than the Cl$^-$ in a spectrochemical series for lanthanides. The symmetry of the compounds described is the highest found for a lanthanide coordination compound with an organic ligand.

REFERENCES AND NOTES

1. M.M. supported by FAPESP and OAS from CNPq. We thank Dr. O. Sala for the Raman instrumentation use and Dr. A.C. Massabni for helpful comments.
2. O.A. Serra and L.C. Thompson, <u>Inorg. Chem.</u>, 15:504 (1976).
3. L.C. Thompson, O.A. Serra, J.P. Riehl, F.S. Richardson, and R.W. Schartz, <u>Chem. Phys.</u>, 26:393 (1977).
4. R.D. Peacock, <u>Struct. Bonding</u>, 22:83 (1975).
5. P. Caro, <u>Structure electronique des elements de transition. L'atome dans le crystal</u>, PUF Paris (1976).
6. F. Choplin and G. Kaufman, <u>Spectrochim. Acta</u>, 26A:2113 (1970).
7. W.J. Geary, <u>Coord. Chem. Rev.</u>, 7:81 (1971).
8. From Ref. 2.: calculations using J-mixing give the value $B_0^4 = 1800$ ($A_4^0 = 220$ cm^{-1}).
9. O.A. Serra, L.C. Thompson, and J.A. Koningstein, <u>Chem. Phys. Letters</u>, 46:253 (1977).

AMINE COMPLEXES OF DIVALENT EUROPIUM

F. A. Hart and Wenxiang Zhu*

Department of Chemistry, Queen Mary College

Mile End Road, London E1 4NS, England

Although binary europium(II) compounds are quite stable in the absence of air, europium(II) complexes have been rarely investigated. We report an investigation of complexes of europium dichloride with 1,10-phenanthroline, and with 2,2'6',2"-tripyridyl.

EXPERIMENTAL

All operations involving europium(II) were carried out in a glove box under a nitrogen atmosphere.

<u>Dichlorobis(1,10-phenanthroline)europium(II)</u>. $EuCl_2$ (0.224g) was added to a solution of 1,10-phenanthroline (0.870g) in acenonitrile (15 ml). The suspension was stirred for 48 hours, after which time all the $EuCl_2$ appeared to have dissolved, and a purple precipitate had formed. The product was collected by filtration, washed and dried <u>in vacuo</u>, with a 90-95% yield.

<u>Dichloro(2,2'6',2"-tripyridyl)europium(II)</u>. Using 2,2'6',2"-tripyridyl and a similar procedure gave the above complex as a dark blue product. Changes in the proportions of reactants gave no different products, with a yield of 75-80%.

Both complexes, which gave satisfactory elemental analyses, decompose in N_2 at 150-250°C, giving a yellow-brown residue. They are air-sensitive, starting to lose their dark colors on exposure to air for about 30 minutes. They are instantly decolorized by air in the presence of acetonitrile. The complexes are sparingly soluble in methanol and ethanol, but dissolve in DMSO to give purple and deep green solutions respectively. UV-VIS spectra of $EuCl_2(phen)_2$ in

---

*On secondment from Peking Normal University, Peking, China.

DMSO show a peak at 500 nm, with a more intense absorption peak beginning at about 435 nm extending to shorter wavelength. Measurements in the same solvent of the 500 nm absorbance at a total concentration of 0.059M gave a Job's plot which showed a sharp maximum at a 1:1 phenanthroline:$EuCl_2$ ratio. The stability constant was evaluated from the plot giving log $\underline{K}$ = 3.9±0.3 M^{-1}. The spectrum of $EuCl_2$ (tripy) in DMSO shows peaks at 620 nm and at 460 nm, with intense absorption at shorter wavelength.

The i.r. spectra of these complexes gave evidence for coordination. The bands due to 1,10-phenanthroline show several changes from the free ligand. For example, the strong CH out-of-plane bending absorptions at 844-855 cm^{-1} and 737 cm^{-1} are lowered to 840 cm^{-1} and 720 cm^{-1}. For 2,2',6',2"-tripyridyl, the sharp α-substituted-pyridine ring-breathing mode at 991 cm^{-1} in the free ligand shifts to 1010 cm^{-1} in the complex leaving no residual peak near 991 cm^{-1}. This evidence favors a coordinated state for all three pyridine rings.

The ^1H n.m.r. spectra in d_6-DMSO of these two complexes each showed a single very broad band. The tripyridyl complex, for example, showed a very broad (∿100Hz at half height) band at 2.9τ. Thus the resonance is unshifted but broadened, as expected for an f^7 configuration.

DISCUSSION

The structure of $EuCl_2(phen)_2$ is rather unlikely to be monomeric and octahedral as six-coordination would be insufficient for Eu^{2+}. More likely is a polymeric chain with bridging chloride ions giving eight-coordination. Similar arguments apply with respect to the tripyridyl complex, but any suggestions are of course speculative in the absence of X-ray data, and single crystals could not be obtained. The Job's plot shows that the bis(phenanthroline) complex dissociates in DMSO to give a stable mono(phenanthroline) complex in which DMSO doubtless completes the coordination sphere. The stability constant value log $\underline{K}$ = 3.9 represents considerable stability and gives ΔG = -22 kJ $mole^{-1}$.

The absorptions of the complexes at 500 nm and 633.5 nm may represent a $4f^7 \rightarrow 4f^6(ligand)^1$ charge-transfer band or a $4f^7 \rightarrow 4f^6 5d^1$ transition. These absorptions can be compared with an absorption presumably $4f^7 \rightarrow 4f^6 5d^1$ at 332 nm (ϵ = 449) given by a solution of $EuCl_2$ in DMSO. The bathocromic shift could then be caused by a σ-donor nephelauxetic destabilization of the 4f shell together with stabilization of a 5d level by overlap with a ligand antibonding orbital.

ACKNOWLEDGMENTS

We thank the Ministry of Education of China and the British Council for support.

HIGH COORDINATION POLYHEDRA OF TRIVALENT RARE EARTH IONS

Jean-Claude G. Bünzli and Bernard Klein

Université de Lausanne, Institut de chimie minérale

et analytique, CH-1005 Lausanne, Switzerland

When they are bonded to multidentate ligands or to small bidentate ions, rare-earth ions can form complexes with high coordination numbers. The resulting coordination polyhedra usually possess a low symmetry. However, close scrutiny of the polytopes very often reveal that they arise from a slight distortion of a more symmetrical entity. This can be expected since the bonding of Ln(III) ions is essentially non-directional; the interesting point is that for Eu(III) compounds, the symmetry of this idealized polyhedron may be estimated from the emission spectra or, more precisely, from the crystal field splitting of the J-levels. To illustrate this point, we discuss here some examples of nitrato complexes in which the Eu(III) ion is 10-, 11-, and 12-coordinate.

An example of 10-coordination is the pentanitrate ion. The most stable arrangement for CN = 10 is the bicapped square anti-prism (BCSAP) (1) and the coordination polyhedron may indeed be described as a distorted BCSAP (2). However, when bidentate ligands having a small 'bite' are involved, a better way of discussing the structure is to use the suggestion of Bergman and Cotton (3) and to consider that these ligands occupy only one coordination site. As a matter of fact, the five N atoms $[Eu(NO_3)_5]^{2-}$ are exactly arranged on the vertices of a trigonal bipyramid, and this D_{3h} symmetry is reflected in the pattern of the crystal field splittings (Figure).

The 1:1 complex between europium nitrate and 15-crown-5 ether is, to our knowledge, the only reported 11-coordinate compound of Eu(III) (4). The coordination polyhedron can be viewed as arising either from a distortion of the most energetically favorable

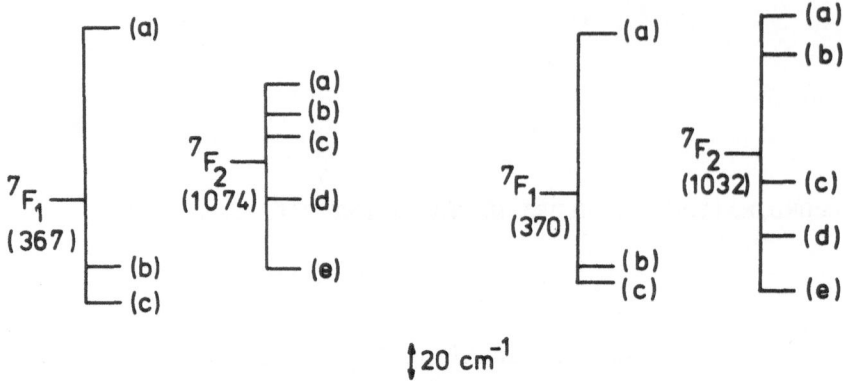

Figure : Crystal field splittings for $[Eu(NO_3)_5]^{2-}$ (left) and $Eu(NO_3)_3 \cdot C_{10}H_{20}O_5$ (right) showing the A+E (J = 1) and 2E+A (J = 2) patterns of pseudo-trigonal symmetry.

polyhedron, the monocapped pentagonal antiprism with C_{5v} symmetry, or from the expansion of the 10-coordinate polyhedron of pseudo-D_{3h} symmetry discussed above. Again, the pattern of the crystal field splittings supports either of these descriptions.

The Eu(III) is 12-coordinate in the hexanitrate ion, which is isostructural with $[Nd(NO_3)_6]^{3-}$. In this latter compound, the coordination polyhedron is a distorted icosahedron, but the arrangement of the N atoms is almost perfectly octahedral (5). We therefore conclude that the Ln(III) ion has the tendency to realize the more closely packed and the more symmetrical arrangement when bonded to small bidentate ligands.

REFERENCES

(1) M.G.B. Drew, Coord. Chem. Rev., 24 : 179 (1977).

(2) J.-C. G. Bünzli, B. Klein, G. Chapuis, and K.J. Schenk, J. inorg. nucl. Chem., 42 : 1307 (1980).

(3) J.G. Bergman Jr. and F.A. Cotton, Inorg. Chem., 5 : 1208 (1966).

(4) B. Klein, Ph. D. Thesis, University of Lausanne (1980).

(5) J.-C. G. Bünzli, B. Klein, D. Wessner, K.J. Schenk, G. Chapuis, G. Bombieri, and G. de Paoli, Inorg. Chim. Acta, 54 : L43 (1981).

ERBIUM CHLORIDE COMPLEXATION IN AQUEOUS DMF

Herbert B. Silber and M. Rebecca Riddle (1)

Division of Earth and Physical Sciences
The University of Texas at San Antonio (UTSA)
San Antonio, Texas 78285

In water or mixed aqueous solvents, $ErCl^{2+}$ complexation occurs via the following multi-step mechanism:

$$Er^{3+}(S) + Cl^-(S) = Er^{3+}(S)Cl^- = ErCl^{2+}(S) \qquad (1)$$

where S represents bound solvent, $Er^{2+}(S)Cl^-$ is an outer-sphere complex and $ErCl^{2+}(S)$ is an inner sphere complex. The ultrasonic adsorption, α/f^2, equals:

$$\alpha/f^2 - B = A_{os}/(1+(f/f_{os})^2) + A_{is}/(1+(f/f_{is})^2) \qquad (2)$$

where A_{os} and A_{is} are the amplitudes of the outer and inner sphere complexes, F_{os} and f_{is} are the respective relaxation frequencies and B is the solvent background. The excess adsorption maxium μ_{max} is equal to the product of one-half of the relaxation frequency, the relaxation amplitude and the solution sound velocity. At constant salt concentration, μ_{max} is proportional to the square of the reaction volume change. If a coordination number change accompanies complexation, a plot of μ_{max} as a function of water mole fraction, X_{H_2O}, has a maximum. In the absence of a coordination number change, μ_{max} decreases with increased water content due to reduced complexation. For $ErCl^{2+}$ formation in aqueous methanol (2) the coordination number change is absent, but is present in aqueous DMSO (3). This study in aqueous DMF was initiated to study the effect of solvent on coordination number in the Er(III) system.

RESULTS

The experimental α/f^2 data are shown in the Table. At low X_{H_2O} a double relaxation is present, which becomes a single relax-

ation upon the addition of more water. The calculated μ_{max} values
are plotted in Figure 1, based upon a single relaxation. Our experience has demonstrated that μ_{max} for inner sphere complexation
is essentially unchanged whether a single or double relaxation calculation is used. A large increase in μ_{max} below X_{H_2O} of 0.25 is
consistent with a coordination number change. The sharp drop in f_{is}
occuring as water is initially added to the DMF solutions is indicative of a change from bis to mono-complexation (2).

A lower amplitude coordination number change appears at high
water mole fraction, a feature not yet observed in any other
lanthanide(III) system we have studied. In order to insure that
this observation is not an artifact of the calculations, Figure 2
is a plot of α/f^2 at two frequencies as a function of X_{H_2O}. The
data at 7.0 MHz are indicative of inner sphere complexation, whereas
the 151.7 MHz data contains a small contribution from inner sphere
complexes coupled to the outer-sphere results. At 7.0 MHz the data
mimic the μ_{max} curve, confirming the results of Figure 1. At
high frequency, this effect is absent and if the results were an
artifact, even if caused by impurities, the effect would be present.
At 151.7 MHz the small increase at low water compositions is a
result of the increasingly high amplitudes of the low frequency
relaxation.

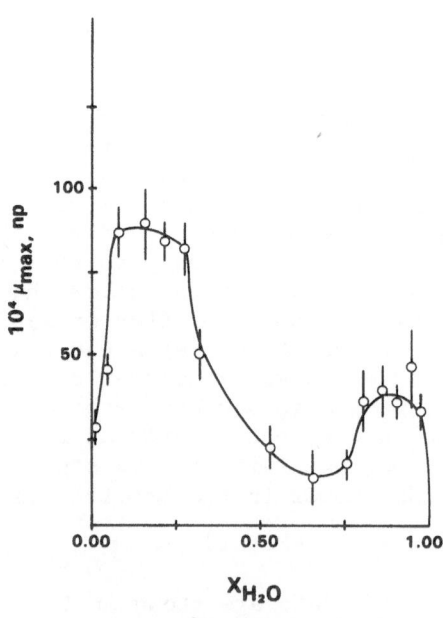

Figure 1. μ_{max} for $ErCl^{2+}$ in aqueous DMF at 25C.

Table. The ultrasonic absorption data for 0.200 M ErCl$_3$ in aqueous DMF at 25C.

water mole fraction

f, MHz	0	0.037	0.082	0.155	0.216	0.272	0.320	0.528	0.656	0.763	0.806	0.866	0.909	0.944	0.974	1.00
4.13±0.05	–	730.9	1499.7	1524.5	1347.8	1441.5	712.9	–	–	–	–	–	–	–	–	
5.15±0.16	–	–	1068.3	1063.5	1042.6	1053.1	–	–	–	–	–	–	–	–	–	
7.01±0.02	–	484.9	881.5	911.6	893.1	746.6	692.1	258.1	201.7	189.1	254.6	291.9	312.9	348.4	280.3	N
9.11±0.02	–	376.5	675.8	700.8	664.1	546.2	405.4	184.7	142.6	155.7	175.7	203.9	220.0	230.8	194.4	O
11.32±0.46	–	307.0	566.6	–	502.2	432.9	356.3*	158.1**	–	–	133.4	156.3	161.8	178.8	141.0	
11.08±0.01	–	305.8	552.4	569.6	525.2	449.7	295.4	153.3	115.9	126.1	133.0	134.0	148.6	–	133.8	E
12.17±0.04	169.1	258.0	478.4	–	438.4	415.7	268.7	141.2	109.3	111.0	117.1	130.7	148.3	159.8	125.9	X
13.09±0.01	–	271.1	458.2	471.1	427.6	374.6	250.4	130.5	98.8	97.1	109.9	122.9	127.1	129.9	106.3	C
15.05±0.03	174.9	226.4	400.4	467.5	366.1	315.8	229.0	121.0	92.1	85.1	92.2	101.7	107.0	97.0	78.4	E
17.06±0.01	–	202.2	336.5	340.3	314.5	272.0	197.6	116.0	89.7	84.7	88.6	90.5	89.2	90.8	78.5	S
19.06±0.01	–	182.0	283.0	291.7	271.2	233.2	170.7	106.1	81.0	76.7	73.9	79.0	81.9	78.0	66.4	S
20.18±0.03	–	172.8	262.5	280.1	263.0	219.2	160.1	103.5	83.3	69.2	73.4	72.2	69.5	66.7	56.1	
24.89±0.06	107.3	145.3	228.4	233.6	225.5	186.9	133.6	86.5	72.3	66.3	63.1	64.2	60.5	55.4	46.1	A
28.19±0.07	–	128.4	201.1	202.7	190.3	169.0	132.6	92.5	74.1	62.9	58.9	57.7	50.8	47.8	42.2	B
30.39±0.15	93.5	124.1	199.3	190.7	180.5	162.6	131.3	87.2	70.1	56.2	55.2	51.2	47.9	48.2	38.9	S
35.23±0.15	88.5	105.6	160.2	163.5	161.0	142.0	116.0	82.5	70.4	59.9	55.8	48.3	45.8	40.2	34.0	O
36.33±0.03	–	112.1	157.8	162.4	156.4	139.4	112.0	81.3	67.2	57.5	49.7	45.9	42.9	38.9	35.2	R
44.34±0.04	–	95.7	136.1	139.9	134.9	118.3	99.4	77.4	66.3	56.3	49.3	45.1	39.1	34.1	31.5	P
45.17±0.11	80.3	96.3	133.1	140.1	132.4	110.3	104.7	76.9	66.5	55.2	53.3	45.8	40.6	35.5	31.1	T
50.28±0.40	75.1	85.5	133.0	134.1	125.9	115.4	104.7	70.9	63.0	53.6	49.2	43.2	38.3	34.3	28.1	I
55.09±0.08	70.1	85.9	120.8	119.1	119.7	107.0	97.3	74.3	63.5	55.5	48.2	43.6	36.2	33.0	28.4	O
65.26±0.07	65.9	76.1	104.5	104.4	107.1	100.8	89.0	72.9	66.5	54.6	47.6	38.5	35.4	30.2	27.3	N
70.17±0.45	58.4	69.7	101.6	100.0	102.4	96.1	–	75.2	61.8	50.3	46.6	40.1	34.1	30.2	27.1	
75.45±0.59	56.9	67.4	94.8	95.5	96.5	96.2	86.8	70.8	61.4	53.6	45.9	40.7	32.9	30.2	25.6	
85.37±0.10	53.4	67.8	90.3	89.8	89.7	88.2	81.1	68.4	61.2	52.0	44.9	38.6	31.0	28.2	26.5	
90.17±0.66	51.8	67.6	85.4	84.8	87.9	86.1	85.4	68.7	57.6	47.7	44.2	37.9	33.1	28.9	26.2	
95.46±0.09	50.5	61.2	80.8	82.9	88.2	83.4	76.8	66.0	59.6	51.3	43.4	37.9	31.6	28.2	25.2	
105.53±0.13	49.5	57.0	75.0	79.5	81.9	82.7	79.7	67.8	59.5	50.8	44.0	38.5	31.3	27.4	24.8	
110.11±0.95	48.4	54.6	75.6	76.2	80.2	79.9	78.6	63.6	60.3	48.9	44.1	39.3	31.9	28.1	24.8	
115.63±0.14	44.2	53.2	71.6	75.2	76.9	77.4	73.7	63.8	55.8	48.4	42.9	37.3	31.2	24.2	24.2	
125.66±0.22	–	54.1	69.3	71.1	72.2	73.8	71.8	62.4	57.0	50.5	43.0	36.9	30.6	26.7	24.5	
130.21±1.36	44.0	54.4	70.6	68.7	72.3	73.7	72.4	59.8	55.1	45.3	43.8	37.1	30.6	26.9	25.2	
135.62±1.10	43.2	50.3	65.9	67.1	70.1	70.4	70.4	63.5	56.2	49.8	44.7	35.7	32.8	28.3	24.5	
145.91±0.13	–	49.4	65.2	63.2	63.9	67.8	68.2	62.4	55.3	48.7	43.5	35.1	32.2	26.6	24.9	
151.73±1.51	40.8	47.9	62.9	55.9	67.0	68.4	62.8	61.1	56.4	47.0	42.2	36.5	30.4	27.3	25.3	
171.04±0.53	39.8	43.3	57.2	59.3	65.1	63.1	–	–	–	–	42.2	34.8	29.7	26.4	23.8	
191.16±0.37	38.0	40.1	56.0	56.1	58.1	59.4	–	–	–	–	41.9	35.7	28.7	26.3	23.9	
211.23±0.27	37.0	37.4	51.7	52.1	53.1	57.4	–	–	–	–	–	–	29.0	26.3	24.6	

Column 1.00: NO EXCESS ABSORPTION

*f = 10.37 MHz   **f = 10.88 MHz

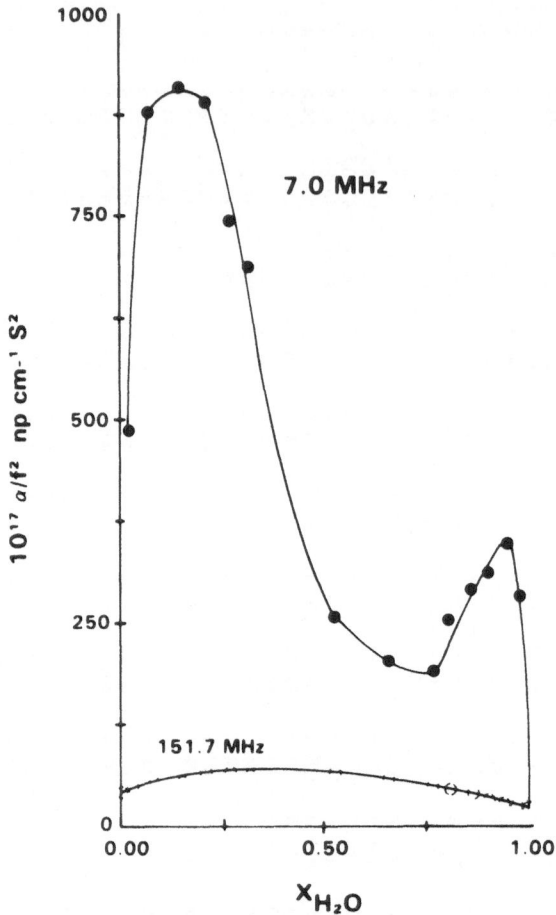

Figure 2.   Ultrasonic absorption for $ErCl^{2+}$ at selected frequencies.

The solvent parameters which may affect the coordination number change are dielectric constant, extent of hydrogen bonding or size. The dielectric constants in aqueous methanol range from 32.6 to 78.3, whereas in aqueous DMF the range is 36.8 to 78.3 (4). Despite the similarities in dielectric constant, the absence of a coordination number change in methanol eliminates this explanation.   The sizes of DMSO (71 ml/mol) and DMF (77 ml/mol) suggest that the coordination number change is steric in nature since both solvent systems act similarly at low water mole fraction.   Insufficient information exists to determine the effect of solvent hydrogen bonding.

The high X_{H_2O} maximum suggest that the environment surrounding Er(III) consists of water plus DMF, an unexpected observation. Since no measurable inner sphere complexation exists in water, the second coordination number change requires that the solvation number of the Er(III) ion differs from the coordination number of the inner sphere complex, even at high water contents.

REFERENCES

1. Supported by a grant from the Robert A. Welch Foundation of Houston, Texas.
2. J. Reidler and H. B. Silber, J. Phys. Chem., 78, 424 (1974).
3. H. B. Silber in The Rare Earths in Modern Science and Technology, Vol. 2, (edited by G. J. McCarthy, J. J. Rhyne and H. B. Silber) pp. 93-98, Plenum Press, N.Y. (1980).
4. J. I. Kim, A. Cecal, H.-J. Born and E. A. Gomaa, Z. Physik. Chem. N. F., 209 (1978).

LANTHANIDE HEXAFLUOROANTIMONATE COMPLEXES OF HEXAMETHYLPHOSPHORAMIDE*

Sérgio M. Melo and Nágila M. P. S. Ricardo

Centro de Ciências da Universidade Federal do Ceará

60.000 - Fortaleza, Brasil

Adducts between hexamethylphosphoramide and lanthanide perchlorates (1,2), clorides (3), nitrates (4,5), hexafluorophosphates (6), perrhenates (7), and trifluoromethanesulfonates (8) have been described previously. Recently, we have prepared the adducts between tetraphenylborates (9,10) and hexafluoroarsenates (11), and have made crystal calculations of the doped lanthanide perchlorates with HMPA (12). Here we describe preparations of adducts of lanthanide hexafluoroantimonates and hexamethylphosphoramide compounds of the general formulas $Ln(SbF_6)_3 \cdot 6HMPA$. The addition compounds were characterized by chemical analyses and by vibrational, electronic absorption and emission spectra.

EXPERIMENTAL

The hexamethylphosphoramide (Aldrich Chem Co.) was used as received. The lanthanide oxides (CERAC and Sigma) were 99.9% purity and the potassium hexafluoroantimonate from Alfa-Ventron was used as received. The solvents and other chemicals were of reagent quality.

The hydrated lanthanide hexafluoroantimonates were obtained from ion exchange reactions in an acetone medium between hydrated perchlorates and potassium hexafluoroantimonate. To this solution, a stoichiometric amount of HMPA in absolute ethanol was slowly added for a few minutes. The resulting precipitates were filtered off in a sintered glass funnel, washed with absolute ethanol and dried under vacuum over anhydrous calcium chloride.

---

*The authors thank the OAS and CNPq for financial support.

The nitrogen content was obtained by a conventional micro-kjel-dahl method (13). The antimony and lanthanide contents and microan-alysis for carbon, nitrogen and hydrogen were done by Galbraith Labor-atories. IR spectra were recorded with a Perkin-Elmer model 283B spectrophotometer as nujol mulls between CsI plates. The fluorescence spectrum of the europium complex were obtained on a Perkin-Elmer model MPF 44B spectrophotofluorimeter, using 394 nm excitation. The absorp-tion of neodymium compounds as fluorolube mulls at room temperature was recorded with a Cary 17 spectrophotometer.

RESULTS AND DISCUSSION

The analytical data are in agreement with the general formula $Ln(SbF_6)_3 \cdot 6HMPA$. The compounds are not hygroscopic and no decomposi-ton was observed when the compounds are heated to 300°C. The com-pounds could not be recrystallized since they are essentially insolu-ble in the common solvents.

Vibrational spectra were used to determine the coordination site of the ligands and whether the anions were coordinated or ionic in nature. The ionic hexafluoroantimonate ion belongs to the point group O_h which has two IR-active vibrations (14). Comparison of the spec-trum of the $KSbF_6$ with those of the complexes indicates that the octahedral symmetry of the hexafluoroantimonate ion has not been re-duced (Table 1). All IR spectra show a very strong band around 645 cm^{-1} (ν_3) and another very strong band around 275 cm^{-1} (ν_3) at-tributed to the SbF_6 anion. This is an indication of the retention of the O_h symmetry for the hexafluoroantimonate ion in the complexes and that they are not coordinated to the lanthanides. No absorption bands in the region 3710 cm^{-1} and 1630 cm^{-1} were observed, indicating the absence of water in the compounds. A considerable shift of the P-O stretching mode to lower frequencies and an enhancement of P-N (+15 cm^{-1}) are indicative that coordination occurred through the phos-phoryl oxygen of the hexamethylphosphoramide. The P-O frequency of the coordinated HMPA in these complexes shows an unusually large shift of -125 cm^{-1} compared to similar complexes with the hexafluoroarsenate anion (11). The ν_{MO} for these compounds were not found in the 4000-200 cm^{-1} region as observed by Berwerth (7) for $Ln(ReO_4)_3 \cdot 6HMPA$.

The number of fluorescent transitions which arise from radiative decay from the ^5D excited levels is defined by the symmetry of the point group. In the fluorescence spectrum of the Eu^{3+} ion, the lines originate mainly from the transitions between the $^5D_0$ excited level and the ^7F_J manifold. The electron-dipole transitions to the $^7F_0$ and $^7F_2$ levels are strictly forbidden in a non-vibrating site with a centrosymmetric environment. However, the $^5D_0 \rightarrow {}^7F_1$ transition is magnetic-dipole allowed.

Table 1. Functional group vibrations of $Ln(SbF_6)_3 \cdot 6HMPA$ (cm^{-1})

Compound	PO	$_{as}PN^{mode}$	$_{s}PN$	$(SbF_6)^-$	
				ν_3	ν_4
HMPA	1212 s	978 vs	740 s	–	–
$Pr(SbF_6)_3 \cdot 6HMPA$	1080 vs	996 vs	753 s	650 vs	278 vs
$Nd(SbF_6)_3 \cdot 6HMPA$	1080 vs	990 vs	754 s	648 vs	276 vs
$Sm(SbF_6)_3 \cdot 6HMPA$	1082 vs	990 vs	754 s	648 vs	276 vs
$Eu(SbF_6)_3 \cdot 6HMPA$	1084 vs	990 vs	755 s	645 vs	275 vs
$Tb(SbF_6)_3 \cdot 6HMPA$	1086 vs	990 vs	755 s	645 vs	275 vs
$Ho(SbF_6)_3 \cdot 6HMPA$	1086 vs	992 vs	754 s	646 vs	276 vs
$Tm(SbF_6)_3 \cdot 6HMPA$	1086 vs	990 vs	756 s	646 vs	276 vs
$Lu(SbF_6)_3 \cdot 6HMPA$	1092 vs	990 vs	755 s	648 vs	278 vs

v = very,   s = strong,   m = medium,   w = weak

Figure 1 contains the fluorescence spectrum of the europium compound. The forbidden $^5D_0 \rightarrow {}^7F_0$ transition is absent in the spectrum and the magnetically allowed $^5D_0 \rightarrow {}^7F_1$ transition occurs as the most intense band at 591 nm. In the region where the $^5D_0 \rightarrow {}^7F_2$ transition is expected to occur, one line of weak intensity at 609.4 nm was observed. Thus, we conclude that the observed splittings of the $^5D_0 \rightarrow {}^7F_{1,2}$ transitions are consistent with a O_h site symmetry (15).

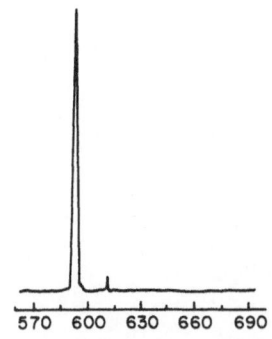

Figure 1. Emission spectrum of $Eu(SbF_6)_3 \cdot 6HMPA$.

REFERENCES

1. E. Giesbrecht and L.B. Zinner, Inorg. Nucl. Chem. Lett., 5:575 (1969).
2. J.T. Donoghue, E. Fernandez, J.A. McMillan, and D.A. Peters, J. Inorg. Nucl. Chem., 31:1431 (1969).
3. J.T. Donoghue and D.A. Peters, J. Inorg. Nucl. Chem., 31:467 (1969).
4. J.A. Sylvanovich, Jr. and S.K. Madan, J. Inorg. Nucl. Chem., 34: 1675 (1972).
5. S.P. Sinha, Z. Anorg. Allg. Chem., 434:227 (1977).
6. O.A. Serra and A.B. Nascimento (private communication).
7. O.A. Serra and E.C. Berwerth (private communication).
8. L.B. Zinner and G. Vicentini, J. Inorg. Nucl. Chem., 43:193 (1981).
9. S.M. Melo and O.A. Serra, Proc. 12th Rare Earth Research Conf., pp. 180-5 (1976).
10. M.K. Kuya, S.M. Melo, and O.A. Serra, An. Acad. Brasil. Ciênc., 51:239 (1979).
11. S.M. Melo and V.L.M. Albuquerque, unpublished work.
12. S.M. Melo and O.A. Serra, in The Rare Earths in Modern Science and Technology, 2, G.J. McCarthy, J.J. Rhyne and H.B. Silber (eds.), Plenum Press, New York, pp. 127-31 (1981).
13. E.P. Clark, "Semicro Quantitative Organic Analysis," Academic Press, New York, pp. 37 (1943).
14. N.B. Colthup, L.H. Daly, and S.E. Wiberley, "Introduction to Infrared and Raman Spectroscopy," Academic Press, New York, pp. 161-2 (1964).

PARAMETRISATION OF THE MCD-SPECTRA OF RARE EARTHS: EXAMPLE OF Pr^{3+} IN PVA-MATRIX

C. Görller-Walrand, N. De Moitie-Neyt and Y. Beyens

Universiteit Leuven, Afdeling Anorganische en Analytische

Scheikunde, Celestijnenlaan 200F, 3030 Heverlee, Belgium

INTRODUCTION AND THEORETICAL BACKGROUND

The aim of this note is to show how a parametrisation of Magnetic Circular Dichroism (MCD) spectra can be given as a function of G_2, G_4 and G_6 in a way similar to the Judd-Ofelt analysis of the absorption spectra. This is illustrated by an example of Pr^{3+} ion spectra.

The theory of forced electric-dipole transitions in trivalent rare-earth ions developed by Judd [1] and Ofelt [2] relates the probability of transition between a ground state $<f^N \psi J|$ and an excited state $|f^N \psi' J'>$ to matrix elements of the tensor operators U^λ,

$$D = 1/d_J \sum_{\lambda=2,4,6} F_\lambda |<f^N\alpha[SL]J||U^\lambda||f^N\alpha'[S'L']J'>|^2 \qquad [1]$$

where D = dipole strength

$$F_\lambda = e^2(2\lambda + 1) \sum_t B^t Z^2(\lambda,t) \qquad [2]$$

$$B^t = \sum_p |A_p^t|^2/(2t+1) \qquad [3]$$

$$Z(\lambda,t) = 2\Sigma(2l+1)(2l'+1)x(-l)^{l'+l} x \begin{Bmatrix} 1 & \lambda & t \\ l & l' & l \end{Bmatrix} x \begin{pmatrix} l & 1 & l' \\ 0 & 0 & 0 \end{pmatrix} \begin{pmatrix} l' & t & l \\ 0 & 0 & 0 \end{pmatrix}$$

$$\qquad [4]$$

$$x <nl|r|n'l'><nl|r^t|n'l'>/\Delta E(n'l')$$

The empirical parameters F_λ are related to certain integrals between appropriate radial wavefunctions, and to the odd-order terms, A_p^t characterizing the ligand field at the rare earth ion.

By analogy, H. Pink (3) developed a theory which describes the A and C, MCD terms by:

$$A = 1/d_J (-1)^{J+J'+1} \sum_{\lambda=2,4,6} G_\lambda |<f^N\alpha[SL]J||U^\lambda||f^N\alpha'[S'L']J'>|^2$$

$$\times \left[\begin{Bmatrix} \lambda & \lambda & 1 \\ J' & J' & J \end{Bmatrix} <J'||\mu^{(1)}||J'> + \begin{Bmatrix} \lambda & \lambda & 1 \\ J & J & J' \end{Bmatrix} <J||\mu^{(1)}||J> \right]$$

$$C = -1/d_J (-1)^{J+J'+1} \sum_{\lambda=2,4,6} G_\lambda |<f^N\alpha[SL]J||U^\lambda||f^N\alpha'[S'L']J'>|^2$$

$$\times \begin{matrix} \lambda & \lambda & 1 \\ J & J & J' \end{matrix} <J||\mu^{(1)}||J>$$

where $G_\lambda = \sqrt{3/2}\, e^2 (2\lambda+1)^2 \sum_t B^t Z^2(\lambda,t) \begin{Bmatrix} 1 & 1 & 1 \\ \lambda & \lambda & t \end{Bmatrix}$

A comparison between the expressions of D and A shows a great similarity. Both of them contain three parameters ($F_{2,4,6}$ and $G_{2,4,6}$) depending on the symmetry of the environment. The sign of G_λ depends on the relative importance of the various terms of odd order t in the expansion of the crystal field potential. Due to the magnetic field in MCD, the expression of A contains the matrix elements of the Zeeman-perturbation $<|\mu^{(1)}|>$.

Pr^{3+} EXAMPLE

Low temperature experiments were performed on PVA-films of $Pr(ClO_4)_3$ (8). The absorption spectra were recorded on a CARY 219 spectrophotometer. MCD was recorded on a CARY 61 dichrometer equipped with an Oxford superconducting magnet using a 40 kG magnetic field.

Table 1 gives the theoretical values for the A- and C- parameters of the transitions in Pr^{3+}. As an example we consider the transition $^3P_0 \leftarrow {}^3H_4$. Using the expression for A one gets:

$$A = -\ 1/9\ G_4 |<^3H_4||U^4||^3P_0>|^2 \left[\begin{Bmatrix} 4 & 4 & 1 \\ 0 & 0 & 4 \end{Bmatrix} \right.$$

$$\left. \times\ <^3P_0||\mu^{(1)}||^3P_0> + \begin{Bmatrix} 4 & 4 & 1 \\ 4 & 4 & 0 \end{Bmatrix} <^3H_4||\mu^{(1)}||^3H_4> \right] \qquad [5]$$

The selection rules on the 3j- and 6j-symbols (4) give a restriction for $\lambda = 4$ for the specific case of $^3P_0 \leftarrow\ ^3H_4$. The reduced matrix elements of U^λ are tabulated by Carnall, Fields and Rajnak (5). The matrix elements in the Zeeman-operator in Russell-Saunders coupling is given by:

$$<LSJ||\mu^{(1)}||LSJ> = -\ g\ \beta\ \hbar\sqrt{(J+1)(2J+1)J}$$

We finally get $A = -\ 2.29\ 10^{-2}\ \beta\ \hbar\ G_4$. The value for C equals A but has an opposite sign. Experimentally the sign-inversion is indeed found (see Fig. 1). The theoretical values for A and C (see Table 1) in combination with experimental results, give an indication of the relative importance of the G_λ-parameters. The results for $^3P_0 \leftarrow\ ^3H_4$ (A(-), C(+)) indicate that G_4 is positive. To explain the experimental sign of $^3P_2 \leftarrow\ ^3H_4$ (A(+), C(-)), G_6

Table 1.   Parametrisation of the MCD-spectra of Pr^{3+} in a PVA-matrix.

Transition	Spectral Region (cm^{-1})	A- and C-Terms and Sign (9) Exper.	
$^3P_0 \leftarrow\ ^3H_4$	20540–21050	A(-)	A = $-2.29\ 10^{-2}\ \beta\ \hbar\ G_4$
		C(+)	C = $+2.29\ 10^{-2}\ \beta\ \hbar\ G_4$
$^3P_1 \leftarrow\ ^3H_4$	21050–21500	neg. dispers. C(+) ?	A = $-2.36\ 10^{-2}\ \beta\ \hbar\ G_4$ C = $+2.15\ 10^{-2}\ \beta\ \hbar\ G_4$
$^3P_2 \leftarrow\ ^3H_4$	21930–23090	A(+)	A = $-5.43\ 10^{-3}\ \beta\ \hbar\ G_4$ $-2.80\ 10^{-2}\ \beta\ \hbar\ G_6$
		C(-)	C = $+4.08\ 10^{-3}\ \beta\ \hbar\ G_4$ $+1.44\ 10^{-2}\ \beta\ \hbar\ G_6$

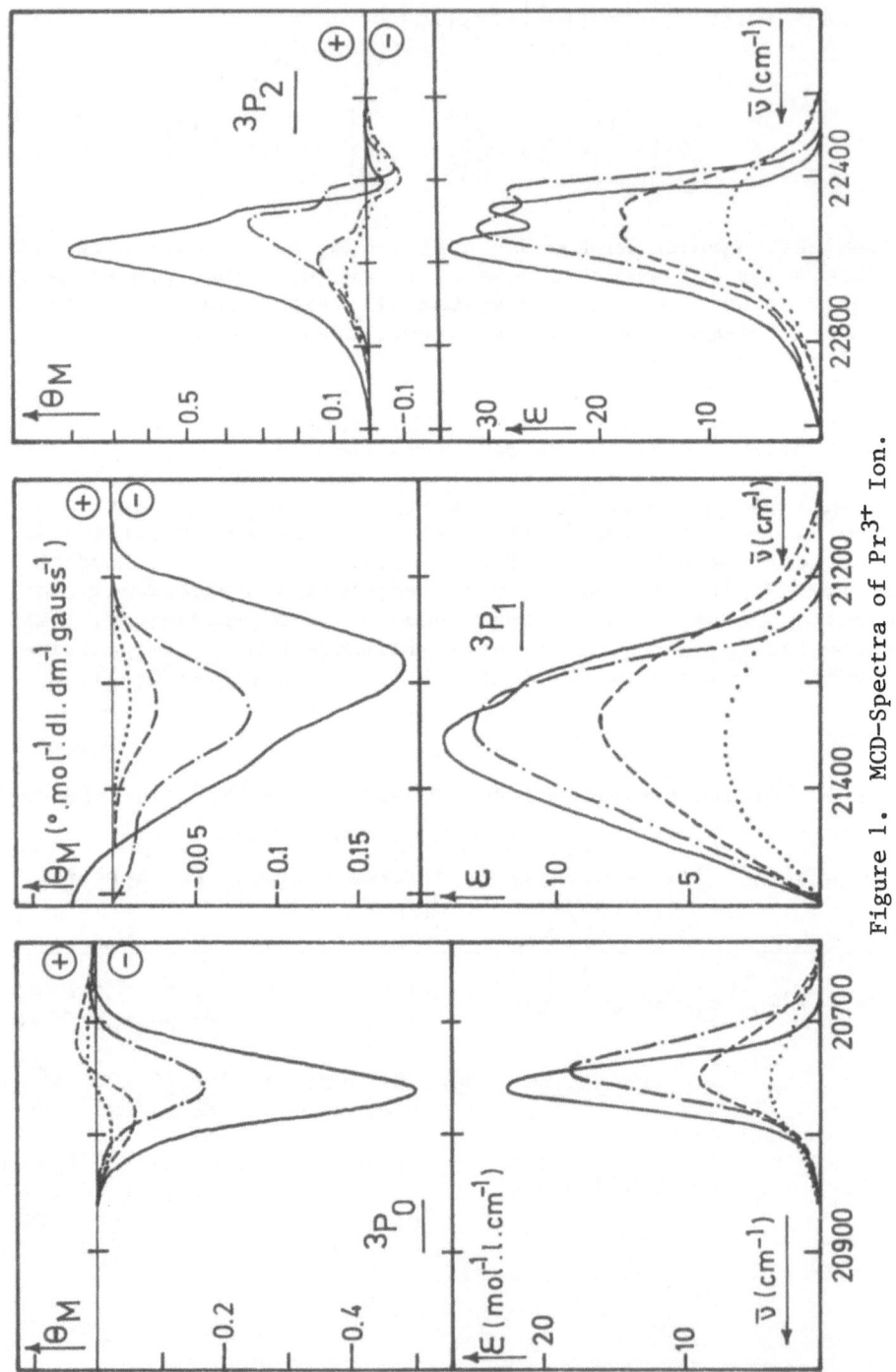

Figure 1. MCD-Spectra of Pr^{3+} Ion.

should be negative. This is in agreement with the results of
Livshits (6). Using the method of moments (7) we can evaluate the
Faraday parameters:

$$A(^3P_2 \leftarrow {}^3H_4)/A(^3P_0 \leftarrow {}^3H_4) = 4.27\ 10^{-3}/-2.69\ 10^{-4} = -15.9$$

From Table 1, it follows that: $A(^3P_2 \leftarrow {}^3H_4)/A(^3P_0 \leftarrow {}^3H_4) =$
$0.24 + 1.22\ G_6/G_4$, so that $G_6 = -13.23\ G_4$. The $^3P_1 \leftarrow {}^3H_4$ transitions depend only on the G_4-parameter. Although partly influenced
by $^1I_6 \leftarrow {}^3H_4$, this transition shows indeed the correct behavior
(C(+)).

DISCUSSION

It is hoped that the parametrisation of the MCD-spectra will
make possible the definition of consistent sets of G_λ-parameters
for the whole series of lanthanides in different media. These
parameters are characterized by a magnitude and a sign. Resulting
from the interplay of the odd crystal field parameters, the G_λ's
by their inner consistency may shed new light on the symmetry
hypothesis made about the environment of lanthanides.

In the specific example briefly illustrated here, the
$^3P_{0,1,2} \leftarrow {}^3H_4$ transitions can be rationalized by G_4 and G_6, respectively positive and negative. Previous results (8) restricted
to the analysis of the $^3P_0 \leftarrow {}^3H_4$ transition supported the hypothesis
of a trigonal symmetry for Pr^{3+} in PVA-matrix. In the framework
of the G-parametrisation of this note, the positive value of G_4
confirms this assumption. It is worthwhile to mention that the
analysis of the Eu^{3+} done by Pink (3) also leads to a sign-inversion
between G_4 and G_6. We hope to find inspiration for further work
in this "coincidence".

ACKNOWLEDGMENTS

One of us (N.N.) thanks the IWONL (Belgium) for financial
support. The laboratory is indebted to the IWONL and NFWO (Belgium)
for the experimental equipment. The authors thank F. Morissens
and M. Behets for technical aid.

REFERENCES

1. B.R. Judd, Phys. Rev., 127:750 (1962).
2. G.S. Ofelt, J. Chem. Phys., 37:511 (1962).
3. H. Pink, Dissertation, Syracuse University (1975).

4. K.M. Howell, Revised tables of 6j-symbols, Research Report
    59-1, University of Southampton, 1959; Rotenberg, Bivins,
    Metropolis, Wooten, The 3j- and 6j-symbols, M.I.T. Press, 1959.
5. W.T. Carnall, P.R. Fields and K. Rajnak, J. Chem. Phys.,
    49:4424 (1968),
6. M.A. Livshits, Opt. Spektrosk. 33:603 (1972).
7. P.J. Stephens, R.L. Mowery and P.N. Schatz, J. Chem. Phys.,
    55:224 (1971).
8. C. Görller-Walrand, Y. Beyens and J. Godemont, J. Chem. Phys.,
    76:190 (1979).
9. For the sign conversion of A- and C-terms see reference (8).

AB-INITIO CALCULATION OF LANTHANIDE CRYSTAL FIELD PARAMETERS AND TRANSITION PROBABILITIES

Michéle Faucher and Oscar Malta*

ER 210, CNRS, 1 Place A. Briand
92190 Meudon-Bellevue, France
*Departamento de Quimica de U.F.PE
Cidade Universitaria, Recife, PE. 50 000
Brasil

INTRODUCTION

We discuss here two aspects of the influence of quadrupolar moments induced in a crystal containing rare earth ions.

The gradient of the <u>static</u> crystal field induces quadrupoles which in turn produce an <u>extra</u> contribution to the crystal field at the rare earth site. This contribution is evaluated for $LaCl_3:Nd^{3+}$. Recently, it has been argued (1,2) that a mechanism (the Pseudo-Multipolar Field), based on the fact that oscillating dipoles induced on the ligands by the external radiation produce a field which is effective in inducing electronic transitions within the $4f^N$ configuration, can explain the main features of hypersensitivity (3). We shall analyze the contribution due to <u>oscillating quadrupole</u> moments induced on the ligands by the external radiation. The transition probability due to this mechanism is evaluated for the $^5D_0 \rightarrow {}^7F_2$ and $^5D_0 \rightarrow {}^7F_4$ transitions of the Eu^{3+} ion doped in Y_2O_3.

QUADRUPOLAR CONTRIBUTION TO CRYSTAL FIELD PARAMETERS (CFP)

Until now, most electrostatic calculations utilized the point charge model which neglects the spatial aspect of charge distribution. Morrison (4) in 1976 calculated electrostatic cfp including the contribution of point charges and point dipoles at the ions' site, the dipole moments being determined in a consistent way. To our knowledge, few attempts have been made in the same direction (5,6,7). In the case of Nd_2O_3 (6) and $LaCl_3:Nd^{3+}$ (7), the dipolar correction improved the agreement between experimental and calculated values. We have investigated the following term in the expansion (quadrupolar)

115

to check if the agreement is better. The calculation is made in two steps: first, the evaluation of induced dipolar and quadrupolar moments; followed by the evaluation of the correction term to the cfp.

Induced Dipolar and Quadrupolar Moments

The dipolar and quadrupolar polarizabilities α_D and α_Q are assumed to be scalar so that:

$$P(j,v) = \alpha_D(j) \cdot E(j,v) \quad Q(j,v,v') = \alpha_Q(j) \cdot \partial E(j,v)/\partial v' \quad (v,v' = x,y,z) \tag{1}$$

where $P(j,v)$ is the induced dipolar moment at j in the v direction, $Q(j,v,v')$ the vv' component of the quadrupolar moment at j and $E(j,v)$ the v component at j of the total electric field produced by monopoles, dipoles and quadrupoles. $P(j,v)$ and $Q(j,v,v')$ are determined by the consistency equation expressing their equilibrium value at each lattice site in presence of the total fields and field gradients.

Quadrupolar Correction

The point charge expression for the cfp is well known. The dipolar correction is given in reference 6. The quadrupolar correction is written as:

$$(B_q^k)_Q = -\left(\frac{4\pi}{2k+1}\right)^{1/2} e \langle r^k \rangle \sum_{j,v,v'} Q(j,v,v') \nabla_{vv'}^2 (Y_q^{k*}(\alpha_j,\beta_j))/ R_j^{k+1} \tag{2}$$

Once the derivations are achieved, the final expression is:

$$
\begin{aligned}
(B_q^k)_Q = -\left(\frac{4\pi}{2k+5}\right)^{1/2} e \langle r^k \rangle \Bigg\{ &(Q_{xx}-Q_{yy}+2iQ_{xy})\frac{}{4}\left[(k+q+1)(k+q+2)(k+q+3)(k+q+4)\right]^{1/2} Y_{k+2}^{*q+2} \\
&+ \frac{(Q_{xx}-Q_{yy}-2iQ_{xy})}{4}\left[(k-q+1)(k-q+2)(k-q+3)(k-q+4)\right]^{1/2} Y_{k+2}^{*q-2} \\
&- (Q_{zx}+iQ_{zy})\left[(k+q+1)(k+q+2)(k+q+3)(k-q+1)\right]^{1/2} Y_{k+2}^{*q+1} \\
&+ (Q_{zx}-iQ_{zy})\left[(k-q+1)(k-q+2)(k+q+1)(k-q+3)\right]^{1/2} Y_{k+2}^{*q-1} \\
&+ \frac{(2Q_{zz}-Q_{xx}-Q_{yy})}{2}\left[(k-q+1)(k+q+1)(k+q+2)(k-q+2)\right]^{1/2} Y_{k+2}^{*q} \Bigg\} / R^{k+3}
\end{aligned} \tag{3}
$$

Table 1. Monopolar, dipolar, and quadrupolar contributions to the crystal field parameters of $LaCl_3:Nd^{3+}$ and Nd_2O_3.

cm^{-1}	$LaCl_3:Nd^{3+}$					Nd_2O_3					
	B_0^2	B_0^4	B_0^6	B_6^6 R	I	B_0^2	B_0^4	B_3^4	B_0^6	B_3^6	B_6^6
Exper.	163	-336	-713	462		-836	634	1606	752	-237	672
PC	-767	-182	-61	29	-21	-1685	518	-1438	344	205	317
Dipole	1547	0	-43	-8	-23	-507	21	27	-37	10	15
Total 1	780	-182	-104	21	-44	-2192	539	-1411	307	215	332
Dipole	3927	-20	-103	-23	-51	1865	-84	-50	139	-28	-56
Quadr.	20188	-4646	806	39	726	2690	-1355	-200	-147	-189	-193
Total2	23348	-4848	642	45	654	2870	-921	-1688	336	-12	68
Dipole	1428	3	-41	-7	-22	-828	36	35	-61	15	25
Quadr.	143	4	-5	20	5	-444	-222	-114	-10	40	-8
Total 3	804	-175	-107	42	-38	-2957	332	-1517	273	260	334

Application to $LaCl_3:Nd^{3+}$ and Nd_2O_3

For $LaCl_3$ the X-ray data from reference 8 were utilized. We set $\alpha_D(Cl^-)= 3\text{Å}^3$. Table 1 reports the experimental results from Gruber et al. (7) and the calculated B_q^k obtained with the radial integrals of Freeman and Desclaux (9). Total 1 is the sum of the point charge and dipolar contributions only ($\alpha_Q(Cl^-)=0$). Once multiplied by $1/<r^k>$. $(0.5292)^k$, these values well agree with the lattice sums A_{kq} of reference 7. Total 2 reports the total results with $\alpha_Q(La^{3+})=0.9 \text{ Å}^5$ (which is the value for Y^{3+} in reference 10) and $\alpha_Q(Cl^-) = 5.5 \text{ Å}^5$. The calculated B_0^2 and B_0^4 are rather enormous. A last calculation was performed with quadrupolar polarizabilities equal to a tenth of the above quoted values and Total 3 is the sum of the three contributions.

Analogous calculations were undertaken for Nd_2O_3 and the results are reported in Table 1 as well as the experimental values of Caro et al. (11). The dipolar polarizabilities were set to 2 and 1 Å^3 for oxygen and neodymium respectively. The quadrupolar polarizabilities were first set to zero values (Total 1), then to 3 and 2 Å^5 respectively following reference 10 (Total 2), at last to a tenth of these values (Total 3). In neither of these cases is the agreement between experimental and calculated values improved by the quadrupolar contribution, but probably some care must be taken concerning the polarizabilities which are utilized. As it stands, until further investigation, with usual polarizability values, the successive electrostatic contributions to crystal field parameters look like a diverging expansion.

CONTRIBUTION TO TRANSITION PROBABILITIES DUE TO QUADRUPOLE
MOMENTS INDUCED BY THE EXTERNAL RADIATION

We start from Equation 4 which is the tensorial form of (3):

$$W = e \, (5/6)^{1/2} \sum_{\substack{k,q,q',p \\ \mu \, j}} (-1)^{k+q+p} \left[4\pi(k+1)(k+2)(2k+1) \right]^{1/2}$$

$$\times \, (2k+3) \frac{\langle r^k \rangle}{R_\mu^{k+3}} Y_{q'}^{k+2 \, *}(\Omega_\mu) \begin{Bmatrix} 1 & k+2 & k+1 \\ k & 1 & 2 \end{Bmatrix} \begin{pmatrix} k & k+2 & 2 \\ q & -q' & p \end{pmatrix} M_p^{(2)}(\mu) C_q^{(k)}(j) \tag{4}$$

The static moments $M_p^{(2)}(\mu)$ are replaced by:

$$M_p^{(2)}(\mu) = \alpha_Q(\mu) \, (30)^{1/2} \sum_{m \, n} (-1)^P \begin{pmatrix} 1 & 1 & 2 \\ m & n-p \end{pmatrix} \left[\nabla_m^{(1)} E_n^{(1)} \right](\mu) \tag{5}$$

where $(\nabla_m^{(1)} E_n^{(1)})(\mu)$ expresses the gradient of the external electric field, at the μ-th ligand's position $R\mu$. We put this electric field, for a given component (M= 0, ±1) of a given polarization in the form

$$E_n^{(1)} = (-1)^M \, E_0 \, e^{i(\vec{K} \cdot \vec{R} - \omega t)} \, \delta_{n,-M} \tag{6}$$

where $\vec{K}$ is the wave vector and ω the angular frequency of the radiation field. We then use the expansion of the exponential in terms of Bessel functions and approximate $j_{k'}(KR)$ by $(KR)^{k'}/(2k'+1)!!$, so that the lowest order contribution to $M_p^{(2)}$ is given by:

$$M_p^{(2)}(\mu) = -(1)^{M+p+1} \, i \, \left(\frac{160\pi}{3} \right)^{1/2} \alpha_Q(\mu) \, E_0 \, e^{i\omega t} \begin{pmatrix} 1 & 1 & 2 \\ M+p & -M & -p \end{pmatrix} K \, Y_{M+p}^1(\Omega_K) \tag{7}$$

The interaction energy W(t)=Re(W) is written W(t)=W+e^{-i\omega t}+W-e^{i\omega t}. We use Fermi's golden rule (12) to get the total probability S for emission from a manifold ΨJ to Ψ'J'. We integrate over $d\Omega_k$, sum over the assumed equally thermally populated Stark components, and finally obtain:

$$S = \frac{64\pi e^2}{27hc^5} \frac{\omega^5}{(2J+1)} \alpha_Q \chi \sum_{k \, q \, \mu\mu'} (1-\sigma_k)^2 \frac{(k+1)(k+2)(2k+3)}{(2k+5)} \langle r^k \rangle^2$$

$$\times \langle f \| C^{(k)} \| f \rangle^2 \frac{Y_q^{k+2}(\Omega_\mu)}{R_\mu^{k+3}} \frac{Y_q^{k+2}(\Omega_{\mu'})}{R_{\mu'}^{k+3}} \langle (4f^N)\psi J \| U^{(k)} \| (4f^N)\psi'J' \rangle^2 \tag{8}$$

where we have included the Lorentz local field correction χ and the shielding factors (1). In this equation k spans over the values 2, 4 and 6. We applied Eq. (8) to the $^5D_0 \to {}^7F_2$ and $^5D_0 \to {}^7F_4$ transitions of the Eu^{3+} ion doped in the Y_2O_3 host which have been studied by Krupke (13). α_Q was set equal to 3.694 $Å^5$ for the O^{--} ion (14). The other appropriate values for the quantities in Eq. (8) were taken from references (15,16,17). We find a contribution to the $^5D_0 \to {}^7F_2$ intensity which is of the order of $10^{-2}s^{-1}$, and a contribution to the $^5D_0 \to$ $^7F_4$ intensity which is smaller than $10^{-3}s^{-1}$. These values differ from those quoted by Krupke (13) by four orders of magnitude showing that oscillating quadrupole moments give a negligible contribution to the Pseudo-Multipolar Field. These contributions are indeed of the same order of magnitude as those arising from the term $i\vec{K}.\vec{R}$ in the treatment of oscillating dipoles if the long wavelength approximation is not assumed. This is rather a fortunate point since it shows the fast convergence of the multipolar series and its contributions to the Pseudo-Multipolar Field.

Two remarks can be made. The first one is that Eq. (8) is non-vanishing even in the case where the rare-earth ion occupies a site with an inversion center since it depends on lattice sums of even rank. The second one is that the contribution due to each multipole moment does not seem to decrease exactly as powers of KR as one could think apriori. For example, from the present calculation the contribution due to oscillating quadrupoles is 10^4 times smaller than that due to oscillating dipoles instead of 10^6 as it could be estimated from the value of $(KR)^2$ which is around 10^{-6}.

ACKNOWLEDGMENTS

This work was done in the framework of the cooperation between the CNRS (France) and the CNPq (Brazil).

REFERENCES

1. O.L. Malta, <u>Mol. Phys.</u>, 42:65 (1981).
2. O.L. Malta, G.F. de Sá, <u>Phys. Rev. Lett.</u>, 45:890 (1980).
3. C.K. Jörgensen and B.R. Judd, <u>Mol. Phys.</u>, 8:281 (1964).
4. C.A. Morrison, <u>Solid State Commun.</u>, 18:153 (1976).
5. C.A. Morrison and R.P. Leavitt, <u>J. Chem. Phys.</u>, 71:2366 (1979).
6. M. Faucher, J. Dexpert-Ghys and P. Caro, <u>Phys. Rev. B</u>, 21:8 (1980).
7. J.B. Gruber, R.P. Leavitt and C.A. Morrison, <u>J. Chem. Phys.</u>, 74: 2705 (1981).
8. B. Morosin, <u>J. Chem. Phys.</u>, 49:3007 (1968).
9. A.J. Freeman and J.P. Desclaux, <u>J. Magn. Magn. Mat.</u>, 12:11 (1979).
10. P.C. Schmidt, K.D. Sen, T.P. Das and A. Weiss, <u>Phys. Rev. B</u>, 22: 4167 (1980).
11. P. Caro, J. Derouet and L. Beaury, <u>J. Chem. Phys.</u>, 70:2542 (1979).
12. A.S. Davidov, "Quantum Mechanics," (Pergamon Press, Oxford, 1965).
13. W.F. Krupke, <u>Phys. Rev. A</u>, 145:325 (1966).
14. P.C. Schmidt, A. Weiss, T.P. Das, <u>Phys. Rev. B</u>, 19:5525 (1979).
15. B.R. Judd, <u>J. Chem. Phys.</u>, 70:4830 (1979).
16. W.T. Carnall, H. Crosswhite, H.M. Crosswhite, "Energy Level Structure and Transition Probabilities of the Trivalent Lanthanides in LaF_3," (Argonne National Laboratory, Argonne, IL 60439).
17. M. Faucher, J. Dexpert-Ghys, to be published in Phys. Rev. B.

ON THE ORIGIN OF RESONANCE LINES IN 3d EMISSION

SPECTRA OF RARE EARTHS

Janusz Kanski

Department of Physics, Chalmers University of
Technology
S-412 96 Gothenburg, Sweden

Radiative decay of 3d holes in rare earth metals leads to complex spectra, the structures in which can be divided into three
classes (1,2): i) $M_{\alpha,\beta}$ emission lines, ii) resonance lines (R) and
iii) satellite lines (S). $M_{\alpha,\beta}$ emission represents the decay of
$3d^9 4f^n$ states to $3d^{10} 4f^{n-1}$ and can thus be regarded as "normal"
characteristic transitions. The resonance lines are identified
through coincidence with lines in the corresponding absorption spectra and are interpreted as reversed photoabsorption processes, i.e.
$3d^9 4f^{n+1} \rightarrow 3d^{10} 4f^n + h\nu_R$. The origin of the satellite lines (S) is not
well understood, but it has been suggested that they represent
resonant-like transitions in excited systems (2). The purpose of
this paper is to reexamine by which mechanism the resonance lines
are produced.

Emission of resonance lines requires as a first step excitation of $3d^9 4f^{n+1}$ states. Figure 1 illustrates three ways in which
such states can be obtained.

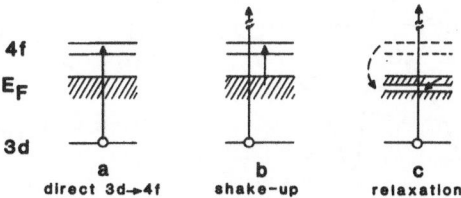

4f			
E_F			
3d			
	a	b	c
	direct 3d→4f	shake-up	relaxation

Figure 1. Different mechanisms by which $3d^9 4f^{n+1}$ states can be excited.

121

In processes 1b and 1c the $4f^{n+1}$ state is populated in consequence of ionization, and can therefore be excited both by photons
and electrons of any energy in excess of the ionization threshold.
Process 1a on the other hand cannot occur under photon irradiation,
unless the photon energy is just equal to the $3d \rightarrow 4f$ transition
energy. Divergencies between photon and electron excited emission
spectra can thus in principle be used to decide whether mechanism
1a is more important than mechanisms 1b and 1c. Unfortunately,
emission data obtained in fluorescence are very scarce, and the ones
available are distorted by self absorption (3).

A significant observation that can be made in electron excited
emission spectra, not heavily distorted by self absorption, is that
the resonance lines appear as intense as the "normal" $M_{\alpha,\beta}$ lines.
From this one can conclude that the $3d^9 4f^{n+1}$ states must be produced
at approximately the same rate as $3d^9 4f^n$ states. To account for the
high relative intensities of resonance lines, mechanism 1a requires
that the probability to form an excited $3d^9 4f^{n+1}$ state is as large
as the integrated ionization probability. A rough estimate considering only the dipole part of the primary electron beam indicates that
the $3d^{10} 4f^n \rightarrow 3d^9 4f^{n+1}$ partial cross section constitutes only about
20% of the total oscillator strength at a primary energy 700 eV in
excess of ionization threshold (4). Inclusion of monopole and quadrupole excitations would reduce this factor further. The intense resonance lines are therefore unlikely to be produced through mechanism
1a.

In Fig. 2 we show Electron excited X-ray Appearance Potential
Spectra (EXAPS) of La, Eu and Er covering the energy range of $3d_{5/2}$
excitations. The spectra are set on a reduced energy scale obtained

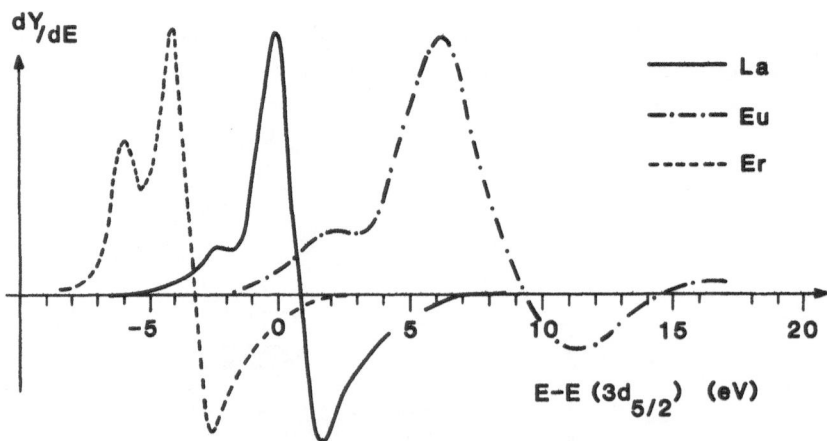

Figure 2. $3d_{5/2}$ EXAPS spectra of La, Eu and Er. The excitation
energies are in each case reduced with the XPS binding energy obtained from literature (refs. 8, 6 and 9 respectively).

by subtracting the $3d_{5/2}$ binding energy, as given by X-ray Photoelectron Spectroscopy (XPS), from the EXAPS energy in each case.
From this figure the possibility of mechanism 1c can be appreciated.
For all rare earths that have been investigated, with the notable
exception of Eu, the $3d_{5/2}$ excitation thresholds occur well below
the $3d_{5/2}$ binding energies. Thus, upon removal of a 3d electron,
the lowest energy of the remaining system is obtained by increasing
the 4f occupancy. This effect is also manifested in XPS via high
kinetic energy satellites on 3d emission peaks (5,6). In the case
of Eu no such satellite is observed (6), and the EXAPS spectrum is
located fully above the $3d_{5/2}$ binding energy. Therefore, if mechanism
1c is the important one, resonance lines should be absent in the
emission spectrum of Eu. (Shake-up processes (1b) and direct excitations (1a) are of course not precluded by the particular position of
the 4f level in Eu.) In Fig. 3 we have reproduced from ref. 1 the
XAS and XES spectra of Eu and Gd. It is quite clear that the R-line
in Eu, if at all distinguishable, is substantially less intense than
in Gd.

    From the above we conclude that mechanism 1c is likely to be the
one producing the intense R-lines in XES.

    The origin of satellite lines (S) seems to be different from
that of the resonance lines, since the S-lines are present with comparable intensities in the Eu and Gd spectra. One can note that at
least in the $3d_{5/2}$ spectra of Ce-, Eu-, Gd- and Dy-oxides no "S"-
lines are identified (1) and that in many cases the S-lines in
metallic spectra coincide with R-lines in oxide spectra. It can
therefore not be excluded that the S-lines present in the spectra
from metallic samples originate from thin oxide layers.

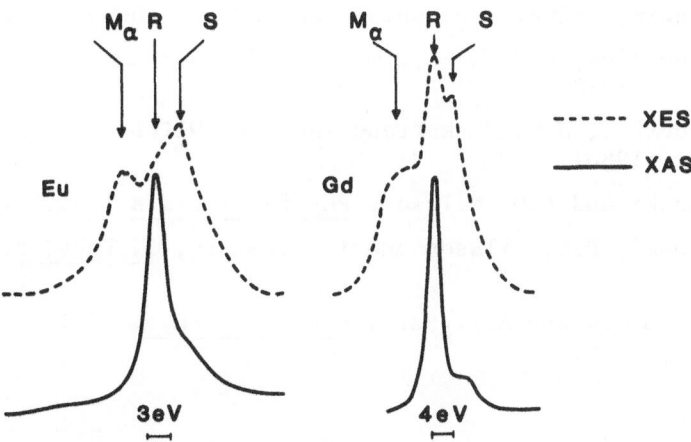

Figure 3. XES and XAS spectra of Eu and Gd from ref. 1. The spectra
have been approximately normalized at the M_α intensities.

Returning finally to the EXAPS results in Fig. 2 we note another very interesting feature. While the spectrum of Eu is located fully above the $Eu(3d_{5/2})$ XPS energy, the spectrum of Er is located fully below the $Er(3d_{5/2})$ XPS energy. Within an independent particle model the EXAPS signal is proportional to the self convolution of the density of unoccupied states (7). For an element with a peak in the density of states, the EXAPS spectrum is then expected to be dominated by a peak-minimum structure, showing excitation of a state in which the orbitals within the high density of states are doubly occupied. The energy required to excite such a state is given by zero transition between the maximum and the minimum. Although the independent particle model is not applicable quantitatively in the case of rare earths, it can be used for qualitative description. In Fig. 2 we see that in all three spectra one can distinguish the dominant peak-minimum structure. For Er we see that the zero transition occurs approximately 3.5 eV below the $3d_{5/2}$ XPS energy. Thus, by ionizing a 3d level it is energetically favourable populate not only the $3d^9 4f^{n+1}$ state, as in the case of La (8), but also the $3d^9 4f^{n+2}$. (This state is of course screened by a hole in the valence charge.) It is obvious that radiative decay of the $3d^9 4f^{n+2}$ state should give a spectrum different from those obtained by decay of $3d^9 4f^n$ or $3d^9 4f^{n+1}$ states and must be considered in detailed analysis of the emission spectrum.

REFERENCES

1.   R.C. Karnatak, Ph.D. thesis, Paris (1971)

2.   C. Bonnelle and R.C. Karnatak, J. de Physique, 32:C4:230 (1971)

3.   R.E. LaVilla, Phys. Rev. A, 9:1801 (1974)

4.   G. Wendin, Private Communication, CTH, Gothenburg (1981)

5.   G. Crecelius, G.K. Wertheim and D.N. Buchanan, Phys. Rev. B, 18:6519 (1978)

6.   J.F. Herbst, J.M. Burkstrand and J.W. Wilkins, Phys. Rev. B, 22:531 (1980)

7.   J. Kanski and P.O. Nilsson, Physica Scripta 12:103 (1975)

8.   J. Kanski, P.O. Nilsson and I. Curelaru, J. Phys. F, 6:1073 (1976)

9.   J.A. Bearden and A.F. Burr, Rev. Mod. Phys. 39:125 (1967)

TRANSITION INTENSITIES FOR Nd^{3+} IN CRYSTALS*

A. A. S. da Gama and Gilberto F. de Sá

Departamento de Física
Universidade Federal de Pernambuco, Recife, Brazil

The intensities of the transitions in rare earth spectra have been obtained in many cases by fitting the phenomenological parameters, Ω_λ. However, few calculations have been performed using "ab initio" parameters. Recently a modification of the model of Jørgensen and Judd has been shown to give better correlation for the intensities, especially for the hypersensitive transitions of Nd^{3+} (1) and Pr^{3+} (2). In this work the intensities of the f-f transitions of Nd^{3+} have been calculated using this pseudo-multipolar field. The results of the calculations are in agreement with the experimental intensities for Nd^{3+} in different host systems.

The well-known Judd-Ofelt expression for the oscillator strength, f, can be written

$$f = \left[\frac{8\pi^2 mc\sigma}{3h}\right] \chi \sum_\lambda \Omega_\lambda <4f^N \alpha SLJ||U^\lambda||4f^N \alpha'SL'J'>(2J+1)^{-1}$$

The principal problems in using this expression lie in determining the crystal field parameters, A_{Kq}, and the integrals $<4f|r^K|nl>$. The pseudo-multipolar field is an extension (3) of the pseudo-quadrupolar field proposed by Judd and Jørgensen (4). Although it approaches the problem in a different way, it yields the same expression for the quadrupolar term (Ω_2) and is extended to include the contributions from the 2^K-polar terms (K = 4 and 6).

$$f = \frac{8\pi^2 mc\sigma}{3he^2} \frac{(n^2+2)^2}{9n} \frac{\alpha_D^2}{g^2(2J+1)} \sum_{Kq}(1-\sigma_K)^2 (K+1)\left|A_q^{K+1}\right|^2$$

*This work was supported by FINEP, CNP$_q$ and OAS.

Table 1. Oscillator strengths (x 10^6).

	LiYF$_4$		CaWO$_4$	Y$_2$O$_3$		LaF$_3$	
	theor	exp	theor	theor	exp	theor	exp
$^4G_{5/2}$	4.93 ⎤		5.66	48.90 ⎤		2.11 ⎤	
		⎬ 6.01			⎬ 41.64		⎬ 5.60
$^2G_{7/2}$	2.33 ⎦		2.72	8.43 ⎦		5.60 ⎦	

$(4f|r^K|4f)^2 \ <1|C^K|1>^2 \ <4f^N \alpha SL||U^K||4f^N \alpha'SL'>^2$

$[\alpha_p$ = polarizability of ligand$]$

This mechanism yields a contribution to f which can be written in the same form as the Ω_λ values in the Judd-Ofelt theory:

$$\Omega_\lambda = \frac{\alpha_p^2}{e^4 g^2} \Sigma (1-\sigma_\lambda)^2 \ (\lambda+1) < 4f|r^\lambda|4f>^2 \ \underset{q}{\Sigma} \ |A_{\lambda+1,q}|^2 <f||C^\lambda||f>^2$$

The main difficulty in evaluating this expression lies in determining the values of α_p and $A_{\lambda+1,q}$. The results obtained for the hypersensitive transition $^4I_{9/2} \rightarrow {}^4G_{5/2}, \ {}^2G_{7/2}$ are given as typical results (Table 1).

The agreement between the theoretical and experimental values is satisfactory. Consideration of the cross-terms (which can have opposite signs to those of the squared terms) should improve the agreement.

REFERENCES

1. O.L. Malta and G.F. de Sá, <u>Chem. Phys. Lett.</u>, 74:101 (1980).
2. O.L. Malta and G.F. de Sá, <u>Phys. Rev. Lett.</u>, 45:890 (1980).
3. O.L. Malta, <u>Mol. Phys.</u>, 38:1347 (1979).
4. C.K. Jørgensen and B.R. Judd, <u>Mol. Phys.</u>, 8:281 (1964).

TRUNCATION EFFECTS ON CRYSTAL FIELD CALCULATIONS FOR THE

$4f^6(Eu^{3+})$ CONFIGURATION IN SOLIDS

Genevieve Teste de Sagey, Pierre Porcher, Guy Garon
and Paul Caro
ER 210 du C.N.R.S., 1, place A. Briand
92190 Meudon-Bellevue (France)

Trivalent europium is an important rare earth element which has many applications in luminescent materials. It is also widely used as an optical structural probe for solid state studies because of its simple sequence of energy levels and well established set of selection rules for transitions. The fluorescence can also be excited easily in the upper $^5D_0$ level with the relatively new Rhodamin 6G dye laser technology. Despite the very extended use of europium, crystal field problems are usually treated for this ion on the limited basis of the $49|SLJM_J>$ kets in the ^7F_J multiplet. This of course is sufficient to determine the crystal field parameters, but it is not precise enough when wave vectors are needed to parametrize some important properties such as intensities of fluorescence lines. Then, one should correct the vector composition in a way which may be arbitrary (1).

It is desirable to have a crystal field calculation which takes into account ^5D_J, and higher excited levels, components, in order to reproduce, as exactly as possible, the positions in the energy scale of the practically important ^7F_J and ^5D_J levels. Moreover, a score of other excited Stark levels can be experimentally determined in the UV through excitation spectra. It will be interesting to see if the same set of crystal field parameters which reproduce the ^7F_J Stark sequence works as well for the excited states first of all the ^5D_J.

We have recently tried to extend the crystal field calculation by introducing kets for the other multiplets. There are 3003 $|SLJM_J>$ kets in the configuration. It is not easy to handle the crystal field problem on such a large basis. The free ion problem can however be easily treated because the size of the matrix can be broken by separating the free ion configuration components (295 $^{2S+1}L_J$ states)

127

according to their J values. Then one can get for the J levels,
positions in good agreement with the barycenters of the experimentally observed crystal field levels in solids provided that Racah's
parameters and spin orbit coupling constant have realistic values
(2,3). The examination of the free ion results already shows some
interesting characters. For instance it can be seen that the energies for the sequence of the ^7F_J levels (Table I) does not follow the
Landé interval rule in the free ion calculation, whereas, of course,
it is followed for the calculation involving ^7F_J levels alone. However the ^7F_J in the free ion calculation are 93 to 97% pure in ^7F
when J increases from 0 to 6. Nevertheless the discrepancy is so
large, 500cm^{-1} for the barycenter of $^7F_6$) that it is usual, when
doing crystal field calculation on the 49 F_J basis, to add "corrections" to take into account that effect. Another almost pure
state in the configuration, the ^5L multiplet, does not behave better.
For instance we have for the free ion the sequence: 23298cm^{-1} ($^5L_6$),
24370cm^{-1} ($^5L_7$), 25274cm^{-1} ($^5L_8$), 26076cm^{-1} ($^5L_9$), 26629cm^{-1} ($^5L_{10}$).
It is easy to see that Landé interval rule (the proportionality of
the spacings to J "superior") is not followed although the lowest
purity is 90% for $^5L_6$ and the highest 96% for $^5L_9$. The other components in the wave vectors are tiny ones from exotic terms such as
^5K, ^3K (1), ^3K (2), ^3M (1), ^3M (3). For the ^7F the largest minor
components in the wave vectors are from the three ^5D terms by order
of importance $^5D(1)$, $^5D(3)$ and $^5D(2)$ but one notices also numerous
components of the various ^5F, ^5G and ^3P terms. It is then clear that
the $^{2S+1}L_J$ levels energy spacings are <u>very sensitive</u> to small admix-
tures from the numerous other terms to which they can be crossed in
such a large configuration.

     The first truncation which is natural to try involves the
22 $^{2S+1}L_J$ levels of the ^7F multiplet together with the three ^5D mul-
tiplets. The result is found in Table I. The $^7F_6$ is too high with
respect to the free ion (the free ion calculation is basically in
agreement with the barycenters, deduced from experimental crystal
field splittings in solids) and the $^5D_0$ is much too high (the gap is
2552cm^{-1} !). Such a large discrepancy makes this apparently simple
truncation of little value for practical calculations. If the ^5F
are added, raising the number of $^{2S+1}L_J$ levels to 32, no improvement
is obtained. Adding the ^5G (we have then 47 $^{2S+1}L_J$ levels) does not
improve $^5D_0$ but markedly affects the positions of $^7F_{4,5,6}$ bringing
them more in line with the experimental data. If instead of ^5G one
adds the six ^3P levels (that is 50 $^{2S+1}L_J$ levels), we still have bad
$F_{4,5,6}$ positions but a very clear amelioration for the ^5D. $^5D_0$ drops
to 17332cm^{-1}, much closer to the experimental values. It should be
noted that this is obtained only if one takes into account <u>all</u> of the
six ^3P terms. To get both an improvement on the highest J, $\overline{^7F}$ and
on the ^5D it is then obvious that both ^5G and ^3P terms should be
included making a total of 65 $^{2S+1}L_J$ levels. Then the lowest free
ion levels are reproduced in a way which may be deemed satisfactory,
at least for crystal field calculations. The arbitrary constants

Table I. Computed Barycenters for the ^7F_J and Lowest ^5D_J Levels of the $4f^6$ Depending on Different Truncations (Free Ion Parameters (cm^{-1}) E° = 3154 E^1 = 5995, E^2 = 29.4, E^3 = 680, ζ = 1306, α = 20, β = -640, γ = 1750)

Level	Free ion (all levels)	65 $^{2S+1}L_J$ Levels from Terms $^7F,^5D,^5F,^5G,^3P$	50 $^{2S+1}L_J$ Levels from Terms $^7F,^5D,^5F,^3P$	47 $^{2S+1}L_J$ Levels from Terms $^7F,^5D,^5F,^5G$	35 $^{2S+1}L_J$ Levels from Terms $^7F,^5D,^5F$	22 $^{2S+1}L_J$ Levels from Terms $^7F,^5D$	7 $^{2S+1}L_J$ Levels from ^7F
$^7F_0$	27	27	27	27	27	27	27
$^7F_1$	409	410	410	374	374	405	245
$^7F_2$	1079	1087	1099	1018	1030	1119	680
$^7F_3$	1941	1956	2016	1875	1935	2097	1333
$^7F_4$	2922	2939	3121	2857	3039	3256	2203
$^7F_5$	3969	3989	4409	3907	4328	4519	3292
$^7F_6$	5039	5078	5906	4997	5825	5825	4598
$^5D_0$	17180	17332	17332	19732	19732	19732	
$^5D_1$	18892	18892	18892	20474	20474	20539	
$^5D_2$	21492	21467	21470	22013	22017	22183	
$^5D_3$	24351	24525	24541	24444	24459	24711	
$^5D_4$	27515	27417	27984	27336	27903	28150	

are no longer needed for the $^{7}F_{J}$. $^{5}D_{0}$ is still off by 150cm^{-1}, but the other ^{5}D levels are on line with the free ion calculation.

It is clear that this matrix, of size 383x383 in the crystal field $|SLJM_{J}>$ basis, will simulate reasonably well the Stark levels for $^{5}D_{0}$, $^{5}D_{1}$, $^{5}D_{2}$ and yields wave functions reasonably accurate for practical calculations of intensity parameters from the classical fluorescence lines. Unfortunately, for the great maze of levels between 23000 and 30000 cm^{-1} which are easily collected from excitation spectra, this truncation is still unsufficient. The crystal field splittings of some individual levels, reasonably pure such as the ^{5}L ones, are well reproduced if the barycenter position is not taken into account (4). This is because the ^{5}L are not mixed with their neighbors by free atom operators. This is not the case of the other terms ^{5}G, ^{5}F, ^{5}H, etc. which will, moreover, be intensely J-mixed by the crystal field. A correct detailed interpretation of the excitation spectrum of Eu^{3+}4f^{6} above 23000cm^{-1} obviously depends on a future 3003 by 3003 crystal field calculation !, a size which may be somewhat reduced however, if symmetry allows it.

REFERENCES

1. P. Porcher and P. Caro. J. Chem. Phys., 68: 4176 (1978).
2. H.H. Caspers, H.E. Rast, and J.L. Fry. J. Chem. Phys., 47: 4505 (1967).
3. G.S. Ofelt. J. Chem. Phys., 38: 2171 (1963).
4. J. Hölsä and P. Porcher, J. Chem. Phys., to be published.

STUDY OF $M_{4,5}$-LEVEL SOFT X-RAY APPEARANCE POTENTIAL SPECTRA OF SELECTED RARE EARTHS*

D. Chopra and G. Martin

Physics Department, East Texas State University
Commerce, Texas 75428, USA

INTRODUCTION

Soft x-ray appearance potential spectroscopy (SXAPS) is recognized as an important technique for the study of core electron binding energies and the density of unoccupied states above the Fermi level.[1] The purpose of this paper is to show a comparative study of the $M_{4,5}$-level SXAPS spectra of the rare earth metals Ce^{58}, Pr^{59}, Tb^{65}, Dy^{66}, Ho^{67}, Er^{68}, Tm^{69} and Yb^{70}. We have attempted to compare the spectra in terms of total x-ray fluorescence, width and height of positive peaks and the fine structural features, and to correlate the findings with increasing Z. The spectrometer used for the measurements, has been described in detail elsewhere.[2] In each case, the samples were high purity polycrystalline metal foils which were argon ion sputtered and annealed in an attempt to obtain a clean surface for examination.

RESULTS AND DISCUSSION

We have made measurements on heavy rare earths and have also included two light elements, Ce and Pr to enlarge the scope of the investigation. The spectra are plotted on the same energy scale and are aligned for purposes of making systematic comparisons of their features (Fig. 1). As shown the M_4 and M_5 spectral region consists of two main peaks superimposed with secondary structure. The complexity of the structure is rich for medium rare earths and decreases for light and heavy elements. The intensity of the M_4-level peak decreases with Z. We have also recorded the M_2 and M_3-level SXAPS of the same elements and found these to be more than an order of magnitude weaker relative to

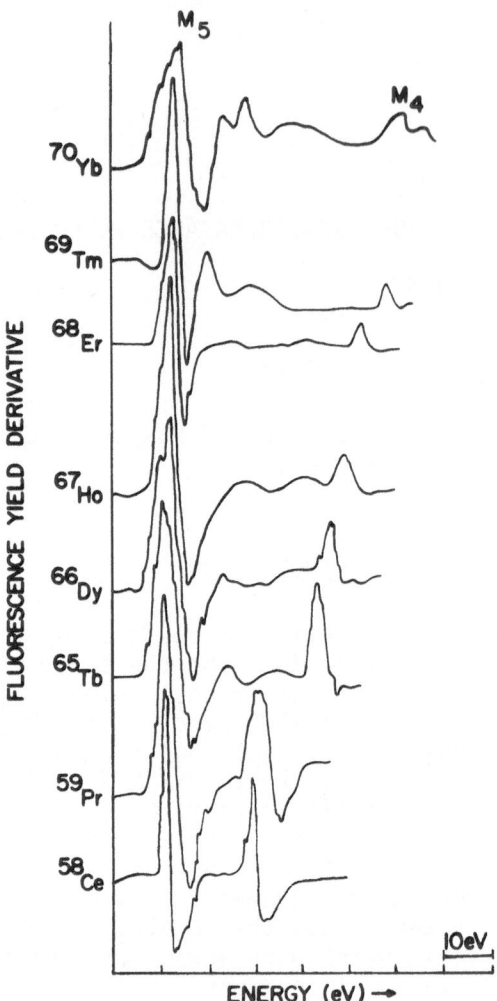

FIG. 1. $M_{4,5}$ appearance-potential spectra of rare earths.

the M_4 and M_5 peaks. This indicates that rare earths exhibit intense SXAPS structures only for those core levels which have
proper symmetry to satisfy the dipole selection rule for transitions to the final states. The M_4 and M_5-level spectra represent
the excitation of 3d electrons to the final f states.
    Park et al[1] have successfully employed the one-electron and
one-density of states theory to explain the spectral line shapes
of 3d transition elements. They have observed systematic changes
in threshold peak widths and L_3/L_2 peak intensity ratios in their
L_3 and L_2-level SXAPS with Z, demonstrating reasonable accord with
the rigid-band model. According to Dev and Brinkman[3] the total

transition probability is determined by the product of the transition probabilities of the incident and core electrons. The density of the final states is assumed to be nearly the same for both electrons in the case of simple metals and 3d transition elements. The density of the conduction band states above the Fermi level for rare earths should be similar to that of 3d transition elements, consisting of a strong peak of 4f density of states crossed by a Fermi level. However, the spectra of rare earths are not amenable to such description. In the case of rare earths the scattered incident electron probes the true density of conduction band states and the core electron performs a transition to the localized 4f orbitals. The excitation process is definitely atomic-like and a simple one-electron theory is not considered valid. Wendin[4] has suggested a two-density-of-states model, localized states for the core electron and conduction band states for the incident electron for explaining the SXAPS of Ba.

Liefeld et al[5] have observed the resonant behavior in a series of bremsstrahlung isochromat spectra of La and Ce at energies spanning the binding energies of their 3d core levels. The excitation of this singularity depended both on the energy of the incident electron and the angular momentum of core state vacancy. Some authors[6,7] have proposed the resonance scattering interaction in which both the incident and a 3d core electron scatter into the atomic 4f level according to the $3d^{10}4f^n + e \rightarrow 3d^94f^{n+2}$ transition where n varies from 0 for La to 14 for Yb. On the other hand soft x-ray absorption (SXA) results from $3d^{10}4f^n \rightarrow 3d^94f^{n+1}$ transition. The presence of the additional electron in the atomic 4f state accounts for the difference between SXAPS and SXA. This implies that SXAPS are not predicted for rare earths with less than two 4f vacancies and SXA with less than one. The exchange interaction between the 4f electrons and 3d hole, splits the final state configuration into a multiplet. The multiplicity of these levels, however, depends entirely on the number of 4f electrons.[6] The 3d $\rightarrow$ nf (n $\geq$ 5) channels are not considered explicity because high energy f orbitals remain outside the centrifugal barrier.[8] Some of these multiplet levels (bonding) remain below the ionization threshold while others (antibonding) autoionize to various channels having the core configuration $3d^94f^n$. The autoionizing levels decay 100 times more strongly to f channels than to the p channels.[8] The exchange interaction in the case of 3d levels of rare earths is weaker than the spin-orbit interaction. The observed line structure is, therefore grouped around the two spin-orbit components and constitutes a small fraction of total oscillator strength of the 3d excitation. This line structure is then expected to get quite complex for medium rare earths and then tend to decrease towards the end of the rare earth series. This interpretation is supported by the data of the present investigation. The systematic investigation of the rare earths along the periodic table reveals some clues in understanding the excitation mechanism for their SXAPS. In the case of heavy rare earths when the 4f level is nearly full, the importance of various proposed transition channels can be estimated. Thulium[69] (ground state configuration of $3d^{10}4f^{12}$) may assume either of the final state configurations; $3d^94f^{14}$ or $3d^94f^{13}\varepsilon f$ depending on the scattering modes of incident and core electrons. Here $4f^{13}\varepsilon f$ are the continuum channels. Thulium SXAPS yields a strong M_5 peak which is not rich in complex line structure. The next element, Yb[70] is divalent and has no 4f vacancy. However, Yb_2O_3 has been reported to be trivalent with $4f^{13}$ configuration. We have measured the SXAPS of an oxidized Yb surface. The M_5 peak is quite broad and complex, and is approximately 15-20 percent in magnitude relative to the Tm M_5 peak. Soft x-ray absorption measurements also give a strong absorption in the Yb_2O_3 corresponding to the transition of single d electron to the 4f state. It is obvious that the resonance process which scatters both electrons into the 4f state is not valid in Yb_2O_3 which has only one 4f vacancy. The SXAPS of Yb_2O_3, therefore results from the second resonance process of the scattering of two electrons to the excited configuration $4f^{14}\varepsilon f$. As a result of comparison of M_5 level SXAPS of Yb_2O_3 with that of Tm, it is estimated that the second resonance process is relatively weak. If the transition of two electrons to the 4f and εf states play an important role in the SXAPS of Yb_2O_3, where it makes the major contribution, then it is expected to be of significant importance in the SXAPS of other rare earths.

In conclusion, we propose that the oscillator strength of the M_5 and M_4-level SXAPS peaks in the rare earths (La through Tm) is divided between the $3d^94f^{n+2}$ and $3d^94f^{n+1}\varepsilon f$ configuration states in the ratio of $\sim$ 5:1, based on the peak intensities. The SXAPS of Yb_2O_3 is interpreted to result principally from the $4f^{14}\varepsilon f$ final state configurations.

REFERENCES

1. R.L. Park and J.E. Houston, Phys. Rev. B 6, 1073 (1972).
2. D. Chopra, H. Babb and R. Bhalla, Phys. Rev. B 14, 5231 (1976).
3. B. Dev and H. Brinkman, Ned. Tydschr Vacuum Techn. 8, 176 (1970).
4. G. Wendin, Proc. Int. Conf. Vacuum Ultraviolet Radiation Phys., Hamburg 1974 (Vieweg Pergamon, 1974).
5. R.L. Liefeld, A.F. Burr and M.B. Chamberlain, Phys. Rev. A 9, 316 (1974).
6. W.E. Harte and P.S. Szcezepanek, Jpn. J. Appl. Phys. 17-2 [Suppl.], 305 (1978).
7. R.J. Smith, M. Piacentini, J.L. Wolf and D.W. Lynch, Phys. Rev. B 14, 3419 (1976).
8. J. Sugar, Phys. Rev. B 5, 1785 (1972).

*   Supported by grants from Robert A. Welch Foundation and East Texas State University.

THE EFFECT OF MULTIPLET SPLITTING OF THE 4d LEVELS IN RARE EARTH ELEMENTS ON THEIR X-RAY SPECTRA

S. I. Salem

Department of Physics-Astronomy, California State
University, Long Beach, California 90840

The rare earth elements are characterized by partially filled
4f shells which give rise to an exchange interaction between the
4f and 4d levels. This interaction results in the splitting of
the $4d_{3/2}$, $4d_{5/2}$, $4f_{3/2}$ and $4f_{7/5}$ atomic levels, thus generating
electronic subshells which are found in the atoms of the rare
earth elements only. As a result, the x-ray spectra of these elements exhibit emission lines which are not to be found in any
other element. The energy and the relative intensity of these
lines, which are observed on the low energy side of the $L\gamma_1$ and
$L\beta_2$ diagram x-ray lines, have been measured and compared with the
results of a simple theory based on the exchange integral and the
total spin of the 4f electrons.

The rare earth elements $58 \leq Z \leq 69$ have partially filled 4f
shells and display multiplet splitting of both the 4d and the 4f
levels, [1,2] Fig. 1. Elements with atomic numbers smaller than 58
or larger than 69; that is, elements having the configuration $4f^0$
or $4f^{14}$, exhibit only spin-orbit splitting of the 4d level.

The photoabsorption spectra of the rare earth elements show
relatively broad peaks ten to twenty eV above the 4d absorption
edge[3]. The energy and the relative intensity of those peaks were
qualitatively described, using matrix elements evaluated with central field wave functions[4]. The x-ray emission spectra involving
transitions from the $4d^9 4f^{N+1}$ configuration exhibit complex structures which are characteristics of the rare earth elements alone[5].

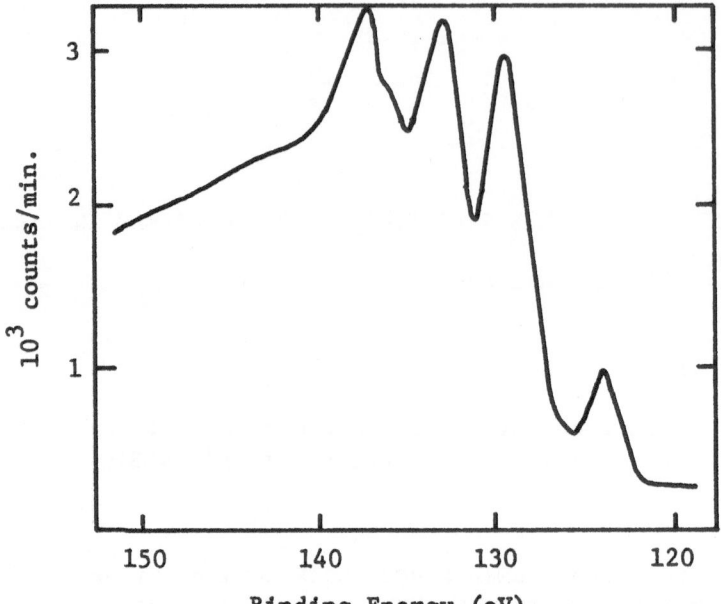

Figure 1. The x-ray photoemission spectra
of the 4d levels of $_{62}$Sm
exhibiting multiplet splitting.
[From Ref. 2.]

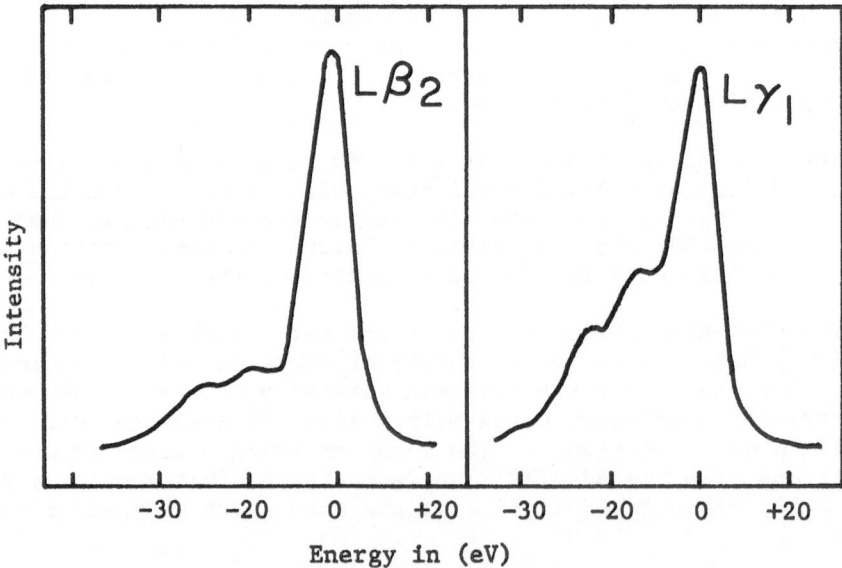

Figure 2. Energy distribution of Lβ_2 and Lγ_1 complex
of $_{62}$Sm.

Fig. 2 shows the $L\gamma_1$ [L_2 - N_4] and the $L\beta_2$ [L_3 - N_5] complexes from pure samarium metal (99.9% pure). The clean, polished sample was secured to a water-cooled anode, and inserted in a continuously evacuated chamber. The element was excited by a highly regulated (better than 99.7%) electron beam. The x-ray emission spectra was analyzed using a single-crystal, high-angle goniometer whose resolution, $d\lambda/\lambda$, is about 10^{-5} in this energy range, and whose instrumental width at λ = 1.6A° is about 9 x 10^{-4} A° = 4.4 eV.

It should be noted that the transitions L_2 - N_5 and L_3 - N_4 are very weak and could not account for any part of the structure shown in Fig. 2. It is proper to indicate that had it not been for the complex exchange interaction, the spin-orbit splitting between the $N_4(4d_{3/2})$ and the $N_5(4d_{5/2})$ would vary in this region from about 4 eV for $_{58}$Ce to about 8 eV for $_{69}$Tm.

By comparing Fig. 1 and Fig. 2, the discrepancy in the energy separation between the observed peaks becomes evident. Simple theory based on the exchange interaction of the 4d hole and the partially filled 4f level gives the energy separation in the form,

$$\Delta E = J(2S + 1) \qquad\qquad -1-$$

where for $_{62}$Sm, the total spin of the 4f electrons, S = 3, and the exchange integral J is given by

$$J = \frac{3}{35} G^1(4d4f) + \frac{4}{105} G^3(4d4f) + \frac{10}{231} G^5(4f4d) \qquad -2-$$

Putting the values of Hartree Fock Slater integrals as computed by Mann[7] in equation 2, gives J = 2.3, and an energy separation ΔE = 16.1 eV for $_{62}$Sm; a value which lies between the two observed extremes, Fig. 1, and Fig. 2, and which predicts only one peak in addition to the diagram line.

The measured radiative transition probability of the satellite structure relative to that of the main (diagram) line from $_{62}$Sm is 0.20 ± 0.03 for the $L\beta_2$ and 0.35 ± 0.05 for the $L\gamma_1$. These values were obtained by a computer program which was devised to unfold the complex structure and compare the areas under the unfolded curves. Calculations based on level multiplicity give the ratio of the transition probability as S/S + 1 = 0.75 for both the $L\beta_2$ and $L\gamma_1$ of $_{62}$Sm.

It is evident from the previously stated disagreement between measured results and calculated values that we still do not fully apprehend the true mechanism of the interactions that give rise to the multiplet splitting of the 4d4f configuration in the rare earth elements, and that effects other than exchange interaction must be present.

This lack of understanding becomes more disturbing, when one realizes that the problem is not confined to the rare earth elements, where it is the most severe, but it also exists in the transition elements $21 \leq Z \leq 28$ where the 3d shell is partially full, and in elements having $39 \leq Z \leq 45$, where the 4d shell is partially full. There are reasons to believe that similar interaction exists in elements with $91 \leq Z \leq 101$ where the 5f shell is partially full.

REFERENCES

1. G. K. Wertheim, A. Rosencwaig, R. L. Cohen, and H. J. Guggenheim; Phys. Rev. Lett. 27, 505 (1971).

2. S. P. Kowalezyk, N. Edelstein, F. R. McFelly, L. Ley, and D. A. Shirley; Chem. Phys. Lett. 29, 491 (1974).

3. R. Haensel, P. Rabe, and B. Sonntag; Solid State Commun. 8, 1845 (1970).

4. J. L. Dehmer, A. F. Starace, and U. Fano; Phys. Rev. Lett. 26, 1521 (1971).

5. S. I. Salem, C. W. Schultz, B. A. Rabbani, and R. T. Tsutsui; Phys. Rev. Lett. 27, 477 (1971).

6. K. Tsutsumi, J. Phys. Soc. Jap. 25, 1418 (1968).

7. J. B. Mann, Los Alamos Report No. LA-3690 (1967) (Unpublished).

8. S. I. Salem, G. M. Hockney, and P. L. Lee; Phys. Rev. A 13, 330 (1976).

OPTICAL SPECTRA, ENERGY LEVELS, AND CRYSTAL-FIELD ANALYSIS OF

TRIPOSITIVE RARE EARTH IONS IN Y_2O_3:   I.   KRAMERS IONS IN C_2 SITES

N. C. Chang, Aerospace Corporation, Los Angeles, CA 90009;
J. B. Gruber, Portland State University, Portland,
OR 97207; R. P. Leavitt and C. A. Morrison, Harry Diamond
Laboratories, Adelphi, MD 20783

INTRODUCTION

The optical spectra of trivalent rare earth ions (R^{3+}) doped into single crystal yttrium oxide (Y_2O_3) have received considerable attention over the last two decades. A survey of the literature reveals that there is still disagreement over the interpretation of the crystal-field splittings and subsequent wave functions of R^{3+}:Y_2O_3 in both C_2 and C_{3i} sites (1-3). Several calculations have appeared in the past for the optical spectra of $Nd^{3+}(4\underline{f}^3)$, $Er^{3+}(4\underline{f}^{11})$ and $Tm^{3+}(4\underline{f}^{12})$ in C_2 sites of Y_2O_3 and for R^{3+} in other crystals with C_2 sites (4-7). Even less information is available regarding crystal-field splitting calculations for R^{3+} in C_{3i} sites (2,3,8).

In this paper, we report unpublished optical spectra for Ce^{3+}, Sm^{3+}, Dy^{3+}, and Yb^{3+} in the C_2 sites of Y_2O_3. A crystal-field analysis of this new data together with previously published data (4,5) on Nd^{3+} and Er^{3+} is performed by means of a crystal-field Hamiltonian of C_2 symmetry. Parameters in this Hamiltonian are varied to obtain a best rms fit of calculated and experimental levels for Nd^{3+}, Sm^{3+}, Dy^{3+}, and Er^{3+}. Resultant crystal-field parameters are used in conjunction with the three-parameter theory of crystal fields (7) to obtain a smoothed set of crystal-field parameters for the entire lanthanide series. Levels computed with these parameters are compared with experimental results, where available.

EXPERIMENTAL SECTION

The R^{3+}:Y_2O_3 crystals have the cubic C (bixbyite) structure (Ia3), with sixteen formula units per elementary cell. The thirty-two cations are distributed 24 in C_2 sites and 8 in C_{3i} sites (9), and the R^{3+} ions appear to enter these sites randomly during crystal growth. Single crystals of Y_2O_3 containing rare earths in concentrations ranging between 0.05 and 20% at. wt. were grown using a flame fusion technique. Data were recorded covering the ultraviolet, visible, and infrared regions using a Cary 14 spectrophotometer, a Bausch and Lomb dual geating spectrograph, a Perkin-Elmer 301 spectrometer, and a Beckman FS-720 Fourier transform spectrophotometer. Absorption and fluorescence spectra covering a wide spectral range were recorded at liquid helium and liquid nitrogen temperatures.

OBSERVED SPECTRA AND ENERGY LEVELS

The details of our experimental results will be given elsewhere (10). For Ce^{3+}, absorption spectra recorded at 80K yielded the four Stark levels of the excited $^2F_{7/2}$ multiplet as 2182, 2750, 3061, and 3880 cm^{-1}. For Sm^{3+}, absorption and fluorescence measurements allowed the assignment of 59 out of 63 levels between the ground state and 20,076 cm^{-1}. For Dy^{3+}, 73 Stark components were assigned, covering the range from 0 to 23,780 cm^{-1}. For Yb^{3+}, the three Stark levels of $^2F_{5/2}$ were identified as 10,219, 10,640, and 11,044 cm^{-1}.

A crystal-field analysis of the data for Nd^{3+}, Sm^{3+} (this work), Dy^{3+} (this work), and Er^{3+} (5) was undertaken assuming a crystal-field Hamiltonian of C_2 symmetry:

$$H_{CEF} = \sum_{km} B_{km}^{\dagger} \sum_{i} C_{km}(\hat{r}_i) , \qquad (1)$$

where the B_{km} are crystal-field parameters and C_{km} are spherical tensors. In Eq. (1), the sums run over k = 2, 4, and 6 and m = 0, ± 2, ..., $\pm k$; the sum on i runs over the electrons of the $4f^n$ configuration. All B_{km} are complex except B_{k0}, which are real.

The crystal-field Hamiltonian is diagonalized together with an effective free-ion Hamiltonian described elsewhere (7) in a truncated basis covering the lowest several levels of the $4f^n$ configuration. The B_{km} in Eq. (1) are varied starting with values obtained from point-charge lattice sums until a minimum rms deviation between calculated and experimental energy levels is found. The problem of multiple minima is eliminated by requiring that the set of B_{km} found for one ion be consistent with those found for the other ions. The best-fit B_{km} are presented in Table 1, together with the corresponding rms deviations.

THEORETICAL PREDICTIONS

According to the three-parameter theory, the B_{km} are related to

Table 1. Best-fit B_{km} for Kramers ions in C_2 sites.

Ion	Nd	Sm	Dy	Er
B_{20}	−106	−312	−278	−150
B_{22}	−831	−773	−669	−678
B_{40}	−1564	−1436	−1190	−1389
RB_{42}	−1772	−1228	−1409	−1061
IB_{42}	14	381	270	239
RB_{44}	753	682	799	712
IB_{44}	−1006	−776	−902	−812
B_{60}	133	348	58	252
RB_{62}	255	277	281	271
IB_{62}	−115	−88	−40	119
RB_{64}	784	690	385	180
IB_{64}	−512	−600	−319	−218
RB_{66}	−132	−113	−97	−24
IB_{66}	−73	−44	−7	−42
rms	7.0	5.1	6.7	7.9

point-charge lattice sums, A_{km}, by (7)

$$B_{km} = \rho_k A_{km}, \tag{2}$$

where

$$\rho_k = \tau^{-k} <r^k>_{HF} (1-\sigma_k), \tag{3}$$

and where τ is a host-independent, ion-dependent radial expansion parameter, $<r^k>$ are Hartree-Fock expectation values (11), and σ_k are shielding factors (12). Values of the ρ_k are given (13) in Ref. 7. Values of the A_{km} are given by an effective point-charge model in which the optimum values of the charges are a factor of three lower than the valence charges, so that $q_y = +1$ and $q_0 = -2/3$.

A smoothed set of B_{km} for the entire lanthanide series is obtained by dividing the best-fit B_{km} for each rare earth by the appropriate ρ_k and averaging the results. This procedure gives a set of phenomenological A_{km}, and the smoothed set of B_{km} is obtained by multiplying by the ρ_k for each lanthanide ion. As a check on the validity of this procedure, rms deviations between theoretical and experimental levels were calculated for the four ions Nd^{3+}, Sm^{3+}, Dy^{3+}, and Er^{3+}; results for the rms deviation are approximately twice the best-fit values. Detailed predictions and comparisons are given in Ref. 10.

REFERENCES

1. R.M. Moon, W.C. Koehler, H.R. Child and L.J. Raubenheimer, Phys. Rev., 176:722 (1968); see also J.B. Gruber, USAEC Progress Report RLO-2221-T6-20, pp. 171-204 (1974).
2. D. Bloor and J.R. Dean, J. Phys., C5:1237 (1972); J.R. Dean and D. Bloor, ibid, C5:2921 (1972).
3. D.W. Forester and W.A. Ferrando, Phys. Rev., B14:4769 (1976).
4. N.C. Chang, J. Chem. Phys., 44:4044 (1966).
5. P. Kisliuk, W.F. Krupke and J.B. Gruber, J. Chem. Phys., 40:3606 (1964).
6. J.B. Gruber, W.F. Krupke and J.M. Poindexter, J. Chem. Phys., 41:3363 (1964).
7. C.A. Morrison and R.P. Leavitt, J. Chem. Phys., 71:2366 (1979) and references therein.
8. G. Schaak and J.A. Koningstein, J. Opt. Soc. Am., 60:1110 (1970); see also A.M. Lejus and D. Michel, Phys. Stat. Sol., (b)84:K105 (1977).
9. R.W.G. Wyckoff, Crystal Structure, Vol. 2, 2nd Edition (Interscience, New York), pp. 2 ff.
10. N.C. Chang, J.B. Gruber, R.P. Leavitt and C.A. Morrison, J. Chem. Phys., (in press).
11. A.J. Freeman and R.E. Watson, Phys. Rev., 127:2058 (1962).
12. P. Erdos and J.H. Kang, Phys. Rev., B6:3393 (1972).
13. Note that ρ_4 for Tb^{3+} should be 0.4490 rather than 0.4990 in Ref. 7.

Sm^{2+} ACTIVATED MIXED FLUORIDES : SYNTHESIS AND FLUORESCENCE

F. Gaume, A. Gros and J.C. Bourcet

Physico-Chimie des Matériaux Luminescents, ER N° 10

University Lyon I, 43 Bd du 11 Nov. F 69622 Villeurbanne

INTRODUCTION

The intense red fluorescence of Sm^{2+} has been used in fluoride single crystals for laser applications (1). It seemed of interest to test this fluorescence in a series of polycristalline mixed fluorides to obtain new phosphors for lighting.

Among rare-earth ions giving a stable divalent state, Sm^{2+} is one of the most difficult to obtain as the pure fluoride. This is due to its reactivity towards oxidation agents and to the stability of SmOF oxyfluoride. We have chosen to dope the fluoride matrices with SmF_3 (1 %) and to reduce Sm^{3+} "in situ". A relatively low melting point (900 - 1000°C) and the possibility to stabilize Sm^{2+} on a weak field site have been the two principal criteria in selecting the host lattices (the importance of a weak field has been previously pointed out in the case of Eu^{2+} (2)). In preliminary tests, seven pure mixed fluorides have been prepared from a stoichiometric mixing of the simple fluorides and analyzed using thermal and radiocrystallographic analyses (Table 1).

SYNTHESIS

A preliminary study of the thermodynamic constants of the $Sm^{3+} + e^- \rightarrow Sm^{2+}$ reduction in fluoride phases points out two essential conditions for the Sm^{2+} stabilization : completely anhydrous starting fluorides and the use of a powerful reducing agent acting in a molten phase. Samarium metal may be used as reducing agent. From these initial results, a preparation technique has been designed in two distinct stages, a dehydration followed by a reduction in which the host matrix is also prepared.

Two dehydration methods have been set up and compared. In the first method, each fluoride is dehydrated inside a pyrex tube under vacuum (10^{-6} torr) at 300° for a minimum of two days. In the second technique, NH_4HF_2 is used as a fluorant agent at a relatively high temperature (880°C) in a nickel reactor previously evacuated.

The dessicated Sm^{3+} activated mixed fluorides are stored in a glove-box containing a balance. The argon atmosphere in the glove-box is continuously dehydrated and deoxygenated by passing through epoxy resins (the H_2O and O_2 contents are less than 1 v.p.m.). The reaction mixture (starting fluorides, SmF_3 and Sm metal) is weighed in the glove-box, mixed in an agate mortar then introduced into a small nickel tube previously sealed at one end. The free end of the nickel tube is obtured with a tin brazing, removed from the glove-box, then definitively sealed with the oxyacetylene torch. The sealed nickel tube is heated in an oven according to a thermal cycle varying with the compound (Table 1). An important condition for the complete reduction is to obtain a melt between 5 and 8 h. The melting is not congruent for $BaLiF_3$, BaY_2F_8 and $BaMgF_4$. In these cases however, a pure phase can be obtained by annealing. After cooling, the nickel tube is sawed in order to recover the Sm^{2+} doped mixed fluoride. This fluoride is stable and free of nickel (3).

FLUORESCENCE

Emission and excitation spectra – under λ = 365 nm excitation, the phosphors exhibit emission lines assigned to the $4f^6$ $^5D_J \rightarrow ^7F_J$ transitions. This proves the Sm^{2+} stabilization in the mixed

Table 1

Matrix	Site	Thermal cycle maximum °C	Sm total content ($2SmF_3$ + Sm) %
$BaLiF_3$	Ba^{2+}	950	1,5 and 3
BaY_2F_8	Ba^{2+}	1 060	1,5 and 3
$KMgF_3$	K^+	1 190	1,5 and 3
KY_3F_{10}	K^+ // Y^{3+}	1 050	0,5
$NaMgF_3$	Na^+	1 100	3
$BaMgF_4$	Ba^{2+}	1 100	0,5
$LiYF_4$	Y^{3+}	950	

fluorides. The excitation spectra consist of a band characteristic of the $4f^5$-5d configuration and of a few $4f^6$ lines. The lower energy of the $4f^5$-5d states is easily determined at T=77K (Table 2). This energy depends on the substituted cation and on the number of ligands. With a weak crystal field (substitution on Ba or K, large coordination number) this energy is relatively high and transitions arising from $^5D_0$, $^5D_1$, $^5D_2$ and even $^5D_3$ are observed in emission at low temperature. The $4f^5$-5d excitation band width increases with temperature, so that it can partially overlap the ^5D_J excited states. However in a few cases, a weak emission arising from $^5D_1$, $^5D_2$ and $^5D_3$ is still observed at room temperature. As a result, the fluorescence varies from deep red (BaLiF$_3$:Sm^{2+}, BaMgF$_4$:Sm^{2+}) to orange-red (KY$_3$F$_{10}$:Sm^{2+}).

Table 2

Matrix	Principal emission levels T=77K	$4f^5$-5d lower energy (cm^{-1}) T=77K	Fluorescence T=300K
BaLiF$_3$	$^5D_0$	19 600	middle
BaY$_2$F$_8$	$^5D_0$	17 850	strong
KMgF$_3$	$^5D_0$	20 000	strong
KY$_3$F$_{10}$	$^5D_0$, $^5D_3$ // $^5D_0$	21 700 // 16 600	strong
NaMgF$_3$	$^5D_0$	20 400	weak
BaMgF$_4$	$^5D_0$, $^5D_1$, $^5D_2$	17 250	very strong
LiYF$_4$	$^5D_0$, $4f^5$-5d	14 000	very weak

Quantum yield – For three phosphors BaY$_2$F$_8$:Sm^{2+}, KY$_3$F$_{10}$:Sm^{2+} and BaMgF$_4$:Sm^{2+}, the apparent quantum yield η (fluorescence photon number/exciting photon number) has been determined versus λ according to a previous technique (4) using the following excitation lamps : i) a deuterium lamp between 230 and 370 nm ii) a calibra-

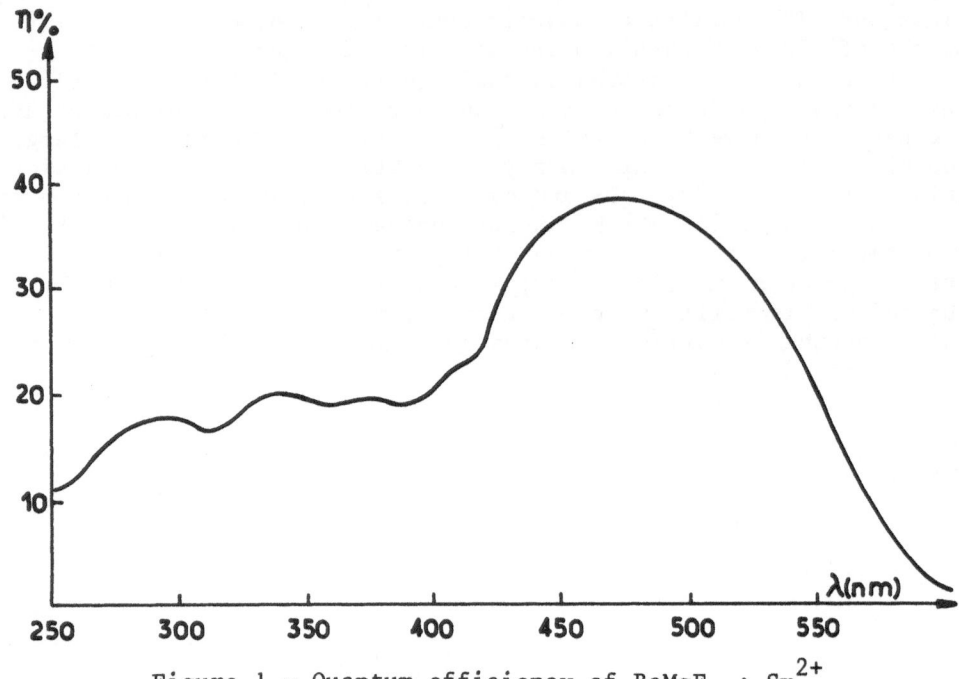

Figure 1 - Quantum efficiency of BaMgF$_4$: Sm^{2+}

ted halogen lamp above 370 nm. The yields are low between 230 and 370 nm especially for BaY$_2$F$_8$:Sm^{2+} and KY$_3$F$_{10}$:Sm^{2+}. Above 370 nm, the BaMgF$_4$:Sm^{2+} yield exhibits a peak (λ = 475 nm, η = 40 %) (Figure 1). This last phosphor looks interesting. Actually, BaMgF$_4$:Sm^{2+} is not efficient enough for lamp applications but the yield can certainly be improved by a study of the optimum samarium concentration.

REFERENCES

(1) G.J. Goldsmith and H.L. Pinch. Technical Report AFML-TR-65-115, RCA Laboratories (1965).

(2) C. Fouassier, B. Latourette, J. Derouet and P. Hagenmuller. Spectroscopie des éléments de transition et des éléments lourds dans les solides, Coll. Intern. CNRS n° 255, Lyon, 83 (1976).

(3) A. Gros. Thèse de Docteur-Ingénieur, Lyon, July 15 (1980).

(4) J. Janin, J.C. Bourcet and J.P. Jorus. Revue d'Optique, 44 : 393 (1965).

ELECTRIC AND MAGNETIC DIPOLE STRENGTHS OF f-f TRANSITIONS IN CUBIC Cs$_2$NaYCl$_6$:Ln^{3+} SYSTEMS

Frederick S. Richardson

Department of Chemistry, University of Virginia

Charlottesville, Virginia 22901

INTRODUCTION

Optical spectra of trivalent lanthanide ions (Ln^{3+}) doped into the cubic elpasolite host, Cs$_2$NaYCL$_6$, have proved to be of special value in theoretical studies of Ln^{3+}-crystal field interactions (1). In the Cs$_2$NaYCl$_6$:Ln^{3+} systems, the Ln^{3+} ions are each surrounded by six nearest-neighbor chloride (Cl$^-$) ions and each resides at a site of exact octahedral (O$_h$) symmetry. The O$_h$ site symmetry allows the lanthanide crystal field to be described entirely in terms of just two crystal field coefficients (excluding the spherically symmetric components of the crystal field), and it precludes any electric dipole contributions to the intensities of no-phonon (origin) crystal field transitions associated with lanthanide f-f excitations and de-excitations. In general, the optical absorption and emission spectra of the Cs$_2$NaYCl$_6$:Ln^{3+} systems may be analyzed entirely in terms of pure magnetic dipole lines associated with f-f (crystal field) origins, and sets of one-phonon vibronic lines which acquire electric dipole intensity via a vibronic coupling mechanism. Invariably, the most intense vibronic lines are associated with the three ungerade vibrational modes localized within the LnCl$_6^{3-}$ chromophoric clusters. Vibronic lines associated with non-cluster vibrational modes (e.g., lattice modes) are frequently observed, but with lesser intensities.

The relative ease with which magnetic dipole origins may be located and identified in the Cs$_2$NaYCl$_6$:Ln^{3+} optical spectra facilitates crystal field energy level analyses, and permits quantitative determinations of the magnetic dipole strengths associated with f-f crystal field transitions. This information may then be used to determine the octahedral crystal field coefficients and to

147

characterize the 4f-electron crystal field wave functions. The
isolation of nearly all of the electric dipole intensity into vibronic lines associated with $LnCl_6^{3-}$ cluster modes makes a vibronic
analysis of the electric dipole parts of the spectra theoretically
tractable (2).

The accuracy with which f-f magnetic dipole strengths can be
calculated depends upon (a) the quality of the "free-ion" wave
functions being used, and (b) the accuracy to which the crystal
field coefficients are known. The requirements for calculating
good vibronically-induced electric dipole strengths for f-f transitions are considerably more stringent. Not only are good 4f-elec-
tron crystal field wave functions needed, but one also requires an
accurate description of the lanthanide-ligand vibrational force
fields and an accurate representation for the mechanism whereby
odd-parity states of the system are mixed into the 4f-electron
even-parity states. A model for calculating the vibronically-
induced electric dipole strengths of f-f transitions in $LnCl_6^{3-}$ systems has been presented previously (2), and calculations based on
this model have been of considerable value in interpreting the vibronic spectra of $Cs_2NaYCl_6:Ln^{3+}$ systems. In the present paper we
summarize some of the theoretical results obtained previously on
magnetic dipole and vibronically-induced electric dipole strengths,
and present several sets of new results.

CALCULATIONS

The model and methods used in calculating the dipole strength
data reported here have been described elsewhere (2), so we shall
only summarize them in the present paper. The "free-ion" 4f-electron
basis functions used in our crystal field calculations are obtained
in an intermediate-coupling (IC) representation by diagonalizing a
parameterized "free-ion" Hamiltonian in the appropriate $4f^N$ Russell-
Saunders (RS) basis set. The parameters included in the free-ion
Hamiltonian are (a) the Slater-Condon electrostatic radial functions F_2, F_4, and F_6, (b) the spin-orbit coupling constant ζ_{so}, and
(c) the configuration-interaction parameters α and β (3). The 4f-
electron crystal field (CF) wave functions are obtained by diagonalizing the octahedral crystal field Hamiltonian, H_{cf}, in the free-
ion IC basis set. The H_{cf} operator is defined by

$$H_{cf} = B_0^{(4)}[U_0^{(4)}+(5/14)^{\frac{1}{2}}(U_4^{(4)}+U_{-4}^{(4)})] + B_0^{(6)}[U_0^{(6)}-(7/2)^{\frac{1}{2}}(U_4^{(6)}+U_{-4}^{(6)})],$$

where the $U_q^{(k)}$ are intraconfigurational unit tensor operators and
the $B_q^{(k)}$ (k = 4 or 6) are the socalled crystal field coefficients
(3). The crystal field wave functions obtained by this procedure
are then used in calculating f-f transition dipole strengths. The
free-ion parameters (F_2, F_4, F_6, ζ_{so}, α, and β) are "optimized" by
determining "best fits" between calculated and experimentally ob-

served baricenter energies for the $4f^N$ term levels. The crystal field parameters, $B_0^{(4)}$ and $B_0^{(6)}$, are optimized by determining "best fits" between calculated and empirically observed crystal field splittings in the $Cs_2NaLnCl_6$ optical spectra.

The f-f magnetic dipole strengths are calculated directly from the 4f-electron crystal field wave functions, and they have the same parameter dependence as do the 4f-electron energy levels. The vibronically-induced electric dipole strengths, however, require for their calculation an additional set of parameters. These include ligand (Cl^-) charges and dipolar polarizabilities, 4f-electron radial integrals $<r^k>$ (k = 2, 4, or 6), the Ξ parameters of the Judd-Ofelt f-f intensity theory (4,5), and force constants for the $\nu_3(t_{1u})$, $\nu_4(t_{1u})$, and $\nu_6(t_{2u})$ $LnCl_6^{3-}$ (cluster) vibrational modes. How these paramters enter into our electric dipole intensity model has been described previously (2). For all of the calculations reported here we have used q_L(ligand charge) $=-1.0e$ and $\bar{\alpha}_L$(mean ligand polarizability) = 3.0 A^3. The Ξ parameters were taken from Krupke (6) and the values of $<r^k>$ were taken from Freeman and Watson (7). We used a common set of vibrational force constants for all of the $LnCl_6^{3-}$ systems. The force-field parameters employed in our vibronic calculations are identical to those used previously (2). The free-ion parameters and crystal field coefficients used in the present study are listed below:

	Pr^{3+}	Eu^{3+}	Tb^{3+}	Ho^{3+}
F_2/cm^{-1}	304	401	417	426
F_4/cm^{-1}	48.2	55.4	57.6	67.2
F_6/cm^{-1}	4.84	6.06	6.30	7.39
ζ_{so}/cm^{-1}	750	1,320	1,840	2,127
α/cm^{-1}	23.8	16.8	17.5	23.5
β/cm^{-1}	-500	-640	-655	-811
$B_0^{(4)}/cm^{-1}$	2,800	2,030	1,900	1,850
$B_0^{(6)}/cm^{-1}$	-600	-306	-200	-155

Our calculated dipole strengths were found to be insensitive to ±10% variations in all of the free-ion parameters except ζ_{so}. However, they were considerably more sensitive to variations in ζ_{so}, $B_0^{(4)}$, and $B_0^{(6)}$.

RESULTS

Our computational model calculates magnetic dipole strengths for each of the no-phonon crystal field components associated with a given term-to-term f-f transition. It also calculates the electric dipole strengths associated with the $0-1(\nu_3)$, $0-1(\nu_4)$, and $0-1(\nu_6)$ one-phonon vibronic lines associated with each crystal field transition (where ν_3, ν_4, and ν_6 denote ungerade $LnCl_6^{3-}$ cluster modes) within a given term-to-term manifold. Simulated optical absorption and emission spectra based on these calculated magnetic dipole and vibronically-induced electric dipole strengths have been used with considerable success in performing detailed analyses of the experimentally obtained spectra for the Pr^{3+}(8), Eu^{3+}(8), Tb^{3+}(2), Ho^{3+}(8), Er^{3+}(9), and Tm^{3+}(10) ions doped into Cs_2NaYCl_6. With very few exceptions, semiquantitative (and sometimes fully quantitative) agreement between theory and experiment is achieved for the intensity distributions among the crystal field and vibronic components of term-to-term transitions.

The results we present here are for term-to-term or "partial" term-to-term transitions. That is, we present magnetic and electric dipole strengths which represent unweighted sums of crystal field or vibronic dipole strengths associated with components of various term-to-term transitions. These results reflect, then, the overall magnetic dipole or vibronically-induced electric dipole character of a given term-to-term transition. Results calculated for the $EuCl_6^{3-}$, $TbCl_6^{3-}$, $HoCl_6^{3-}$, and $PrCl_6^{3-}$ complexes are presented in Tables 1-3. Where it is possible to make comparisons, the calculated results for the $EuCl_6^{3-}$, $TbCl_6^{3-}$, and $HoCl_6^{3-}$ systems are in nearly quantitative agreement with observed relative intensity data (2,8). The results calculated for the $PrCl_6^{3-}$ system are in less

Table 1. Total electric and magnetic dipole strengths calculated for selected $^5D_J \leftrightarrow {}^7F_J$ term-to-term transitions in $EuCl_6^{3-}$ and $TbCl_6^{3-}$.

Complex	Transition	Electric[a,b]	Magnetic[a,c]
$EuCl_6^{3-}$	$^5D_0 \leftrightarrow {}^7F_0$	0.87	forbidden
	$^7F_1$	2.07	39,800
	$^7F_2$	15,800	forbidden
	$^7F_3$	737	490
	$^7F_4$	598	757
	$^5D_1 \leftrightarrow {}^7F_0$	1.92	11,000
	$^7F_1$	14,600	34.2
	$^7F_2$	4,900	93,700
	$^7F_3$	20,500	5,840
	$^7F_4$	1,300	650
	$^5D_2 \leftrightarrow {}^7F_0$	6,010	forbidden
	$^7F_1$	1,000	6,180
	$^7F_2$	11,300	4,880
	$^7F_3$	14,400	130,000
	$^7F_4$	12,300	3,820
$TbCl_6^{3-}$	$^5D_4 \leftrightarrow {}^7F_6$	10,700	1,900
	$^7F_5$	156,000	306,000
	$^7F_4$	6,010	1,650
	$^7F_3$	22,100	42,800
	$^7F_2$	19,500	3,210

(a) Expressed in units of 10^{-10} D^2.

(b) Summed over the ν_3, ν_4, and ν_6 vibronic components of all crystal field transitions within a given term-to-term transition.

(c) Summed over all crystal field (origin) transitions.

Table 2. Total electric and magnetic dipole strengths calculated for transitions originating in the aE_g, aT_{1g}, and A_{1g} crystal field levels of the $^5I_8$ term level in $HoCl_6^{3-}$.

Transition	Electric[(a,b)]	Magnetic[(a)]
$^5I_8(aE_g, aT_{1g}, A_{1g}) \rightarrow {}^5I_7$	74.5	42,700
$^5F_5$	166	55.9
$^5F_4$	198	46.4
$^5F_3$	83.8	1.62
$^5F_2$	24.1	0.11
$^3K_8$	255	2,660
$^5G_6$	22,300	29.9
$^5G_5$	293	20.4
$^3K_7$	476	98.2

(a) Expressed in units of 10^{-8} D^2.

(b) Summed over all vibronic components.

Table 3. Electric and magnetic dipole strengths calculated for selected emissive transitions in $PrCl_6^{3-}$.

Transition	Electric[(a,b)]	Magnetic[(a)]
$^3P_0 \rightarrow {}^3H_4$	1,760	69.5
$^3H_5$	882	2.24
$^3H_6$	4,590	6.69
$^3F_2$	27,600	forbidden
$^3F_3$	63.3	4.46
$^3F_4$	947	110
$^1D_2(T_{2g}) \rightarrow {}^3H_4$	238	24.8
$^3H_5$	102	20.6
$^3H_6$	1,310	16.2
$^3F_2$	1,190	98.8
$^3F_3$	2,980	1,830
$^3F_4$	36,200	51.2

(a) Expressed in units of 10^{-8} D^2.

(b) Summed over all vibronic components.

good agreement with experimental observation, but they do reproduce the semiquantitative and qualitative aspects of the Pr^{3+} emission spectra (8).

In general, the relative magnetic dipole strengths strongly reflect the free-ion selection rules for magnetic dipole transitions ($\Delta J = 0, \pm1$, excluding $J = J' = 0$). Furthermore, certain transitions listed in Tables 1-3 acquire a major portion of their electric dipole strength via a transition quadrupole(Ln^{3+})-transition dipole(Cl^-) coupling mechanism (a component of the socalled "dynamic coupling" or "ligand polarization" f-f intensity mechanism). These transitions are characterized by $\Delta J = \pm2$ and they include $^5D_0 \rightarrow {}^7F_2$, $^5D_1 \rightarrow {}^7F_3$, $^5D_2 \rightarrow {}^7F_0$, and $^5D_2 \rightarrow {}^7F_4$ for $EuCl_6^{3-}$, $^5D_4 \rightarrow {}^7F_2$ for $TbCl_6^{3-}$, $^5I_8 \rightarrow {}^5G_6$ for $HoCl_6^{3-}$, and $^3P_0 \rightarrow {}^3F_2$ and $^1D_2 \rightarrow {}^3F_4$ for $PrCl_6^{3-}$. Elimination of the dynamic coupling mechanism from our intensity model leads to gross underestimates of the electric dipole strengths of these transitions.

CONCLUSIONS

Vibronically-induced electric dipole strengths for f-f transitions in $Cs_2NaYCl_6:Ln^{3+}$ systems can be calculated with semiquantitative accuracy using a model based on isolated $LnCl_6^{3-}$ chromophoric clusters. Inclusion of both static coupling (Judd-Ofelt) and dynamic coupling (or ligand polarization) mechanisms in this model is essential to its general applicability.

ACKNOWLEDGMENTS

This work was supported by the National Science Foundation (NSF Grant CHE80-04209).

REFERENCES

(1) C.A. Morrison, R.P. Leavitt, and D.E. Wortman, J. Chem. Phys., 73:2580 (1980).
(2) T.R. Faulkner and F.S. Richardson, Mol. Phys., 35:1141 (1978); Mol. Phys., 36:193 (1978); Mol. Phys., 38:1165 (1979).
(3) B.G. Wybourne, Spectroscopic Properties of Rare Earths, Interscience, New York (1965).
(4) B.R. Judd, Phys. Rev., 127:750 (1962).
(5) G.S. Ofelt, J. Chem. Phys., 37:511 (1962).
(6) W.F. Krupke, Phys. Rev., 145:325 (1966).
(7) A.J. Freeman and R.E. Watson, Phys. Rev., 127:2058 (1962).
(8) J.P. Morley, Ph.D. Dissertation, University of Virginia (1981).
(9) Z. Hasan and F.S. Richardson, unpublished results.
(10) R.W. Schwartz, T.R. Faulkner, and F.S. Richardson, Mol. Phys., 38:1767 (1979).

ABSORPTION SPECTROPHOTOMETRIC CHARACTERIZATION OF Sm(II), Sm(III),

AND Sm(II/III) BROMIDES AND Sm(III) OXYBROMIDE IN THE SOLID STATE*

A. B. Wood, J. P. Young[†], and J. R. Peterson, University
of Tennessee, Knoxville, TN 37916 and †Oak Ridge National
Laboratory, Oak Ridge, TN 37830;
J. M. Haschke, Rockwell International, Golden, CO 80401

ABSTRACT

Absorption spectra obtained from $SmBr_3$, $SmBr_2$, and SmOBr were
used in identifying the samarium species in several mixed-valence
Sm(II/III) compounds produced by H_2 reduction of $SmBr_3$. The nature
of the absorption of SmOBr made it possible to detect even traces
of SmOBr in the Sm bromides.

INTRODUCTION

The results of a crystallographic study of phase equilibria in
the samarium-bromine system have identified three intermediate or
mixed-valence bromide compounds between the di- and trivalent ones
(1,2). It is of interest to determine if the individual oxidation
states, Sm(II) and (III), can be identified in these vernier compounds. If so, it would imply that separate crystallographic sites
are occupied by individual Sm(II) and (III) ions; if not, the
implication would be that electron exchange is occurring so rapidly
that only an average species is present. The presence or absence of
such individual Sm(II) and (III) ions cannot be determined by X-ray
diffraction techniques but can be determined by absorption spectrophotometry if their individual spectra are known. In a recent
crystallographic study of single crystals of U_2F_9, it was shown that
the uranium atoms were all equivalent within the precision of the

---
*Research sponsored by the Division of Chemical Sciences, U. S.
Department of Energy under contracts DE-AS05-76ER04447 with the
University of Tennessee (Knoxville) and W-7405-eng-26 with Union
Carbide Corporation.

structural determination (3). Spectrophotometry was then applied to samples of U_2F_9, UF_4, and UF_5 to show that U_2F_9 exhibits spectral characteristics ascribable only to U(IV) and U(V) (3). In this paper we present absorption spectroscopic data for $SmBr_3$, $SmBr_2$, and two mixed-valence compounds denoted here as $SmBr_x$ and $SmBr_{2+y}$. The absorption spectrum of SmOBr, a likely contaminant in samarium bromide samples, is also reported.

EXPERIMENTAL

All spectral studies described in this paper were made at room temperature by the use of a microscope-spectrophotometer of local design (4). Samples of $SmBr_2$ and $SmBr_x$ (1.67<x<2.20) were provided by J. M. Haschke (1). Although preliminary spectral data were obtained for both compounds, the $SmBr_2$ and $SmBr_x$ samples were relatively thick and so were opaque at wavelengths below 600 nm. Therefore, in-house preparation of Sm bromides was undertaken to produce thinner samples on which more complete spectral studies could be made.

Our methods of preparation of tri-, di-, and oxybromides on the μg scale are described elsewhere (4). All samples prepared at Oak Ridge were derived from the same source of Sm_2O_3, kindly supplied by R. G. Haire. Following hydrobromination of the oxide, cyclic hydrogen reductions of product $SmBr_3$ yielded $SmBr_2$. SmOBr was prepared by treatment of $SmBr_3$ with an HBr/H_2O gas mixture. The mixed-valence compound, $SmBr_{2+y}$, was produced by three consecutive thermal treatments (10 seconds each at about 500 °C) of a $SmBr_3$ sample sealed under 575 torr H_2.

One sample of $SmBr_3$ was also examined by X-ray powder diffraction techniques. The LCR-2 program (5) was used to calculate the least-squares lattice parameters ($\pm\sigma$) including the Nelson-Riley correction. Line intensities were compared to those calculated by program POWD (6) assuming the atomic coordinates and thermal parameters of $CmBr_3$ (7) but not including a correction for sample absorption.

RESULTS AND DISCUSSION

The absorption spectra of $SmBr_2$, $SmBr_{2+y}$, and $SmBr_3$ are shown in Fig. 1. The spectrum of red $SmBr_2$ (top) is characterized by one broad band centered at 17×10^5 m^{-1} and a shoulder at 20×10^5 m^{-1}. No peaks are seen in the spectral region of 9 to 14×10^5 m^{-1}. In contrast the spectrum of light yellow $SmBr_3$ (bottom) exhibits a high energy tail of an absorption peak at 9×10^5 m^{-1} and a band of lower intensity at 10.3×10^5 m^{-1}. Additional features include three weak bands between 18 and 20×10^5 m^{-1} and a unique pattern of f-f

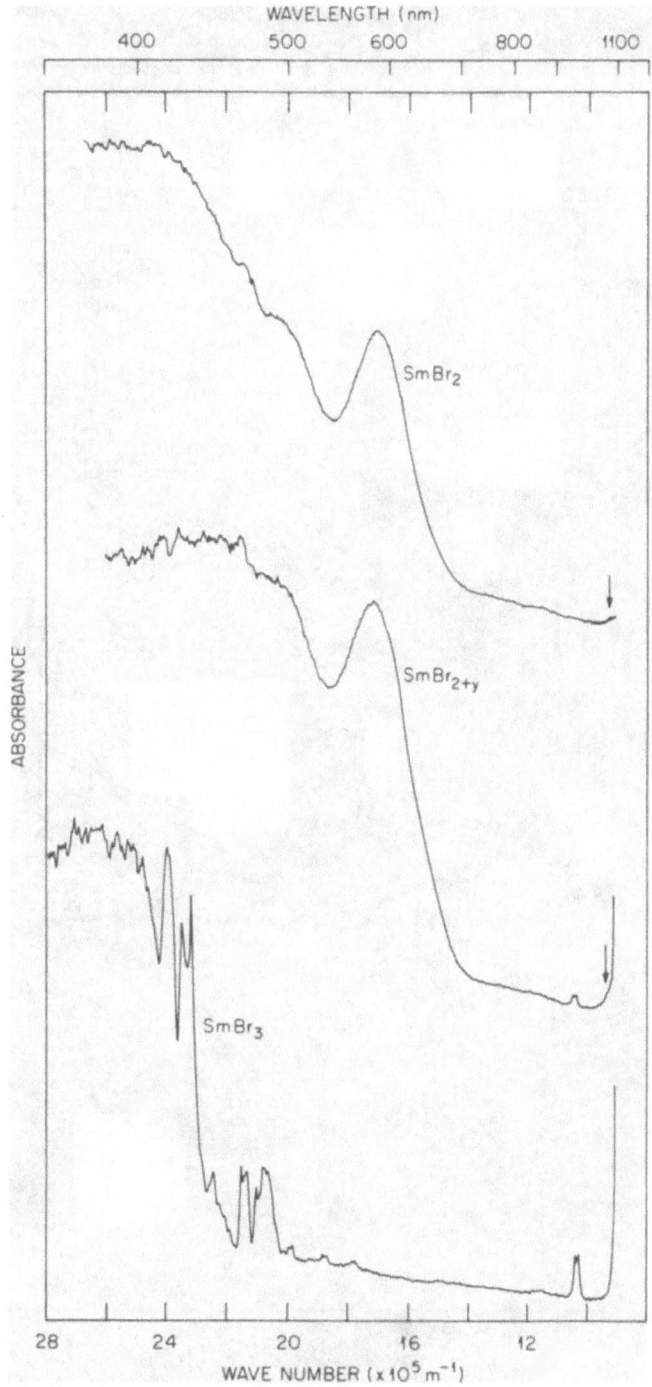

Figure 1. Absorption spectra of
SmBr$_2$, SmBr$_{2+y}$, and SmBr$_3$.

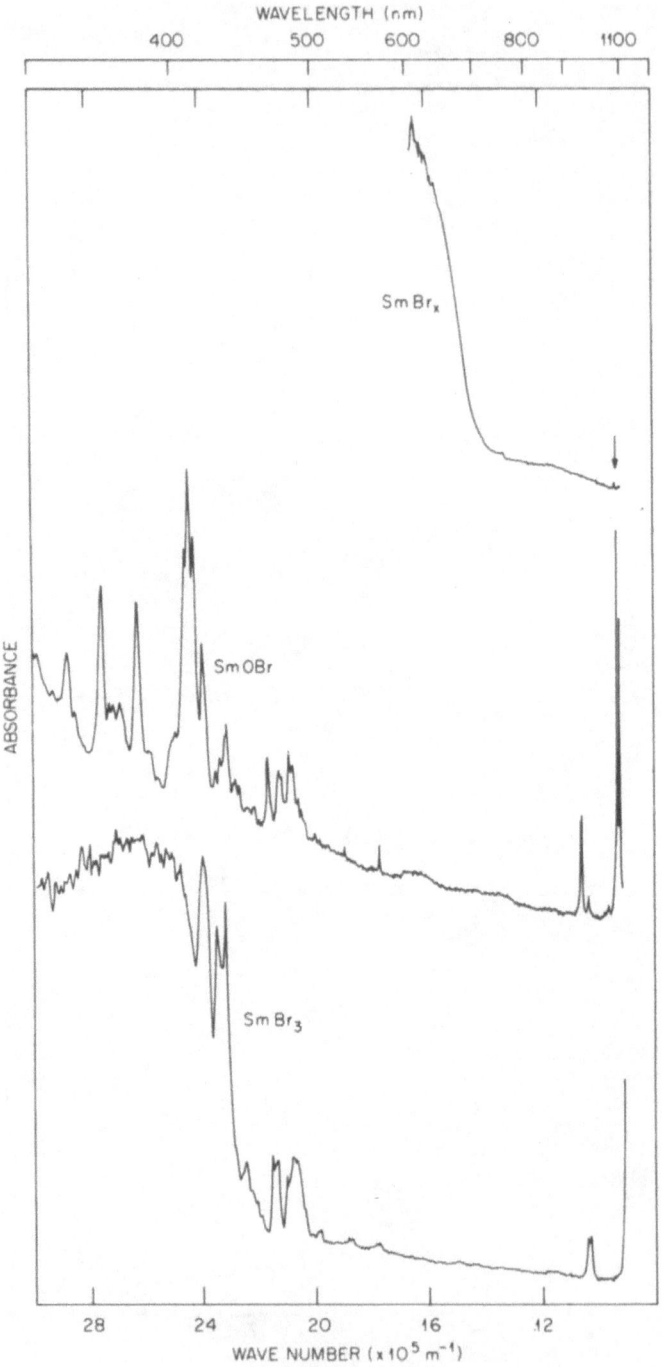

Figure 2.   Absorption spectra of
SmBr$_x$, SmOBr, and SmBr$_3$.

transitions at still higher energies. The spectrum of $SmBr_{2+y}$
(middle) exhibits the tribromide peaks at low wave numbers and the
intense Sm(II) absorption peaks above 12 x 10^5 m^{-1}. This broad and
intense absorption of Sm(II) prevents the observation of the weak f-f
Sm(III) absorption peaks in the higher energy spectral region.

The absorption spectra of $SmBr_x$, SmOBr, and $SmBr_3$ are presented
in Fig. 2. The spectrum of SmOBr can be distinguished easily from
that of $SmBr_3$ in the spectral region of 9 to 11 x 10^5 m^{-1} on the
basis of the number, position, and extreme sharpness of the SmOBr
absorption peaks. Samples of the $SmBr_x$ compound did not yield useful
spectra at energies greater than 17.5 x 10^5 m^{-1}. The more intense
set of SmOBr peaks around 9.3 x 10^5 m^{-1} are present in the spectrum
of $SmBr_x$. Haschke also pointed out that SmOBr was present (1).

It is noteworthy that no spectral evidence for $SmBr_3$ is seen in
the spectrum of $SmBr_x$ in contrast to its obvious presence (Fig. 1)
in the spectrum of $SmBr_{2+y}$. The absence of the $SmBr_3$ peak at 10.3 x
10^5 m^{-1} and its high energy tail at 9 x 10^5 m^{-1} in the $SmBr_x$ spectrum
cannot be attributed to the opacity of the sample. This absence might
mean that no uniquely $SmBr_3$ sites are present in the $SmBr_x$ compound,
but more evaluation of such vernier compounds will be necessary to
confirm this point. It has been reported that only Cf(II) spectral
peaks are seen in $Cf_4Gd(III)X_{11}$ halide compounds exhibiting the
M_5X_{11} vernier structure (8). It is also possible that the presence
or absence of individual cationic species in these compounds is not
related to the vernier structure but only to the relative likelihood
of electron transfer between the two cations. Additional work is
necessary to resolve this question.

Not only can the sharp absorption peaks of SmOBr at 9.3 x 10^5 m^{-1}
be seen in the $SmBr_x$ spectrum, but also they can be detected in the
spectra of many of the samples of $SmBr_3$, $SmBr_2$, and mixed Sm(II/III)
bromides (note the arrows in Figs. 1 and 2 pointing to this characteristic peak of SmOBr). This ability to detect even minor amounts
of SmOBr attests to the sensitivity of this type of spectrophotometric analysis. The stability of $SmBr_2$ with respect to reaction with
SiO_2 to form SmOBr has also been demonstrated; $SmBr_2$ sealed in SiO_2
was repeatedly heated, even to melting, with no resultant increase
in the amount of SmOBr detected.

The powder pattern of one sample of $SmBr_3$ exhibited 36 lines out
to $2\theta = 161°$. Following indexing on the basis of the $PuBr_3$-type
orthorhombic structure, the lattice parameters were calculated to be
$a_o = 0.4039(2)$ nm, $b_o = 1.2729(6)$ nm, and $c_o = 0.9140(3)$ nm. These
values are in agreement with the recent values of Haschke (1).

The X-ray data obtained from the $SmBr_3$ sample confirmed its
stoichiometry. Similar data collected from the $SmBr_2$ and $SmBr_{2+y}$
samples were not able to be indexed completely on the basis of the

SrBr$_2$-type tetragonal and M$_5$X$_{11}$/M$_{11}$X$_{24}$/M$_6$X$_{13}$ vernier structures, respectively. These samples were macrocrystalline in character and produced low quality diffraction patterns. Thus, we have not confirmed the stoichiometries of these latter compounds.

REFERENCES

1. J.M. Haschke, Inorg. Chem., 15:298 (1976).
2. H. Bärnighausen and J.M. Haschke, Inorg. Chem., 17:18 (1978).
3. P.G. Eller, A.C. Larson, J.R. Peterson, D.D. Ensor, and J.P. Young, Inorg. Chim. Acta, 37:129 (1979).
4. J.P. Young, R.G. Haire, R.L. Fellows, and J.R. Peterson, J. Radioanal. Chem., 43:479 (1978).
5. D.E. Williams, "LCR-2, A Fortran Lattice Constant Refinement Program," IS-1052, Ames Laboratory, Ames, Iowa (1964).
6. D.K. Smith, "A Fortran Program for Calculating X-Ray Powder Diffraction Patterns," UCRL-7196, Lawrence Radiation Laboratory, Livermore, CA (1963).
7. J.H. Burns, J.R. Peterson, and J.N. Stevenson, J. Inorg. Nucl. Chem., 37:743 (1975).
8. R.G. Haire, J.P. Young, J.R. Peterson, and R.L. Fellows, "Absorption Spectrophotometric and X-Ray Diffraction Evidence for Mixed-Valence Compounds in Anhydrous Halides of Lanthanide-Actinide Mixtures," in The Rare Earths in Modern Science and Technology, G. J. McCarthy and J. J. Rhyne (eds.), Plenum Press, New York, pp 501-506 (1978).

CASCADE LASER ACTION IN Tm^{3+}:YLF

L. Esterowitz, R. Allen, and R. Eckardt

Naval Research Laboratory
Washington, D. C. 20375

Among current systems used for atmospheric aerosol remote sensing is the ruby laser at 0.69 μm. Due to uncertainties in the complex index of refraction and its dependence with wavelength it would be very useful to expand the current ruby lidar system and include transmission measurements at longer IR wavelengths.

Using a ruby laser together with a thulium doped LiYF$_4$ single crystal we have obtained laser action at 2.30 μm corresponding to the $^3H_4 \rightarrow {}^3H_5$ transition in trivalent thulium (see Fig. 1). This wavelength is in a region of excellent atmospheric transmission. The 0.69-μm ruby wavelength is resonant with the $^3H_6 \rightarrow {}^3F_3$ thulium absorption transition. The energy thus deposited in the $^3F_3$ manifold quickly relaxes to the $^3H_4$ level. The lifetime of the $^3H_4$ level in our sample of 5% Tm:YLF is 80 μsec. The ruby laser pump was used in the long pulse mode (600 μsec) and threshold for the 2.3-μm thulium laser in these preliminary measurements was 1.5 J with the multimode ruby beam. Aperturing and utilizing a TEM$_{oo}$ beam should drop the threshold by an order of magnitude. Q-switching which will be tried later will also substantially reduce the energy threshold. The laser cavity for the thulium laser consisted of a 1 meter mirror with reflectivity of >99% between 1.8 and 2.4 μm and a flat mirror with reflectivity of >99% between 1.8 and 2.4 μm. The thulium crystal was pumped longitudinally with the ruby laser through the flat mirror which is 90% transparent at 0.69 μm. With this same experimental setup we also observed stimulated emission at 1.88 μm corresponding to the $^3F_4 \rightarrow {}^3H_6$ transition (see Fig. 1). This cascade laser action is achieved by lasing the $^3H_4 \rightarrow {}^3H_5$ transition and subsequent decay to the $^3F_4$ level. The 1.88-μm laser transition occurs 200 μsec after termination of the 2.3-μm laser pulse indicating the sequential laser emission. Of much greater

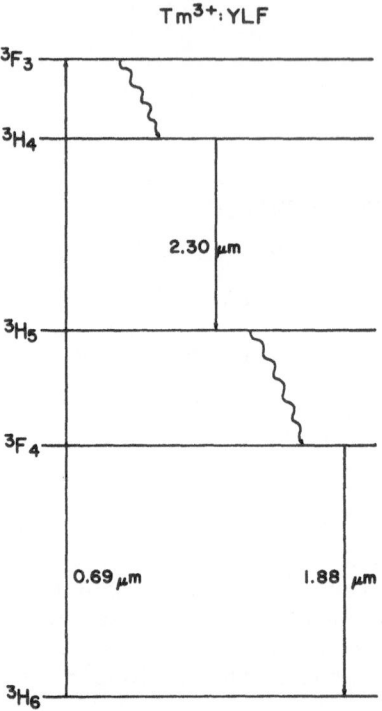

Figure 1.   Energy level diagram of Tm³⁺:LiYF showing relevant
            levels and transitions for ruby laser pumping of
            cascade laser transitions.

interest and of more practical use would be the laser transition at
around 3.3 μm corresponding to the $^3H_5 \rightarrow ^3F_4$ transition. Current
plans are underway to obtain the appropriate laser mirrors and to
Q-switch the ruby laser to attempt lasing this transition via
sequential cascade laser action. A multiple LIDAR system with
wavelengths at 0.69 μm, 2.3 μm and 3.3 μm should prove extremely
useful for measuring the backscatter cross-sections and extinction
coefficients of atmospheric aerosols.

LASER INDUCED LUMINESCENCE OF Pr^{3+} IN CaF_2 ATTRIBUTED TO DIFFERENT LOCAL SITE SYMMETRIES

J. Chrysochoos[a], M.J. Stillman[b] and P.W.M. Jacobs[b]
(a) Department of Chemistry, University of Toledo, Toledo
Ohio 43606; (b) Centre for Interdisciplinary Studies in
Chemical Physics, University of Western Ontario, London
Canada N6A 3K7

INTRODUCTION

The incorporation of trivalent lanthanides into CaF_2 is accompanied by a charge-compensation process brought about by interstitial anions, F_i^- or 0^{2-}. This process gives rise to such site symmetries as cubic (0_h), trigonal (C_{3v}), tetragonal (C_{4v}) and others. The optical properties of trivalent lanthanides in such sites are very characteristic. The type of the site-symmetry can be inferred, at least in principle, from the Stark splitting (1,2), and the intensity of the crystal spectrum. The unambiguous identification of the local symmetry of a Pr^{3+}-site requires a combination of techniques such as optical spectroscopy, Zeeman spectroscopy and EPR (3).

Although site symmetries have been reported for $CaF_2:Er^{3+}$ (4), $CaF_2:Gd^{3+}$ (5) and $CaF_2:Nd^{3+}$ (6) only one early report on $CaF_2:Pr^{3+}$ claimed the existence of tetragonal-sites (7), in conflict with another early report based upon the EPR spectrum of Pr^{3+} in CaF_2, which concluded the existence of trigonal sites of Pr^{3+} attributed to 0^{2-}-impurities (8). EPR spectroscopy has been applied to several trivalent lanthanides in CaF_2 leading to tentative site-symmetry assignments (9).

EXPERIMENTAL

The crystals used in this study were grown and kindly supplied by Dr. A.V. Chadwick of the University of Kent at Canterbury Chemical Laboratory. The concentrations of Pr^{3+} reported are nominal. An Argon ion laser was used.

RESULTS AND DISCUSSION

Typical room and low temperature emission spectra of 1.0 mole % Pr^{3+} in CaF_2, excited with 4765.05 Å, are shown in Fig. 1. The spectra shown consist of several narrow bands attributed to the $^3P_0 \rightarrow {}^3F_2$; $^3H_{6,5,4}$ transitions. The corresponding assignments were made by grouping the values of $E(^3P_0)-\bar{\nu}_{em}$, $E(^3P_1)-\bar{\nu}_{em}$ and $E(^1D_2)-\bar{\nu}_{em}$, where $^3P_0$, $^3P_1$ and $^1D_2$ are the radiative states of Pr^{3+}. Two groups of emission lines at about 19080 cm^{-1} attributed to the $^3P_1 \rightarrow {}^3H_5$ transition and at 16300 cm^{-1}, due to the $^3P_1 \rightarrow {}^3F_2$ transition are present only at room temperature.

The emission spectrum of Pr^{3+} is concentration dependent (Fig. 2). In the first place it is apparent that there are more emission lines than the maximum number expected via the Stark splitting of Pr^{3+} in a single site. Furthermore, the emission lines appear to be divided into three groups, namely Group (I), (II) and (III). Group (II) which is the most intense, consists of several Stark components. Group (I) consists of fewer components and it is weaker. It gains intensity at lower [Pr^{3+}]. Finally, Group (III) is quite weak. The appropriate grouping of the emission lines was determined based upon the dependence of the intensity ratios on [Pr^{3+}]. Some of these plots are depicted in Fig. 3. Lines whose intensity ratio was found to be independent of [Pr^{3+}] were assumed to belong to the same group.

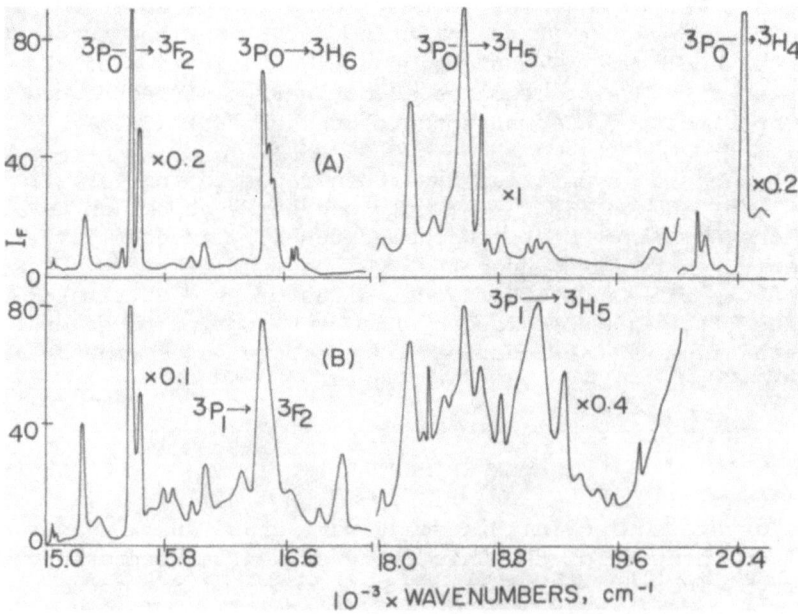

Figure 1.   Luminescence spectra of $CaF_2:Pr^{3+}$, 1.0 mole % Pr^{3+}; (A) 125 K (B) Room temperature. λ_{exc}=4765 Å.

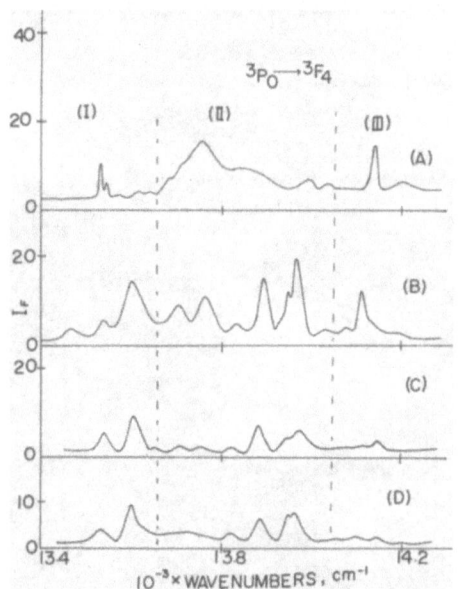

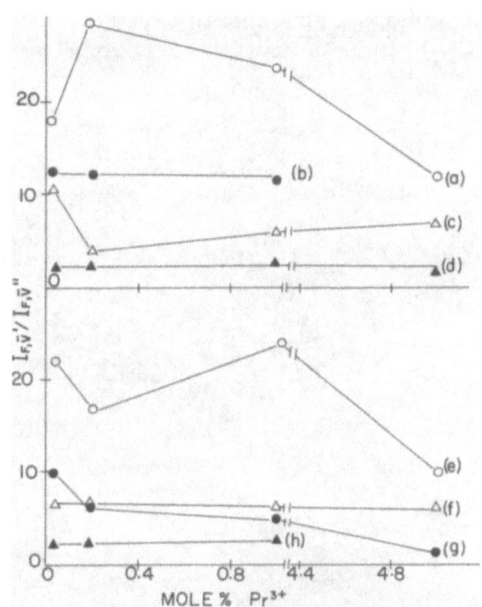

Figure 2. Emission of CaF_2:Pr^{3+} at 125 K; 4765 Å; (D) 0.04 (C) 0.2 (B) 1.0 and (A) 5.0 mole % Pr^{3+}.

Figure 3. I_F, $\bar{\nu}'/I_F,\nu''$ vs mole % of Pr^{3+} at 125 K. (a) I_{16440}/I_{16750}; (b) I_{16440}/I_{16460}; (c) I_{16440}/I_{16280}; (d) I_{16280}/I_{16090}; (e) I_{15560}/I_{15980}; (f) I_{15235}/I_{15860}; (g) I_{15980}/I_{15860}; (h) I_{15560}/I_{15860}.

Groups (I), (II) and (III) can be selectively excited using different laser lines leading to the population of the $^3P_0$-state only. Excitation at 4765.05 Å gives rise to all three Groups, shown in Fig. 4., whereas excitations at 4880 and 4965 Å are more selective. Light excitation at 4880 Å gives rise to Groups (II) and (III), whereas excitation at 4965 Å enhances Group (I) relative to Groups (II) and (III).

The energy of the emission lines of 1.0 mole % of Pr^{3+} in CaF₂ together with literature values in several hosts are given elsewhere (10). The energies of Groups (II) and (III) are in fair agreement with the fluorescence spectrum of Pr^{3+} in LaF₃ at 77 K (11,12). Since the symmetry of Pr^{3+} in this host is considered to be orthorombic, C_{2v} (11) the symmetry of Groups (II) and (III) is likely to be a distorted trigonal (C_{3v}). Group (I) is in fair agreement with the fluorescence spectrum of Pr^{3+} in Yttrium Aluminum Garnet (11) leading to either a tetragonal (C_{4v}) or a cubic (O_h) symmetry. Finally, Group (II) and part of Group (III) are in fair agreement with the fluorescence spectrum of Pr^{3+} in $La_{1-x}Pr_xP_5O_{14}$ (monoclinic or pseudo-orthorhombic (13)). Thus, Group (II) is closer to trigonal (C_{3v}) than cubic or tetragonal.

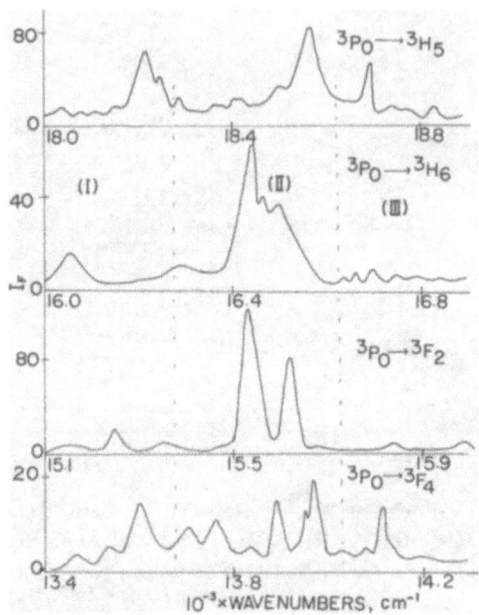

Figure 4. Luminescence spectra of $CaF_2:Pr^{3+}$; 1.0 mole % Pr^{3+}, 4765 Å; 125 K.

REFERENCES

1. J.M. O'Hare and V.L. Donlan, Phys. Rev., 185:416 (1969).
2. J.M. O'Hare, Phys. Rev., B3:3603 (1971).
3. R.E. Bradbury and E.Y. Wong, Phys. Rev., B4:690,694,701 (1971).
4. D.R. Tallant and J.C. Wright, J. Chem. Phys., 63:2074 (1975).
5. J. Makovsky, J. Chem. Phys., 46:390 (1967).
6. J.C. Toledano, J. Chem. Phys., 57:1046,4468 (1972).
7. W.A. Hargreaves, Phys. Rev., B6:3417 (1972).
8. S.D. McLaughlan, Phys. Rev., 150:118 (1966).
9. W. Hayes "Crystals with the Fluorite Structure" W. Hayed, (Ed)
    Calrendon Press, Oxford, pp. 378-387 (1974).
10. J. Chrysochoos, M.J. Stillman and P.W.M. Jacobs, Chem. Phys.,
    submitted.
11. E.Y. Wong, D.M. Stafsudd and D.R. Johnson, J. Chem. Phys. 39:
    786 (1963).
12. W.M. Yen, W.C. Scott and A.L. Schawlow, Phys. Rev. 136:A271 (1964).
13. H. Dornauf and J. Heber, J. Luminescence, 20:271 (1979).

PHOTOCONDUCTIVITY DUE TO AUTOIONIZATION OF DIVALENT RARE EARTH

IMPURITIES IN CRYSTALS HAVING THE FLUORITE STRUCTURE

Christian Pedrini and Françoise Gaume-Mahn

University of Lyon I, 69622 Villeurbanne, France

Donald S. McClure

University of Princeton, New Jersey 08544

INTRODUCTION

Photoionization of impurity ions in solids can lead to photo-conductivity or oxidation-reduction within the crystal. In the present case of divalent rare earth ions in alkaline earth fluorides, the photoconductivity occurs by exciting in the 4f→5d adsorption bands of the impurities. A good correlation exists between photo-conduction signals and absorption spectra, indicating that some of the transitions lead to the release of electrons from the centres. Therefore, we are dealing with an autoionization process where impurity-induced discrete electronic states are in resonance with the conduction energy band, as confirmed by low temperature photoconductivity measurements. The emphasis is on a systematic study of photoionization thresholds in CaF_2, SrF_2 and BaF_2 host crystals for the divalent rare earth ions. The threshold energy is interpreted as the energy mismatch between the ground state of the impurity ion and the bottom of the conduction band. A simple electrostatic model including the third ionization potential of the free rare earth ion and the influence on its surrounding in the crystal permits an explaination of the threshold shift upon changing the nature of the crystal and the nature of the impurity.

EXPERIMENT

The materials used in these studies were grown by using a modified Bridgman technique. The typical concentration of rare earth dopant was around 0.05 % and made up a maximum of 0.1 mass % of the crystal. To reduce the rare earth impurity and consequently to introduce broad absorption bands throughout the visible and UV, the technique of additive coloration was used which can result in almost complete reduction and in stable conversion.

The measuring system and the sample mounting geometry were
described previously (1,2). The samples are placed in a cryostat
permitting low-temperature photoconductivity measurements.
The theory of the experiments is explained in detail in (2).

RESULTS AND DISCUSSION

The room temperature results concern the following divalent
rare earth impurities : Tm, Ho, Dy, Pr and Ce. As an example of curves representing the spectral dependence of both the photoresponse
and the absorption constant, we have choosen to show CaF_2:Ce (Fig. 1).
The very rapid rise of photoelectron yield is typical of these systems and permits the identification of the threshold to a good approximation. For all the samples studied, the order of magnitude of
the photosensitivity is around 10^{-12} cm^2 x v^{-1}. The experimental
results are contained in the last column of table 1. When the same
host lattice CaF_2 is doped with different ions, it can be seen that
the onsets of the photoconductivity are found in the same range of
energy for most rare earth guest ions excepted for Tm^{2+} which exhibits a threshold shift of about 1eV towards higher energies. For the
same type of impurity, the change of the host lattice leads to a
strong displacement of the threshold towards lower energies in going
from CaF_2 to SrF_2, then to BaF_2.

In order to explain the experimental results, it is necessary
to construct a model. The photoionization energies of the rare earth
ions should depend primarily on the ionization potential I_3 of the
free rare earth ion, but it is expected that their values will be
modified by the crystalline surroundings. So the threshold energy
can be described by the expression

$$E_{PI} = I_3 - C \quad , \tag{1}$$

where C represents the effect of the crystal. Since we are dealing
with ionic crystals, C may be evaluated as the Madelung energy at
a metal site in the crystal and it takes into account the electronic
polarization of the host lattice during the ionization process. At
first sight, it is expected that E_{PI} will vary like I_3 when the
impurities are embedded in the same crystal. Furthermore, for a same
impurity in the same crystal, E_{PI} should follow the trend of Madelung
energy, the polarization energy being about constant. However, it is
not generally the case, due to the fact that the impurity-fluorine
distance is different from the host crystal metal-fluorine (3). Consequently, it is necessary to take into account these variations,
so that the threshold energy must be calculated as :

$$E_{PI} = I_3 + E_{M^{++}} + \Delta E_{M^{++}} - E_{PM^{++}} \quad . \tag{2}$$

It can be seen in table 1 that eq. (2) fits very well the experimental results. Then it seems that these crude calculations are adequate
for assigning the threshold which are correctly interpreted as the
energies between the impurity level in the band gap and the edge of
the conduction band of the crystal.

Low temperature photoconductivity measurements were done on

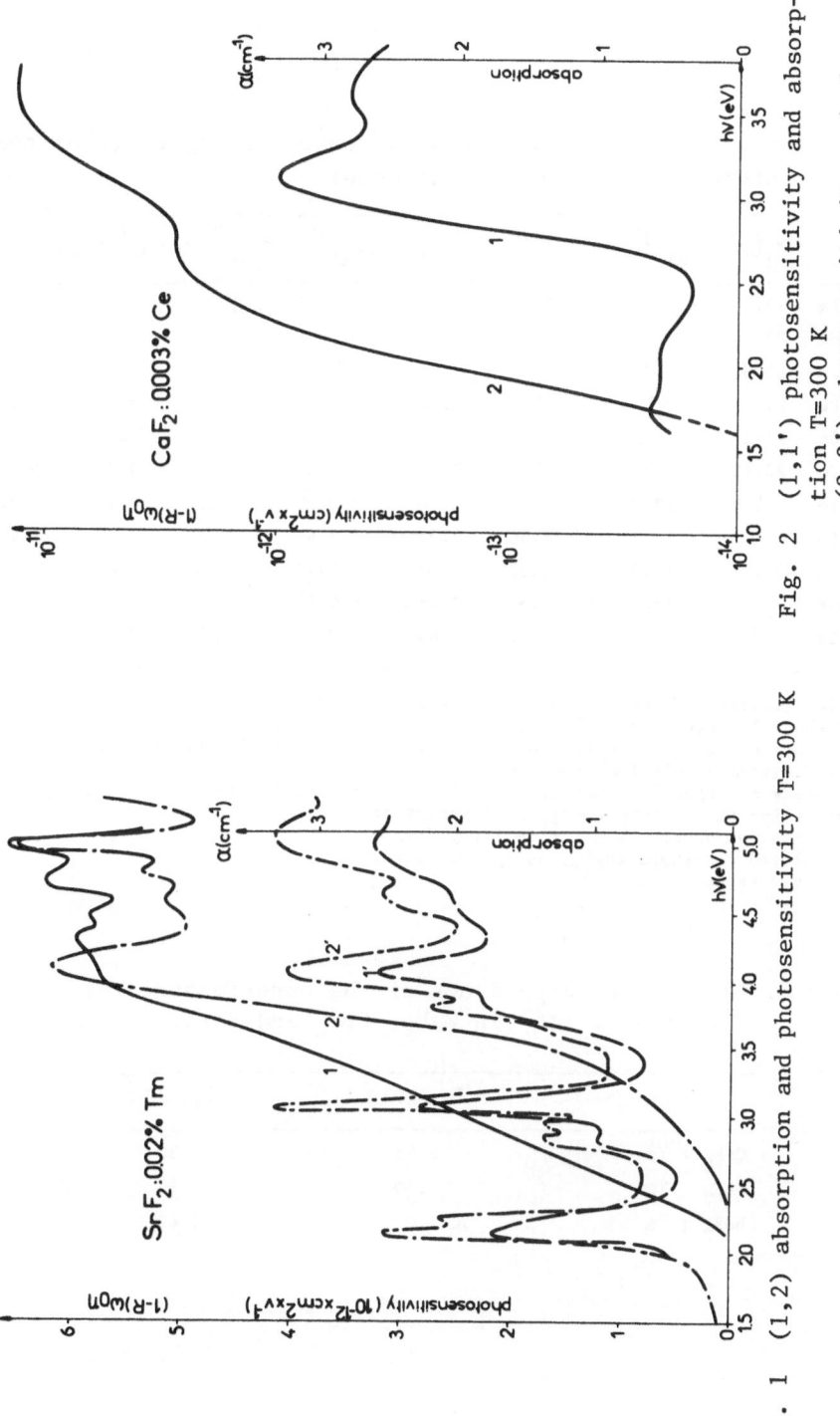

Fig. 1   (1,2) absorption and photosensitivity T=300 K

Fig. 2   (1,1') photosensitivity and absorption T=300 K
(2,2') photosensitivity and absorption T=77 K

Table 1. Comparison of experimental threshold energies (room temperature) with ionic crystal model.

	I_3(eV) [a]	$E_{M^{++}}$(eV) [b]	$r(\AA)$ [c]	$r_{corr}(\AA)$ [d]	$\Delta E_{M^{++}}$(eV) [e]	$E_{PM^{++}}$(eV) [f]	Calc. E_{PI} [g]	Exp. E_{PI} [h]
CaF_2:Tm	23.68	− 19.94	2.37	2.42	+ 1.21	1.73	3.22	2.95
CaF_2:Ho	22.84	"	"	"	"	"	2.38	1.75
CaF_2:Dy	22.80	"	"	"	"	"	2.38	1.76
CaF_2:Pr	21.62	"	"	2.47	+ 2.17	"	2.12	1.30
CaF_2:Ce	21.1	"	"	"	"	"	1.60	1.60
CaF_2:Tm	23.68	"	"	2.42	+ 1.21	"	3.22	2.95
SrF_2:Tm	"	− 18.78	2.50	2.46	− 0.75	2.01	2.14	2.10
BaF_2:Tm	"	− 17.59	2.68	2.51	− 2.91	2.02	1.16	~1.0
CaF_2:Ce	21.1	− 19.94	2.37	2.47	+ 2.17	1.73	1.60	1.60
SrF_2:Ce	"	− 18.78	2.51	2.56	+ 0.99	2.01	1.30	1.30
BaF_2:Ce	"	− 17.59	2.68	2.66	− 0.39	2.02	1.10	1.10

(a) Third ionization potential of rare-earth (ref. 7). For cerium, it was added 0.9 eV to the value given in Ref. 4 (see ref. 5).
(b) Madelung energy per electron at M^{++} site of host crystal (ref. 6).
(c) Host crystal metal-fluorine distance.
(d) Impurity-fluorine distance (estimated distance excepted for thulium (ref. 3)).
(e) Correction to Madelung energy at impurity site.
(f) Polarization energy at M^{++} site (ref. 7).
(g) Calculated threshold energy using relation (2).
(h) Observed threshold energy.

Table 2. Temperature dependence of the experimental threshold energies of Tm^{2+} in CaF_2, SrF_2 and BaF_2.

	E_{PI} 300 K	E_{PI} 77 K
CaF_2 : Tm	2.95	3.30
SrF_2 : Tm	2.10	2.30
BaF_2 : Tm	~1.0	~1.60

the series $MF_2:Tm^{2+}$ (M=Ca,Sr,Ba). Typical curves are shown in Fig.2.

The "plateau" photoconductivity (above 4eV) is approximately temperature independent, indicating that the parameters which effect the photoresponse, such as the mobility μ, the lifetime τ and the photogeneration efficiency η, have temperature dependences which cancel each other. The temperature dependence of the initial rise of the photoconductivity (table 2) is assigned not only to a sharpening and a displacement of the absorption bands, but more especially to thermally assisted currents. Probably the principal contribution comes from the temperature dependence of η which can be written

$$\eta = \frac{k_A}{k_A + k_r} = \frac{k_o \exp(-E_A/kT)}{k_o \exp(-E_A/k_T)+k_r} , \qquad (3)$$

where k_A is the autoionization rate, k_r is the relaxation rate for the electron to return to the ground state and E_A is the activation energy of a level just below the conduction band.

REFERENCES

1. C. Pédrini, D.S. McClure and C.H. Anderson, J. Chem. Phys., 70 : 4959 (1979).
2. C. Pédrini, P.O. Pagost, C. Madej and D.S. McClure, J. Phys., 42 : 323 (1981).
3. C.H. Anderson, P. Call, J. Stott and W. Hayes, Phys. Rev., B11 : 3305 (1975).
4. W.C. Martin, L. Hagan, J. Readler and J. Sugar, J. Phys. Chem., Ref. Data 3 : 771 (1974).
5. R.C. Alig, Z.J. Kiss, J.P. Brown and D.S. McClure, Phys. Rev., 186:276 (1969)
6. R.T. Poole, J. Szajman, R.C.G. Leckey, J.C. Jenkin and J. Liesegang, Phys. Rev., B12 : 5872 (1965).
7. N.V. Starostin and V.A. Ganin, Sov. Phys. Solid State, 15 : 2265 (1974) − 16 : 369 (1974).

THE EFFECT OF STRUCTURAL ENVIRONMENT ON THE ABSORPTION SPECTRA OF SELECTED LANTHANIDE SESQUIOXIDES*

R.G. Haire, J.P. Young, and J.R. Peterson†

Oak Ridge National Laboratory, Oak Ridge, TN   37830

†University of Tennessee, Knoxville, TN   37916

INTRODUCTION

Solid-state absorption spectra and X-ray powder diffraction have been used to study the transplutonium elements in several oxidation states. During the course of these studies it was found that a compound's structure, or its local environment, can be deduced from its solid-state spectrum, once the spectrum has been correlated with a particular structure (1,2). One application of this technique would be in examining the chemical and structural consequences of radioactive decay in the transplutonium oxides. The object of the present study was to ascertain if room-temperature absorption spectral data could be used to determine the structural environment of a lanthanide sesquioxide. To evaluate this technique several experiments were carried out using pure lanthanide oxides or lanthanide oxide mixtures. Described here are the absorption spectra obtained from neodymium, samarium, dysprosium and erbium sesquioxides in matrices which exhibit three different crystal forms.

EXPERIMENTAL

The lanthanide oxides and oxide mixtures (6-8 mole%) were prepared by precipitating their oxalates from nitrate solutions and

*Research sponsored by the Division of Chemical Sciences, U.S. Department of Energy under contracts W-7405-eng-26 with the Union Carbide Corporation and DE-AS05-76ER04447 with The University of Tennessee (Knoxville).

calcining the oxalates in platinum containers to 1000°C. Subsequent heat treatments of the oxides were done on a microscale to simulate conditions to be used for the transplutonium oxides and consisted of heating multimicrogram (50-200μg) pieces of the oxide material in miniature platinum coils. The oxide samples were sealed in quartz capillaries for X-ray and spectrophotometric analysis. The techniques used for obtaining and analyzing the X-ray and absorption spectrophotometric data have been described (3,4). All spectra were obtained at room temperature with a single-beam spectrophotometer.

RESULTS AND DISCUSSION

The lanthanide sesquioxides possess one of three structural forms (designated classically as A, B or C) depending on the size of the cation and on the thermal history of the oxide. The actinide sesquioxides also form these same structures, the phase relationship of a particular actinide being similar to that of the lanthanide of comparable ionic radius (5). In the A-structure (hexagonal) and the B-structure (monoclinic) the metal ion has a coordination number of seven. The C-structure (body centered cubic) provides a coordination number of six for the metal ion. In the A-form the metal sites are equivalent, but in the B-form there are three kinds of non-equivalent sites for the metal ion. In the C-structure there are two different six-coordinated metal ion sites (6). The structural environments of the metal ions are different in these three oxide forms, and their absorption spectra might also exhibit differences.

To generate the desired structural environment a metal ion would have in each sesquioxide structure, it was necessary in some instances to use mixtures of sesquioxides in which the metal ion of interest was a minor constituent. In this study neodymium, samarium, dysprosium and erbium oxides were prepared using the same experimental approach to be used with the transplutonium elements. Lanthanum and europium were selected as oxide hosts to provide a particular structural environment: lanthanum was used for the A-structure and europium for both the B- and C-forms (treatment at 1450°C for the B-form; at 1000°C for the C-form).

The absorption spectra of neodymium in the three sesquioxide structures are shown in Fig. 1. The absorption spectrum of pure Nd_2O_3 (A-form) is shown, and it can be compared to the spectra of neodymium in the monoclinic Eu_2O_3 host and in the cubic Eu_2O_3 host. The contribution of europium to the absorption spectra in the latter materials is not significant. The most striking differences in the neodymium spectra occur between the spectrum for the cubic form and the spectra from the other two structures. The differences in the absorption peaks can be classified into two categories:

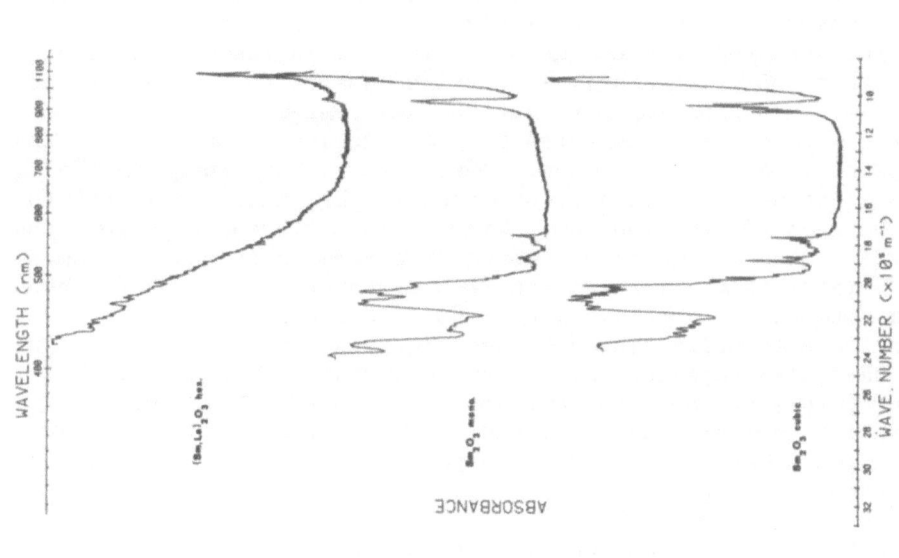

Figure 2. Absorption Spectra of Sm_2O_3 and $(Sm,La)_2O_3$.

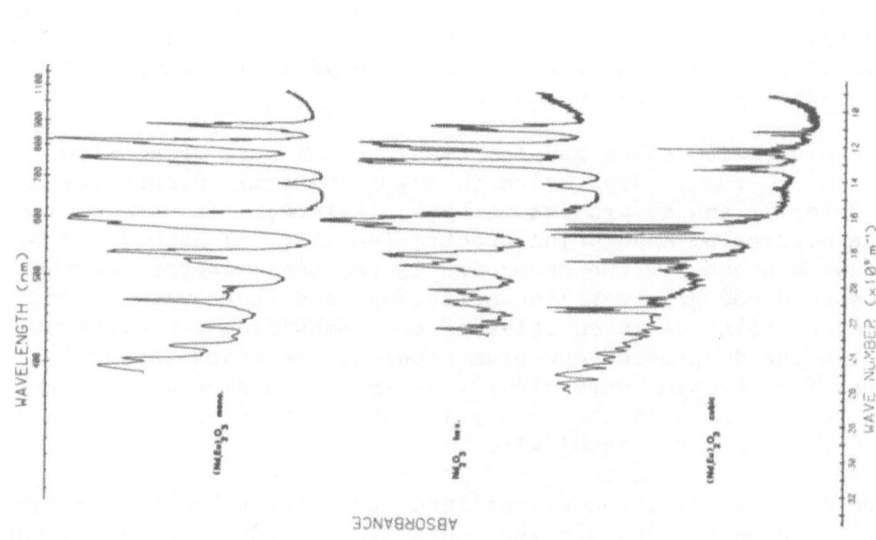

Figure 1. Absorption Spectra of Nd_2O_3 and $(Nd,Eu)_2O_3$.

(1) the degree of splitting; and (2) the relative intensities. The
spectrum for cubic (Nd, Eu)$_2O_3$ exhibits a high-degree of splitting,
whereas the spectra for the hexagonal and monoclinic forms exhibit
unsplit absorption envelopes likely due to unresolved transitions.
Since the spectrophotometer is a single-beam instrument, absolute
intensities cannot be readily determined due to the irregular path
lengths of the solid samples; but the relative intensities of peaks
in one spectrum can be compared to the relative intensities of the
same peaks in another spectrum. The most obvious change in the
relative intensity of absorption peaks in Fig. 1 occurs in the four
bands between 680 and 1000 nm. The subtle differences between the
spectra for the hexagonal and monoclinic forms must result from
small changes in the interaction (M-O distances, etc.) of the neodymium cations and oxygen anions. No effort has been made to
assign the absorption lines to specific transitions. Such assignments have been reported for neodymium in oxide crystals (7,8).
For our purposes, the important aspect is that the reproducible
spectra obtained at room temperature can be recognized as originating from a Ln(III) species in a particular sesquioxide
polymorph.

In the case of samarium, it was possible to prepare pure samarium sesquioxide in both the C-form (950°C) and the B-form
(1450°C). To obtain samarium in the hexagonal structure, a lanthanum sesquioxide host was used. The absorption spectrum of
samarium (see Fig. 2) between 300 and 1100 nm is not as rich as
that of neodymium. Examination of the samarium spectra from these
different oxide environments also reveals differences in the
absorption peaks similar to those discussed for neodymium. The
spectrum of the C-form of samarium sesquioxide appears to exhibit
greater splitting than that of the other two forms. Variations in
the relative intensities of the absorption peaks in the different
samarium samples are also noted.

The dysprosium oxide samples consisted of pure dysprosium
sesquioxide (C-form), dysprosium in Eu$_2O_3$ (C-form), dysprosium in
Eu$_2O_3$ (B-form), and dysprosium in La$_2O_3$ (A-form). It was concluded
that the dysprosium absorption spectra (see Fig. 3) showed the same
trends noted above for the other two lanthanides; spectra of the
C-form showed the greatest line splitting, and there were alterations in the relative intensities of the peaks from the different
forms. In the dysprosium-europium samples, the major absorption
envelopes for europium were clearly visible, and distinct differences were also seen in the absorption peaks of europium in the
B- and C-forms of the sesquioxide.

The spectral differences outlined above for the other oxides
were again noted for the samples containing erbium. In Fig. 4 the
absorption spectra of erbium in different environments are shown.
The spectra of pure erbium sesquioxide (C-form) and erbium (6 mole%)

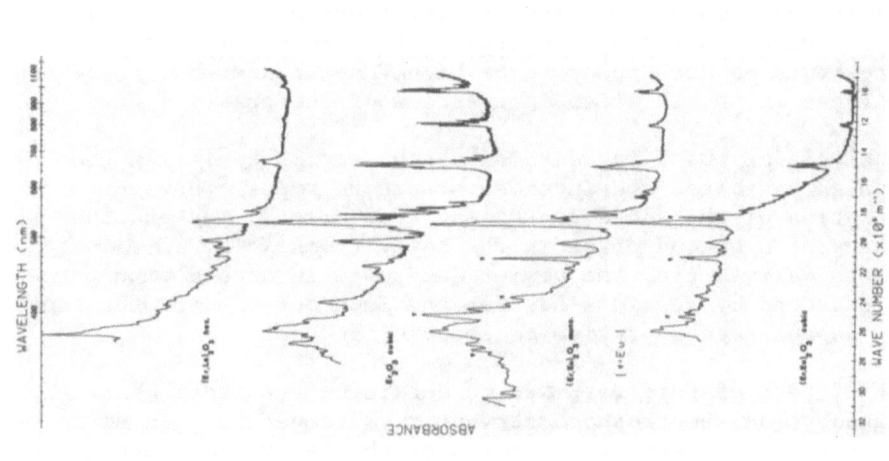

Figure 4. Absorption Spectra of Er_2O_3, $(Er,Eu)_2O_3$ and $(Er,La)_2O_3$.

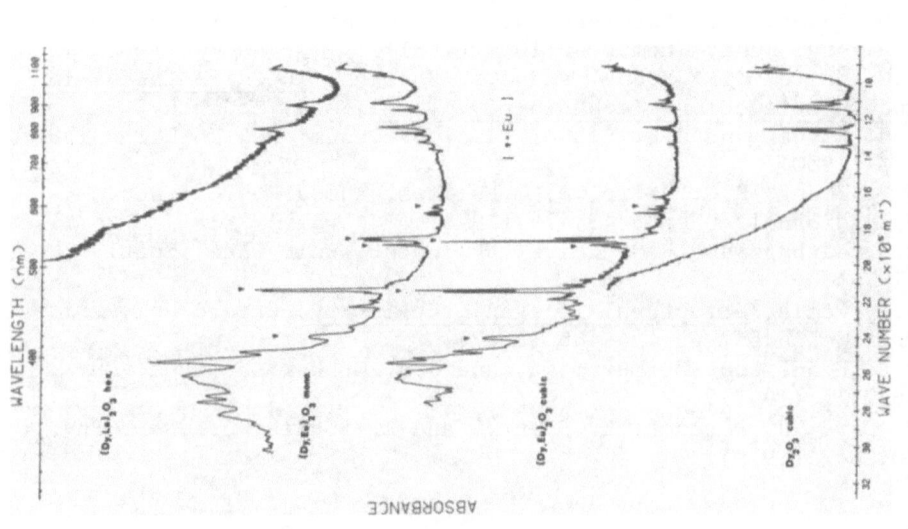

Figure 3. Absorption Spectra of Dy_2O_3, $(Dy,Eu)_2O_3$ and $(Dy,La)_2O_3$.

in cubic Eu_2O_3 are essentially the same. These spectra are characterized by multiple sharp peaks (greater splitting), especially in the 700-1100 nm region. The spectrum of erbium in the monoclinic form of Eu_2O_3 differs in the degree of splitting and in the relative intensities of some peaks. The sample of erbium in La_2O_3 (A-form) exhibits a weak absorption spectrum, but it is sufficient to see that the peaks do not appear to be highly split and that there are also changes in the relative intensities of the peaks.

Analysis of the X-ray data on these samples confirmed that all the samples exhibited the intended structure types. However, the concentration of the added lanthanide (6-8 mole %) may preclude the observance of a second phase in the X-ray films. The lattice parameters derived from the powder data were in accord with parameters expected by Vegard's Law for the incorporation of the lanthanide dopant into a particular sesquioxide host.

The intent of this work was to determine the feasibility of using absorption spectrophotometry at room temperature to differentiate between the structural forms of the lanthanide sesquioxides in anticipation of similar studies of transplutonium sesquioxides. The conclusion is that such absorption spectra can be used to distinguish between the sesquioxide polymorphs.

REFERENCES

1. J.P. Young, R.G. Haire, J.R. Peterson, D.D. Ensor, and R.L. Fellows, Inorg. Chem., 19:2209 (1980).
2. D.D. Ensor, J.R. Peterson, R.G. Haire, and J.P. Young, J. Inorg. Nucl. Chem., 43:1001 (1981).
3. R.G. Haire, J.P. Young, and J.Y. Bourges, The Rare Earths in Modern Science and Technology, Vol. II, G.J. McCarthy, J.J. Rhyne, and H.B. Silber (eds.), Plenum, New York, pp 159-165 (1980).
4. J.P. Young, R.G. Haire, R.L. Fellows, and J.R. Peterson, J. Radioanal. Chem., 43:479 (1978).
5. R.D. Baybarz and R.G. Haire, J. Inorg. Nucl. Chem. Suppl.:7 (1976).
6. A.F. Wells, Structural Inorganic Chemistry, Oxford University Press, London, pp 464-466 (1962).
7. J.R. Henderson, M. Muramoto, and J.B. Gruber, J. Chem. Phys., 46:2515 (1967).
8. P. Caro, J. Derouet, L. Beaury, and E. Soulie, J. Chem. Phys., 70:2542 (1979).

THE $Ce^{3+} \rightarrow Tb^{3+}$ TRANSFER IN PHOSPHATE HOST LATTICES

Philippe Bochu, Claude Parent, Abdelhamid Daoudi,*
Gilles Le Flem and Paul Hagenmuller, Laboratoire de
Chimie du Solide du CNRS, Université de Bordeaux I 351,
cours de la Libération, 33405 Talence Cedex, France

Phosphors showing $Ce^{3+} \rightarrow Tb^{3+}$ energy transfer are industrially used as green components in low pressure mercury vapor lamps (1).

The best yield for this transfer is obtained in oxides involving a good overlapping between terbium excitation and cerium emission, which is related indeed to the covalent character of Ce-O bonding (2,3).

A $Ce^{3+} \rightarrow Tb^{3+}$ transfer has been shown to appear in various phosphates belonging to different structural types and particularly in some new compounds: $Na_3Ln(PO_4)_2$:Ce, Tb; $Na_xSr_{3-2x}Ln_x(PO_4)_2$: Ce, Tb and $KCaLn(PO_4)_2$:Ce, Tb.

The presence of $[PO_4]$ groups, in which the P-O bonds are strongly covalent, leads to a relatively weak crystal field at the rare-earth sites. The best·overlapping between cerium emission and terbium excitation spectra is obtained for phosphates containing alkali ions. This result can be explained by a lowering of the emitting level of the cerium 5d configuration due to stronger covalency of the Ce-O bond and to an increasing crystal field at the rare-earth sites.

The full text of this paper is published in the <u>Materials Research Bulletin</u> (4).

---

*Permanent Address: Departement de Chime, Faculté des Sciences de l'Université Mohamed V, Avenue Ibn Batouta, Rabat, Morocco.

REFERENCES

1. J.M. Verstegen, Internationales Kongress fur Reprographie und
    Information (1975).
2. C. Parent, J. Fava, R. Salmon, G. Le Flem and P. Hagenmuller,
    Solid State Comm., 35:451 (1980).
3. C. Parent, G. Le Flem, M. Et-Tabirou and A. Daoudi, Solid
    State Commun. (in press).
4. P. Bochu, C. Parent, A. Daoudi, G. Le Flem and P. Hagenmuller,
    Mat. Res. Bull. 16:883-886 (1981).

LASER EXCITED LUMINESCENCE OF Tb^{3+}, Eu^{3+}-ACTIVATED FERROELECTRIC GADOLINIUM MOLYBDATE (GMO)

B.K. Chandrasekhar and William B. White

Materials Research Laboratory
The Pennsylvania State University
University Park, PA   16802

INTRODUCTION

Gadolinium molybdate, $Gd_2(MoO_4)_3$ (GMO), a ferroelectric-ferroelastic material exhibits interesting electrical and optical properties. Stimulated emission (1) has been obtained in this material with Nd^{3+} ions. However the spectroscopic properties of this material activated with Tb^{3+} and Eu^{3+}, are not well known. In this paper some anomalous luminescence properties of a single domain GMO crystal activated with Tb^{3+} and Eu^{3+} are discussed. The excitation was accomplished directly by the 4f-4f absorption lines of the rare earth ions using a linearly polarized argon laser.

EXPERIMENTAL

The experimental set-up consists of a double spectrometer mainly used in Raman spectroscopy. The emitted light was detected at right angles to the direction of the exciting light. The sample was rectangular with its edges parallel to the crystallographic axes. The sample was mounted on the cold end of an Air Products Heli-Tran system. The emitted light passed through an analyzer which can be rotated through 90 degress. The detector was a cooled RCA C31034 GaAs photomultiplier tube. The temperature of the sample was monitored directly by a digital thermometer.

GMO belongs to the orthorhombic system with space group Pba2 (C_{2v}^8) and contains four formula units per unit cell. The optic plane is parallel to (100) and by convention (2) the X,Y,Z, indicatrix axes coincide with the b,a,c, crystallographic axes

respectively. All spectral measurments were relative to the
indicatrix axes and Porto's notation (3) was used to denote the
orientation.

RESULTS AND DISCUSSION

Fig. 1 shows the emission spectrum of the GMO (Eu, Tb activated) crystal at 81.5 K when excited by the 488 nm (20492 cm^{-1})
laser line. The emission peaks are assigned as shown; the intense
peaks are electric dipole transitions.

The strong lines around 16525 cm^{-1} are assigned to the $^5D_1 \rightarrow$
$^7F_4$ transition of Eu^{3+} and will be discussed later. The total
number and/or width of the peaks are due to the low symmetry
environment of the activators which substitute for Gd^{3+} on both
sites (4). The relative intensity and shape of the strong peaks
were studied for different orientations of the crystal. It was
found that the intensity was maximized when the incident beam was
polarized along the Z direction.

The other group of intense lines around 18,425 cm^{-1} was
assigned to the $^5D_4 \rightarrow ^7F_5$ transition of Tb^{3+}. In order to further
resolve the spectra, measurements were taken below liquid nitrogen
temperature. However, we obtained the unexpected result that the
intensity drastically decreased and it was not possible to obtain
the polarized spectra. In Fig. 2 we have plotted the relative
integrated intensity of the $^5D_4-^7F_5$ transition of Tb^{3+} as a function of temperature without the analyzer. Also the other peaks
including the ones around 16525 cm^{-1} disappear in this temperature
range. Such a decrease was not observed when GMO was excited by
the 254 nm mercury line (5) which indicates that the observed
behavior is excitation wavelength-dependent. The peaks around
16525 cm^{-1} are not prominent under 254 nm excitation. The emission
intensity was very small when other lines of the argon laser were
used as excitation sources. In part, mis-match between the laser
line and the absorption level could result in loss of intensity
as the absorption line narrows at low temperature. However, a calculation of the intensity-temperature relation, assuming a gaussian
line shape for the absorption gives a function with opposite curvature to that observed in Fig. 2. The positions of the Stark levels
of the $^7F_6$ ground state are not known in the GMO structure and one
of them could be the level that is thermally populated. An alternative explanation for this low temperature behavior, preferred by
us, is that the excitation energy of 20492 cm^{-1} is slightly less
than needed to excite the $^5D_4$ state of Tb^{3+}. At higher temperatures, the thermal energy makes up for the deficiency. Candidate
phonons for this phonon-assisted transition are the A_1 symmetry
modes that appear at 43 and 50 cm^{-1} at k = 0 and produce intense

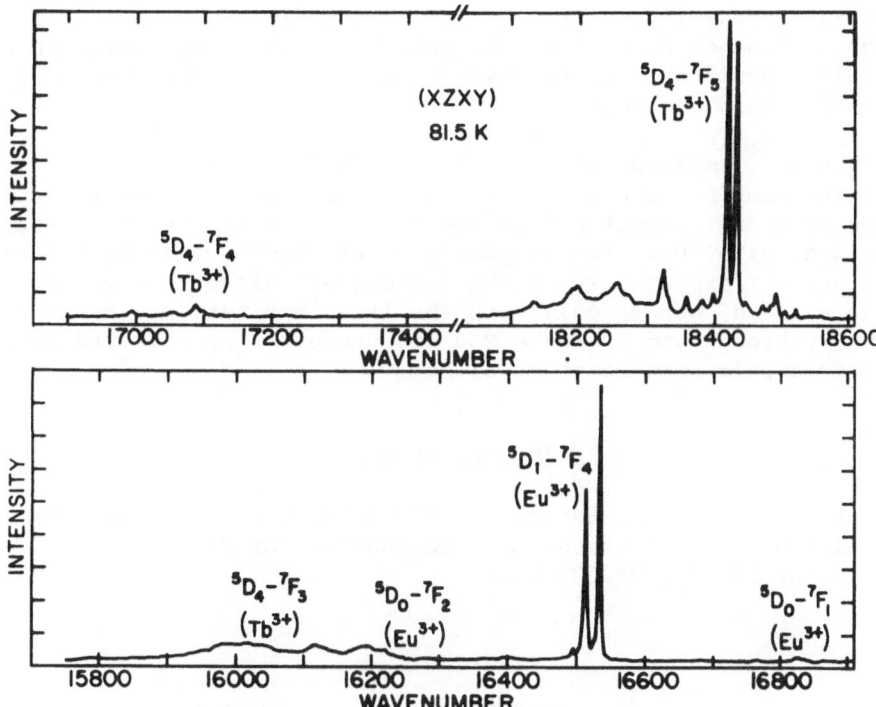

Figure 1.   Emission spectrum of gadolinium molybdate (Tb, Eu activated) excited by the 20,492 cm^{-1} Ar$^+$ laser line.

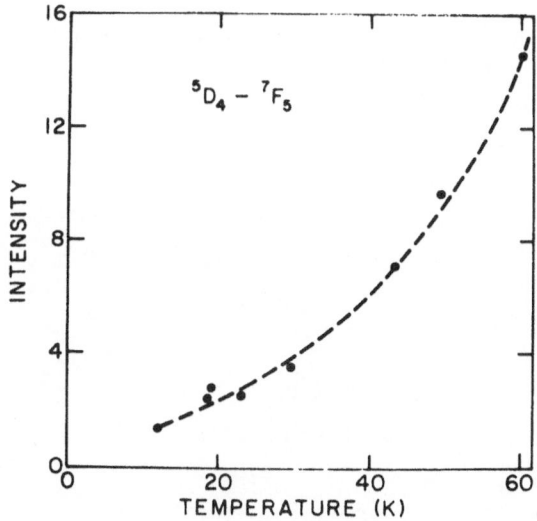

Figure 2.   Variation of intensity of the Tb^{3+} $^5D_4 \rightarrow {}^7F_5$ transition in GMO with temperature.

Raman scattering (6). Fitting a Boltzmann distribution to Fig. 2 yields a value of 23 cm^{-1} as the height of the energy level to be thermally populated. Experiments with a tunable laser are needed to confirm this conclusion.

The Eu^{3+} emission lines at 16,525 cm^{-1} also decrease at very low temperatures. These anomalous Eu^{3+} emissions appear to be pumped by energy transfer from the Tb^{3+} $^5D_4$ level which is nearly coincident with $^5D_1$. The dominance of the Eu^{3+} $^5D_1 \rightarrow {}^7F_4$ transition over the expected $^5D_0 \rightarrow {}^7F_2$ transition, although electric dipole allowed, occurs only with the laser excitation. Excitation of GMO by broad band UV (Hg – 254 nm) produces the expected forced electric dipole transition at 610 nm.

ACKNOWLEDGEMENTS

We thank Dr. A.S. Bhalla of this laboratory for providing the crystal and the National Science Foundation for financial assistance (Grant No. DMR-74-00340).

REFERENCES

1. A.A. Kaminskii, Soviet Physics-Crystallography, 17:194 (1972).
2. Akio Kumuda, Ferroelectrics, 3:115 (1972).
3. T.C. Damen, S.P.S. Porto and B. Tell, Phys. Rev. 142:570 (1966).
4. W. Jeitschko, Acta Cryst. B28:60 (1972).
5. B.K. Chandrasekhar, W.B. White and A.S. Bhalla, Mat. Res. Bull (in press).
6. F.C. Ullman, B.J. Holden, B.N. Ganguly and J.R. Hardy, Phys. Rev. B 8:2991 (1973).

LUMINESCENCE OF HIGH-PRESSURE PHASES OF Eu^{2+} ACTIVATED ALKALINE EARTH BORATES AND SILICATES

Gin-ya Adachi, Ken-ichi Machida and Jiro Shiokawa
Department of Applied Chemistry,
Faculty of Engineering, Osaka University
Yamadakami, Suita, Osaka, 565 Japan

ABSTRACT

Luminescence properties of the high-pressure phases of Eu^{2+} activated alkaline earth borates ($MB_2O_4:Eu^{2+}$) and silicates ($MSiO_3:Eu^{2+}$) were studied. Their quantum efficiencies increased strikingly with transformation from low pressure to high pressure phases. One of the high pressure forms of the metaborates, cubic $Sr_{1-x}Eu_xB_2O_4$ (x=0.01), was found to be an efficient photoluminescent material. The quenching effect owing to the interaction between the neighboring Eu^{2+} ions in the high-pressure phase is expected to be small compared with that in an atmospheric pressure phase. The same phase transformation effect was observed for Eu^{2+}-activated silicates, $Ca_{1-x}Eu_xSiO_3$ and $Sr_{1-x}Eu_xSiO_3$.

INTRODUCTION

Alkaline earth metaborates crystallize in several structures with varying pressure (Table 1). Metasilicates ($MSiO_3$) also undergo several high-pressure transformations. These crystallographic modifications cause deformation in the anion environment around Eu^{2+} ions which results in profound changes in the optical properties of the compounds. We report here the luminescence properties of the high pressure phases of Eu^{2+} doped SrB_2O_4, CaB_2O_4, $SrSiO_3$, and $CaSiO_3$.

RESULTS AND DISCUSSION

The atmosphere stable phases of the metaborates, SrB_2O_4 and

Table 1. High-pressure Polymorphism of CaB_2O_4 and SrB_2O_4

		Lattice parameters (Å)	
Phase	Symmetry	CaB_2O_4*	SrB_2O_4*
I	Orthorhombic	a= 6.214 b=11.604 c= 4.284	a= 6.589 b=12.018 c= 4.337
	12kb		
II	Orthorhombic	a= 8.369 b=13.816 c= 5.007	8kb ———
	15kb		
III	Orthorhombic	a=11.380 b= 6.382 c=11.304	a=12.426 b= 6.418 c=11.412
	25kb		15kb
IV	Cubic	a= 9.008	a= 9.222

*Treatment temperature = 600°C for SrB_2O_4 and 900°C for CaB_2O_4.

CaB_2O_4, are easily transformed into several high-pressure phases. The triangularly coordinated borons in the borates are partially or completely changed into tetrahedrally coordinated borons by the high-pressure treatment. Table 1 shows the lattice parameters and symmetry for the resulting phases of the metaborates (1).

High-pressure treatments and luminescence measurements were made on the Eu^{2+} activated metaborates, $SrB_2O_4:Eu^{2+}$ and $CaB_2O_4:Eu^{2+}$. The phases of $Sr_{0.99}Eu_{0.01}B_2O_4$ resulting from treatments at 15-30 kbar and 600°C and their luminescence data are shown in Table 2. The treatment pressure dependence of the quantum efficiency of $Sr_{0.99}Eu_{0.01}B_2O_4$ is shown in Fig. 1.

It is notable that the quantum efficiency increases by a factor of about 40 under 254 nm excitation and about 60 under the optimum excitation (313 nm) for phase IV with transformation from phase III to phase IV. In Fig. 2, we summarize the Eu^{2+} concentration quenching effect on the quantum efficiency of $Sr_{1-x}Eu_xB_2O_4$ (IV). These results are interpreted in terms of crystal structures. The phases I, III, and IV consist of (BO_2) chains, a (B_6O_{12}) network and a (B_3O_6) network, respectively (Fig. 3). Packing of the (B_3O_6) network in phase IV becomes denser by the high-pressure treatment so that anions vibrations allow the excitation energy loss by radiationless processes to become small. The second effect responsible for the increase in the luminescence intensity may be due to

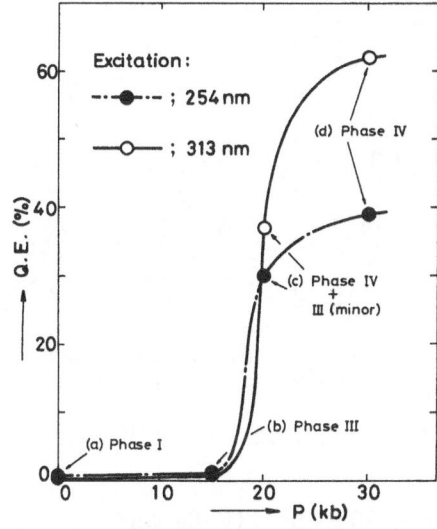

Fig. 1. Quantum efficiency vs
         treatment pressure for
         $Sr_{0.99}Eu_{0.01}B_2O_4$.

Fig. 2. Quantum efficiency vs
         Eu^{2+} content (x) for the
         phases of $Sr_{1-x}Eu_xB_2O_4$:
         ▲, I; ● and O, IV.

Table 2. Phase Formation and Luminescence Data of $Sr_{1-x}Eu_xB_2O_4$
          from High-Pressure Treatments

	Phase	p kbar	λ max nm	QE %
a	I		367	1
b	III	15	367	1
c	III+IV	20	404	30
d	IV	30	395	39

Treatment temperature = 700 C.   QE = quantum efficiency under
254 nm excitation at 300 K.

a Sr^{2+} (Eu^{2+}) arrangement in phase IV, where the number of neighboring Sr^{2+} (Eu^{2+}) acting like a cation pair in a magnetic
interaction is small. In phase I, cations (Sr^{2+} and Eu^{2+}) are in
a line and the Eu^{2+}-Eu^{2+} energy transfer can occur easily. The
third effect is based on dissociation of Eu^{2+} clusters which seem
to exist in phase I.

Strontium silicate, $SrSiO_3$, crystalizes with three structures,
α´, δ and δ´ according to applied pressures and temperatures (2).

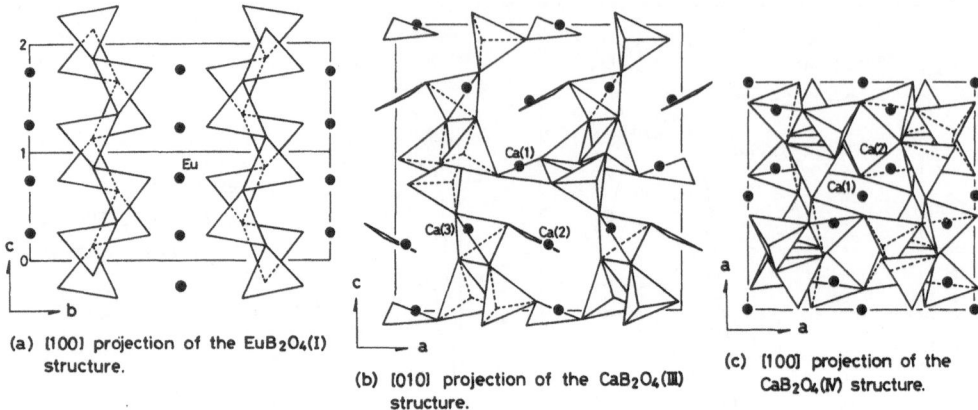

(a) [100] projection of the EuB_2O_4(I) structure.

(b) [010] projection of the CaB_2O_4(III) structure.

(c) [100] projection of the CaB_2O_4(IV) structure.

Fig. 3. Structures for the high-pressure phases of $M^{II}B_2O_4$.

Table 3. Luminescence Properties of $MSiO_3:Eu^{2+}$ (M=Ca and Sr)

	Treatment P(kb)	Temp.(°C)	Phase Symmetry		λ max (nm)	$\lambda I/2$ (nm)	QE (%)
$CaSiO_3$	40	1300-800	α	monoclinic	507	70	weak
			δ	triclinic	472	83	14
$SrSiO_3$	40	1300-800	α	monoclinic	-	-	-
			δ	triclinic	506	67	6
	55	1400-800	δ'	monoclinic	466	59	27

Quantum efficiencies under optimized Eu^{2+} concentrations and excitation.

The luminescence properties of these phases are summarized in Table 3. Apparently the phase transformation cause the relative quantum efficiencies of the emission to increase. The increase in the intensities is related to the structural framework and the Eu coordination to oxgyen.

REFERENCES

1. M. Marezio, J.P. Remeika and P.D. Dernier, "The crystal structure of the high pressure phase, CaB_2O_4(IV), and polymorphism in CaB_2O_4", Acta Cryst., B25:965 (1969).
2. A.E. Ringwood and A. Major, "Some high-pressure transformations of geophysical significance", Earth and Planetary Sci. Lett., 2:106 (1967).

PMR SPECTROSCOPIC STUDIES OF LANTHANIDE AMINOCARBOXYLATE COMPLEXES

G. R. Choppin, P. A. Baisden and E. N. Rizkalla

Department of Chemistry, Florida State University

Tallahassee, FL 32306

INTRODUCTION

Based on PMR spectra at 60 MHz in which doublets were observed for the acetate protons at low temperature (1,2), it was proposed that EDTA is pentachelated for the lighter lanthanides and hexachelated for the heavier ones. We had observed (3) a single AB quartet pattern for the acetate protons of YEDTA and LuEDTA, confirming that all 4 acetate groups were equivalently bonded. Such an AB pattern is associated with short (on NMR scale) lifetimes for the metal-carboxylate bonds and long lifetimes for the metal-nitrogen bonds (4) confirming the bonding of both nitrogens in EDTA. The simplest interpretation is hexachelation.

A somewhat broadened acetate singlet, indicating short M-O and short M-N bond lifetimes was observed at 90 MHz and room temperature for the acetate protons in $LaEDTA^{-1}_{(aq)}$. This singlet is not in agreement with the reported doublet for LaEDTA (1) using a 60 MHz instrument nor it it explained by the proposal of pentachelation involving an unbonded acetate group in LaEDTA.

To resolve this uncertainty in the chelation by EDTA of lighter lanthanides we measured the PMR spectra of $CaEDTA^{-2}$, $LaEDTA^{-1}$, $YEDTA^{-1}$ and $LuEDTA^{-1}$ at 0°, room temperature (ca. 30°) and 85 °C. We also studied the PMR spectra of the ternary complexes $La(HEDTA)(IDA)^{-2}$ and $Lu(HEDTA)(IDA)^{-2}$ to learn if the binding of IDA affects the chelation of the LnHEDTA complex.

187

RESULTS AND DISCUSSION

The preparation of the solutions in D_2O at pH 7-8 (La, Y, Lu) and 8-9 (Ca) followed that in ref. 3. The spectra were measured using the pulsed Fourier transform mode with a deuterium lock on the Florida State University Bruker 270 MHz Spectrometer with DDS and acetone as internal standards. Solutions with Na^+ and with Li^+ as the counter cations were measured. No significant dependence on the counter cation was observed.

For the EDTA complexes, an AB quartet pattern was observed for the acetate protons in all the solutions at all three temperatures (Figure 1(a)) except for CaEDTA at 85 °C which had a broadened singlet. The ethylenic protons all showed a singlet and that of the baricenter of the acetate quartet agreed with our earlier room temperature values (3) and showed no significant temperature dependence. The separation of the two inner peaks of the quartets were, at room temperature (Hz): Ca, 4.9; La, 5.4; Y, 43; Lu, 49. For all the complexes, this separation decreased with increasing temperature.

The observation of a quartet in these high resolution spectra confirms that both nitrogen donors are bonded and all 4 carboxylate groups are equally bonded in CaEDTA as well as in all lanthanide-EDTA complexes. There is no evidence for non-equivalency in the carboxylate bonding as expected for true pentachelation. The quartet separation and the temperature dependence are consistent with decrease in the metal-nitrogen bond lifetimes with increase in temperature. The relative values for the metal-nitrogen bond lifetimes are:

$$Lu \approx Y >> La \approx Ca$$

For all the systems the M-N bond lifetime is longer than the NMR time window except for CaEDTA at 85 °C. Misidentification of AB quartets as doublets due to inadequate resolution at 60 MHz led to the earlier proposal of pentachelation in the LnEDTA complexes.

In the study of the ternary complexes, spectra of La(HEDTA) (IDA) and Lu(HEDTA)(IDA) were obtained in a slight excess of lanthanide (La/HEDTA $\geq$ 1) at 1:1 and 1:2 mole ratios of HEDTA;IDA. The pH was varied in 0.5 units from 6.3 to 9.3. Using the reported stability constants (5), we estimate the ratio Ln(HEDTA): Ln(HEDTA)(IDA) varied from 6.8 at pH 6.3 to 0.05 at pH 9.3. Spectra of La-HEDTA-IDA solutions are shown in Figure 1(b) with the three quartets, X,Y,Z, associated with the acetate protons in La(HEDTA) (3) diagrammed in (b). Little pertubation of the La(HEDTA) spectra are observed as ternary complexation occurs. For the protons of IDA, a single shift as observed which shifts

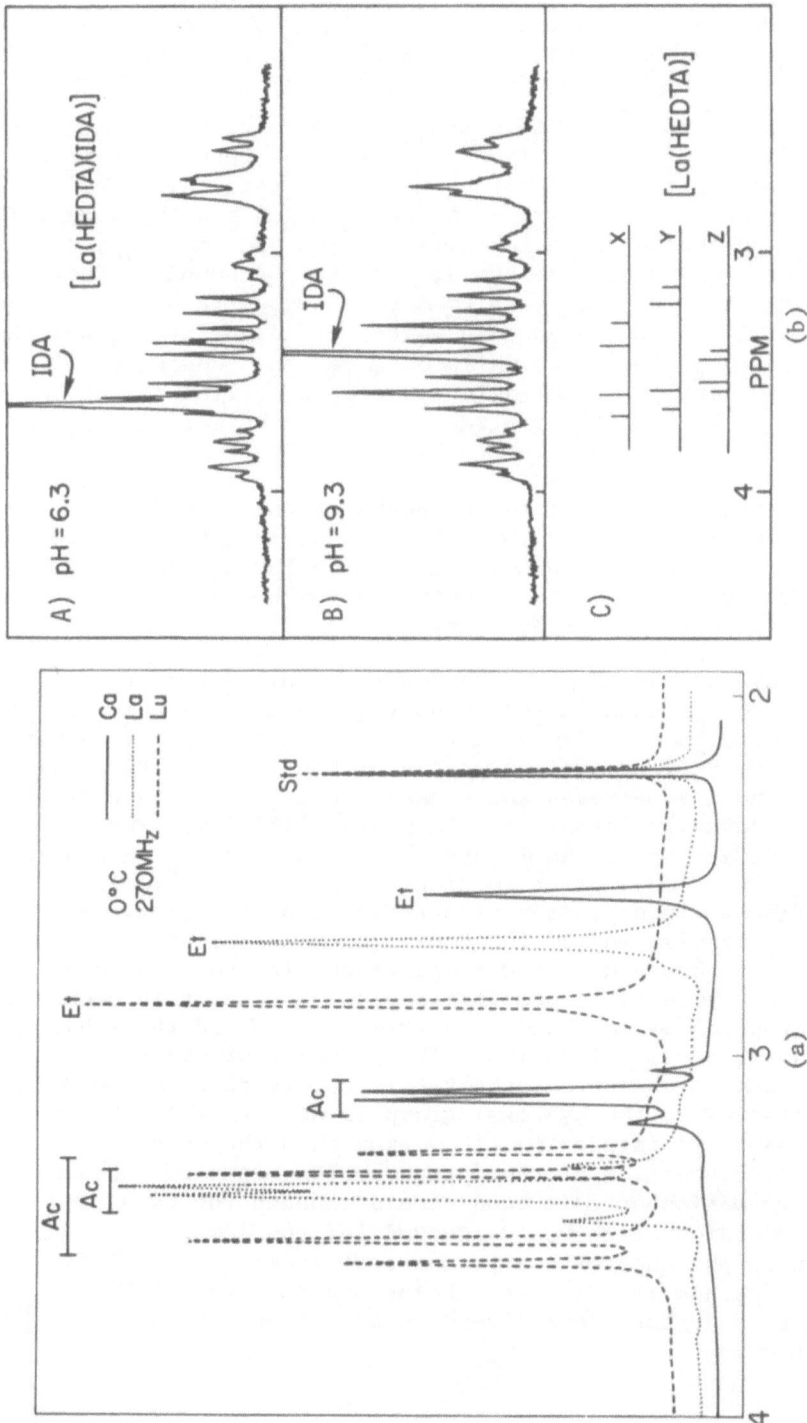

Figure 1. (a) Spectra of CaEDTA, LaEDTA and LuEDTA at 0°C. (b) Spectra of 1:1:1 solutions of La(III), HEDTA and IDA at pH 6.3 (a) and 9.3 (b). The quartet pattern of La(HEDTA) is shown in (c).

from 3.64 ppm at pH 6.3 (the same shift was observed for uncomplexed IDA at this pH) to 3.43 ppm at pH 9.3 (compared to 3.35 ppm for free IDA). Computer analysis gives a limiting shift of 3.40 ppm for IDA bound in the ternary complex.

The spectra of Lu(HEDTA) + IDA solutions are different than those of the La solutions. There is a loss of resolution in the HEDTA spectra due to broadening of the X quartet. Also, two peaks are observed due to IDA - an $IDA_{(f)}$ which has the same shift values with pH as the uncomplexed (free) IDA and an $IDA_{(b)}$ which is invariant with pH at 3.35 ppm and is the IDA bound in the ternary complex. This was confirmed by the spectra of the solutions of 1:2 stoichiometry in which the intensity but not the shift of the $IDA_{(f)}$ increased while both remained constant for $IDA_{(b)}$. The presence of a single, pH dependent line in the La+ HEDTA + IDA systems reflects a rapid exchange between $IDA_{(f)}$ and $IDA_{(b)}$.

By contrast, the separate IDA peaks of the Lu + HEDTA + IDA solutions are due to a slow (by PMR timescale) exchange. Study of the Ln (paramagnetic)-EDTA-IDA system by PMR (6) gave similar results with the order of IDA exchange rates being: Pr (too fast to measure) > Eu > Yb.

Ternovaya and Kostromina (7) reported collapse of the quartet pattern of EDTA in the ternary La(EDTA)(IDA) complex which would indicate libility of the La-N bonds of EDTA. However, our spectra of HEDTA as well as those of the Ln-EDTA-IDA complexes show sharp quartet patterns which means that the presence of IDA does not increase noticeably the Ln-N bond lability. The most noticeable effect in the HEDTA spectra is the loss of resolution of the X quartet lines by broadening in the ternary complex of Lu(HEDTA)(IDA). This quartet is attributed to the single acetate group on the ethyoxy end of the HEDTA molecule. Such a loss of resolution was observed at low temperatures in the paramagnetic Ln(EDTA)(IDA) systems. A possible explanation for this broadening is the disruption of the bonding between the Lu and the ethanolic group upon ternary complexation. The presence of IDA does not affect the analogous acetate quartet in the lanthanum complex where no bonding to the hydroxyl group is assumed (3). Since the metal-hydroxyl interaction is weaker than the metal-carboxylate, disruption of the former by IDA seems reasonable. Further, disruption of this bond should enhance the rate of nitrogen inversion which would accound for the loss of resolution in the quartet due to the acetate protons. These effects strengthen the earlier conclusions (3) that HEDTA functions as a pentachelate ligand to La but as a hexachelate one to Y and Lu.

This research was supported by USDOE through a contract with F.S.U. P.A.B. is presently at Lawrence Livermore National Laboratory. E.N.R. is on leave from Ain Shams U., Cairo, Egypt.

REFERENCES

1. E. Nieboer, Structure and Bonding, 22, 1(1975).
2. G.A. Elgavish and J. Reuben, J. Am. Chem. Soc., 98, 4775(1976)
3. P.A. Baisden, G.R. Choppin and B.B. Garrett, Inorg. Chem., 16, 1367(1977).
4. R.J. Day and C.N. Reilley, Anal. Chem. 36, 1073(1963), 1367(1977).
5. L.C. Thompson and J.A. Loraas, Inorg. Chem. 2, 89(1963).
6. R.V. Southwood-Jones and A.E. Merbach, Inorg. Chim. Acta, 30, 135(1978).
7. T.V. Ternovaya and N.A. Kostromina, Russ. J. Inorg. Chem., 18, 1266(1973).

THE INTERACTION OF 1:1 AND 1:2 LANTHANIDE-EDTA CHELATES WITH ALKALI

CATIONS: AQUEOUS RELAXATION REAGENTS FOR METAL NUCLIDE NMR

Gabriel A. Elgavish, Isotopes Department, The Weizmann
Institute of Science, Rehovot, Israel. Present Address:
Comprehensive Cancer Center, University of Alabama in
Birmingham, Birmingham, AL 35294

Nuclear Magnetic Resonance (NMR) Spectroscopy of alkali-metal nuclides, in particular of ^{23}Na, has increasingly become a valuable tool to probe into the solution chemistry and biology of the mono-valent alkali cations (1,2). Sodium is highly abundant in biological fluids and is essential to many life sustaining processes. Its selective transport across membranal barriers plays a fundamental role in cellular activity. Therefore, studies of sodium transport in such systems is essential for the understanding of many physiological phenomena. In this domain ^{23}Na NMR may have a unique contribution due to its noninvasive nature, the limits of accessible time scale which are lower than by other methods and the straightforward interpretation of the data in terms of kinetic constants. There was, however, a serious obstacle in utilizing this potential because a differentiation in the magnetic environments of the observed nuclide inside and outside the cells was needed. Here, the special magnetic properties of the lanthanides to enhance greatly the NMR relaxation rates of nuclei in their vicinity might be the solution. For this purpose an appropriate cation-binding lanthanide complex has to be used. A method based on the utilization of the Gd-EDTA chelate to follow ionophore mediated sodium and lithium transport across the membrane of phospholipid vesicles was therefore employed (3). In those experiments the solution outside the vesicles was doped with Gd-EDTA which by virture of its binding of sodium cations broadened out the extravesicular ^{23}Na signal. The intravesicular sodium signal, however, was not affected and therefore could be separated from the now broad one from the outside. The effect of an ionophore on the exchange kinetics could then be conveniently monitored via this separated sharp signal by conventional NMR methods.

It is evident from the above that for the Gd-EDTA to fulfill

optimally the role of relaxation reagent for metal cations the
equilibrium conditions of complexation in terms of stoichiometries,
affinities and pH dependence better be characterized. To accomplish
this, measurements of longitudinal relaxation rates $(1/T_1)$ of ^{23}Na
and ^{7}Li in aqueous solutions of NaCl and LiCl have been carried out
as a function of the EDTA–Gd concentration ratio and also as a
function of solution pH. Analogous experiments have also been done
on the same systems with the diamagnetic lanthanum to separate
between the paramagnetic dipolar and the diamagnetic quadrupolar
contributions of the gadolinium chelates to the enhancement of the
sodium and lithium relaxation rates. Figure 1 shows the observed
$1/T_1$ of ^{7}Li versus EDTA/Gd with the Gd^{3+} to Li^+ ratio held constant
at 0.025. There is an abrupt increase of slope at EDTA/Gd = 1.0
and a sharp leveling off at EDTA/Gd = 2.0. The formation of a
$Gd(EDTA)_2^{5-}$ chelate following the step of Gd(EDTA) formation is indicated. It seems that the overall relaxation enhancement effect

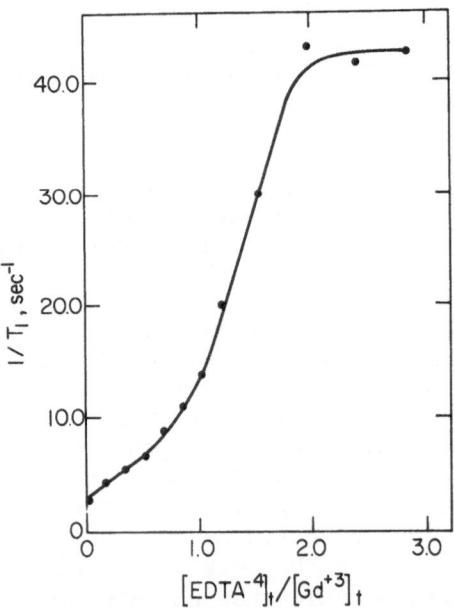

Figure 1   The observed longitudinal relaxation rate $(1/T_1)$ of ^{7}Li
            at 34.98 MHz in aqueous solution of 0.56 M LiCl with
            0.014 M GdCl$_3$ versus the EDTA to GdCl$_3$ total concentra-
            tion ratio. The pH was held constant at 5.4. The smooth
            curve has no theoretical meaning.

Table 1. The normalized gadolinium induced longitudinal relaxation
rates (R_i) of ^{23}Na and ^{7}Li.

gadolinium complex	R_i, sec^{-1}	
	^{23}Na	^{7}Li
Gd^{3+}_{aq}	180[a]	110[a]
$Gd(EDTA)^-$	430[a]	450[a]
$Gd(EDTA)_2^{5-}$	1320[a] 980[b]	1800[a] 1700[b]

[a]Values calculated from the slopes in Fig. 1 and from similar
slopes for the sodium data.

[b]Values calculated from the plateau in Fig. 1 and from a similar
plateau for the sodium data.

of the former on Li$^+$ is significantly larger than of the latter.
This may be caused either by the increase of the intrinsic dipolar
effect, or by a higher fraction of Li$^+$ bound to the 1:2 chelate, or
both. An experiment on the sodium system has given similar results.
Since in these systems the alkali cations exchange fast among their
unbound and various bound forms, Eq. 1 would give the total observed
relaxation rate,

$$1/T_1 = 1/[Na^+]_t \cdot \sum_i [Na^+]_i R_i \qquad i = 1,2,3,4 \qquad (1)$$

where $[Na^+]_i$ are the equilibrium concentrations of sodium as the
unbound aquoion and as the fractions that interact with Gd^{3+},
Gd(EDTA)$^-$ and Gd(EDTA$_2$)$^{5-}$. $[Na^+]_t$ is the total sodium concentration
and R_i are the corresponding normalized relaxation rates of ^{23}Na in
its various forms. The same expression is valid for the Li system.
Analysis of the data of Fig. 1 and similar data of the sodium system
yields the normalized relaxation rates of sodium and lithium induced
by the various species of gadolinium. These values are compiled in
Table 1. For comparison note that the relaxation rates of free
aquosodium and aquolithium are 19 sec^{-1} and 0.04 sec^{-1}, respectively.
The sizable effects induced by the aquogadolinium cation are probably due to nonspecific interactions since one can assume that
direct cation-cation binding is highly improbable. When these
nonspecific effects are subtracted from the relaxation rates induced
by each of the chelates, an approximately four-fold larger effect is
evident for the 1:2 chelate relative to Gd(EDTA)$^-$ both for sodium
and for lithium. Previous results on the interaction of Ln(EDTA)$^-$
with substituted ammonium cations (4) were interpreted in terms of
a single acetate arm out of the total of four which were not bound
to the lanthanide metal ion and therefore free to bind other
cations. On the other hand, extraneous acetate did readily bind to

$Ln(EDTA)^-$. Therefore it seems that this expansion of the lanthanide coordination sphere is favored thermodynamically. It is due to steric effects that the fourth arm of EDTA in the 1:1 chelate cannot be accommodated by the lighter lanthanides with their larger ionic radius. Freely moving outside carboxylates, however, would bind. Based on these considerations and on the data presented in this paper we would like to suggest that a second EDTA molecule interacts with $Ln(EDTA)^-$ in the above manner to form $Ln(EDTA)_2^{5-}$. In such a 1:2 chelate one would expect, based on symmetry considerations, that the two identical EDTA molecules rearrange positions around the lanthanide cation each binding to the metal through two acetate arms. This con-figuration would leave a total of four acetate arms, two from each EDTA, free to bind additional cations. If the distance of these four presumably equivalent cation-binding sites from the Gd^{3+} ion is not different from that of the single site in the 1:1 chelate, then one could expect the relaxation rate enhancement induced by the latter to be one fourth of the effect caused by the 1:2 chelate. Since this has indeed been the experimental ratio, it is therefore highly suggestive that the above configuration of the $Gd(EDTA)_2^{5-}$ chelate and its mode of interaction with cations are correct. The dissociation constant for the $Gd(EDTA)_2^{5-} \rightleftarrows Gd(EDTA)^- + (EDTA)^{4-}$ equilibrium has been calculated from the type of data presented in Fig. 1 and found to be 2 mM from the sodium results and 9 mM from the lithium data. Therefore, these values should be regarded as order of magnitude indication only.

The pH dependence of the ^7Li relaxation rate affected by $Gd(EDTA)_2^{5-}$ is depicted in Fig. 2. Alongside with it the effect of the analogous lanthanum chelate is also shown. The low pH plateau (below ∿pH 5) and the high pH plateau (∿pH 9-11) are separated by a sharp titration with pK ∿8 in the lanthanum system while the gadolinium analog reveals a titration spread out over four pH units. The source of this difference is yet unresolved. In general, the shape of these pH profiles is not fundamentally different from those of the 1:1 chelate interacting with Na^+ and Li^+ and with tetramethyl ammonium (4). From this similarity of the pH profiles of the 1:2 and the 1:1 chelate systems one can conclude that there is no pH dependence of the binding of the second EDTA molecule. Therefore it seems that as in the 1:1 chelate, the higher pH plateau in the 1:2 system is also due to the formation of the hydroxo complex of $Ln(EDTA)_2^{5-}$ thereby increasing the overall negative charge density of the chelate. This in turn might well increase the chelate-bound fraction of the monovalent cation and thus bring about the 3-4 fold larger relaxation rate enhancement by the high pH chelate relative to that of the low pH region. Due to this larger effect the pH range of 9-11 is recommended for optimal effect, whenever it is attainable.

Comparison of the lanthanum pH profile with that of gadolinium also reveals some information about the difference of mechanisms by which these two enhance the alkali metal relaxation rates. The gadolinium effect on [7]Li, as shown in Fig. 1, is more than two orders of magnitude larger than the one caused by lanthanum. A similar ratio has been obtained with the 1:1 chelates and [7]Li. With [23]Na, however, the Gd/La ratio of the 1:1 chelates has been only about a factor of three. This difference is quite expected if one assumes that the lanthanum chelates merely distort the symmetry of the bound alkali cations' coordination sphere relative to the aquocations.

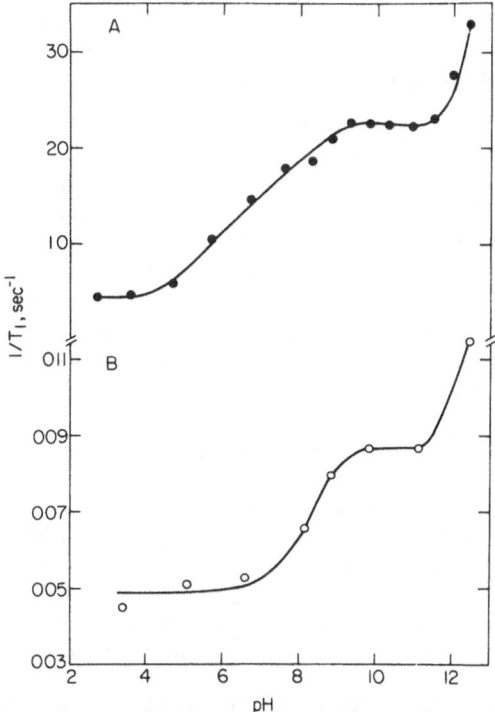

Figure 2   The pH dependence of the longitudinal relaxation rate
           $(1/T_1)$ of [7]Li at 34.89 MHz in aqueous solutions of 0.56
           M LiCl with 0.005 M $GdCl_3$ (A) and with 0.005 M $LaCl_3$ (B).
           In both cases a threefold concentration of EDTA was
           present. The smooth lines have no theoretical meaning.

This would increase the electric field gradient of the latter causing an enhancement of their already existing quadrupolar relaxation mechanism by a certain factor. Since the relaxation rate of [23]Na as the free aquocation is already almost three orders of magnitude larger than that of [7]Li, the La-EDTA enhanced relaxation rate of sodium is larger than that of lithium by about the same ratio. The paramagnetic dipolar mechanism would be affected by the replacement of sodium by lithium in an entirely different manner. Assuming equal distances from Gd of the two alkali cations, the ratio of their dipolar relaxation rate enhancement would be the ratio of their gyromagnetic ratios squared. For ^{23}Na/^{7}Li this is 0.46. Therefore the purely dipolar mechanism might cause an enhancement which is in fact two fold larger for lithium. The experimental ^{23}Na/^{7}Li factor for Gd(EDTA)$^-$ has been 0.7-0.8. Therefore, it is concluded that while the Gd effects on ^{7}Li are predominantly dipolar, there is a substantial quadrupolar contribution to the ^{23}Na relaxation rate. Thus in the event of employing gadolinium enhanced relaxation rates for distance determinations, ^{7}Li data may be straightforwardly utilized. Sodium data, on the other hand, may have to be corrected for the quadrupolar contribution by lanthanum measurements.

In summary, lanthanide-EDTA chelates bind alkali metal cations. The 1:2 chelates are readily formed with an excess of EDTA present and this increases the number of cation binding sites from one to four. The enhancements in the NMR relaxation rates of the alkali metals are concomitantly increased. The pH range of 9-11 is recommended for maximum effect. When qualitative usage of the relaxation reagent is sought, e.g. lifting the magnetic degeneracy between two compartments (3), Gd-EDTA is sufficient both for sodium and lithium. For possible structural studies, however, La-EDTA corrections are needed for sodium systems.

REFERENCES

1. P. Laszlo, Angew. Chem., 17:254 (1978).
2. B. Lindman and S. Forsen, "The Alkali Metals," in NMR and the Periodic Table, R.K. Harris and B.E. Mann (eds.), Academic Press, London, pp. 129-181 (1973).
3. H. Degani and G.A. Elgavish, FEBS Lett., 90:357 (1978).
4. G.A. Elgavish and J. Reuben, J. Amer. Chem. Soc., 99:1762 (1977).
5. G.A. Elgavish and J. Reuben, J. Amer. Chem. Soc., 98:4755 (1976).

LUMINESCENCE OF L-(-)-TRYPTOPHAN IN DMSO AT 77 K IN THE PRESENCE

OF Tb^{3+}

V. Anantharaman and J. Chrysochoos

Department of Chemistry, The University of Toledo
Toledo, Ohio 43606

INTRODUCTION

The luminescence properties of tryptophan and its derivatives exhibit a strong dependence upon the nature of the microenvironment of the tryptophan moiety. Such a dependence can be exploited leading to information regarding the structural characteristics of systems containing tryptophan residues. The fluorescence and phosphorescence spectra of tryptophan and derivatives in solution depend upon the polarity and the viscosity of the solvent as well as upon the excitation wavelength (1-4). Due to the strong interaction of tryptophan with its microenvironment, the observed luminescence properties are often different from those of the unperturbed molecule.

Unperturbed optical properties such as energies of excited singlet and triplet states as well as radiative lifetimes and quantum yields are necessary for the use of tryptophan as a probe of the microenvironment. Such unperturbed values may be obtained if tryptophan is coupled strongly with a spectroscopic probe characterized by well-defined optical properties unaffected by the microenvironment. Lanthanide ions provide such a probe since their narrow and well-defined spectra are virtually independent of the environment. Furthermore, the magnetic properties of the lanthanide ions render them even more suitable as probes, because appropriate magnetic perturbations upon the molecule under consideration can be exploited.

Preliminary studies dealing with interactions of Eu^{3+} with excited singlet and triplet states of tryptophan have been reported (5). The present report deals with interactions of the triplet state(s) of tryptophan with Tb^{3+}.

EXPERIMENTAL

The luminescence spectra and lifetimes of tryptophan were measured using the Aminco Bowman spectrophotofluorimeter at room temperature and the phosphoroscope at liquid N_2 temperature. $TbCl_3.6H_2O$, of 99.9% purity was converted to anhydrous $TbCl_3$ and was dissolved in spectroscopic grade DMSO. Spectroscopic grade L-(-)-Tryptophan (Gold label) was used with no further purification.

RESULTS AND DISCUSSION

Typical luminescence spectra of L-(-)-Tryptophan in DMSO as 77 K in the absence and presence of Tb^{3+} are shown in Fig. 1. The spectra consist of a fluorescence band with a maximum at $340^{\pm}$ 1 nm (the narrow peak at 360 is scattered light), a phosphorescence envelope with maxima at $415^{\pm}1, 437^{\pm}1$ and $465^{\pm}1$ nm and several sensitized fluorescence bands (in the presence of Tb^{3+}) monitored at $490^{\pm}1$ nm $(^5D_4 \rightarrow {}^7F_6)$ and $546^{\pm}1$ nm $(^5D_4 \rightarrow {}^7F_5)$.

Light excitation of tryptophan at $261^{\pm}1$ nm $(^1L_a \leftarrow {}^1A_0)$ leads to an increased phosphorescence emission compared to that under excitation at $291^{\pm}1$ nm $(^1L_b \leftarrow {}^1A_0)$. Both the fluorescence and phosphorescence intensities decrease in the presence of Tb^{3+}. The luminescence spectrum of tryptophan exhibits similar characteristics at room temperature. The phosphorescence envelope is completely absent under such conditions.

Fluorescence quenching of tryptophan by Tb^{3+} expressed in terms of the ratio $(I_F)_0/(I_F)_{Tb^{3+}}$, as well as the intensity of the sensitized fluorescence of Tb^{3+}, in terms of $(I_F)_{545}/(I_F)_{340}$, are plotted against the concentration of Tb^{3+} in Fig. 2 at 77 K. Similar results obtained at room temperature are not shown here. The fluorescence intensity of tryptophan decreases initially in the presence of low $[Tb^{3+}]$ at both room temperatures and 77 K. After this initial reduction the fluorescence of tryptophan exhibits no further dependence upon $[Tb^{3+}]$.

The phosphorescence lifetime of tryptophan varies randomly from 5.8 to 7.8 sec. for several tryptophan concentrations ranging from 5×10^{-5} to 2.5×10^{-4} M. An average value of τ_{ph} of about 7.3 sec. was obtained in 1×10^{-4} M tryptophan (at 261 and 291 nm), compared with a corresponding phosphorescence lifetime of tryptophan in EG-W of 6.8 sec. (5). Furthermore, the value of τ_{ph} of 1×10^{-4} M tryptophan in the presence of up to 5×10^{-3} M Tb^{3+} also varies randomly from 5.8 to 8.0 sec. Therefore, the phosphorescence lifetime of tryptophan appears to be independent of $[Tb^{3+}]$ (Fig. 3). It is also independent of the concentration of tryptophan which rules out quenching of the triplet by the ground state of tryptophan.

The apparent independence of τ_{ph} upon $[Tb^{3+}]$ in conjunction with the significant phosphorescence quenching observed (Fig. 4), indicates the existence of two excited triplet states. A lower energy triplet state which emits phosphorescence but does not interact with Tb^{3+}, and a higher energy one which does not emit, feeds the lower triplet and interacts with Tb^{3+}. The initial quenching of fluorescence by Tb^{3+} indicates an enhanced intersystem crossing induced by Tb^{3+}

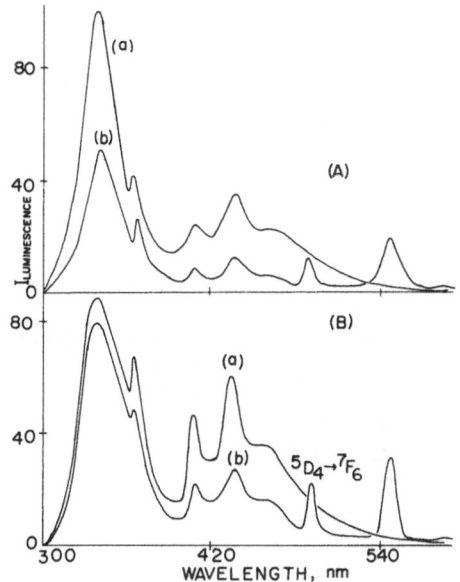

Figure 1. Luminescence spectra of 1×10^{-4} M L-(-)-Tryptophan in DMSO at 77 K. (A) $\lambda_{exc}=291^{\pm}1$ nm; (B) $\lambda_{exc}=261^{\pm}1$ nm; (a) No Tb^{3+} (b) 4×10^{-3} M Tb^{3+}.

Figure 2. Fluorescence quenching of 1×10^{-4} M Tryptophan by Tb^{3+} at RT (B) and sensitized fluorescence of Tb^{3+} (A); ● λ_{exc} $261^{\pm}1$ nm, ○ λ_{exc} 291-1 nm.

Excitation of tryptophan at $291^{\pm}1$ nm leads to results which may be accounted for via the following spectroscopic mechanism.

$$^1A_0 \longrightarrow 291^{\pm}1 \text{ nm} \longrightarrow {}^1L_b$$
$$^1L_b \rightarrow {}^1A_0 + h\nu_{fl} \qquad\qquad ;k_{fl}$$
$$^1L_b \rightsquigarrow {}^3L_a \qquad\qquad ;k_{isc}$$
$$^1L_b + Tb^{3+} \rightarrow {}^3L_a + Tb^{3+} \qquad ;k_{isc}^{ind}$$
$$^3L_a \rightsquigarrow {}^3L_b \qquad\qquad ;k_{ic}$$
$$^3L_a + Tb^{3+} ({}^7F_6) \rightarrow {}^1A_0 + Tb^{3+} ({}^5D_j) \qquad ;k_{ET}$$
$$^3L_b \rightarrow {}^1A_0 + h\nu_{ph} \qquad\qquad ;k_{ph}$$

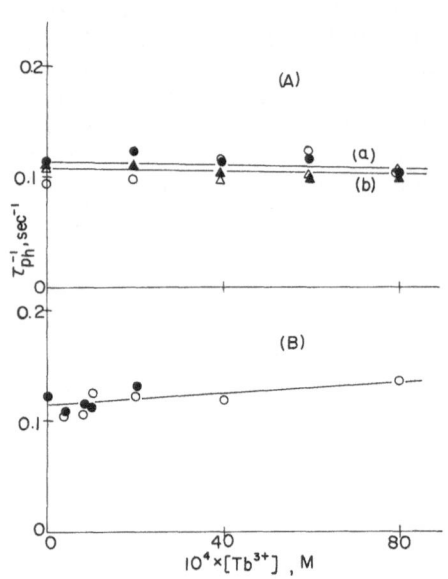

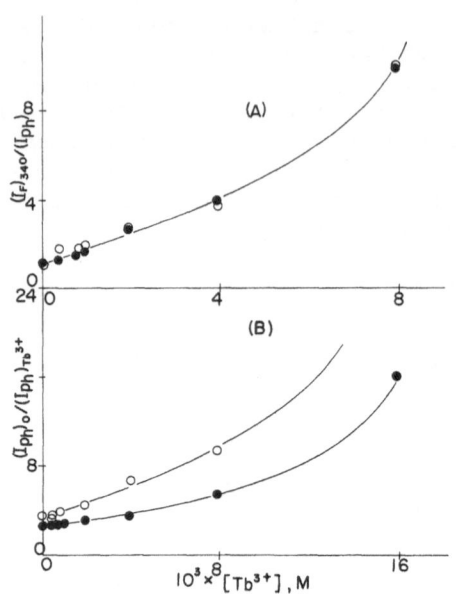

Figure 3. Variation of τ_{ph}^{-1} with [Tryptophan] (A) and with [Tb^{3+}] (B) at 77 K; (a) 4×10^{-3} M Tb^{3+} (b) No Tb^{3+}; ●, ▲ $\lambda_{exc} = 261 \pm 1$ nm, ○, △ $\lambda_{exc} = 291 \pm 1$ nm.

Figure 4. Phosphorescence quenching of 1×10^{-4} M Tryptophan by Tb^{3+} (B) and variation of $(I_F)/(I_{ph})$ with [Tb^{3+}] (A) at 77 K; ● $\lambda_{exc} = 261 \pm 1$ nm, ○ $\lambda_{exc} = 291 \pm 1$ nm.

A kinetic analysis leads to the following equations:
$$(I_F)_0/(I_F)_{Tb^{3+}} = 1 + (k_{isc}^{ind}/k_{fl} + k_{isc})\,[Tb^{3+}] \qquad (1)$$
for the intersystem crossing process induced by Tb^{3+},
$$(I_{ph})_0/(I_{ph})_{Tb^{3+}} = (k_{fl} + k_{isc}(k_{isc} + k_{isc}'))/k_{fl} + k_{isc})(k_{isc} + k_{isc}')$$
$$+ (k_{ET}/k_{fl} + k_{isc})\,[Tb^{3+}] \qquad (2)$$
describing the phosphorescence quenching by Tb^{3+}, and
$$(I_F)_{Tb^{3+}}/(I_{ph})_{Tb^{3+}} = (k_{fl}/k_{isc} + k_{isc}') + (k_{fl}k_{ET}/k_{isc} + k_{isc}')\,[Tb^{3+}] \qquad (3)$$
where k_{isc}' stands for k_{isc}^{ind} [Tb^{3+}] at low [Tb^{3+}]. Results plotted according to Equations (2) and (3) are shown in Fig. 4.

REFERENCES

1. J. Chrysochoos, Mol. Photochem., 5:1 (1973).
2. N. Mataga, Y. Torihasi and K. Ezumi, Theoret. Chim. Acta, 2: 158 (1964).
3. I. Weinryb, Biochim. Biophys. Res. Comm., 34:865 (1969).
4. P.S. Song and W.E. Kurtin, J. Am. Chem. Soc., 91:4892 (1969).
5. J. Chrysochoos and V. Anantharaman, "The Rare Earths in Modern Science and Technology", Vol. 2:475 (1980).
6. M.X. Balcavage and T. Alvager, Mol. Photochem., 7:309 (1976).

HYPERSENSITIVITY IN THE $4f$-$4f$ ABSORPTION SPECTRA OF Er^{3+} AND Ho^{3+} COMPLEXES

Scott A. Davis and Frederick S. Richardson

Department of Chemistry, University of Virginia

Charlottesville, Virginia 22901

The absorption intensities of certain lanthanide ion (Ln^{3+}) 4f→4f transitions exhibit an especially strong sensitivity to the detailed nature of the ligand environment about the Ln^{3+} ion. Transitions showing this behavior are referred to as "hypersensitive" transitions. The absorption spectra of these transitions are of value in following changes in the ligand environment brought about by complex formation, chelate:substrate adduct formation, or ligand structural alterations in solution media. Hypersensitivity has been exploited extensively as a structure probe even though a full understanding of the phenomenon, from a theoretical point of view, has remained elusive (1,2).

We have measured the absorption spectra of Er^{3+} and Ho^{3+} ions in aqueous solution with four different terdentate ligands under variable pH conditions. The ligands were: oxydiacetic acid (ODA), iminodiacetic acid (IDA), N-methyliminodiacetic acid (MIDA), and dipicolinic acid (DPA). In solution at pH>8 and with [Ln^{3+}]:[ligand] = 1:3, each of these ligands form tris-terdentate $Ln(ligand)_3^{3-}$ complexes having trigonal symmetry. Evidence for these tris-terdentate, trigonally symmetric structures is provided by magnetic-field-induced circularly polarized luminescence (MCPL) spectra obtained on the 1:3 [Eu^{3+}]:[ligand] systems in aqueous solution at pH≥8(3). In the case of ODA and DPA, further evidence is provided by x-ray crystallographic data obtained on crystals grown from neutral solutions of 1:3 [Ln^{3+}]:[ligand] in water (4). Although each complex has trigonal symmetry, each presents a slightly different ligand environment to the Ln^{3+} ion. Furthermore, each complex is expected to show a different coordination-versus-pH behavior in aqueous solution. To probe the differences between these complexes we exploited the hypersensitivity behavior of the $^4I_{15/2} \rightarrow {}^2H_{11/2}$, $^4G_{11/2}$ transi-

203

tions of Er^{3+}, and the $^5I_8 \to ^5G_6$ transition of Ho^{3+}. Although the
structural and chemical differences between the ODA, IDA, MIDA, and
DPA complexes are small (especially with regard to the LnL_9 coordination cluster), their absorption spectra in the hypersensitive
transition regions exhibit both qualitative and quantitative differences. Furthermore, the pH dependence of the absorption spectra
was found to vary significantly from complex to complex.

To help rationalize the experimental results cited above, intensity calculations were carried out and simulated absorption
spectra were computed for each of the complexes in each hypersensitive transition region. These calculations were based on an intensity model which includes both the static-coupling (Judd-Ofelt)
intensity mechanism and the dynamic-coupling (or ligand polarization)
intensity mechanism for the electric dipole strengths of f-f transitions in noncentrosymmetric complexes. This intensity model has been
described previously (5). In calculating the spectra for each of
the $Ln(ligand)_3^{3-}$ complexes, we assumed a tricapped trigonal prism
structure for the coordination cluster (4). For the $Ln(ODA)_3$ and
$Ln(DPA)_3$ systems we assumed a mer isomeric form for the chelate
rings, as is suggested by available crystallographic data (4). For
the $Ln(IDA)_3$ and $Ln(MIDA)_3$ systems we assumed a fac isomeric form
for the chelate rings, as is dictated by the stereochemistry at the
nitrogen donor atoms of the IDA and MIDA ligands.

Excellent agreement was achieved between the simulated (theoretically calculated) and experimental absorption spectra for each of
the complexes (in their pH>8 forms), showing that the intensity
model is quite sensitive to very small differences in ligand properties (i.e., it reflects the experimentally observed hypersensitivity).
Significantly, it was found that in each case at least 40% of the
calculated electric dipole strength was contributed by the transition quadrupole (Ln^{3+})-transition dipole (ligand) terms of the dynamic-coupling mechanism. Magnetic dipole strength contributions to
the intensities of the hypersensitive transitions investigated here
were found to be very small (generally less than 0.1%). For comparison with the tris-terdentate $Ln(ligand)_3^{3-}$ complexes, dipole strengths
and simulated absorption spectra were also calculated for $Ho(H_2O)_9^{3+}$
and $Er(H_2O)_9^{3+}$ complexes of exact D_{3h} symmetry. For the $^5I_8 \to ^5G_6$
transition of Ho^{3+}, the total (integrated) intensity for the $Ho(ODA)_3^{3-}$
complex was calculated to be about 25 times greater than that for
the $Ho(H_2O)_9^{3+}$ complex. Although this calculated result closely
mimics the experimentally observed intensity differences between the
$Ho(ODA)_3^{3-}$ and Ho^{3+}(aquo) $^5I_8 \to ^5G_6$ absorption spectra, comparisons
must be made with some circumspection since it is not yet known what
structural forms are dominant in Ho^{3+}(aquo). Indeed, it is likely
that both $Ho(H_2O)_8^{3+}$ and $Ho(H_2O)_9^{3+}$ structures are present, in which
case our $Ho(H_2O)_9^{3+}$ model structure provides only an approximate
representation for the Ho^{3+}(aquo) system.

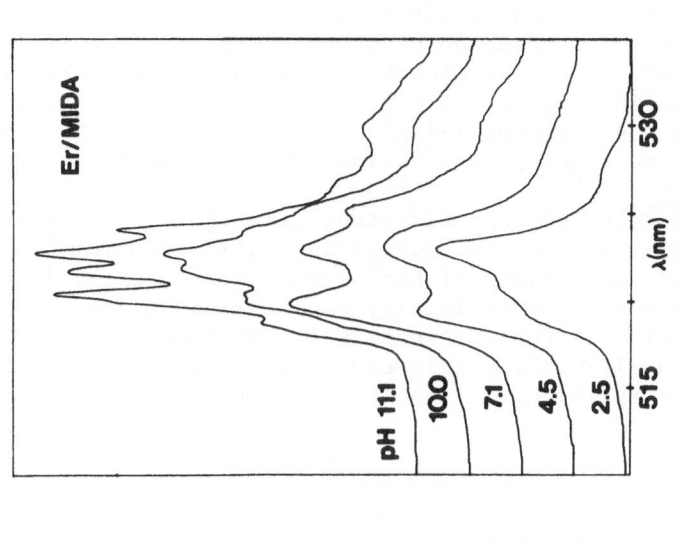

Figure 2. Absorption spectra recorded in the $^4I_{15/2} \rightarrow {}^2H_{11/2}$ transition region for 1:3 [Er³⁺]:[MIDA] in aqueous solution at various pH values.

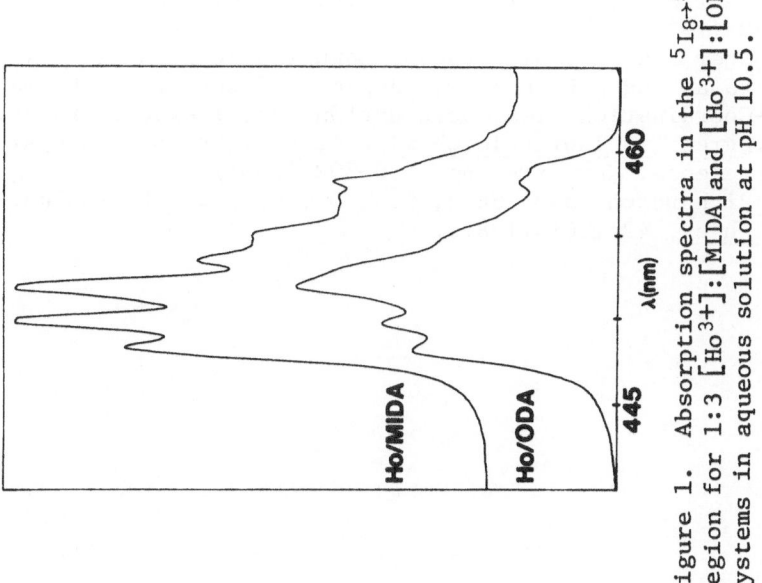

Figure 1. Absorption spectra in the $^5I_8 \rightarrow {}^5G_6$ region for 1:3 [Ho³⁺]:[MIDA] and [Ho³⁺]:[ODA] systems in aqueous solution at pH 10.5.

Examples of our experimental results are presented in Figures 1 and 2. Absorption spectra for Ho(ODA) and Ho(MIDA) at pH = 10.55 are shown in Figure 1 (for the $^5I_8 \rightarrow {}^5G_6$ transition region). These spectra illustrate the sensitivity of the $^5I_8 \rightarrow {}^5G_6$ Ho^{3+} transition to ligand type. The pH dependence of the Er(MIDA) $^4I_{15/2} \rightarrow {}^2H_{11/2}$ absorption spectra is illustrated by the data in Figure 2. The changes observed in these spectra can be attributed to the stepwise binding of MIDA donor groups (eliminating water molecules from the inner coordination sphere) as solution pH is raised from 2.5 to 11.1.

The results obtained in this study illustrate the utility of hypersensitive f-f absorption spectra in probing subtle changes in the coordination environment of Ln^{3+} ions. Furthermore, it is shown that observed hypersensitivity can be rationalized, in large part, in terms of an intensity model which includes both static-coupling (Judd-Ofelt) and dynamic-coupling (ligand polarization) contributions to f-f electric dipole strengths (5).

ACKNOWLEDGEMENTS

This work was supported by the National Science Foundation (NSF Grant CHE80-04209). We also acknowledge the valuable help of J.D. Saxe (in carrying out the intensity calculations).

REFERENCES

(1) R.D. Peacock, Struct. Bonding, 22:83 (1975).
(2) D.E. Henrie, R.L. Fellows, and G.R. Choppin, Coord. Chem. Rev., 18:199 (1976).
(3) D. Foster and F.S. Richardson, unpublished results.
(4) M.C. Favas and D.L. Kepert, "Aspects of the Stereochemistry of Nine-Coordination, Ten-Coordination, and Twelve-Coordination", in Progress in Inorganic Chemistry, Vol. 28, S.J. Lippard (ed.), Interscience, New York, pp. 239-308 (1981).
(5) F.S. Richardson, J.D. Saxe, S.A. Davis, and T.R. Faulkner, Mol. Phys., 42:1401 (1981).

POSSIBLE SYMMETRIES OF THE COORDINATION SPHERE OF Eu^{3+} IN DMSO VIA A CORRELATION OF MAGNETIC CIRCULAR DICHROISM (MCD) WITH ABSORPTION AND EMISSION SPECTRA

J. Chrysochoos[a], M.J. Stillman[b] and P.W.M. Jacobs[b]

(a) Department of Chemistry, University of Toledo, Toledo, Ohio 43606; (b) Centre for Interdisciplinary Studies in Chemical Physics, University of Western Ontario, London, Canada N6A 3K7

INTRODUCTION

Energy level assignments of lanthanides in crystals can be made via a combination of absorption and emission spectra at low temperatures with appropriate Zeeman spectra (1,2), provided the appropriate absorption bands are narrow. In systems with relatively broad absorption bands, the information supplied by Zeeman spectroscopy can be obtained via magnetic circular dichroism (MCD) (3-5). This technique leads to the determination of molar ellipticities, $[\theta]_M$, which relate to the difference between the absorption coefficients of l.c.p and r.c.p light in the presence of a longitudinal magnetic field.

Although most MCD spectra are quite complex, they are easier to analyze in systems which possess nondegenerated lower ground states like in the case of $Eu^{3+}(^7F_0)$. In such systems the experimental ellipticities consist mainly of A-terms with minor contributions from B-terms, attributed to the mixing of the final state of the transition under consideration with nearby states in the presence of a magnetic field. Systems with a degenerate ground state give rise to ellipticities which consist of A, B and C terms.

Information regarding the local symmetry of the system under consideration may be obtained by comparing both the values and the signs of the experimental and theoretical MCD terms as well as from the splitting of the MCD spectra. Such attempts led to possible local symmetries of Eu^{3+} in aqueous solutions (6,7) in aqueous solutions and in ethanol (8) and in crystalline $Eu(C_2H_5SO_4)_3 \cdot 9H_2O$ (9). The purpose of this brief paper is to consider the possible local symmetries for Eu^{3+} in DMSO using a combination of MCD, absorption and emission spectra.

207

EXPERIMENTAL

Magnetic circular dichroism spectra were recorded on a JASCO-UV-5 spectrometer using a superconducting Oxford Instrument magnet, at 5.5 T or at lower fields where necessary.

RESULTS AND DISCUSSION

Typical MCD spectra of $EuCl_3$ in DMSO along with corresponding absorption bands extending from 3000 to 6200 Å are shown in Figs. 1 and 2. The appropriate MCD spectra are plotted as negative counts vs. wavelength because the maximum number of counts corresponds with the lowest value of $[\theta]_M$(negative). Several MCD signals appear to have symmetrical A-forms. Molar ellipticities of $EuCl_3$ in DMSO vs. λ are shown in Figs. 3 and 4. A complete list of the observed MCD signals, in terms of $\bar{\nu}_{max}$, $\bar{\nu}_{min}$, $\bar{\nu}_0$ and appropriate signs defined as $\bar{\nu}_{max}-\bar{\nu}_{min}$, as well as corresponding absorption and emission energies are given in Table 1. The emission spectra of Eu^{3+} in $LaAlO_3:Eu^{3+}$ (D_3-symmetry) at RT (10) are in fair agreement with the emission spectra of $EuCl_3$ in DMSO. Based on the number of absorption, emission and MCD components shown in Table 1, the symmetry of Eu^{3+} in DMSO cannot be higher than D_{3h} (C_{3h} in the presence of a magnetic field), although a lower symmetry is also possible.

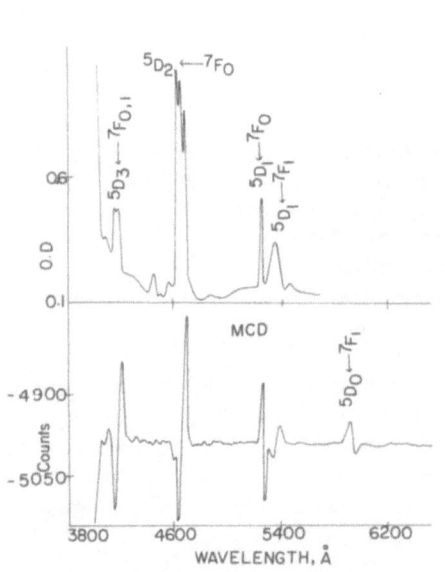

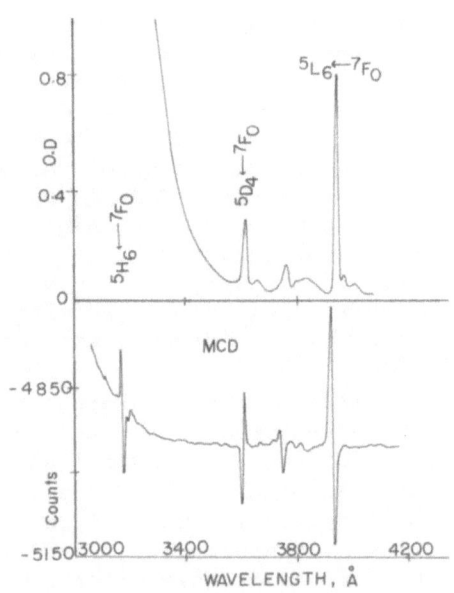

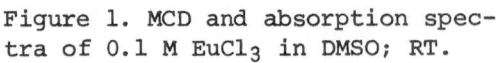

Figure 1. MCD and absorption spectra of 0.1 M $EuCl_3$ in DMSO; RT.

Figure 2. MCD and absorption spectra of 0.1 M $EuCl_3$ in DMSO.

Table 1. Energies of MCD, Absorption and Emission Spectra of $EuCl_3$ in DMSO (cm^{-1})

MCD			sign	Absorption	Emission	(10)	Transition
$\bar\nu_{max}$	$\bar\nu_{min}$	$\bar\nu_o$	$(\bar\nu_{max}-\bar\nu_{min})$	$\bar\nu_{abs}$	$\bar\nu_{em}$	$\bar\nu_{em}$	
16118	-	-	-		16155	16158	$^5D_0 \leftrightarrow {}^7F_2$
16160	16245	16210	Negative		16233	16233	$^5D_0 \leftrightarrow {}^7F_2$
16290	16375	16325	Negative	16339	16339	-	$^5D_0 \leftrightarrow {}^7F_2$
16680	16540	16560	Positive	16766	-	-	$^5D_0 \leftarrow {}^7F_{1,2}$
16980	16841	16895	Positive	16891	16863	16858	$^5D_0 \leftarrow {}^7F_1$
17030	16890	16960	Positive	16949	16949	16907	$^5D_0 \leftrightarrow {}^7F_1$
-	-	-	-	-	17123	17029	$^5D_1 \rightarrow {}^7F_3$
-	-	-	-	-	-	17084	$^5D_1 \rightarrow {}^7F_3$
-	-	-	-	-	-	17149	$^5D_1 \rightarrow {}^7F_3$
	No MCD			17271	17281	17281	$^5D_0 \leftrightarrow {}^7F_0$
17989	17119	18054	Negative	18075	17950	17924	$^5D_1 \leftrightarrow {}^7F_2$
-	-	-	-	18115	18115	17992	$^5D_1 \leftrightarrow {}^7F_2$
-	-	-	-	18242	18150	18015	$^5D_1 \leftrightarrow {}^7F_2$
18572	18764	18668	Negative	18656	18640	18625	$^5D_1 \leftrightarrow {}^7F_2$
18645	18825	18695	Negative	18708	18736	-	$^5D_1 \leftarrow {}^7F_1$
19065	18975	19025	Positive	19011	18990	18984	$^5D_1 \leftarrow {}^7F_0$
19015	19055		Negative?	19029	19047	18995	$^5D_1 \leftrightarrow {}^7F_0$
19920	19860	19910	Positive	19576	-	-	$^5D_2 \leftarrow {}^7F_2$
20240	20050	20200	Positive	21110	-	-	$^5D_2 \leftarrow {}^7F_1$
21400	21300	21350	Positive	21365	-	-	$^5D_2 \leftarrow {}^7F_0$
21508	21601	21554	Negative	21439	-	-	$^5D_2 \leftarrow {}^7F_0$
-	21764	-		21520	-	-	$^5D_2 \leftarrow {}^7F_0$
22525	22627	22576	Negative	23255	-	-	$^5D_3 \leftarrow {}^7F_2$
23610	24307	23958	Negative	23896	-	-	$^5D_3 \leftarrow {}^7F_1$
-	-	-	-	23950	-	-	$^5D_3 \leftarrow {}^7F_1$
23926	24110	23980	Negative	24066	-	-	$^5D_3 \leftarrow {}^7F_1$
24415	24465	24450	Negative	24390	-	-	$^5D_3 \leftarrow {}^7F_0$
24535	24580	24550	Negative	-	-	-	$^5D_3 \leftarrow {}^7F_0$
25303	25153	25255	Positive	25345	-	-	$^5L_6 \leftarrow {}^7F_0$
25535	25480	25507	Positive	-	-	-	$^5L_6 \leftarrow {}^7F_0$
-	-	-	-	26004	-	-	$^5L_8 \leftarrow {}^7F_1$
-	-	-	-	26247	-	-	$^5L_8 \leftarrow {}^7F_1$
26515	26390	26430	Positive	26540	-	-	$^5L_8 \leftarrow {}^7F_0$
26780	26696	26738	Positive	26680	-	-	$^5L_8 \leftarrow {}^7F_0$
-	-	-	-	26760	-	-	$^5D_4 \leftarrow {}^7F_0$
27280	27350	27310	Negative	27280	-	-	$^5D_4 \leftarrow {}^7F_0$
27400	27470	27410	Negative	27563	-	-	$^5D_4 \leftarrow {}^7F_0$
27715	27780	27745	Negative	27624	-	-	$^5D_4 \leftarrow {}^7F_0$
30909	30729	30819	Positive	-	-	-	$^5H_7 \leftarrow {}^7F_1$
31220	31280	31250	Negative	-	-	-	$^5H_6 \leftarrow {}^7F_1$?
31132	30920	31026	Positive	-	-	-	$^5H_6 \leftarrow {}^7F_0$
31525	31456	31490	Positive	-	-	-	$^5H_6 \leftarrow {}^7F_0$?

An analysis of the most certain MCD spectra based upon a D_{3h}-symmetry is given below.

a. $^5D_1 \longleftrightarrow ^7F_0$

   Absorption: 19011 cm^{-1} $(^5D_1(1) \leftarrow \; ^7F_0(0))$; MD (π)

                        19025 cm^{-1} $(^5D_1(0) \leftarrow \; ^7F_0(0))$; MD (σ)

   Emission: 18990 cm^{-1} $(^5D_1(1) \rightarrow \; ^7F_0(0))$; MD (π)

                        19045 cm^{-1} $(^5D_1(0) \rightarrow \; ^7F_0(0))$; MD (σ); Mixing?

   MCD: 19025 cm^{-1} $(^5D_1(\pm1) \leftarrow \; ^7F_0(0))$; MD (π)

b. $^5D_0 \longleftrightarrow ^7F_1$

   Absorption: 16949 cm^{-1} $(^5D_0(0) \leftarrow \; ^7F_1(1))$; MD (π)

                        16891 cm^{-1} $(^5D_0(0) \leftarrow \; ^7F_1(0))$; MD (σ)

   Emission: 16949 cm^{-1} $(^5D_0(0) \rightarrow \; ^7F_1(1))$; MD (π)

                        16863 cm^{-1} $(^5D_0(0) \rightarrow \; ^7F_1(0))$; MD (σ)

   MCD: 16895 cm^{-1} $(^5D_0(0) \leftarrow \; ^7F_1(\pm1))$; MD ($\pi$)

c. $^5D_1 \longleftrightarrow ^7F_1$

   Absorption: 18708 cm^{-1} $(^5D_1(0,1) \leftarrow \; ^7F_1(1))$; MD, ED (σ)

                        18656 cm^{-1} $(^5D_1(0,1) \leftarrow \; ^7F_1(0))$; MD (π)

   Emission: 18736 cm^{-1} $(^5D_1(0,1) \rightarrow \; ^7F_1(1))$; MD, ED (σ)

                        18640 cm^{-1} $(^5D_1(0,1) \rightarrow \; ^7F_1(0))$; MD (π)

   MCD: 18695 cm^{-1} $(^5D_1(0,1) \leftarrow \; ^7F_1(\pm1))$; MD, ED; Mixing?

                        18668 cm^{-1} $(^5D_1(0,1) \leftarrow \; ^7F_1(0))$; MD? Mixing?

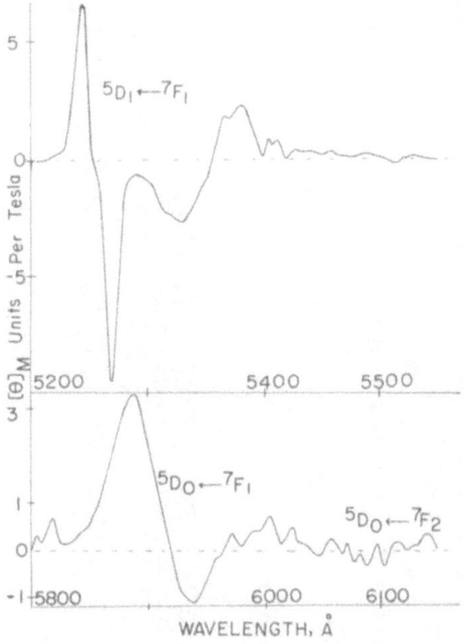

Figure 3. Molar ellipticity of 0.1M EuCl$_3$ in DMSO vs wavelength.

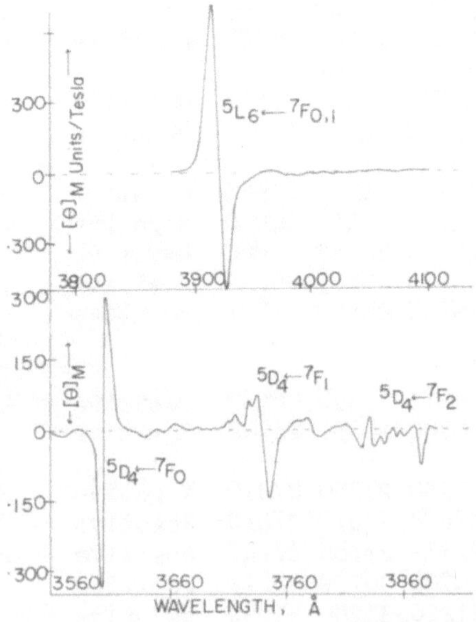

Figure 4. Molar ellipticity of 0.1M EuCl$_3$ in DMSO vs wavelength.

REFERENCES

1. G.H. Dieke, "Spectra and Energy Levels of Rare Earth Ions in Crystals", Interscience Publishers, New York (1968).
2. D.N. Olson and J.B. Gruber, J. Chem. Phys., 54:2077 (1971).
3. J. Margerie, Physica, 33:238 (1967).
4. A.C. Boccara and B. Briat, J. Phys. (Paris) 30:445 (1969).
5. J. Ferre, T. Nagai and T. Nakaya, Chem. Phys. Lett., 39:183 (1976).
6. C. Gorller-Walrand and J. Godemont, J. Chem. Phys., 66:48 (1977).
7. C. Gorller-Walrand and J. Godemont, J. Chem. Phys., 67:3655 (1977).
8. M.L. Sage, M.H. Buenocore and H.S. Pink, Chem. Phys., 36:171 (1979).
9. Y. Kato and S. Nakano, Chem. Phys. Letts., 45:359 (1977).
10. M. Faucher and P. Caro, J. Chem. Phys., 63:446 (1975).

DIPHENYL-PHOSPHINYL-MORPHOLIDE (DPPM) LANTHANIDE SALT ADDUCTS:

Nd^{3+} and Eu^{3+} SPECTRA

G. Vicentini and L.R.F. Carvalho*

Instituto de Química, Universidade de São Paulo
Caixa Postal: 20.780, São Paulo, Brazil

INTRODUCTION

This paper reports the absorption and emission spectra of several adducts with DPPM: $Ln(NO_3)_3 \cdot 3DPPM$, $Nd(NCS)_3 \cdot 4DPPM$, $Eu(NCS)_3 \cdot 3DPPM$ (1), $Ln(ClO_4)_3 \cdot 4DPPM$ (2), $LnCl_3 \cdot 3DPPM$ and $LnBr_3 \cdot 3DPPM$ (3) (Ln = Nd, Eu). Structural information concerning bond character, symmetry, and geometry around the central lanthanide ions, as recently done with N,N-dimethyl-diphenylphosphinamide (DDPA) (4) and diphenylphosphinamide (DPPA) (5), was obtained.

RESULTS AND DISCUSSION

From the neodymium absorption spectra (Figs. 1 and 2 and Table 1) the nephelauxetic parameter, β (6,7), was determined. This parameter was related to $b^{1/2}$ (7), which measures the amount of mixing of the 4f-ligand orbitals, or to Sinha's parameter, δ (8), which also gives the relative covalent character of the bond. The data indicate that the participation of the 4f orbitals in bonding is rather small, which means a Ln^{3+}-ligand interaction almost completely electrostatic in character. The number of hypersensitive bands in the spectra, at 77K, is indicative of the existence of non cubic sites around Nd^{3+}, but octahedral symmetries, with some distortion, may be suggested for the chloride and bromide adducts. Spectral investigations in solution were not made due to the slight solubility of the adducts.

---

*Postdoctoral fellowship from FAPESP.

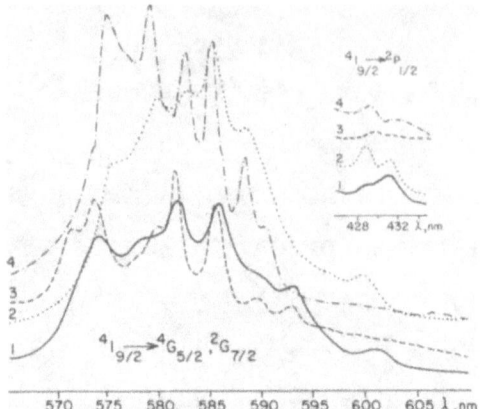

Figure 1.   Absorption spectra of neodymium DPPM adducts in Kel-F
            mulls: a - perchlorate, 1 - room temperature, 2 - 77K;
            b - nitrate, 3 - room temperature, 4 - 77K. (Cary 17
            spectrophotometer).

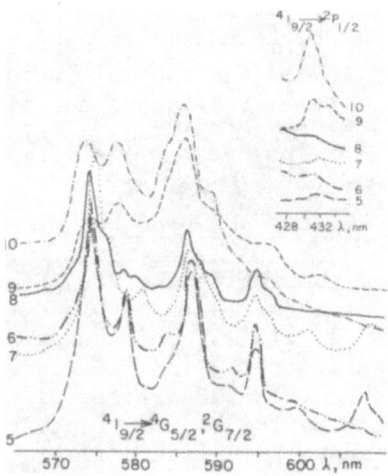

Figure 2.   Absorption spectra of neodymium DPPM adducts in Kel-F
            mulls: a - chloride, 5 - room temperature, 6 - 77K;
            b - bromide, 7 - room temperature, 8 - 77K; c - isothio-
            cyanate, 9 - room temperature, 10 - 77K. (Cary 17
            spectrophotometer).

     Figure 3 contains the emission spectra of the europium adducts,
at 77K and presents the proposed symmetries and geometries (9,10).
For the perchlorate adducts (2) we deduce that there are two

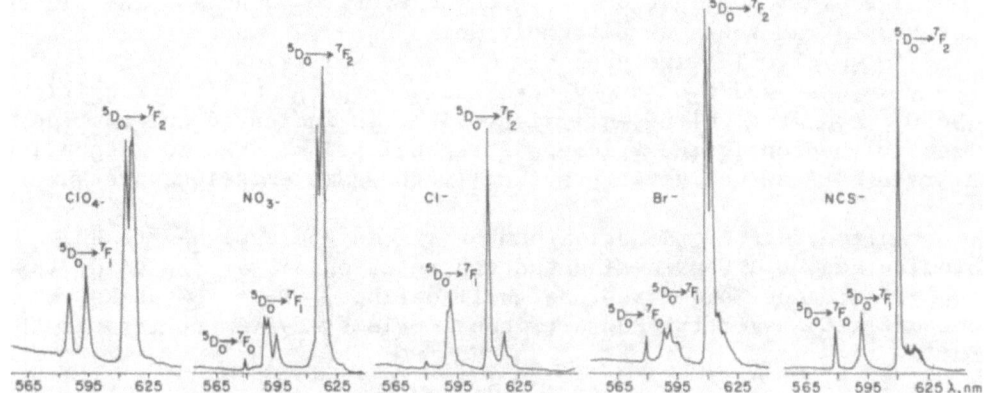

Figure 3. Emission spectra of the europium compounds at 77K (Perkin-Elmer MPF-4 spectrofluorimeter).

bidentate ions and that the coordination number is eight. The emission spectrum, which is identical to that of the DDPA adduct (4), shows one band and a shoulder for the $^5D_0 \rightarrow ^7F_1$ transition and two $^5D_0 \rightarrow ^7F_2$ bands, one of which presents a shoulder, attributed to a splitting of an E species. This is consistent with a D_{2d} symmetry of an $Eu(O)_8$ chromophore. The band at 17152 cm^{-1} is attributed to a $^5D_1 \rightarrow ^7F_J$ transition.

Table 1. Spectroscopic data for diphenyl-phosphinyl-morpholide neodymium salt adducts.

Compounds[a]	Transition $^4I_{9/2} \rightarrow ^4G_{5/2}, ^2G_{7/2}$		β	$b\frac{1}{2}$	δ
	λ (nm)	$\bar{\nu}$			
Nd(ClO$_4$)$_3 \cdot$4DPPM(2)	583.1	17148	0.9895[c]	0.075[d]	1.14[d]
Nd(NO$_3$)$_3 \cdot$3DPPM(1)	584.1	17118	0.9878	0.078	1.23
Nd Cl$_3 \cdot$3DPPM(3)	581.4	17198	0.9924	0.061	0.76
Nd Br$_3 \cdot$3DPPM(3)	582.8	17159	0.9902	0.070	0.99
Nd(NCS)$_3 \cdot$4DPPM(1)	584.7	17103	0.9869[b]	0.079[b]	1.26[d]

[a]Purity checked by EDTA titrations and carbon and hydrogen by microanalysis; [b]$\beta^4I_{9/2} \rightarrow ^2P_{1/2}$ = 0.9880; $\bar{\beta}$ = 0.9875; [c]β $^4I_{9/2} \rightarrow ^2P_{1/2}$ = 0.9879, $\bar{\beta}$ = 0.9887; [d]Calculated from $\bar{\beta}$.

The nona coordination is given to nitrate adducts. The fluorescence spectrum shows an extremely weak $^5D_0 \to {}^7F_0$ band; three $^5D_0 \to {}^7F_1$ bands, two of which are probably due to a splitting of E species, and one very weak and two strong peaks due to the $^5D_0 \to {}^7F_2$ transition. The D_{3h} symmetry, with small distortion, is indicated and the idealized polyhedron is the tricapped trigonal prism. The corresponding nitrate DDPA adduct gives practically the same emission spectrum (4).

An apparent coordination number six may be assigned to the bromide adduct. Considering that there is one $^5D_0 \to {}^7F_0$ band, three $^5D_0 \to {}^7F_1$ bands, and three, not well defined, $^5D_0 \to {}^7F_2$ peaks, we suggest a C_{3v} symmetry and a trigonal prismatic geometry around the Eu^{3+}.

The chloride and isothiocyanate adducts have apparently hexacoordination. Both emission spectra are very similar and they show one $^5D_0 \to {}^7F_0$ band, one $^5D_0 \to {}^7F_1$ band with a shoulder and two $^5D_0 \to {}^7F_2$ bands, one of which presents a splitting, attributed to an E species. The C_{3v} symmetry is suggested for the complex species, with the ligands octahedrally surrounding the central ion.

REFERENCES

1. L.B Zinner, G. Vicentini and L. Rothschild, J. Inorg. Nucl. Chem., 36:2499 (1974).
2. G. Vicentini, L.B. Zinner and L. Rothschild, An. Acad. brasil. Ciênc., 46:207 (1974).
3. G. Vicentini and J.E. Viana, An. Acad. brasil. Ciênc., 48:709 (1976).
4. L.R.F. Carvalho and G. Vicentini, J. Coord. Chem. (in press).
5. L.R.F. Carvalho, G. Vicentini and K. Zinner, J. Inorg. Nucl. Chem., 43:1088 (1981).
6. P. Caro and J. Derouet, Bull. Soc. Chim. (France), 1:46 (1972).
7. D.E. Henrie and G.R. Choppin, J. Chem. Phys., 49:477 (1968).
8. S.P. Sinha, Spectrochim. Acta, 22:57 (1966).
9. J.H. Forsberg, Coord. Chem. Rev., 7:81 (1971).
10. S.P. Sinha, Structure and Bonding, 25:69 (1976).

ADDUCTS OF LANTHANIDE TRIFLUOROMETHANESULFONATES AND TETRAMETHYLENE SULFOXIDE (TMSO)

L.B. Zinner and F.A. Araujo

Instituto de Química, Universidade de São Paulo

Caixa Postal: 20.780, São Paulo, Brazil

INTRODUCTION

Adducts between lanthanide (Ln) perchlorates, nitrates, isothiocyanates, chlorides, hexafluorophosphates, bromides, perrhenates, iodides and methanesulfonates with TMSO have been previously described (1). In this paper, the synthesis, characterization and spectroscopic properties of Ln trifluoromethanesulfonate complexes with TMSO, having the following stoichiometry $Ln(CF_3SO_3)_3 \cdot 7.5TMSO$ (Ln = La-Lu,Y), are reported.

EXPERIMENTAL

The hydrated trifluoromethanesulfonate (2) was dissolved in an excess of warm TMSO (Aldrich) and the adduct precipitated by addition of triethyl-orthoformate (teof). The crystals were filtered, washed with teof and dried *in vacuo*, over anhydrous calcium chloride. The lanthanides were titrated with EDTA. Carbon and hydrogen were determined by microanalytical procedures (Table 1). Conductance, X-ray diffraction patterns, absorption and emission spectra were obtained as described in (2).

RESULTS AND DISCUSSION

Molar conductance values (Table 1) indicate a 1:2 electrolyte behavior (3) probably due to ion-pair formation. X-ray powder patterns show only one isomorphous series.

217

Table 1. Summary of analytical results and electrolytic conductance
          data for the compounds of formula $Ln(CF_3SO_3)_3.7.5TMSO$.

Ln	Analysis (%)						Conductance			
	Lanthanide		Carbon		Hydrogen		Acetonitrile		Nitromethane	
	Theor.	Exp.	Theor.	Exp.	Theor.	Exp.	Conc.,mM	Λm^*	Conc.,mM	Λm^*
La	10.16	10.18	28.97	28.67	4.42	5.01	1.02	278	1.01	137
Ce	10.24	10.16	28.96	29.28	4.42	4.74	1.00	289	0.99	144
Pr	10.29	10.17	28.94	29.34	4.42	4.81	1.00	283	1.00	146
Nd	10.51	10.53	28.87	29.15	4.41	4.27	1.02	286	1.00	149
Sm	10.90	10.97	28.75	27.98	4.39	4.83	1.01	292	0.99	152
Eu	11.01	10.91	28.71	28.73	4.38	4.94	1.03	289	1.00	156
Gd	11.35	11.27	28.00	28.72	4.36	4.71	1.03	297	1.00	159
Tb	11.45	11.45	28.57	28.71	4.36	4.92	1.00	300	1.01	172
Dy	11.68	11.52	28.49	28.49	4.35	4.44	1.02	306	1.00	169
Ho	11.84	11.60	28.45	28.35	4.34	4.53	1.02	300	0.99	173
Er	11.98	11.92	28.40	28.48	4.33	4.56	0.98	311	0.99	176
Tm	12.09	12.04	28.36	28.42	4.33	5.24	1.05	301	0.96	180
Yb	12.35	12.19	28.28	28.76	4.32	4.62	1.03	297	0.94	177
Lu	12.47	12.35	28.24	28.81	4.31	4.45	1.00	304	1.00	175
Y	6.75	6.59	30.09	30.94	4.59	4.87	0.97	295	1.00	164

Λm^* $(\Omega^{-1}.cm^2.mol^{-1})$.

A shift of the S = O stretching mode (1026 to 970 cm^{-1}) is a clear indication of coordination through the oxygen. In all the compounds, the bands assigned to the stretching and bending modes of the SO_3 group ($\nu_{as}\sim1275vs$, $\nu_s\sim1030m$-s, δ_{as} 636m-s and $\delta_s\sim515w$ cm^{-1}) (4), are unsplit indicating that all the trifluoromethanesulfonate ions in the complexes have a C_{3v} symmetry and are not coordinated.

From the absorption spectra of the Nd^{3+} complexes (Fig. 1) the following observations were made: a) the nephelauxetic parameter, $\beta(5,6)$, covalent factor, $b^{1/2}$ (7) and Sinha's parameter, $\delta(8)$ were determined. The values (Table 2) are compatible with a minor bonding contribution of 4f orbitals and an essentially electrostatic interaction between Nd^{3+} and the ligand; b) the number of bands in the $^4I_{9/2}\rightarrow$ $^4G_{5/2}, ^2G_{7/2}$ transition, at 77K, indicates a non cubic site around Nd^{3+}. The shapes of the spectra in acetonitrile and nitromethane solutions are identical, and very similar to that in the solid state

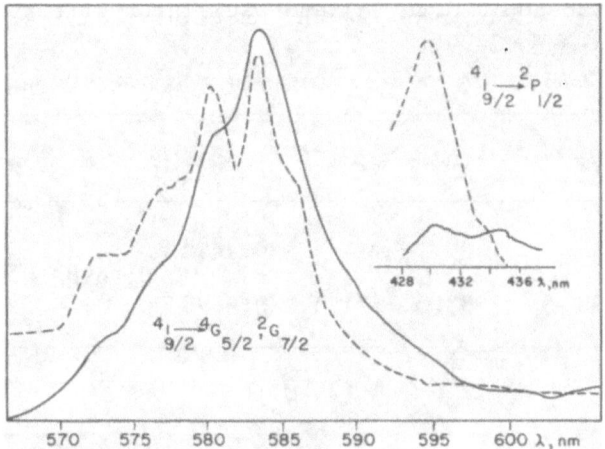

Figure 1.   Absorption spectra of Nd(CF$_3$SO$_3$)$_3$.7.5TMSO, (Kel-F mull)——
room temperature ---- 77K.

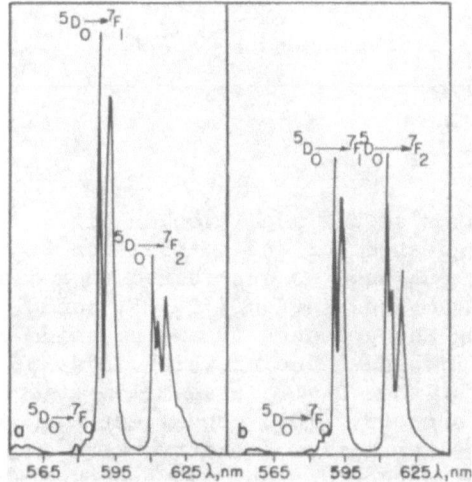

Figure 2.   Emission spectra, at 77K; a ⁻ Eu(CF$_3$SO$_3$)$_3$.7.5TMSO   b ⁻
Eu(ClO$_4$)$_3$.7.5TMSO.

at room temperature, with practically the same P values (9), which
implies the same environment. It is interesting to mention that the
spectra are very similar to those obtained for Nd(PF$_6$)$_3$.7.5DTMSO (10)
and Nd(PF$_6$)$_3$.7.5TSO (11).

Table 2. Spectroscopic data (A) and oscillator strengths (B) for the
          neodymium adduct.

A (Kel-F mull, 25°C)

Transition	λ,nm	$\bar{\nu}$,cm^{-1}	β	$\bar{\beta}$	$b^{1/2}$	δ
$^4I_{9/2} \rightarrow {}^2P_{1/2}$	430.5	23229	0.9898			
				0.9898	0.0714	1.03
$^4I_{9/2} \rightarrow {}^4G_{5/2},{}^2G_{7/2}$	583.0	17153	0.9898			

B (25°C)

Transition	Solvent	Conc.,M	η	$P \times 10^6$ cm^2.mol^{-1}.ℓ
$^4I_{9/2}, \rightarrow {}^4G_{5/2},{}^2G_{7/2}$	Acetonitrile	0.0191	1.3474	12.7
"	Nitromethane	0.0191	1.3846	12.8

Emission spectra at 77K are given in Fig. 2. The coordination
number eight is suggested for the central ion in both complexes. The
extremely weak $^5D_0 \rightarrow {}^7F_0$ band is attributed to a distortion of the polyhedron. The spectrum shows strong $^5D_0 \rightarrow {}^7F_1$ bands, as compared with
$^5D_0 \rightarrow {}^7F_2$, suggesting the presence of an inversion center. The spectrum
presents two $^5D_0 \rightarrow {}^7F_1$ bands, one of which is split in two peaks. The
same is observed for the $^5D_0 \rightarrow {}^7F_2$ transition (two E species are expected). The D_{3d} symmetry (10), consistent with a bicapped octahedral
or bicapped trigonal antiprism geometry, with one ligand acting as a
bridge between two lanthanide ions, is proposed for the complex species. Finally, we conclude that all lanthanide complexes with compo-
sition $LnX_3.7.5TMSO$ |X = ClO_4, PF_6, I, CF_3SO_3| present the same sym-
metry and geometry.

ACKNOWLEDGMENTS

The authors are much indebted to G. Vicentini for helpful discussion. One of us (F.A.A.) acknowledges the fellowship from PICD-CAPES.

REFERENCES

1. Last paper in this series: L.B. Zinner and G. Vicentini, <u>J. Inorg. Nucl. Chem.</u>, 42:1349 (1980).
2. G. Vicentini and L.B. Zinner, <u>J. Inorg. Nucl. Chem.</u>, 42:1510 (1980).
3. W.J. Geary, <u>Coord. Chem. Rev.</u>, 7:81 (1971).
4. A.L. Arduini, M. Garnett, R.C. Thompson and T.C. Wong, <u>Can. J. Chem.</u>, 53:3812 (1975).
5. P. Caro and J. Derouet, <u>Bull. Soc. Chim. Fr.</u>, 1:46 (1972).
6. H.H. Caspers, H.E. Rast and R.A. Buchanam, <u>J. Chem. Phys.</u>, 42:3214 (1965).
7. D.E. Henrie and G.R. Choppin, <u>J. Chem. Phys.</u>, 49:477 (1968).
8. S.P. Sinha, <u>Spectrochim. Acta</u>, 22:57 (1966).
9. W.T. Carnall, P.R. Fields and B.G. Wybourne, <u>J. Chem. Phys.</u>, 42:3797 (1965).
10. G. Vicentini and G. Chiericato, Jr., <u>An. Acad. brasil. Ciênc.</u>, 51:217 (1979).
11. A.M.P. Felicissimo, G. Vicentini, L.B. Zinner and Y. Shimizu, <u>An. Acad. brasil. Ciênc.</u>, 52:283 (1980).

PHOTOREDUCTION OF YTTERBIUM AND SAMARIUM

Terence Donohue

Laser Physics Branch
Naval Research Laboratory
Washington, D. C.   20375

ABSTRACT

Two lanthanides, ytterbium and samarium, have been photoreduced from their normal stable states of +3 to their unstable +2 states in neat methanol solution, using a pulsed KrF excimer laser at 248 nm. No photochemical reactions are observed when a continuous low-pressure mercury lamp at 254 nm is employed, implying that multi-photon processes are occurring when using the pulsed laser source. In the case of Sm, complexation with a macrocyclic ether (18-crown-6 or 2.2.2 cryptand) stabilizes the reduced state, and the apparent lifetime of Sm(II) is increased from seconds to hours.

INTRODUCTION

There have been virtually no reports on the redox photochemistry of samarium and ytterbium (1). This is due to the powerful reducing properties of the reduced states of these ions which react with most common solvents. While it is possible to stabilize reactive states in molten salts or glasses, it has also been shown that unstable oxidation states of metal ions can be partially protected by stabilization with simple inorganic complexing ligands, such as carbonates (2) or halides (3). Furthermore, recently discovered (poly)macrocyclic ligands, including the crown ethers and cryptands, display even more powerful properties in the stabilization of metal ions (5,6) and are relatively inert in photochemical reactions at photolysis wavelengths longer than about 250 nm. The stability of an ion complexed by a macrocyclic ligand is given in part by the match between ion size and ligand cavity diameter.

EXPERIMENTAL

Chloride salts of either Sm(III) or Yb(III) were dissolved in neat methanol. No attempt was made to remove the water of hydration that is found in rare earth halides. Concentrations employed were in the range of 0.01-0.05 M in Ln(III), with a slight excess over stoichiometric of either 18-crown-6 polyether or 2.2.2 cryptand added in certain experiments. Solutions were thoroughly deoxygenated by sparging with argon before photolysis. No photoreduction would occur until most or all atmospheric oxygen was excluded. Photolyses and spectrophotometric analyses were carried out in 1 cm fused silica spectrophotometer cells. The photolysis sources included a low-pressure Hg lamp with Vycor filter so that the only significant photolysis wavelength would be 254 nm. Power was about 7 mW into the 3 cm^3 samples. The other source was a pulsed rare gas-halide excimer laser, operating on KrF at 248 nm (7). The beam was not focussed, but masked to fill the cylindrical sample cells uniformly, and peak power levels were about 10^7 W/cm^2, with repetition rates adjusted to 20-30 Hz so that the average power absorbed was 2-3 W.

RESULTS AND DISCUSSION

The lineshapes and intensities of both the CT and f-f transitions in Sm(III) were affected by addition of the crown ether or cryptand. This is sufficient evidence that these macrocyclic ligands are binding Sm(III) (6). No such unambiguous spectral changes were observed with Yb(III), however, but the absence of any accessible f-f transitions in this ion makes conclusions regarding the existence of macrocyclic Yb(III) complexes more difficult. Photolysis of the neat SmCl$_3$ solutions with the pulsed excimer laser produced a transient blue color with a lifetime of 3 hours. Complexation with 2.2.2 cryptand caused an even more pronounced effect, where the photolytically produced species lived for about 4 hours. Comparison of the spectra produced (Fig. 1) indicates that the product is a Sm(II) species, most likely SmCl$_2$ (macrocyclic ether). The shifts in the f-d bands of Sm(II) are expected and similar changes have been observed in Eu(II) and Ce(III) (6), as well as U(III) (11,12). Laser photolysis of Yb(III) solutions produced an absorption band centered at 370 nm; addition of a macrocyclic ether ligand did not appear to change the spectrum but did cause minor increases in the lifetime of the species, the lifetime being about one hour. The observed band is similar to that reported for aqueous Yb(III) (8-10).

In all cases, attempts at photoreduction of Ln(III) using a mercury lamp were unsuccessful. Furthermore, photoreduced samples produced using the excimer laser were slowly destroyed (photooxidized) when photolyzed with the Hg lamp. The CT bands of Sm(III) and Yb(III) at these photolysis wavelengths are quite broad, with no structure distinguishable between 248 and 254 nm. Thus the only

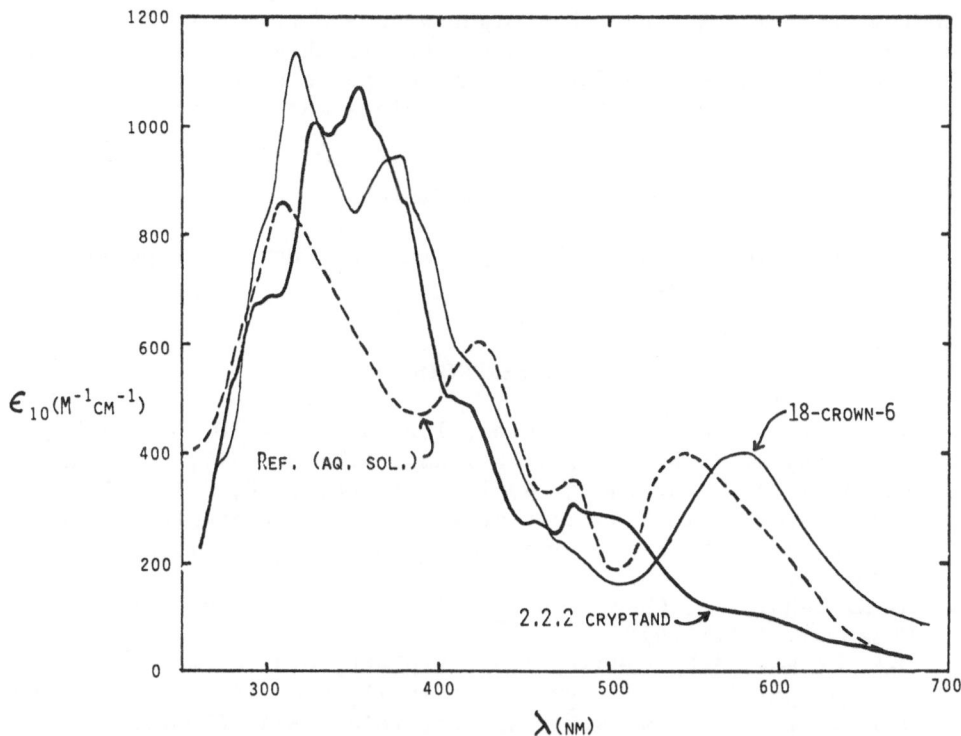

Figure 1. Comparison of several Sm(II) spectra in solution. The
reference spectrum was produced by pulse radiolysis in dilute aqueous
solution, from Ref. 10. The solid lines indicate photochemically
produced Sm(II) in methanol, using a KrF laser at 248 nm. Complexation by the two macrocyclic polyethers indicated causes the large
shifts and structure changes in the f–d spectra of Sm(II). Absolute
extinction coefficients (ε) are for the reference spectrum only;
others are relative intensities.

significant difference between the two photolysis sources used is
peak power. The excimer laser produces a power density approximately
10^9 times larger than the mercury lamp. With this ratio and the peak
power produced by the pulsed laser, the presence of multiphoton
processes during laser photolysis is not surprising (11, 12). Such
multiphoton reactions in liquid phase inorganic studies are just now
being reported (11, 13) and we have inferred such processes from
quantum yield measurements in the photoreduction of the similar actinide ion U(IV) (11, 12). While we have earlier noted that a cryptand
enhanced the stability of Sm(II) produced chemically by reduction
with magnesium (6), the photochemical synthesis of this species
appears to be simpler, cleaner and give higher yields than any chemical preparation. Whether the macrocyclic ligands described here complex Yb(III) or Yb(II) is not clear at this stage, there are probably

other macrocyclic ligands that might give a better match between
cavity size and ion size than those used in these studies. We will
be continuing the search for ligands which can enhance the photochemistry of unstable states of lanthanides and actinides.

ACKNOWLEDGEMENT

I would like to thank the Department of Energy for partial
support of this research.

REFERENCES

1. Y. Haas, G. Stein and R. Tenne, Isr. J. Chem. 10, 529 (1972).
2. D.E. Hobart, K. Samhoun, J.P. Young, V.E. Norvell, G. Mamantov
    and J.R. Peterson, Inorg. Nucl. Chem. Lett. 16, 321 (1980).
3. J.F. Endicott, in "Concepts of Inorganic Photochemistry," A.W.
    Adamson and P.D. Fleischauer, eds. (Wiley-Interscience, New York,
    1975), p. 81.
4. J.-M. Lehn, Acc. Chem. Res. 11, 49 (1978); I.M. Kolthoff, Anal.
    Chem. 51, 1R (1979).
5. O.A. Gansow, A.R. Kausar, K.M. Triplett, M.J. Weaver and E.L.
    Lee, J. Am. Chem. Soc. 99, 7087 (1977); R.M. Izatt, J.D. Lamb,
    J.J. Christensen and B.L. Haymore, ibid., 8344 (1977).
6. T. Donohue, in "The Rare Earths in Modern Science and Technology,
    Vol. 2," G.J. McCarthy, J.J. Rhyne and H.B. Silber, eds. (Plenum
    Press, New York, 1980), p. 105.
7. J.G. Eden, R. Burnham, L.F. Champagne, T. Donohue and N. Djeu,
    I.E.E.E. Spectrum 16, 50 (April, 1979).
8. F.D.S. Butement, Trans. Faraday Soc. 44, 617 (1948).
9. M. Farragi and Y. Tendler, J. Chem. Phys. 56, 3287 (1972).
10. A.K. Pikaev, G.K. Sibirskaya and V.I. Spitsyn, Dokl. Akad. Nauk.
    SSSR 209, 339 (1973).
11. T. Donohue, in "Chemical and Biochemical Applications of Lasers,
    Vol. V," C. B. Moore, ed. (Academic Press, New York, 1980),
    p. 239.
12. T. Donohue, to be published.
13. K.M. Cunningham and J.F. Endicott, Chem. Comm., 1024 (1974);
    R. Sriram, J.F. Endicott and S.C. Pyke, J. Am. Chem. Soc. 99,
    4824 (1977).

CeN: PHASE RELATIONSHIPS AND ENTHALPIES OF SOLUTION

E. Kaldis, B. Steinmann, B. Fritzler,
E. Jilek, and A. Wisard

Laboratorium für Festkörperphysik ETH
8093 Zürich, Switzerland

ABSTRACT

Single crystals and polycrystalline samples of CeN with nitrogen contents 47.0<x<50.5 at %N, do not show inclusions of cerium metal either in metallographic or in X-ray investigations. A T-x plot of the growth or nitridation temperature (2100 C>T>900 C) and the nitrogen concentration show a temperature dependent nitrogen-rich phase boundary. Variation of the nitrogen partial pressure during nitridation of cerium turnings from 1140 to 57 Torr, showed only a slight change of the nitrogen content. Attempts to locate the cerium-rich phase boundary are now concentrated on annealing CeN+Ce mixtures in sealed W-crucibles.

The lattice constants of CeN vary by 0.14% in the range of 47.0-50.5 at%N, with the value for stoichiometric CeN being 5.018 $\overset{o}{A}$, as compared with 0.23% for the corresponding change of δ-NbN. Most instructive is a histogram of the lattice constants vs. nitrogen concentration of 70 samples. Comparison with the lattice constant, $\underline{a}$, of samples which have been doped with oxygen, shows clearly that $\underline{a}$ >5.023 A belong to oxygen contaminated samples. Carefully measured lattice constants are therefore a sensitive criterion in differentiating between contamination and nonstoichiometry. This was supported by chemical analyses both of nitrogen (Kjeldahl; accuracy +0.1%) and cerium (complexometrically; accuracy $\pm$ 0.25 %) in aliquot parts of the samples used for measurements of the lattice constants and enthalpies of solution.

Enthalpies of solution ΔH_S in 4n HCl have been measured at 25 C as a function of nonstoichiometry for 50 undoped and 20 oxygen doped (and alloyed) CeN samples. The proposed value for pure, stoichiometric CeN is ΔH_s=122 kcal/mol. The dependence of ΔH_s on nonstoichiometry indicates an instability of the mixed valent CeN at 49.2%N. The most striking effect however, is the strong decrease of the heat of solution i.e. increase of the stability of samples doped with not more than 1% oxygen. This is explained as the result of destabilisation of the delicate balance of electronic states in mixed valent CeN.

INTRODUCTION

Based on metallographic investigations on pressed powder and sintered samples, Brown and Clark (1) have concluded that in their samples cerium metal coexists with CeN and that the latter does not have an appreciable homogeneity range. X-ray powder diagrams however showed no additional lines of cerium metal. These authors suggested that the small change of the room temperature lattice constant as a function of the chemical composition of CeN supports the above assumption, the slope of the $\underline{a}$=f(x) curve ($\underline{a}$: lattice constant, x: composition) being due only to the solubility of oxygen in CeN resulting from the contamination of the samples.

Since the growth of large single crystals of CeN is now possible (2), we have performed metallographic investigations on large single crystals (5-12 mm dimensions) grown by sublimation, and crystallites (2-3mm dimensions) grown by annealing in nitrogen atmosphere, at temperatures between 2200 and 1500 C, with various nonstoichiometric compositions. Using an optical microscope no evidence of metallic cerium could be found either on cleaved or on highly reflective as grown golden-yellow faces of the CeN single crystals. This result triggered a systematic investigation of the homogeneity range of CeN (3). The extreme sensitivity of this compound, its surface turns green-yellow after only 20 mins. in a well-gettered glove box, forced us to use for this investigation the most advanced preparative methods available in our laboratory.

SYNTHESIS AND CHEMICAL ANALYSIS

Details of the nitridation and crystal growth of CeN (2) and of the crystal growth at T<2600 C in tungsten crucibles sealed by electron beam welding (4) have been published elsewhere. To keep contamination at a minimum, a train of three glove boxes made of stainless steel has been used. These contained among other facilities an optical microscope, a precision balance and a lathe (for preparation of cerium turnings as starting material for the nitridation). The argon atmosphere of the boxes was continuously recycled above hot cerium turnings so that an oxygen- and H_2O-content of approximately 1 ppm was maintained throughout this work.

50 samples of CeN from 15 crystal growth and nitridation experiments of CeN and 25 samples of CeO_xN_{1-x} have been used in order to
establish the dependence of the lattice constant, the heat of solution
in 4n HCl and the T-x phase diagram on the nonstoichiometry. The nonstoichiometry of CeN was determined by chemical analysis both of
cerium and nitrogen. Cerium was determined complexometrically with
an accuracy of $\pm$ 0.25 wt%. Nitrogen was determined by the Kjeldahl
method, using the Büchi 320 semiautomatic apparatus (Büchi Labora-
toriumstechnik AG, Flawil, Switzerland), with a $\pm$ 0.1 wt% accuracy.
Therefore we estimate an accuracy of the chemical analysis for a given
sample of 0.7 wt%. Concentration gradients and slight contamination of
the samples require a sophisticated sampling process in order that
the four sets of measurements mentioned above become compatible. Each
sample, a small piece of a single crystal or a bunch of crystallites,
was pulverized and aliquot parts were used for chemical analysis,
lattice constant and calorimetric investigations. This procedure decreased the accuracy of the complete chemical analysis to 1.0%
($\pm$ 0.5 wt%). Oxygen contents of CeO_xN_{1-x} were deduced from the sum
of the metal and nonmetal contents (difference from 100 wt%) for x
values higher than 0.01. For lower concentrations the indication of
the lattice constant was used, because concentration variations were
within the error limits of the chemical analysis.

3. THE HOMOGENEITY RANGE

Figure 1 shows some first results of the T-x phase diagram near the
stoichiometric compound. The temperatures shown correspond to the
highest temperature at which the sample was grown or annealed. Subsequent annealing at lower temperature changes the nitrogen concentration according to the shape of the phase boundary. Thus annealing
for 24 hours of single crystals (A) at 1600 C is enough to adjust
the nitrogen content at this temperature. The existing data suggest
a preliminary location of the nitrogen-rich phase boundary at p=1140
Torr N_2. Variation of the partial pressure of nitrogen using nitrogen-
argon mixtures did not show large changes down to p=57 Torr N_2. For
this reason, work is under way now with mixtures of Ce+CeN in sealed
W-crucibles, in order to locate the metal-rich phase boundary. The
present form of the phase boundary can be explained by the interplay
of an increasing (catalytic) dissociation rate of the nitrogen molecule on the surface of CeN coupled with higher thermal diffusion of
nitrogen with increasing temperature (T>1000 C) and an increasing
formation of nitrogen vacancies at higher temperatures (T>1300 C).The
lower nitrogen content of the high temperature samples (on the left
of the phase boundary at (T>2000 C) is probably due to oxygen contamination, as indicated by the slightly higher values of the lattice
constant.

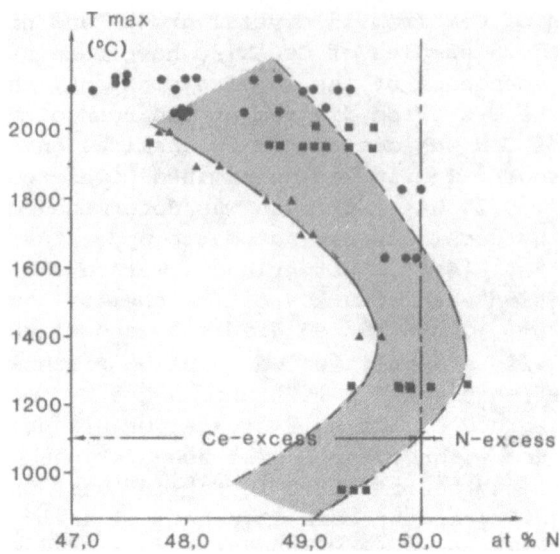

Figure 1. Part of the T-x phase-diagram of Ce-N showing the nitrogen-rich phase boundary of CeN. ● denote single crystals and poly-cristalline material grown in sealed tungsten crucibles (vacuum furnace). ■ denote material nitridated and annealed in nitrogen atmosphere at p=1140 Torr; p denotes reduced nitrogen pressures down to 57 Torr N_2.

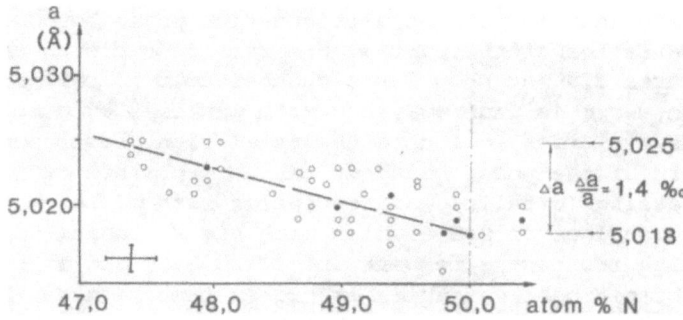

Figure 2. Lattice constant as a function of nitrogen content for CeN. ●● denote two overlapping measurements. Suggested value of stoichiometric CeN single crystals a=5.018 Å. Maximal error of the measurements is denoted by the crossed bar.

The variation of the lattice constants at room temperature as a function of composition (Fig.2) shows a change of 0.14% as compared to 0.23% for the corresponding change of a NaCl phase like δ=NbN(5). Very instructive also is the histogram of Fig.3. It clearly shows that lattice constants with values a>5.022 Å (corresponding to less than 48.5% N) belong to samples doped with oxygen. Doped and mixed-crystal (CeO_xN_{1-x}) samples have higher lattice constants. For x<0.01 we have to rely on the stoichiometric ratio of the starting materials (CeN+CeO$_2$+Ce). Based on this we could find via the lattice constants, which of the CeN samples were contaminated (full lines with a < 5.022 in Fig.3). Strong evidence for the contamination of these samples is given by the calorimetric measurements discussed below.

The above measurements show clearly that contamination is a severe hindrance in establishing the properties of CeN. A homogeneity range of CeN probably exists but seems to be rather narrow, possibly in the range 48.5 - 50.5 at% N at p(N$_2$)=1.5 atm. We hope to establish the location of the metal-rich phase boundary by heating mixtures of Ce+CeN in sealed W-crucibles. With increasing Ce concentration, however, difficulties may arise at high temperatures due to corrosion of the crucibles.

ENTHALPY OF SOLUTION AS A FUNCTION OF NONSTOICHIOMETRY

Although a deep insight in the structural aspects of nonstoichiometry has been established due to the progress in electromicroscopy in the last decades (6), there have been almost no investigations of the energetic aspects of this phenomenon. A possible way to strike the bridge between structure (superstructures) and energy content of nonstoichiometric compounds is in principle the study of the enthalpy of formation, solution, or reaction, as a function of nonstoichiometry. If calculation of the energy of interaction of vacancies (which stabilizes large nonstoichiometry ranges and induces super-structures) were possible from such measurements, a most important step towards a thermodynamic theory of nonstoichiometry would be made. In addition to these aspects, enthalpy measurements as a function of nonstoichiometry are very important for compounds with valence instabilities. Measurements of microhardness (Fig.4) show for example with TmSe (7) a lattice instability associated with the valence instability of this compound (8,9). The question arises however, which energy changes are responsible for these instabilities, i.e. what is the energy difference between the state of normal valence and that of mixed valence. XPS-measurements are not reliable to answer this question, because under the highly energetic beam the final nonintegral valence state dissociates to the initial integral valence states (10). In view of this situation, enthalpy

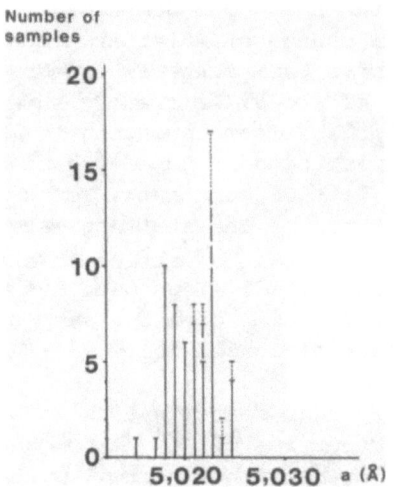

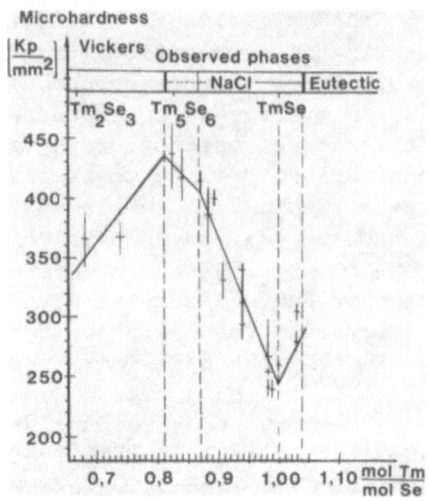

Figure 3. Histogram of the frequency of CeN and CeO$_x$N$_{1-x}$ samples with a given value of the
lattice constant.

Figure 4. Microhardness of Tm$_x$Se
single crystals as a function of
nonstoichiometry. The valence
fluctuation starts at the minimum of microhardness.

measurements as a function of nonstoichiometry would be expected to
give some information on a relative scale (including defect structure contributions) about the stability of phases containing various
valence states.

Recently we have investigated by solution calorimetry the nonstoichiometric systems Tm$_x$Se, Sm$_2$S$_3$-Sm$_3$S$_4$ and CeN$_x$. CeN$_x$ has the advantage of quantitative dissolution in acid, so that no additional
chemical effects (e.g. precipitation) to the dissolution process
exist. Figure 5 shows the dependence of the heat of solution of
CeN in 4n HCl as a function of nonstoichiometry. Most of the samples
with composition 47.0<at% N<48.0 show higher lattice constants indicating slight contamination(sect.3.). Therefore we believe at this
time that the homogeneity range of CeN is 48.5-50.5 at% N (6). There
is a remarkable reproducibility (sevenfold; from samples grown under
various conditions) of the value of ΔH_s=122 kcal/mol for the enthalpy
of solution of the composition 49.8-49.9 at% N. In view of the error
of the chemical analysis (sect.2) we attribute this ΔH_s value to the
pure, stoichiometric CeN. The dependence of ΔH on nonstoichiometry
in the homogeneity range of CeN should be nearly parabolic (as in
the case of an alloy). This is the case in the range 49.4-50.2 at%N.

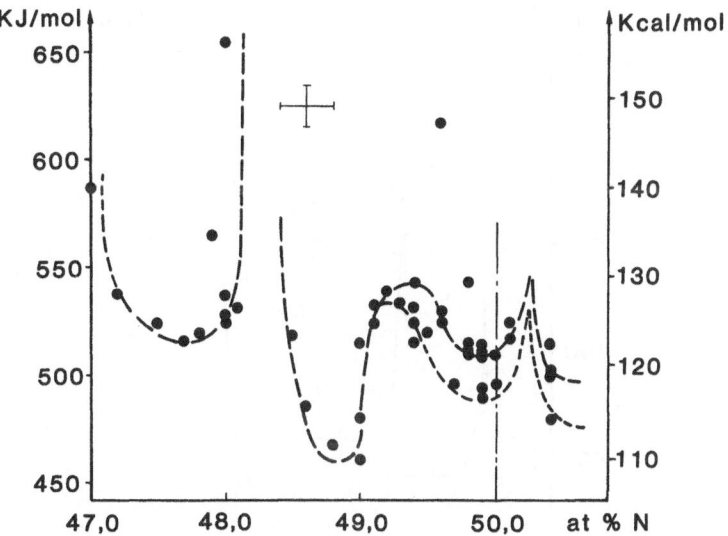

Figure 5. Enthalpies of solution (ΔH_s) of CeN in 4n HCl as a function of stoichiometry. Lower values correspond to more stable compositions. The crossed bar indicates maximal error.

A relative maximum of ΔH_s, indicating an instability of CeN, appears at 49.2 at% N, followed by a pronounced minimum (maximum of stability) at 48.9 at% N. As Fig.5 shows, two stability maxima (48.9 and 49.9) therefore exist in the nonstoichiometry range (48.5-50.5) of CeN. The instability at 49.2 has a form similar to a spinodal decomposition. The explanation for this instability is probably an unstable electronic (valence) state induced by this specific nonstoichiometric composition, similar to the case for TmSe (9). Until the cerium-rich phase boundary is definitely established an alternative explanation for the splitting of the phase at 48-50 at% N could be doping with traces of impurity (such as the 47-48 at% N phase). The shape of the curve in the nitrogen-rich homogeneity range is not clear. More information will be available after experiments at higher nitrogen pressures ($p(N_2) > 1140$ Torr) have been performed.

The main question which arises from these results vis-à-vis the thermodynamics of conventional (integral valence) compounds is the relatively large enthalpy range (approx. 18 kcal/mol, i.e. 14% for the phase at 48,9) in which each of these phases can exist. This is a much higher range than the tangent law for ΔG-curves (of phases in equilibrium) permitts, even if we allow for unusually strong ΔS contributions. This effect is also observed in TmSe (9) and

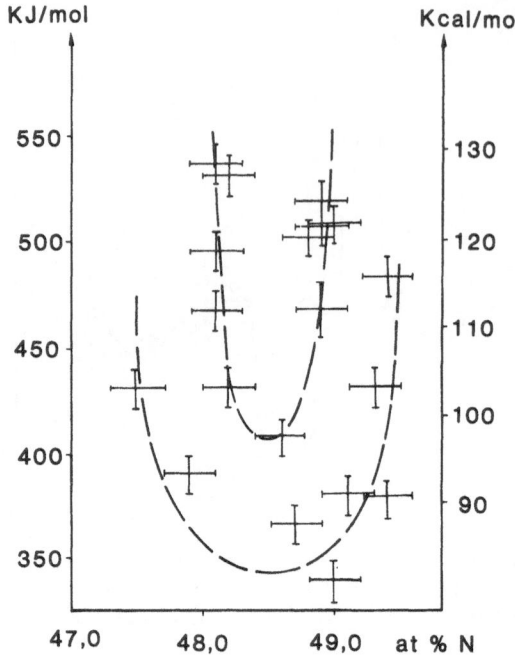

Figure 6. Enthalpies of solution (ΔH_S) of oxygen-doped CeN(<1%wt). Note the increase of stability (decrease of ΔH_S) as compared to pure CeN (Fig.5). The lattice constants of these samples lie in the range 5.022 to 5.025.

$Sm_3S_4-Sm_2S_3$. The explanation which seems more probable at present is, that here metastable states are frozen in the solid with the help of defects, foreign atoms etc. There are several aspects of Fig.5 that support this explanation, e.q. the double curve at 49.5 at% N, and the extremely high ΔH_S values around 48 at% N, which belong to samples quenched from high temperatures.

The most clear indication, however, for the existence of a metastable state which can be destabilized by foreign atoms, is given by the ΔH_S values of oxygen doped samples. Fig.6 shows these results. They belong to samples doped on purpose and to samples doped by contamination during growth. All these samples show distinctly large lattice constants (5.022 to 5.025) in accordance with the histogram

of Fig.3. The sum of the cerium and nitrogen content of these samples is equal to or higher than 1%. As we can see from Fig.6 a
decrease of the heat of solution of CeN from 122 kcal/mol to 98 kcal/
mol (20%) and to 82 kcal/mol (33%) takes place with this doping.
It is difficult to imagine a substance which is so stable that its
admixture by 1% with CeN results in a 20 or 33% increase in its stability. We therefore believe that oxygen doping and the associated defects disturb the unstable balance of the valence states in the mixed
valent CeN, so that stable, possibly integral, valence states of CeN
result with energies 1.04 eV and 1.7 eV lower than that of the pure
unstable phase. On the other hand, as these changes in energy are not
sensed by the lattice constant, we have to assume that the corresponding changes of the 4f-electron configuration does not influence
the orbitals of the outer electrons and therefore the radius of
cerium-ion.

CONCLUSIONS

To date two chemical methods for changing the valence state of
compounds with mixed valence have been available: variation of stoichiometry, illustrated for the first time in the case of TmSe (9),
and mixed crystal formation as in the case of TmSe-TmTe and TmSe-
EuSe (11). The above results of CeN indicate now that doping may also lead to abrupt changes of valence. The theoretical background
for these phenomena will be better understood only when a model for
the 4f-5d hybridization is formulated.

REFERENCES

1. R.C.Brown and N.J.Clark, Mat.Res.Bull, 9:1007 (1974)

2. E.Kaldis and Ch. Zürcher, Proceed. of the 12th Rare Earth Research
   Conference, Denver Research Institute, Denver, pp.915-934 (1976)

3. B.Steinmann, Diploma Thesis, Laboratorium für Festkörperphysik
   ETH, October 1980

4. E.Kaldis, "Principles of the Vapour Growth of Single Crystals",
   in Crystal Growth: Theory and Techniques, Vol.1, C.H.L. Goodman
   (ed.) Plenum Press, New York, pp.49-192 (1974)

5. L.E. Toth, Transition Metal Carbides and Nitrides, Academic
   Press (1971)

6. See e.g. R.R.Roth and S.J.Schneider, Solid State Chemistry,
   Nat. Bureau of Standards, 1972

7. A.Wisard, E.Kaldis, unpublished results

8. B.Batlogg, E.Kaldis and H.R.Ott, Physics Letters, 62 A:270 (1977)

9. E.Kaldis, B.Fritzler, E.Jilek and A.Wisard, Journal de Physique, C5, 40:366 (1979)

10.G.K.Wertheim, W.Eib, E.Kaldis and M. Campagna, Phys.Rev., B 22: 6240 (1980)

11.E.Kaldis and B.Fritzler, Journal de Physique, C5, 41:135 (1980)

12.E.Kaldis, H.Spychiger, B.Fritzler, E.Jilek, this volume p

ENTHALPIES OF SOLUTION AND LATTICE CONSTANT ANOMALIES OF

Sm_3S_4 - Sm_2S_3 SOLID SOLUTIONS

E. Kaldis, H. Spychiger, B. Fritzler, and E. Jilek

Laboratorium für Festkörperphysik ETH

8093 Zürich, Switzerland

The large scattering of the lattice constants, $\underline{a}$, of Sm_3S_4 reported in the literature, recently triggered a systematic investigation as a function of composition in the range of the Sm_2S_3-Sm_3S_4 solid solutions (1), which have the Th_3P_4-structure. This work made obvious that the various values of $\underline{a}$ quoted in the literature, are due to the different stoichiometries of the investigated samples. Lack of trustworthy chemical analyses resulted in the existing confusion. Further, it was shown (1), that in the Sm_2S_3-Sm_3S_4 range the lattice constants deviate abruptly from the expected linear dependence (Vegard's law) towards lower values. This was explained by an anomalous increase of the trivalent contributions in the valence of samarium (1). Recently this investigation was pursued further by a combination of calorimetric, phase diagram and lattice constant investigations, using the chemical composition as a parameter. In this paper we present the calorimetric results and some additional lattice constants.

To avoid the influence of thermal treatment on the scattering of the lattice constants, a series of new samples with various compositions, not covered in the past, were synthesized at 1600 C. The temperature was chosen due to the fact that the preliminary phase diagram shows no liquid phase up to this temperature. Therefore, it was expected that composition changes due to kinetic (constitutional supercooling during solidification) or thermodynamic (peritectic melting) effects would be avoided. These samples were prepared by direct reaction of the elements in quartz ampoules. The reaction product was sealed by electron beam welding in W-crucibles and heated with high frequency to 1600 C for 24 hours. The handling of

237

the samples took place in a train of three glove boxes, made of stainless steel, which contained a balance and a lathe for cutting the samarium turnings. Their argon atmosphere was gettered by hot cerium turnings to a few ppm of oxygen and humidity.

Figure 1 shows the lattice constants of samples which have been used for the calorimetric measurements. The anomaly found previously (1) was further investigated both at high and room temperature. We received information about the high temperature range (800<T< 2400 C) from the phase diagram which we constructed from the DTA data (2). The current results do not indicate any anomaly in the sulfur-rich concentration range of Sm_3S_4 for x>0.42. Solution calorimetry as a function of composition may show the existence of phases appearing near room temperature (T<800 C). Particularly in the case of mixed valence compounds, this method may also show the existence of phases with different valence states within the same structure (3).

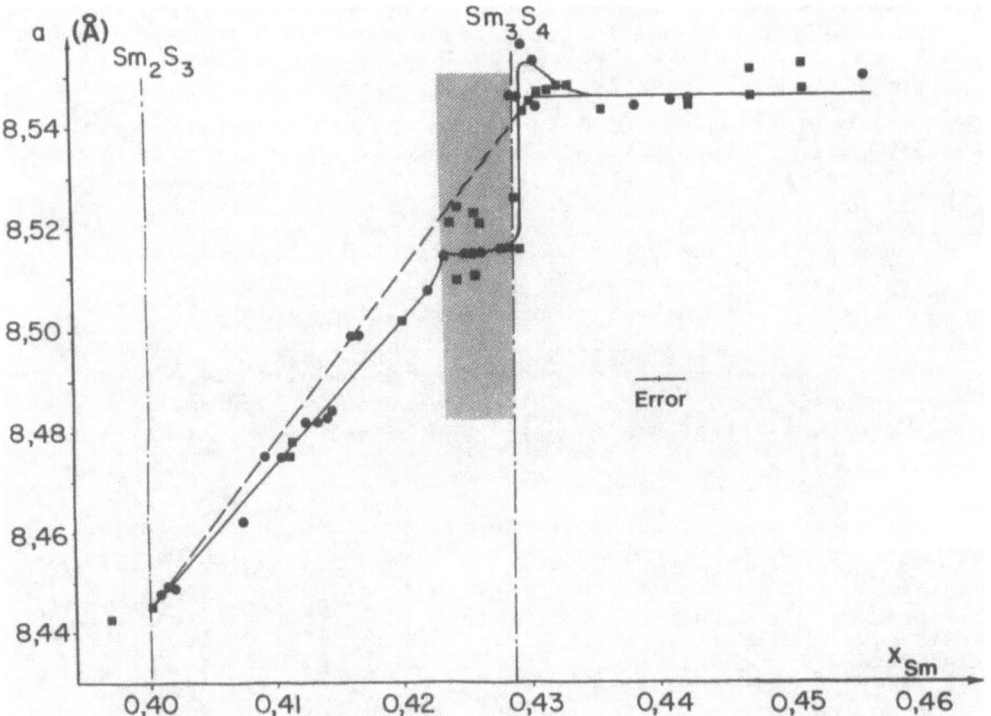

Figure 1. Lattice constant as a function of the mol fraction of Sm. All samples were heated in sealed W-crucibles. ●● denote samples heated up to only 1600 C. The maximum error of the chemical analysis of samarium (x=±0.002) and of the lattice constant are given with the crossed bar. The shadowed region indicates a possible miscibility gap.

Figure 2 shows the variation of the heat of solution ΔH_S in 4n HCl, at 25 C, as a function of the chemical composition. Determination of ΔH_S and chemical analysis of samarium were performed in aliquot parts of the pulverised sample used also for the determination of the lattice constant. The heat of solution was measured using an LKB calorimeter (4). The original LKB glasampoules were slightly modified (funnel neck) in order to be sealed in the glove box with a glass sphere and a drop of molten Apizon (5). From thermodynamic considerations, we expect in a solid solution range a quasi-parabolic form of the free energy of formation as a function of composition, $\Delta G = f(x)$. As the entropy contributions are not large in the solid state, we expect a similar form for the $\Delta H_S = f(x)$ function. In the case of a two-phase region the ΔH_S of the two phases are additive and a linear dependence on the concentration is expected. This

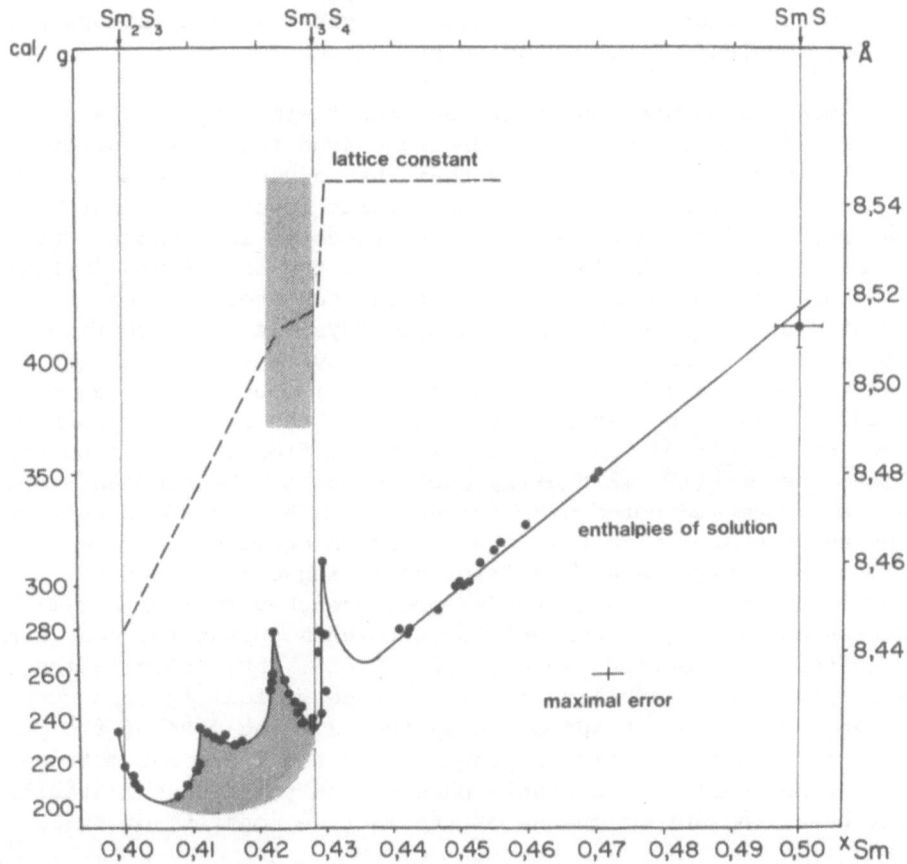

Figure 2. Heat of solution of samarium sulfides in 4n HCl as a function of composition (lower curve, left scale). For comparison the lattice constants are also shown (upper curve, right scale).

condition is very well fulfilled in the range of x>0.44, where according to the phase diagram and metallographic investigations a eutectic of Sm_3S_4-SmS exists.

For the solid solution range, Sm_2S_3-Sm_3S_4, we would expect one wide parabola. The solution calorimetry reveals however a fine structure of three subphases with phase boundaries at $x \simeq 0,399$, 0.411, 0.422, 0.430 and maxima of stability at $x \simeq 0.404$, 0.417, 0.428. It seems probable that the two end-subphases correspond to the homogeneity ranges of Sm_2S_3 and Sm_3S_4. According to the tangent law, the two subphases on the left (0.399<x<0.422) are in equilibrium with each other but not with the Sm_3S_4 phase. Therefore, an instability in the mixed crystal series Sm_2S_3-Sm_3S_4 should be expected in the range of $x \simeq 0.422$. This is about the composition range where the lattice constant anomaly appears in Fig.1. This instability can be due to changes of the valence state as suggested before (1). On the other hand structural and X-ray single crystal investigations have been performed as a function of nonstoichiometry.

It must be pointed out that the reproducibility of the data of Fig.2 is very good. The small deviations from the drawn curves indicate an average error much lower than that given by the maximum error bars in Fig.2. The deviations towards lower Sm-content in the linear part (x>0.45) are due to minute tungsten inclusions. This is also the reason for the large error bar in the ΔH_S of SmS (x=0.5). This faint reaction with the walls of the tungsten crucible at high temperatures makes the growth of single crystals as a function of nonstoichiometry difficult. It is noteworthy that both samples heated to 1600 C and higher temperatures, give compatible calorimetric data. This is not exactly the case for the lattice constants. The set of the a's of Fig.1 give slightly different values to those published before (1). This difference is due to the samples evaporated in vacuum and noted with triangles in the figure of ref.1,which in the short evaporation time did not reach equilibrium. The main new results in Fig.1 are 1) a faint splitting of the lattice constants has been found in two samples quenched from high temperatures and 2) larger composition gradients appear for some samples in the mol fraction range 0.422<x<0.426. They indicate the possibility of a miscibility gap (two-phase region) in this composition range which is not supported by the existing data of the phase diagram (2). An alternative explanation is the increased metastability of thermodynamically unstable phases (due to valence instability). In any case, the abrupt change of the lattice constant of Sm_3S_4

definitely exists. Stoichiometric samples, heated in sealed tungsten crucibles, have a lattice constant in the range of $8.520 > a > 8.516$ Å and a maximum of stability (Fig.2), and overstoichiometric samples are unstable and have $\underline{a} \simeq 8.546$ Å.

REFERENCES

1. E.Kaldis, J.Less-Common Metals, 76:163 (1980)
2. E.Kaldis, H.Spychiger, B.Fritzler, W.Peteler, to be published
3. E.Kaldis, B.Steinmann, B.Fritzler, E.Jilek, this volume pp.
4. H.Spychiger, Diploma Thesis, Laboratorium für Festkörperphysik ETH, April 1981
5. J.Fugger, private communication

HIGH TEMPERATURE PHASE DIAGRAM AND ENTHALPIES OF SOLUTION OF TmSe

B.Fritzler, E.Kaldis, E.Jilek

Laboratorium für Festkörperphysik ETH
8093 Zürich, Switzerland

As we have shown in the past TmSe, a mixed valent compound, has a large homogeneity range (1,2,3,4). Variation of nonstoichiometry leads to a large change of the Tm-valence (approx.25%) and therefore induces fascinating physical properties (1,5). Their interpretation however is not yet complete due to the lack of a theoretical model for the 4f-5d hybridization. In principle, it is also possible that some of the properties attributed to a homogeneous sample of a mixed valent compound result from the existance of several phases. In order to investigate the "homogeneity" of the nonstoichiometric range of TmSe and the phase relationships as a function of temperature, three kinds of investigations have been carried out. Structural, including single crystal presession photographs and electron diffraction, T-x phasediagram up to 2300 C, and calorimetric measurements (heats of reaction of TmSe with 4n HCl) as a function of nonstoichiometry.

The structural investigations showed clearly that only the NaCl-structure appears in the composition range $0.87 < x < 1.05$ (x=mol Tm/mol Se). With increasing Se-excess however, the electron diffraction (5) showed that a short range order (probably vacancies) appears in the range of 100-500 Å. At more Se-rich compositions ($x < 0.87$) the stability range of the Tm_5Se_6 superstructure begins.

T-x PHASE DIAGRAM

The phase diagram Tm Se has been further investigated by DTA using the Mettler Thermoanalyser TA1. Similar to many rare earth chalcogenides with m.p. $>$ 2000 C, a strong incongruent evaporation

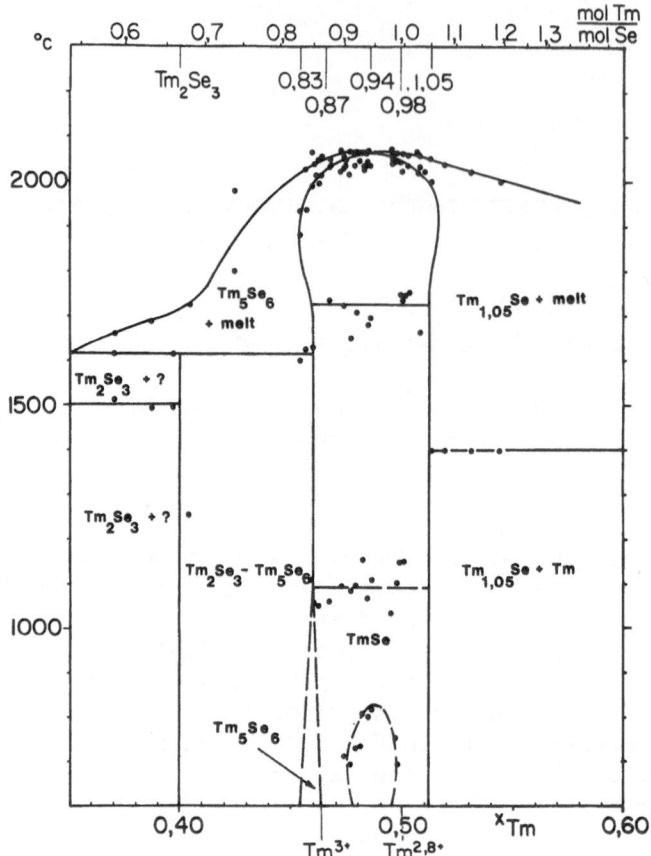

Figure 1. T-x phase diagram of the Tm-Se system.

starts in TmSe already at 1500 C. Meaningful DTA investigations can therefore be carried out only in sealed tungsten capsules. Due to the geometry of the furnace, the capsules must have a diameter of 4 mm and a length between 8 and 10 mm. They are made by electro-erosion of a W-rod. To avoid changes in composition due to incongruent evaporation during sealing of these capsules with an electron beam (T>3600 C at the lid), a special technic was developed which has been described elsewhere (6). More than 40 compositions in the range 0.2<x<0.8 have been investigated by DTA. Each sample was a small part of a single crystal, adjacent parts of which have been used for chemical analysis and lattice constant determination. Fig.1

shows the Tm-Se phase diagram, a modified form of the older version (3,4) according to the latest measurements. TmSe melts congruently at 2065 C. Its large nonstoichiometric range is framed by an entectic and a peritectic. In the Tm-rich range, the eutectic defines the Tm-rich pase boundary of TmSe at $x \simeq 1.05$. In the Se-rich range appear the peritectically melting Tm_2Se_3 and probably an eutectic. The nonstoichiometric range of TmSe shows several characteristic features:

1.The liquidus curve is very flat in a large homogeneity range. As a matter of fact it shows even a slight depression in the middle, which led us to propose in previous publications (3,4) a very narrow eutectic with a phase transition at a slightly lower temperature. We have confirmed the splitting of the peaks which led us to this model in several new compositions. However, the scattering of the data does not allow us at the present stage to decide definitely about the narrow eutectic from the DTA measurements. Due to this reason, we have fitted the data to a conventional maximum solubility diagram. Both narrow eutectic or very flat maximum solubility diagram indicate an instability of TmSe. This conclusion is further supported by other characteristics of this phase diagram and that of TmSe - TmTe.

2.The new measurements support the existence of the phase transitions at approx. 1700 and 1100 C. The nature of these transitions (electronic or structural) is not yet known. The scattering of the measurements is appreciable, indicating a strong metastability of the involved phases. Particularly for the transition at $T \simeq 1700$ C, more complicated shapes could appear after a more detailed investigation has been carried out.

3.Figure 1 shows also a phase transition at $T < 830$ C for $0.94 < x < 0.98$. Based on the current data, we may interpret this transition as a miscibility gap. The existance of a miscibility gap in a nonstoichiometric range needs further explanation as the two phases at the boundaries have the same structure and belong to the same compound. The only difference between these phases known at present is the valence of Tm which changes as a function of nonstoichiometry between approximately 12% to 16% Tm^{2+}-contributions. On the other hand, we may recall that a miscibility gap is consistent with a flat maximum of liquidus (or a narrow eutectic) in a thermodynamic picture of repulsive interactions. We conclude therefore that a miscibility gap in the nonstoichiometric range would be possibly the result of valence segregation i.e. of a repulsive interaction between two different valence states of thulium. The interpretation of this transition as a miscibility gap is based on the fact that it was not possible up to now to synthesize samples in the composition range $0.94 < x < 0.98$. However, two-phase samples were not yet found.

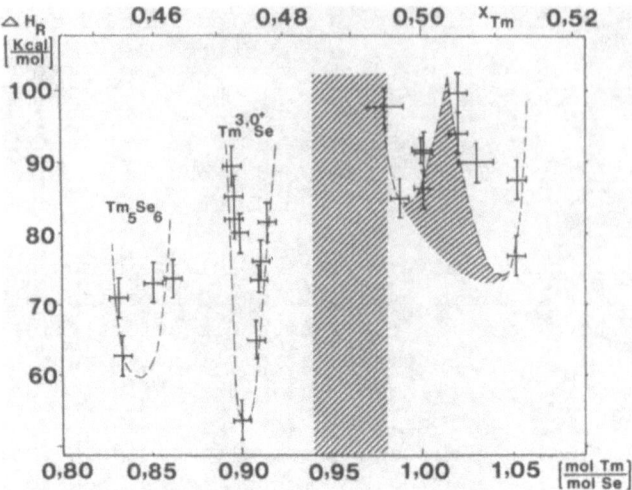

Figure 2. Enthalpies of solution of Tm$_x$Se in 4n HCl as a function of nonstoichiometry. The miscibility gap is shadowed.

CALORIMETRIC INVESTIGATIONS

The reasons why measurements of the heat of solution as a function of nonstoichiometry are interesting have been already discussed (8). Here we note that this method is possibly suitable to trace energetic changes due to the valence change of thulium as a function of non-stoichiometry. Fig.2 shows the dependence of the enthalpy of reaction (ΔH_R) on nonstoichiometry, as determined by the reaction of TmSe with 4n HCl:

$$Tm_x Se_{(s)} + 3x\ HCl \rightarrow x\ TmCl_3 + H_2Se + (3x-1)/2H_2\uparrow$$

Shortly after dissolution the H_2Se is oxidized and a red Se-precipitate appears. Of course this reaction cannot lead to quantitative measurements of the enthalpy of formation (ΔH_f) and therefore to absolute values of the energies of the various valence state configurations of nonstoichiometric Tm$_x$Se. For the determination of quantitative values of ΔH_f, we have built the apparatuses for fluorine combustion calorimetry and purification of commercial fluorine by destillation and we hope to start the measurements soon.

However, under the reasonable assumption that changes of the chemistry of the above reaction as a function of nonstoichiometry do
not cause singularities of the thermal effect, we expect the shape
of the $\Delta H_R = f(x)$ curve to be similar to that of $\Delta H_f = f(x)$. The general
conclusion which can be reached from Fig.2 is that the gross nonstoichiometric range splits up into a number of narrow "subphases".
Their nature can be deduced from their stoichiometries. The phase
at $x \approx 0.85$ corresponds to the Tm_5Se_6 superstructure (3,4). From the
systematic of the lattice constants (3,4) and the physical properties
(5,9) we conclude that the phase at $x \approx 0.91$ is $Tm(3.0+)Se$ and has a
normal (integer valence) state. It is possible that the small changes
of nonstoichiometry in the narrow stability range of this phase do
not influence strongly the 4f-5d hybridization or they may contribute
to a heterogeneous valence state. The twin-form of the $x > 0.98$ phase
indicates the existence of an instability at $x \approx 1.00$. This is also
the composition range where unexpected physical properties (1,5)
indicate the onset of the valence fluctuation (homogeneous mixed
valence). This interpretation indicates that calorimetry reflects
band structure effects. From pure thermodynamic considerations, it
seems astonishing that these phases, particularly $Tm(3.0+)Se$ at
$x \approx 0.91$, can be stable in such large energy ranges ($50 < \Delta H_R < 90 kcal/mol$).
The results of CeN (8) show however, that several valence states
can be frozen in the same range of compositions by defects or impurities. Thus it could be possible that the long parabola of
$Tm(3.0+)Se$ results from superposition of two shorter parabolas:
$70 < \Delta H_R < 90$ and $55 < \Delta H_R < 70$ kcal/mol.

CONCLUSIONS

Calorimetric measurements as a function of nonstoichiometry of compounds with valence instability seem to give interesting information
about the energetic state of the various nonstoichiometric compositions. A further conclusion from the existing data is, that valence
instabilities induce thermodynamic phase instabilities shown in the
phase diagram and the heats of reaction. This is found not only for
TmSe but also for CeN (8) and Sm_3S_4 (10). Characteristic for these
thermodynamically unstable phases is an appreciable degree of metastability.

REFERENCES

1. B.Batlogg, E.Kaldis and H.R.Ott, Phys.Letters, 62A:270 (1977)
2. B.Fritzler, E.Kaldis, W.Peteler and A.Wisard, Proceedings Euchem.
   Conference "Chemistry of the Rare Earths, p.129, L. Niinistö (ed),
   Helsinki Institute of Technology, May 1978

3. E.Kaldis, B.Fritzler and W.Peteler, Z.Naturforsch., 340:55 (1979)

4. E.Kaldis, B.Fritzler, E.Jilek and A.Wisard, J.Physique, 40:C5-366 (1979)

5. B.Batlogg, H.R.Ott, E.Kaldis, W.Thöni and P.Wachter, Phys.Rev., B19:247 (1979)

6. E.Kaldis and W.Peteler, Proceed. 6th Int.Conference on Thermal Analysis, Vol.2, p.67, Birkhäuser Verlag, Basel, 1980

7. E.Kaldis and B.Fritzler, J.Physique, 41:C5-135 (1980)

8. E.Kaldis, B.Steinmann, B.Fritzler, E.Jilek and A.Wisard, this volume pp.

9. G.Wertheim, W.Eib, E.Kaldis and M.Campagna, Phys.Rev.,22:6240 (1980)

10.E.Kaldis,H.Spychiger, B.Fritzler and E.Jilek, this volume pp.

DENSITIES AND CONCENTRATION OF DEFECTS IN TmSe

B.Fritzler, E.Kaldis, B.Steinmann, E.Jilek, A.Wisard

Laboratorium für Festkörperphysik ETH

8093 Zürich, Switzerland

A few years ago we have reported the dramatic dependence of the lattice constant of TmSe on nonstoichiometry (1,2,3,4), which has been used to produce TmSe samples with controlled valence of Tm. Figure 1 shows recent results on more than 40 samples. Chemical analysis of Tm (complexometrically) and lattice constant determination have been made on the samé sample.

Figure 2 shows the results of the density measurements of single crystals of TmSe as a function of nonstoichiometry, using the buoyancy method. Details of the experimental set-up have been given elsewhere (3). The accuracy of the experimental set-up is 190 ppm in density, as determined by measurements of high purity silicon single crystals (7). All experimental densities in Fig. 2 have lower values than 8.95 g/cm^3, which is calculated from the lattice constant for an ideal, stoichiometric crystal with completely occupied unit cell (X-ray density). The measured values in the nonstoichiometric range lie between two curves which converge at ρ=0.87 and have at ρ=1.05 a divergence of density of 1.66%. The scattering of the experimental points ●● in this range (0.87<ρ<1.05) becomes larger with increasing Tm concentration, indicating that metastability increases with ρ. This may be the result of the phase transitions at higher temperatures shown in the T-x phase diagram (6). The discontinuity in density at ρ=0.87 is attributed to the formation of the Tm$_5$Se$_6$ superstructure (doubling of the lattice constant) which is obviously accompanied by a 2% increase of the concentration of the Schottky pairs.

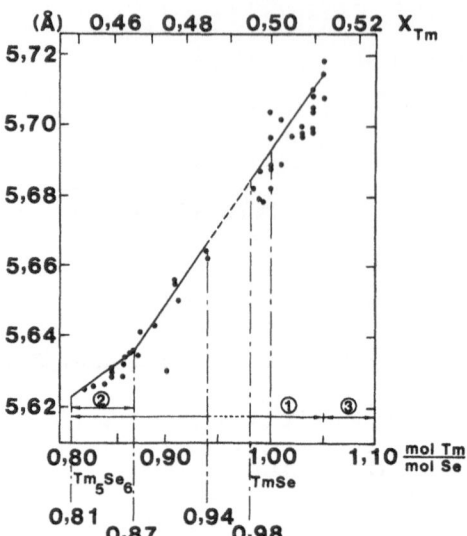

Figure 1. Lattice constants of TmSe as a function of nonstoichiometry. 1 shows the gross nonstoichiometric range of the NaCl structure, containing the superstructure around Tm_5Se_6 ② ③ indicates the eutectic with Tm.

Figure 2 shows also the curves of the theoretical densities as a function of the mol ratio $\rho = n_{Tm}/n_{Se}$ for the vacancy ① interstitial ② and antisite ③ models. These curves are calculated from the following equations (5) respectively:

$$d_{Se}^{exc} = \left[4\rho \ (Tm) + 4(Se) \right] / \ N_L a^3 \tag{1a}$$

$$\text{(vacancies)}$$

$$d_{Tm}^{exc} = \left[4(Tm) \quad + 4(Se)/\rho \right]/N_L a^3 \tag{1b}$$

$$d_{Se}^{exc} = \left[4(Tm) \quad + 4(Se)/\rho \right]/N_L a^3 \tag{2a}$$

$$\text{(interstitials)}$$

$$d_{Tm}^{exc} = \left[4\rho(Tm) + 4(Se) \right] / N_L a^3 \qquad\qquad (2b)$$

$$^{as}d_{Tm/Se}^{exc} = \left[8\rho(Tm) + 8(Se) \right] / \left[(1+\rho)\ N_L a^3 \right] \text{ (antisite)} \qquad (3)$$

where (Tm), (Se) denote the atomic weights of the respective elements, N_L the Avogadro number, $\underline{a}$ the experimentally determined lattice constant and ρ the mol ratio (Tm/Se) as an expression of nonstoichiometry.

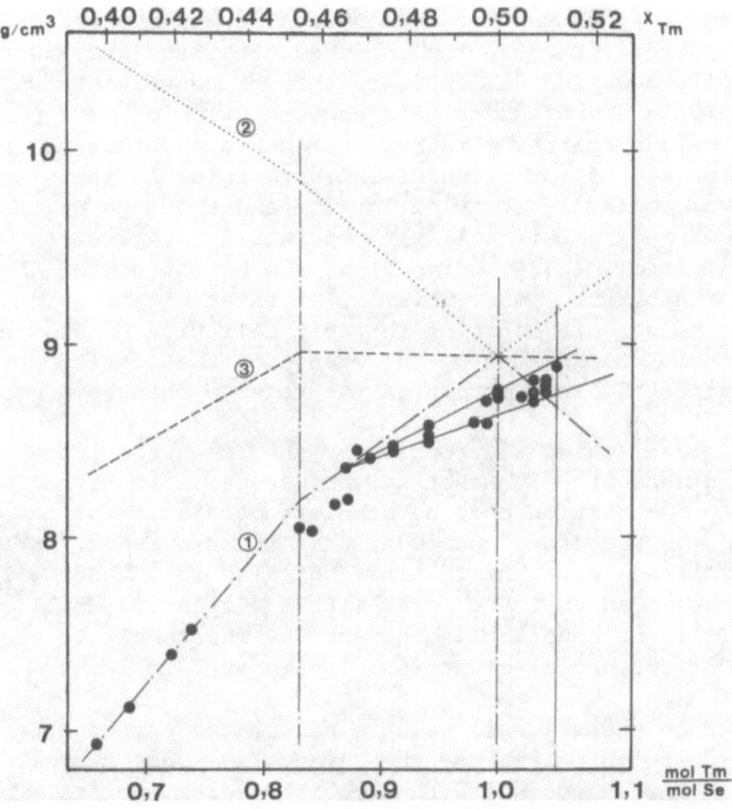

Figure 2. Densities of TmSe as a function of nonstoichiometry.●●● measured densities. Calculated densities according to ① nonstoichiometric vacancy model, ② interstitial model, ③ antisite model.

At $\rho=0.87$ the calculated density for a nonstoichiometric vacancy model with completely occupied Se-sublattice and vacancies in the Tm-sublattice becomes equal to the measured density. The total concentration of vacancies at $\rho=0.87$ ($d_{exp}=8.351$ g/cm^3) is $8.24 \cdot 10^{22}$/mol TmSe, i.e. 6.84%. With decreasing Se-excess (increasing ρ), the curves of calculated and measured densities diverge due to Schottky vacancy-pairs. Stoichiometric crystals ($\rho=1.00$) contain approximately 2% Schottky vacancy-pairs. Therefore a combiantion of nonstoichiometric thulium vacancies (zero at $\rho=1.00$; 6.84% at $\rho=0.87$) and Schottky vacancy-pairs (zero at $\rho=0.87$; 2-3% at $\rho=1.00$) can describe exactly the experimental curve in the Se-excess nonstoichiometric range.

It is more difficult to decide about the Tm-excess range. The measured densities show no maximum at the stoichiometry but continue to increase until the theoretical density of a fully occupied lattice is reached near the phase boundary at $\rho\simeq1.05$. Therefore the formation of nonstoichiometric vacancies similar to the Se excess range is contradictory to the measurements. Also, an interstitial model can be discarded, because the measured densities are lower than that of the fully occupied cell. Due to this, interstitials can be discussed only in the presence of a concentration of some percent of Schottky vacancy-pairs, a rather unprobable combination in view of the high energy of formation of interstitials. An antisite model, assuming a completely occupied lattice and a change of the ratio of Tm/Se atoms in the unit cell with increasing ρ, gives an almost horizontal line (Fig. 2, curve 3) at the height of the X-ray density of stoichiometric TmSe. This is, at first sight, quite astonishing, but the weight change of the unit cell with nonstoichiometry is completely compensated by the change of the lattice constant, which carries the information of the valence change.

The model, which we propose for the Tm-excess range, is based on the presence of 2-3% Schottky vacancy-pairs in crystals of stoichiometric composition and the complete occupation of the lattice at $\rho\simeq1.06$. We therefore conclude, that in this range, Tm ions are not only filled in vacant Tm sites but also in vacant Se lattice sites, that is an antisite model with a decreasing small concentration (<3%) of Schottky vacancy-pairs. The result is the formation of up to 3% octahedral clusters, containing 7 Tm ions.

Based on these ideas, we have calculated the density as a function of nonstoichiometry for the Se and Tm-excess ranges. This model line is in good agreement with the experimental points. The general scheme of the calculation, which will be published in detail elsewhere (5), is the following: As the mechanism of the increase of Schottky pairs with increasing ρ(in the Se-excess range, $0.87<\rho<1.0$, $0.465<x_{Tm}<0.5$) is not easily described, we assume a linear increase

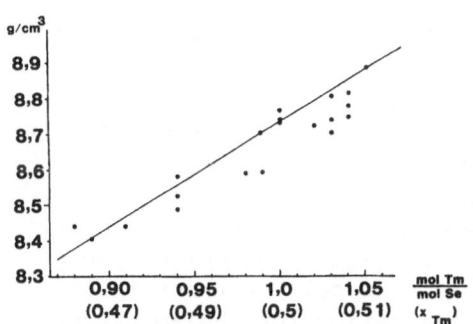

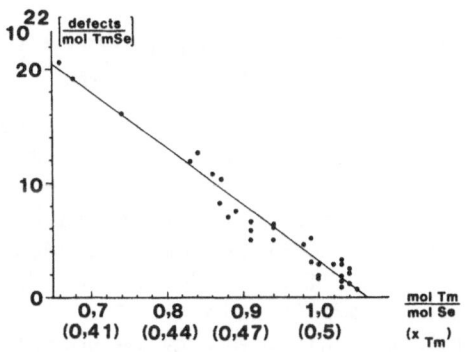

Figure 3. Densities of TmSe (NaCl structure) vs. nonstoichiometry. Full line calculated from eq. (4) and (5).

Figure 4. Vacancy concentration vs. nonstoichiometry of TmSe, calculated from the theoretica densities (eq. 4 and 5). Note that the concentration of vacancies becomes zero at $\rho \approx 1.06$.

between $\rho=0.87$ where it is zero, and $\rho=1.00$. In other words, we fit the density equation to Δd_o, the difference between the measured and X-ray density of stoichiometric TmSe. The corresponding equation based on the mol fraction x_{Tm} as parameter is

$$d_{Se}^{exc} = \frac{1}{N_L a^3} \left[\left(4 - 4\frac{1-2\cdot0.465}{1-0.465} - \frac{x_{Tm}/(1-x_{Tm})-0.87}{0.13} \cdot \frac{\Delta d_o a_o^3}{((Tm)+(Se))/N_L}\right)(Tm) \right.$$

$$+ \left(4 - \frac{x_{Tm}/(x_{Tm}-1)-0.87}{0.13} \frac{\Delta d_o a_o^3}{((Tm)+(Se))/N_L}\right)(Se) + \left[(2x_{Tm}-1)\right.$$

$$\left. \left[4 - \frac{x_{Tm}/(x_{Tm}-1)-0.87}{0.13} \frac{\Delta d_o a_o^3}{((Tm)+(Se))/N_L}\right] - (x_{Tm}-1)\ 4\frac{1-2\cdot0.465}{1-0.465}\right]$$

$$(Tm)/(1-x_{Tm})\right] \tag{4}$$

with Δd_o and a_o taken with their values at $x_{Tm}=0.5$. In the Tm-excess region $1.00<\rho<1.05$ $(0.5<x_{Tm}<0.512)$ the combined antisite-Schottky pairs model leads to the equation (5)

$$d_{Tm}^{exc} = \frac{1}{N_L a^3} \left[\left(4 - \frac{\Delta d_o a_o^3}{((Tm)+(Se))/N_L}\right)\ (Tm)\ +\ \left(4 - \frac{\Delta d_o a_o^3}{((Tm)+(Se))/N_L}\right)\ (Se) \right.$$

$$+ \left(4 - \frac{\Delta d_0 a_0^3}{((Tm)+(Se))/N_L}\right) \left(\frac{2x_{Tm}-1}{1-x_{Tm}}\right) (Tm)\right] \tag{5}$$

Both equations (4) and (5) define a straight line for the whole non-stoichiometric range of TmSe. Figure 3 shows the agreement of the calculated line with the measured densities. If we calculate the concentration of vacancies as a function of the nonstoichiometry using the nonstoichiometric vacancy model for $\rho<1.00$ and the anti-site model for $\rho>1.00$, we can draw a straight line which reaches zero at $\rho \approx 1.06$ (Fig.4). The filling of the cell (without inter-stitials) gives a very reasonable explanation for the appearance of the eutectic of $Tm_{1.05}Se$ with Tm as shown by the T-x phase diagram.

Let us briefly consider the consequences of this simple model for the physical properties of TmSe. If Tm-ions are also filled into Se-vacancies in the Tm-rich range, then a small concentration (<3%) of octahedral clusters will be formed with a 50% smaller Tm-Tm distance, $a/2 \approx 2.85$Å. This is approximately 20% smaller Tm-Tm distance in the hexagonal metallic thulium (≈ 3.5 Å). The strong decrease of the Tm-Tm distance will tend to increase the crystal field. The question arises however, what is the overall effect in view of the small concentration of these clusters. It is interesting to point out, that in this nonstoichiometric range, valence fluctu-ation has been found (1) and the calorimetric measurements show an instability (6).

REFERENCES

1. B.Batlogg, E.Kaldis and H.R.Ott, Phys.Letters, 62A:270 (1977)

2. B.Fritzler, E.Kaldis, W.Peteler and A.Wisard, Proceedings Euchem. Conference "Chemistry of the Rare Earths", p.129, L.Niinistö (ed) Helsinki Institute of Technology, May 1978

3. E.Kaldis, B.Fritzler and W.Peteler, Z.Naturforsch., 34a:55 (1979)

4. E.Kaldis, B.Fritzler, E.Jilek and A.Wisard, J.Physique, 40:C5-366 (1979)

5. B.Fritzler, E.Kaldis (to be published)

6. B.Fritzler, E.Kaldis, E.Jilek, this volume pp.

7. B.Fritzler, Diploma Thesis, Laboratorium für Festkörperphysik ETH Zürich, April 1978

PREPARATION AND THERMOELECTRIC PROPERTIES OF SOME RARE EARTH CHALCOGENIDES

T. Takeshita, B. J. Beaudry and K. A. Gschneidner, Jr.

Ames Laboratory,[*] and Department of Materials Science
and Engineering, Iowa State University
Ames, IA 50011, U.S.A.

Rare earth sesquisulfides were synthesized by the direct combination of the constituent elements (i.e., $2R + 3S \rightarrow R_2S_3$). This
method produces chemically pure products if high purity materials
are used, but often produces mixed phases (1), because this is a gas
(sulfur vapor) - solid (rare earth metal) reaction under the conditions the reaction is normally carried out. The gas-solid reaction
is dictated by not only the reaction temperature and the composition
of the reaction mixture but also the pressure of gaseous species.
As is the case for many gas-solid reactions, the reaction product
(RS_x) forms as a dense hard skin on the metal surface, and sulfur
must diffuse through this sulfide layer into the metal core for the
reaction to proceed. Many times this kinetic factor is difficult to
circumvent (2).

It is known that certain metal halides dissolve chalcogenides
and function as fluxing agents in gas-solid reactions (2). We have
studied the synthesis of rare earth sulfides by a modified flux method, and examined reaction parameters of the reaction, e.g., 1) kind
of halide, 2) the ratio of the reaction mixture and the flux, 3) the
reaction temperature, 4) the vapor pressure of sulfur, and 5) the removal of the halide from the final product.

The rare earth triiodide was chosen as the flux since it has a
low melting point and iodine is easy to handle at ambient temperatures.

[*]Operated for the U.S. Department of Energy by Iowa State University
under contract No. W-7405-ENG-82. This research was supported by the
Assistant Secretary for Nuclear Energy, Office of Nuclear Reactor
Programs, WPAS-AE-15-25.

Table 1. Thermoelectric power (S) and electrical resistivity (ρ) of $La_{3-x}S_4$ and $Pr_{3-x}S_4$. The S:R ratio is based on chemical analysis.

	50°C		500°C	
	S (μV/°C)	ρ (m$\Omega\cdot$cm)	S (μV/°C)	ρ (m$\Omega\cdot$cm)
$LaS_{1.376}$	−20	0.6	−64	0.9
$LaS_{1.489}$	−67	2.6	−153	6.3
$PrS_{1.476}$	−58	2.2	−136	6.2
$PrS_{1.487}$	−86	4.6	−194	11.3

It is not necessary to use a rare earth triiodide as the starting material since the rare earth metal reacts with iodine vapor first to form the rare earth triiodide (3). A desired amount of the rare earth metal, sulfur and iodine was sealed in a quartz ampoule under a vacuum of $\sim10^{-3}$ torr. The composition of the reaction mixture was varied from $R_2S_3/RI_3 = 1$ to 100, where R is a rare earth. It was found that the amount of the triiodide is not critical, and a small amount of iodine (<100 mg) is sufficient to accelerate the reaction. The reaction temperature ranged from ~750°C to ~900°C, and the vapor pressure of sulfur was maintained at about one atmosphere. The reaction time increased with increasing atomic number (La - 2 hrs. to Lu - 2 days). No free sulfur was found and x-ray diffraction only indicated the presence of the sesquisulfide.

The thermoelectric power and the electrical resistivity were measured for some rare earth sesquisulfides prepared by the above method, and the results are given in Table 1. In the composition range studied, $RS_{1.333}$ - $RS_{1.490}$, both the thermoelectric power and the electrical resistivity were seen to increase linearly with the temperature (for the thermoelectric power, this is in the sense of the absolute magnitude). The figure of merit at 500°C of $LaS_{1.489}$ and $PrS_{1.476}$ is 0.5×10^{-3}°C^{-1} and 0.64×10^{-3}°C^{-1}, respectively. These values of the figure of merit are comparable to that of the n-type silicon germanium alloys, which is currently used in radio-isotope heated thermoelectric generators (RTG) in the space missions.

REFERENCES

1.   E. M. Loginova, A. A. Grizik, N. M. Ponomarev and A. A. Eliseev, J. Inorg. Mater. 41, 644 (1975).
2.   W. Kwestroo, Preparative Methods in Solid State Chemistry, ed. by P. Hagenmuller, Academic Press, NY, London (1972), p. 563.
3.   A. W. Sleight and C. P. Prewitt, Inorg. Chem. 7, 2282 (1968).

THERMAL DECOMPOSITION OF RARE EARTH SULFATE AND SELENATE HYDRATES

Lauri Niinistö, Pekka Saikkonen and Risto Sonninen[*]

Department of Chemistry, Helsinki University of

Technology, SF-02150 Espoo 15, Finland

ABSTRACT

The thermal decomposition of $Ln_2(SO_4)_3 \cdot 8H_2O$ (Ln=La-Lu, Y excel. Pm) and $Ln_2(XO_4)_3 \cdot 5H_2O$ (Ln=La-Nd,Sc excl. Pm; X=S,Se) has been investigated. Degradation in air starts with dehydration and proceeds via oxysulfate to oxide. The dehydration reactions are discussed in relation to the crystal structures.

INTRODUCTION

With the exception of scandium, the sulfates of trivalent rare earths crystallize usually with eight molecules of water. The octahydrates form two structurally different series; the first one comprises lanthanum and cerium and the other series extends over the remaining lanthanoids and yttrium. Thus, the octahydrates form an interesting case for comparative thermoanalytical investigations where the effects of crystal structure and the central ion size can be studied. In addition, there exists another series though shorter, namely the pentahydrates (La-Nd,Sc).

Since the early comprehensive study by Wendlandt et al. (1-3), the thermal stability of the sulfate octahydrates have been reinvestigated occasionally by various methods. The most systematic TG study is that by Bukovec et al. (4) who also studied the dehydration reactions with DSC. Recently, Nabar and Paralkar (5) and Hajek et al. (6) have extended the TG and DTA studies to cover the isostructural selenates of the heavier rare earths and yttrium.

---

[*]Present address: Imatran Voima Oy, Helsinki

In connection of our investigation into the structures of rare earth sulfate and selenate hydrates (7-11), a systematic thermoanalytical investigation was initiated which covered, besides the sulfate octahydrates, the sulfate and selenate pentahydrates. Different experimental techniques (TG, DTG, DTA and DSC) as well as kinetic calculations were used in order to obtain additional information of the structure – thermal stability relationships, especially for the dehydration reactions.

EXPERIMENTAL

The TG, DTG and DTA curves were recorded simultaneously from 300 to 1900 K using a Mettler Thermoanalyzer TA-1. Generally two sample weight/heating rate combinations were used: 200 mg/10 K min^{-1} (I) and 20 mg/2 K min^{-1} (II). The sample holder was a standard platinum crucible (diam. 7 mm, depth 19 mm). The effect of sample holder geometry was studied in a series of experiments (III) using a 20 mg/2 K min^{-1} sample weight/heating rate combination and a more open crucible (diam. 5 mm, depth 5 mm). The air atmosphere was a dynamic one with a flow rate of 90 cm^3 min^{-1}. All obtained curves (TG, DTG, DTA) were redrawn with the aid of a computer programme taking care of buoyancy etc. corrections. In DTA runs, alumina was used as a reference sample. The calibrations were performed from 400 to 700 K with high-purity In, Sn, Cd and Zn reference materials.

In order to study the dehydration mechanism in more detail, TG curves were in some instances recorded with a MOM Derivatograph-Q apparatus using so-called quasi-isothermal heating technique. DSC studies were made with a Perkin-Elmer DSC-1B apparatus using a heating rate of 8 K min^{-1} and 5 mg samples.

RESULTS AND DISCUSSION

Decomposition Schemes for the Sulfates

The thermal decomposition of sulfate octahydrates in air starts with dehydration and proceeds via oxysulfate to sesquioxide:

$$Ln_2(SO_4)_3 \cdot 8H_2O(s) \rightarrow Ln_2(SO_4)_3(s) + 8H_2O(g) \qquad /1/$$

$$Ln_2(SO_4)_3(s) \rightarrow Ln_2O_2SO_4(s) + 2SO_3(g) \qquad /2/$$

$$Ln_2O_2SO_4(s) \rightarrow Ln_2O_3 + SO_3(g) \qquad /3/$$

$Ce_2(SO_4)_3 \cdot 8H_2O$ differs by decomposing from anhydrous sulfate directly to oxide CeO_2 without step /2/. Fig. 1 shows as examples

the thermoanalytical curves for Pr and Er sulfate octahydrates.
When comparing the reaction temperatures to the atomic numbers of
the lanthanoids, the most obvious trend is the decreasing stability
of the oxysulfate from lanthanum to lutetium. This trend together
with the slightly increasing stability of the anhydrous sulfate
causes the stability range for the oxysulfate phase to become very
narrow in the Tm-Lu region.

It may be noted that although there are differences in temperatures, the observed U-shape trend in sulfate decomposition temperatures agrees with earlier studies (3,4,12) as does the almost linearly decreasing trend in oxysulfate to oxide conversion temperatures (3,4,12,13). Yttrium takes a position in the Dy-Ho region
in accordance with its ionic radius.

The decomposition scheme for the pentahydrates differs only in
the mechanism and temperatures for the dehydration step /1/ to be
discussed in more detail below.

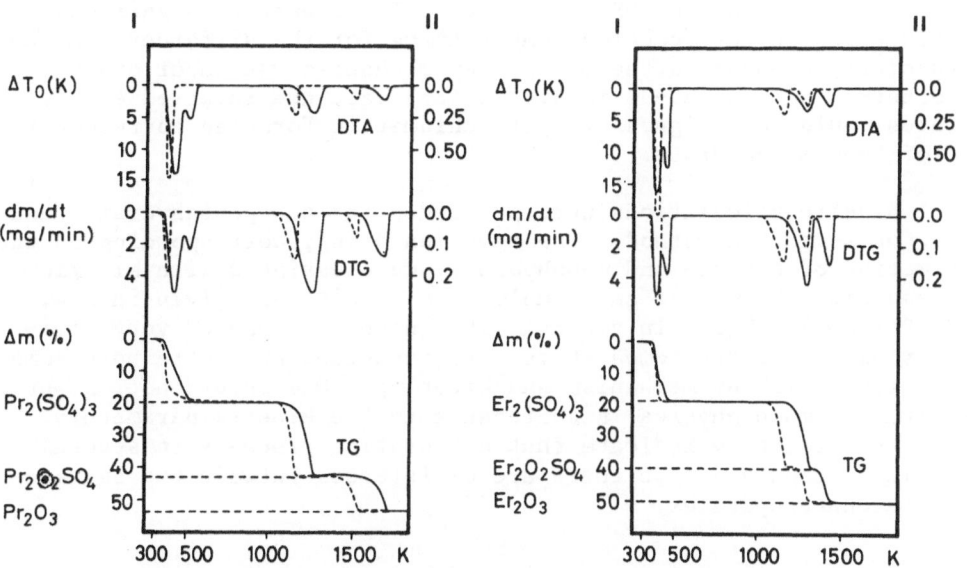

Figure 1. TG, DTG and DTA curves, measured under conditions I and
II, for $Pr_2(SO_4)_3 \cdot 8H_2O$ and $Er_2(SO_4)_3 \cdot 8H_2O$. I: ——————,
II: ------.

Dehydration Mechanism and Enthalpies

With a small amount of sample and a low heating rate (II) the
dehydration of the octahydrates occurred generally in a single step,
the only exceptions being Yb and Lu. The use of other reaction
parameters (I) led into a multistep dehydration, visible clearly
in the DTG and DTA curves. The lanthanum and cerium compounds belonging to other structure type dehydrated in several steps. The
dehydration steps did not results in clear plateaus which could be
assigned with stoichiometric formulae.

Likewise, it was not possible to draw clear conclusions from
the dehydration temperatures, or from the enthalphy values which
showed an irregular but generally increasing trend as observed by
Wendlandt and George (2). The dehydration enthalpies depend on the
experimental conditions but, when comparing values obtained with
the same heating rate (10 K min^{-1}), they agree within the error
limits ($\pm$10 %) with those reported by Wendlandt et al. (2) but are
somewhat lower than those measured with the DSC method (4). For
example; $Pr_2(SO_4)_3 \cdot 8H_2O$: 484, 460 and 628 kJ mol^{-1} for this study
and refs. (3) and (4), respectively. The corresponding values for
$Lu_2(SO_4)_3 \cdot 8H_2O$ are 554, 498 and 586 kJ mol^{-1}. No anomaly (2), however, was observed in the case of ytterbium sulfate hydrate which
is in agreement with the structural study (10).

By contrast to the octahydrates, the dehydration of the pentahydrates was a multi-step process under all conditions studied
(I-III), including the DSC technique. There were no significant
differences in the reaction temperatures for the different compounds.
Especially pronounced the multi-step mechanism was under quasi-
isothermal heating conditions. In that case the intermediate hydrates could be assigned with stoichiometric formulae corresponding
to di- and monohydrate.

Kinetic calculations were performed for the pentahydrates using
the Coats-Redfern method. The reaction steps, corresponding to the
formation of the di- and monohydrate, were assigned with kinetic
parameters. By way of an example, the results are given in Fig. 2
for $Pr_2(SO_4)_3 \cdot 5H_2O$. In general, the selenates behaved very similarly as far as the temperatures are concerned but there were some
differences in the mechanism and kinetics. One should not, however,
assign too much physical significance to the kinetic parameters.
Nevertheless, they indicate that dehydration proceeds in several
distinct steps and that there are differences between the sulfate
and selenate hydrates.

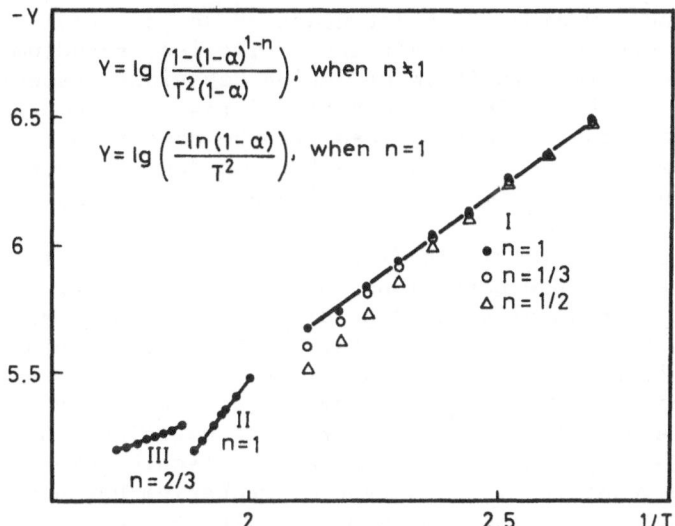

Figure 2.   Determination of apparent reaction order for the dehy-
            dration of $Pr_2(SO_4)_3 \cdot 5H_2O$ using the Coats-Redfern method.
            The steps I, II and III correspond to the formation of
            dihydrate, monohydrate and anhydrous phase, respectively.

Comparison to Structures

Lanthanum and cerium sulfate octahydrates are isostructural
while the sulfates of the heavier lanthanoids and yttrium belong to
another series. The space groups are Pc and $C2/c$, respectively
(14,10). In both cases all water molecules are coordinated to the
metal cation which is nine-coordinated in $La_2(SO_4)_3 \cdot 8H_2O$ but eight-
coordinated in the other structure type.

The pentahydrates, except for scandium, are all isostructural
belonging to space group Cc (7). The trivalent lanthanoid is nine-
coordinated by sulfate and water oxygens but one water molecule remains outside the inner coordination sphere. $Sc_2(SO_4)_3 \cdot 5H_2O$ forms
triclinic crystals where scandium is six-coordinated; all water
molecules are bound to scandium (8,9). $Sc_2(SeO_4)_3 \cdot 5H_2O$ (8) is isostructural with the sulfate but the other selenates differ from the
corresponding sulfates as seen in the case of $Ce_2(SeO_4)_3 \cdot 5H_2O$ (15).

The different structure types are relatively easy to distinguish. Thus, the dehydration mechanism and temperatures for

lanthanum and cerium sulfate octahydrates deviate from the trend
observed for the other octahydrates. Likewise, scandium sulfate
pentahydrate differs in its dehydration behaviour; especially the
enthalphy is higher. The correlation of reaction temperatures with
atomic numbers and ionic radii can also be done in a relatively
straightforward manner although variations due to experimental factors may confuse the trends especially at lower temperatures.

It is more difficult to draw conclusions on the actual mechanism. As seen in the case of octahydrates, the water molecules may
be expelled in several steps depending on the experimental conditions
although in the structure all water molecules are coordinated more
or less similarly. In the pentahydrate series, under quasi-isothermal conditions, two distinct hydrates appeared as intermediate
phases, however. It would be tempting to correlate the monohydrate
as a phase having the coordinated water molecules expelled and only
the lattice water left. This type of mechanism has been suggested
for $CsPr(SO_4)_3 \cdot 4H_2O$ (16). By contrast, it has been reported by
Langfelderova et al. (17) that in $CuSO_4 \cdot 5H_2O$ the coordinated water
is retained in the structure while the lattice water is lost. In
the case of $Ln_2(SO_4)_3 \cdot 5H_2O$ (Ln=La-Nd) the lattice water is probably
fairly tigthly bound as it is locked in a hole in the structure (7)
but additional study by high-temperature X-ray diffraction and other
methods would be needed to corroborate the mechanism.

REFERENCES

1. W.W. Wendlandt, J. Inorg. Nucl. Chem., 7:51 (1958).
2. W.W. Wendlandt and T.D. George, J. Inorg. Nucl. Chem., 19:245
    (1961).
3. M.W. Nathans and W.W. Wendlandt, J. Inorg. Nucl. Chem., 24:869
    (1962).
4. N. Bukovec, P. Bukovec and J. Šiftar, Vestn. Slov. Kem. Drus.,
    22:5 (1975).
5. M.A. Nabar and S.V. Paralkar, Thermochim. Acta, 17:239 (1976).
6. B. Hajek, N. Novotna and J. Hradilova, J. Less-Common Met.,
    66:121 (1979).
7. L.O. Larsson, S. Linderbrandt, L. Niinistö and U. Skoglund,
    Suom. Kemistil. B., 46:314 (1973).
8. L. Niinistö, J. Valkonen and L.O. Larsson, Finn. Chem. Lett.,
    43 (1975).
9. J. Valkonen, L. Niinistö, B. Eriksson, L.O. Larsson and
    U. Skoglund, Acta Chem. Scand. A., 29:866 (1975).
10. L. Hiltunen and L. Niinistö, Cryst. Struct. Commun., 5:561 (1976).
11. L. Hiltunen and L. Niinistö, Cryst. Struct. Commun., 5:567 (1976).

12. A.N. Pokrovskii and L.M. Kovba, Russ. J. Inorg. Chem., 21:305 (1976).
13. M. Leskelä and L. Niinistö, J. Therm. Anal., 18:307 (1980).
14. L.A. Aslanov, I.S.A. Farag, V.M. Ionov and M.A. Porai-Koshits, Zh. Fiz. Khim., 47:2172 (1973).
15. L.A. Aslanov, I.S.A. Farag and M.A. Porai-Koshits, Zh. Fiz. Khim., 47:1057 (1973).
16. N. Bukovec, L. Golič and J. Šiftar, Vestn. Slov. Kem. Drus., 26:377 (1979).
17. H. Langfelderová, M. Linkešová, M. Sérator and J. Gažo, J. Therm. Anal., 17:107 (1979).

RARE EARTH OXYSULFIDE/OXYSULFATE EQUILIBRIA AT 1100-1500K

Ratnesh K. Dwivedi and D.A.R. Kay

Department of Metallurgy and Materials Science
McMaster University, Hamilton, Ontario, Canada

INTRODUCTION

The oxidation of La_2O_2S and Gd_2O_2S to their respective oxysulfates at 1000-1500 K can be represented as:

$$\frac{1}{2} La_2O_2S(s) + O_2(g) = \frac{1}{2} La_2O_2SO_4(s) \qquad (1)$$

and
$$\frac{1}{2} Gd_2O_2S(s) + O_2(g) = \frac{1}{2} Gd_2O_2SO_4(s) \qquad (2)$$

respectively. Under condition of oxysulfide/oxysulfate coexistence at constant temperature, the unique oxygen partial pressure thus established can be conveniently measured using oxygen ion conducting, calcia stabilized zirconica (CSZ), solid electrolytes.

The electrochemical cells used for studying reactions (1) and (2) can be represented as:

$$Pt/La_2O_2S(s), La_2O_2SO_4(s)/CSZ/Air(1 \text{ atm.})/Pt \qquad I$$

$$Pt/Gd_2O_2S(s), Gd_2O_2SO_4(s)/CSZ/Air(1 \text{ atm.})/Pt \qquad II$$

If the cells are reversible and all the solid phases are in their standard states then the emf of cells I and II can be written as:

$$E_I = [RT \ln (Po_2(air)) - \Delta G_1^o]/4F \qquad (3)$$

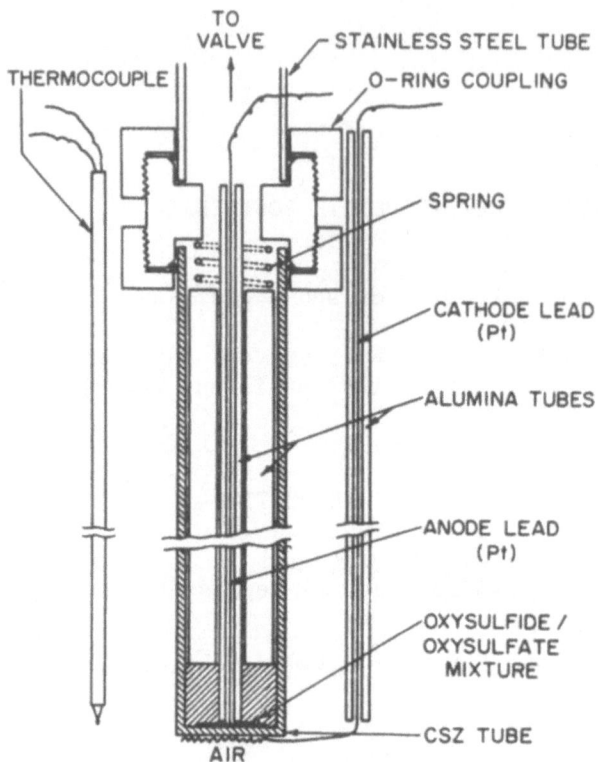

Fig. 1. Cell Assembly

and

$$E_{II} = [RT \ln (Po_2(air)) - \Delta G_2^o]/4F \qquad (4)$$

where ΔG_1^o and ΔG_2^o are the standard free energy changes for reactions (1) and (2), respectively.

EXPERIMENTAL

Starting materials for the preparation of the compounds of lanthanum and gadolinium were La_2O_2S and Gd_2O_3, respectively, each supplied by Cerac Inc. with 99.9% purity. Gd_2O_3 was dissolved in dilute H_2SO_4 to prepare $Gd_2(SO_4)_3$ which was subsequently reduced with H_2 at 923 - 973K. Oxysulfates of both lanthanum and gadolinium were prepared by oxidizing their respective oxysulfides at 1223K. The purity of oxysulfides and oxysulfates was checked using x-ray diffraction and thermogravimetric analysis. In all cases, the purity of compounds was greater than 99.8%, oxide being the main impurity in each case.

The coexistence of an oxysulfide with its corresponding oxysulfate was verified by heating a pelletized mixture of the two in a sealed and evacuated quartz capsule for 120 hours and analysing the products by x-ray diffraction. The coexistence of the compounds of lanthanium was confirmed up to 1373K while that of the gadolinium compounds was tested up to 1223K.

The important portions of the cell assembly are shown in Fig. 1. Calcia stabilized zirconia tubes having a flat closed end, 350 mm length and 7.4 mm outer diameter, were used as electrolytes. The anode consisted of a finely ground mixture of oxysulfide and oxysulfate in a 1:1 molar ratio. Air was used as the reference electrode and platinum wires were used to make all the contacts. The temperature in the platinum wound resistance furnace used to heat the cell assembly, was proportionally controlled within 1K. The temperature of the cell was measured using a Pt-13%Rh/Pt thermocouple which was calibrated against a standard thermocouple, which in turn was calibrated by the Research Council of Canada within 1K of International Practical Temperature Scale of 1968

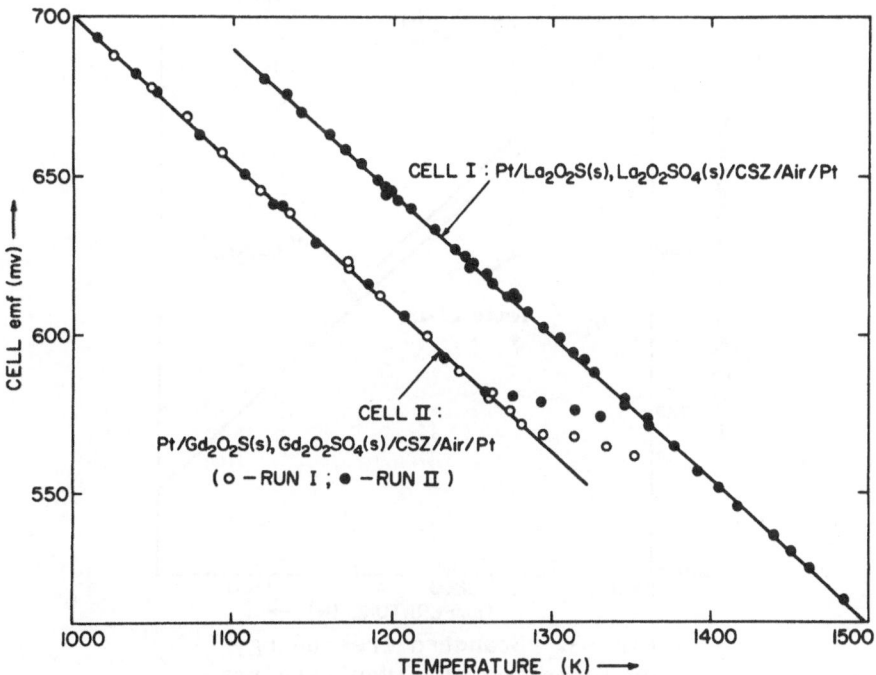

Fig. 2. Cell emf against temperature.

(IPTS–68).   The cell emf was measured using a Keithley 616
electrometer with a precision of $\pm$ 1 mv.

RESULTS AND DISCUSSION

        Figure 2 shows the plots of cell emf against temperature for
cells I and II, respectively.   The data corresponding to the
linear portions of the plots were analysed using linear regression
to obtain:

Cell I : E_I = (1192 – 0.456 T) $\pm$ 1.5 mv   (5)

Cell II: E_{II} = (1154 – 0.455 T) $\pm$ 1.5 mv   (6)

Substituting the values of E_I and E_{II} into the equations (3) and
(4) the following standard free energy changes for the cell
reactions are obtained:

ΔG_1^o (1100 – 1500K) = (–109930 + 38.91T) $\pm$ 150 cals/mole   (7)

ΔG_2^o (1000 – 1280K) = (–106520 + 38.88T) $\pm$ 150 cals/mole   (8)

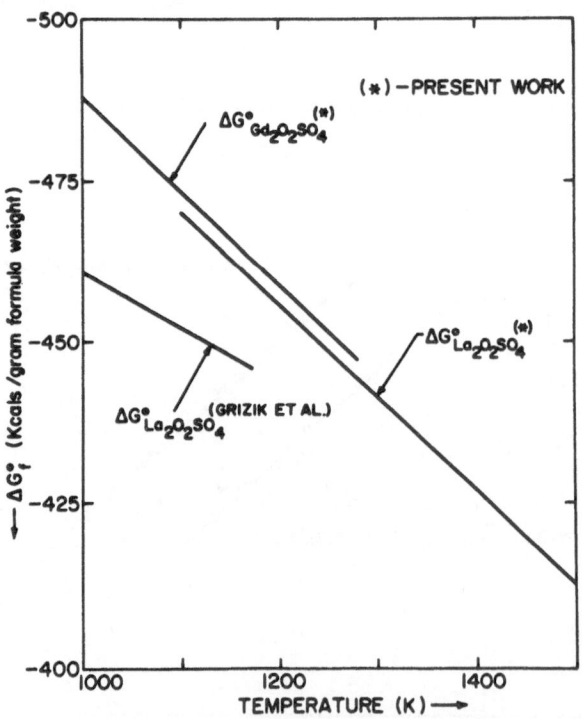

Fig. 3. Standard free energy
of formation of oxysulfate per
gram formula weight against
temperature.

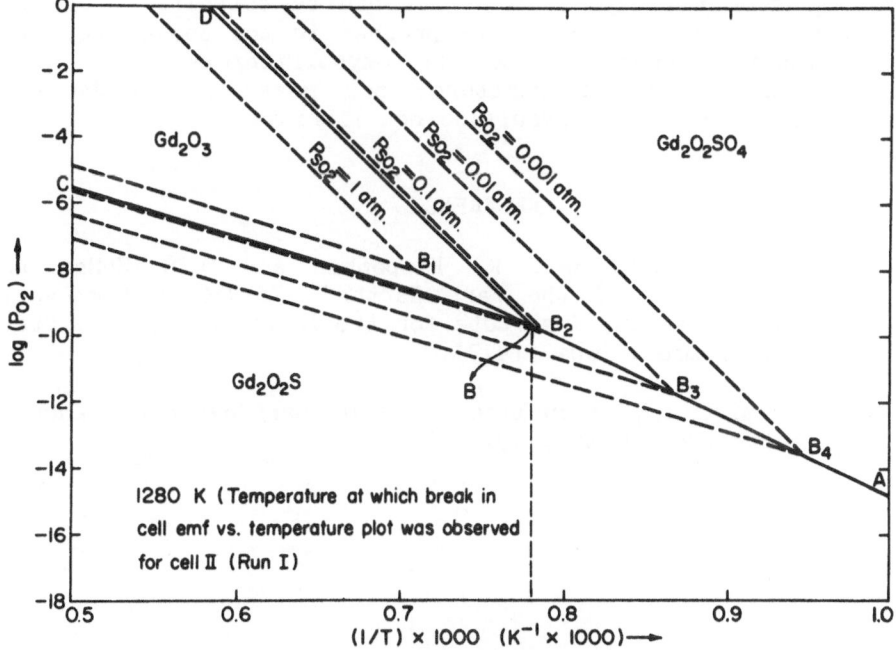

Fig. 4. Oxide-oxysulfide-
oxysulfate equilibria at a
constant partial pressure of
SO_2. ABC is the path followed
by Cell II.

The values of ΔG_1^o and ΔG_2^o can be combined with the values of the
standard free energies of formation of La_2O_2S and Gd_2O_2S, respectively, estimated with large error limits ($\sim \pm 15$ kcal.) by
Gschneidner et al (1), to obtain the standard free energy of
formation of $La_2O_2SO_4$ and $Gd_2O_2SO_4$ from their component elements.
These values, along with the values of the standard free energies
of formation of $La_2O_2SO_4$ obtained on processing the data of Grizik
et al. (2) are plotted in Fig. 3.

Gadolinium oxysulfate decomposes at high temperatures to
gadolinium oxide (c type) giving rise to a break in the cell emf
vs. temperature plot as shown in Fig. 2. No such behaviour is
observed for the cell using lanthanum compounds, presumably,
because lanthanum oxysulfate decomposes to La_2O_3 at temperatures
higher than those examined in the present study. The temperature
at which the break in the cell emf vs. temperature plot occurs for
Cell II, depends upon the partial pressure of SO_2, $p(SO_2)$, in the
anode compartment. This pSO_2 is fixed by the oxide-oxysulfide-
oxysulfate equilibria at any temperature. As Fig. 4 shows, when
the $p(SO_2)$ demanded by the oxide-oxysulfide-oxysulfate equilibrium

becomes greater than the $p(SO_2)$ that can be established by complete decomposition of oxysulfate present in the anode, the cell emf no longer corresponds to the oxysulfide/oxysulfate equilibrium.   Above this temperature, p_{O_2} measured by the cell corresponds to the oxide-oxysulfide equilibrium.

REFERENCES

1.   K.A. Gschneidner, Jr., N. Kippenhan and O.D. McMasters, "Thermochemistry of the Rare Earths", IS-RIC-6, Rare-earth Information Center, Institute for Atomic Research, Iowa State University, Ames, Iowa (1973).

2.   A.A. Grizik, I.G. Abdullina and N.M. Garifdzhanova, Zhurnal Neorganicheskoi Khimii, 19:2586(1974).

PHASE STUDIES AND SELECTIVE OXIDATION IN RARE EARTH-MOLYBDENUM-OXIDE SYSTEMS, Ln_2O_3-MoO_2-MoO_3

H. Prévost-Czeskleba and G. Tourné
Laboratoire de Chimie des Solides, Université des Sciences
et Techniques du Languedoc, 34060 Montpellier Cédex
France

ABSTRACT

Compound formation and subsolidus phase relations in the systems Ln_2O_3-MoO_2-MoO_3 (Ln = Eu, Gd, Tb) have been investigated. New fluorite related compounds of molybdenum VI and V have been characterized. The role of the rare earth element in selective oxidation of molybdenum and crystal chemical properties of the ternary compounds is discussed.

INTRODUCTION

Recently, phase equilibria in the Ln_2O_3-MoO_2-MoO_3 (Ln = Pr, Nd, Sm) systems have been established between 1273 and 1673°K using the controlled atmosphere technique (1,2). Results indicate that molybdenum can be selectively oxidized by fixing the oxygen partial pressure and choosing the rare earth element. In the present study phase equilibria in the Gd_2O_3-MoO_2- MoO_3 system have been established at 1273, 1473 and 1673°K under varying oxygen partial pressure ranging from 10^{-5} to 10^{-15} atm. Published work on phase equilibria in this system by DTA measurements has been limited to the binary line Gd_2O_3-MoO_3 (3) and phases were only reported for the MoO_3-rich portion of the diagram. In the Ln_2O_3-MoO_2-MoO_3 (Ln = Eu, Tb) systems phase equilibria at 1473°K are reported.

On the pseudobinary lines Gd_2O_3-MoO_3, Eu_2O_3-MoO_3 and Tb_2O_3-MoO_3, the following new fluorite-type compounds were identified as stable phases for Ln_2O_3/MoO_3 = 3/1, 5/2, 4/7: Ln_6MoO_{12}, $Ln_{10}Mo_2O_{21}$ and $Ln_{14}Mo_4O_{33}$. Two molybdenum V compounds have been characterized for the ratio Ln_2O_3/Mo = 3/2, 1/2: Ln_3MoO_7 and $LnMoO_4$. In the molybdenum

-rich portion of the phase diagram, two compounds have been found for the molar ratio $Ln_2O_3/Mo = 1/2$ and $1/3$: $Ln_2Mo_2O_7$ and $Ln_2Mo_3O_9$. The role of the rare earth element in selective oxidation of molybdenum to oxidation states IV, V and VI is discussed.

RESULTS AND DISCUSSION

Phase Equilibria

Figure 1 shows phase equilibria in the $Gd_2O_3-MoO_2-MoO_3$ system at $1273°K$. Stable phases under the present experimental conditions were found for the molar ratios Gd_2O_3/Mo: 3/1, 1/1 and 1/3 corresponding to formulas Gd_6MoO_{12}, Gd_2MoO_6 and $Gd_2(MoO_4)_3$ on the pseudobinary line $Gd_2O_3-MoO_3$, and to formulas Gd_2MoO_5 and $Gd_2Mo_3O_9$ on the pseudobinary line $Gd_2O_3-MoO_2$. For the ratio 1/1 a molybdenum V + VI compound, $Gd_{12}Mo_6O_{35}$, forms between Gd_2MoO_5 and Gd_2MoO_6 and for the ratio 1/2 the molybdenum V compound, $GdMoO_4$, has been observed.

Phase equilibria at $1473°K$ are represented in Fig. 2. The molybdenum IV compound $Gd_2Mo_3O_9$ is no longer observed but a new molybdenum VI compound, $Gd_{10}Mo_2O_{21}$, forms for the ratio 5/2. Phase relations and stable phases at $1673°K$ are represented in Fig. 3. The oxygen partial pressures necessary to produce these equilibrium relations have not been determined very precisely at this temperature and are not reported here. For comparison, phase equilibria in the $Tb_2O_3-MoO_2-MoO_3$ system at $1473°K$ are represented in Fig. 4. At $1673°K$ phase relations are identical to those observed in the gadolinium system described in Fig. 3. The numbers on all figures refer to $-\log pO_2$ (atm) and are given with an accuracy of ± 0.1 in terms of $\log pO_2$.

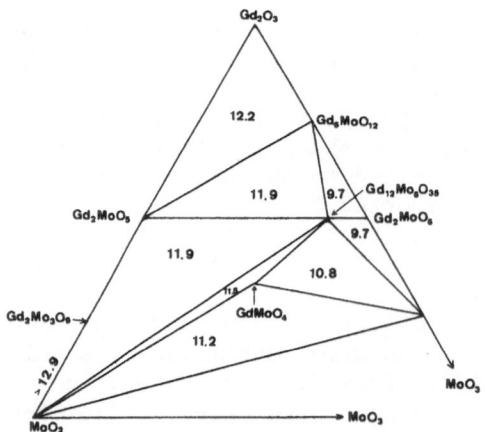

Figure 1. The $Gd_2O_3-MoO_2-MoO_3$ isotherm at $1273°K$.

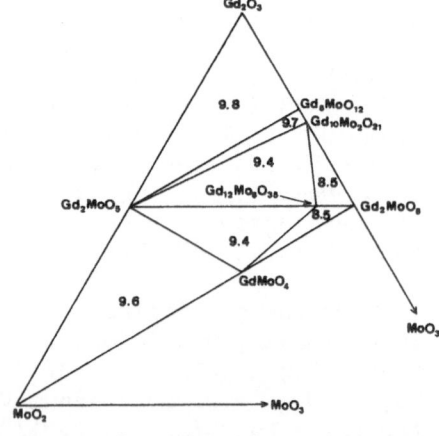

Figure 2. The $Gd_2O_3-MoO_2-MoO_3$ isotherm at $1473°K$.

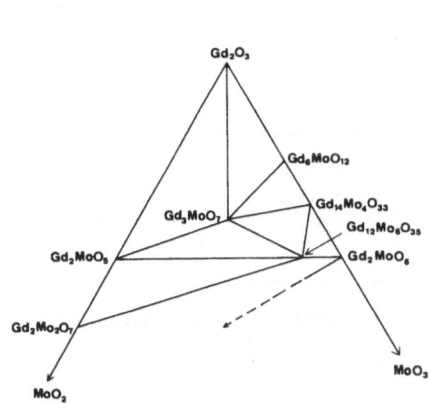

Figure 3. The Gd_2O_3–MoO_2–MoO_3 isotherm at 1673°K.

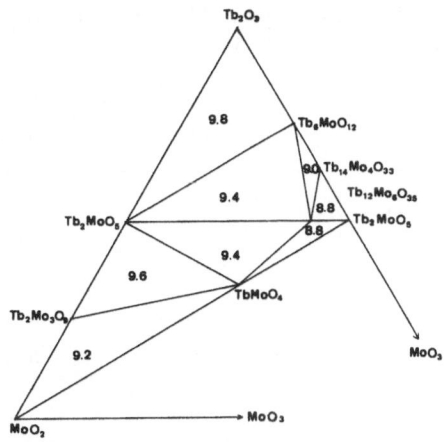

Figure 4. The Tb_2O_3–MoO_2–MoO_3 isotherm at 1473°K.

Phase equilibria results confirm previous observations: the most stable Mo VI compound against reduction is Ln_6MoO_{12}, corresponding to the highest ratio Ln/Mo; the larger the size of the rare earth ion, the more molybdenum, in a given oxidation state, is stabilized toward low oxygen partial pressures; the very narrow stability range in terms of oxygen partial pressure of the scheelite-type compound, $LnMoO_4$, at 1273°K; the particular stability against reduction of the fluorite-related $Ln_{12}Mo_6O_{35}$ compound at low temperatures (compare Fig. 1 and Fig. 2).

The phase diagram observed at 1273°K is characteristic for the Nd, Sm and Gd systems, with the exception that Nd_3MoO_7 is already observed at this temperature and that the MoO_2-rich compound, $Gd_2Mo_3O_9$ forms at much lower oxygen partial pressure than Gd_2MoO_5. Phase relations at 1473°K as represented in Fig. 2 and Fig. 4 are typical for Gd, Tb and heavy rare earth systems in which the scheelite phase exhibits a wide stability range in terms of oxygen partial pressure. Phase relations in the Eu_2O_3-rich portion of the diagram are identical to those observed in the samarium system at 1673°K (1). Eu_3MoO_7 forms at 1473°K in a broad pressure range like the lighter rare earth elements (Ln = Pr, Sm, Nd). The phase diagram at 1673°K is typical for the gadolinium system and those with Ln = Tb, Dy-Er.

X-Ray Crystallographic Analyses

Crystal chemical properties and unit cell parameters were derived from X-ray powder diffraction data. Results for the Mo VI and Mo V compounds are listed in Table 1. The Ln_6MoO_{12} compounds are face-centered cubic and isostructural with Sm_6MoO_{12} (1). No weak extra

Table 1. Cell dimensions for the fluorite type phases of Mo VI and
          Mo V.

Ln	Ln_6MoO_{12}	$Ln_{10}Mo_2O_{21}$	$Ln_{14}Mo_4O_{33}$	Ln_3MoO_7
Eu	a = 5.370(3) V = 155	a = 5.363(2) V = 154	a = 10.22(2) c = 9.54(1) V = 863	a = 10.631(5) b = 7.437(3) c = 7.515(3) V = 594
Gd	a = 5.362(5) V = 154	a = 5.348(2) V = 153	a = 9.89(1) c = 9.51(1) V = 806	a = 5.277(8) V = 147
Tb	a = 5.316(3) V = 150		a = 9.858(5) c = 9.450(5) V = 795	a = 5.290(8) V = 148

a, b, c, in $\overset{\circ}{A}$; V in $\overset{\circ}{A}^3$.

reflections were observed even after prolonged heating at 1673°K.
However, at 1473°K the powder patterns of Gd_6MoO_{12} and Eu_6MoO_{12} showed
a slight broadening of the (311) line. Compounds of formula
$Ln_{10}Mo_2O_{21}$ are of fcc type and have been found from La to Gd. Allow-
ing for the shifts in interplanar spacings due to the lanthanide contraction, one sees that the $Ln_{10}Mo_2O_{21}$ X-ray patterns are in fact
identical to those of the cubuc Ln_6MoO_{12} compounds (Sm to Gd), the
smaller cell edge accounting for the partial substitution of the smaller Mo^{6+} for Ln^{3+}. The $Ln_{14}Mo_4O_{33}$ compounds are of rhombohedral symmetry and isostructural with $Sm_{14}Mo_4O_{33}$. The hexagonal cell parameters are given in Table 1. The molybdenum V compounds Ln_3MoO_7 exhibit
an orthorhombic fluorite-related superstructure from Ln = La to Eu and
have been found isostructural with the analogous Nb V and Ta V compounds. For the smaller Ln cations (Ln = Gd, Tb-Er) a defect fluorite-type structure was observed.

REFERENCES

1. H. Kerner-Czeskleba, Third International Conference on the Chem-
    istry and Uses of Molybdenum, Ann Arbor, August 19-23, 1979.
2. H. Czeskleba-Kerner, B. Cros, and G. Tourné, J. Solid State Chem.,
    37:294 (1981).
3. K. Megumi, H. Yuomoto, S. Ashida, S. Akiyama, and Y. Furuhata,
    Mater. Res. Bull., 9:391 (1974).

CORRELATION OF SPECTRAL AND HEAT-CAPACITY SCHOTTKY CONTRIBUTIONS

FOR Dy_2O_3, Er_2O_3, and Yb_2O_3*

Edgar F. Westrum, Jr. and Robert D. Chirico, University
of Michigan, Ann Arbor, MI 48109; John B. Gruber, Portland State University, Portland, OR 97207

ABSTRACT

A self-consistent interpretation of existing and new Raman- and
infrared spectra on oriented single crystals and heat-capacity measurements on Dy_2O_3, Er_2O_3, and Yb_2O_3, which have the cubic (bixbyite)
structure is presented.

INTRODUCTION

The cubic lanthanide sesquioxides possess two inequivalent cation
sites, C_2 and C_{3i}, the latter with inversion symmetry. The crystal-
field splitting of the low lying [SL]J-states for ions with C_2 site
symmetries has been known in detail for a decade. However, the
absence of corresponding details for those in C_{3i} sites has long
frustrated efforts to fully rationalize the observed thermophysical
property values known for nearly two decades (1,2). Coherent,
polarized, and tuneable excitation sources permit identification
through electronic Raman scattering and double-photon absorption of
crystal-field states of lanthanide ions in sites with inversion symmetry (3-5). However, because Raman scattering experiments sometimes
exhibit vibrational and electronic spectral peaks of similar magnitude
(5,6), ambiguities arise in the spectral assignments. Recently,
Chirico et al. have resolved Schottky contributions to the heat
capacity for lanthanide trihydroxides (7-9) and for hexagonal lanthanide trichlorides (10) through application of a volume-weighted inter-
polated lattice heat-capacity approximation technique. Possible
ambiguity in the interpretation of the spectral data and limitations

---

*Supported in part by the National Science Foundation

275

in the resolution of lattice heat capacities preclude total independence of the two approaches, but as shown, the final results are self-
consistent.

THE SPECTRAL STUDIES

Two photon absorption and scattering experiments on the lanthanide sesquioxides were performed by JBG and coworkers between 1969
and 1980 at Washington State University and North Dakota State University. Details on the experimental design have been described
earlier. The Raman scattering experiments follow the approach
described by Schaack and Koningstein (11) in the use of symmetry types
and choice of coordination axes. A definitive analysis of the vibrational spectra of Yb_2O_3, Dy_2O_3, and Er_2O_3, and comparison with previously published heat-capacity results has been submitted (12).
These data provide the complete set of ground manifold Stark levels
employed in the calculation of the "spectroscopic" Schottky contributions shown in Fig. 1 b-d.

THE SCHOTTKY HEAT-CAPACITY CONTRIBUTIONS

Resolution of Schottky contributions in lanthanide compounds
requires an evaluation of the much larger "lattice" (vibrational) contribution to the total measured heat capacities. In view of its previous success, the volume-weighted lattice-approximation technique
(7-10) has been applied to the lattice heat capacities of the C-type
lanthanide sesquioxides:

$$C_p(\text{lattice, } Ln_2O_3) = (1-f)[C_p(Gd_2O_3^*)] + f[C_p(Lu_2O_3)]. \qquad (1)$$

Here f is defined as:

$$f(Ln_2O_3) = [V(Ln_2O_3) - V(Gd_2O_3)]/[V(Lu_2O_3) - V(Gd_2O_3)] \qquad (2)$$

where V is the molar volume. The asterisk in Eq. (1) indicates that
the cooperative magnetic contribution to the heat capacity of Gd_2O_3
has been deleted. This contribution is usually quite small above 10 K
in Gd^{+3} compounds; however, it is clear from Fig. 1a that it is significant to at least 75 K. Because of this unusually large cooperative magnetic contribution, the lattice contribution calculated from
Eq. (1) must be considered less reliable in this temperature region,
but this is partially offset by the reduced magnitude of the lattice
contribution at low temperatures.

As can be seen in Fig. 1, good agreement between the spectro-
scopically and calorimetrically derived Schottky contributions
obtains. The positive deviations at low temperatures of the calorimetrically derived Schottky contributions from those calculated from

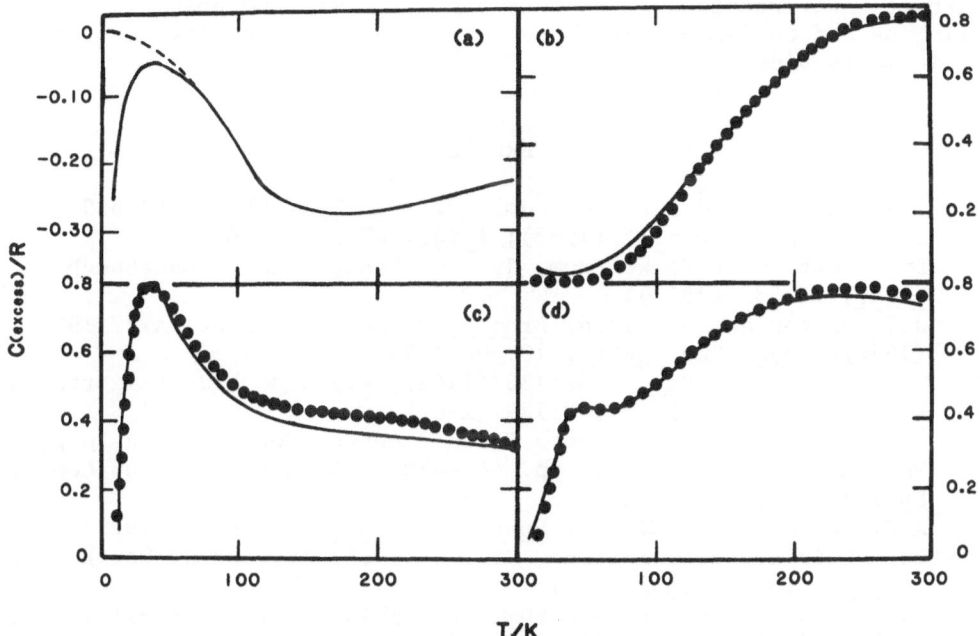

Figure 1. (a). The solid curve represents $\frac{1}{2}\{C_p(Lu_2O_3) - C_p(Gd_2O_3)\}$. The dashed curve represents the estimated difference in the "lattice" heat capacities at low temperature (see text). (b-d). The calorimetric Schottky curves are represented by the solid lines and the spectroscopic Schottky curves are represented by the dotted lines. (b). Yb_2O_3. (c). Er_2O_3. (d). Dy_2O_3.

the spectral data are almost certainly due to cooperative magnetic contributions, which are ignored--except in the case of Gd_2O_3--in the current treatment. At higher temperatures the deviations rarely exceed the estimated experimental uncertainties.

Further confirmation of the assignments results upon comparing the recent crystal-field splitting calculations for these ions in both sites (13) with the reported experimental splittings. The calculations make use of first-order perturbation theory as developed by Judd and Wybourne (14,15) and normalization of CEF parameters follows Wybourne's treatment (15) of CEF theory. A least-squares fitting of calculated to observed levels has established a set of crystal-field parameters $B_{n,m}$ for Dy^{3+}, Er^{3+}, and Yb^{3+} in C_2 sites. These parameters will appear in a separate publication (13). With the agreement between spectroscopic and heat-capacity data relevant to Schottky contributions, it now appears promising with the aid of the forthcoming lattice-sum calculations, CEF parameters, and wave functions that

further interpretation or re-interpretation will be possible for
existing and published EPR, Mössbauer, and heat-capacity data on the
cubic lanthanide sesquioxides.

REFERENCES

1. B.H. Justice and E.F. Westrum, Jr., J. Phys. Chem., 67:339
    (1963); ibid., 67:345 (1963); ibid., 67:659 (1963).
2. B.H. Justice, E.F. Westrum, Jr., E. Chang, and R. Radebaugh,
    ibid., 73:333 (1969).
3. J.T. Hougen and S. Singh, Proc. Roy. Soc., (London) A277:193
    (1964); Phys. Rev. Lett., 10:406 (1963).
4. J.D. Axe, Phys. Rev., A42:136 (1964); see also R.J. Elliott and
    R. Loudon, Phys. Lett., 3:189 (1963).
5. J.B. Gruber, "Prc. 9th Rare Earth Research Conf." Blacksburg,
    VA, October 10-14, 1971, pp. 465-478 NTIS, U.S. Dept. of Commerce,
    Springfield, VA 22151.
6. A.M. Lejus and D. Michel, Phys. Stat. Sol., 84:K105 (1977).
7. R.D. Chirico, E.F. Westrum, Jr., J.B. Gruber, and J. Warmkessel,
    J. Chem. Thermodynamics, 11:835 (1979).
8. R.D. Chirico and E.F. Westrum, Jr., ibid., 11:71 (1979); ibid.,
    12:311 (1980); ibid., 13 (1981) (in press).
9. R.D. Chirico, E.F. Westrum, Jr., and J. Boerio-Goates, ibid.,
    13 (1981) (in press).
10. R.D. Chirico, E.F. Westrum, Jr., and J.B. Gruber, ibid., 12:117
    (1980).
11. G. Schaack and J.A. Koningstein, J. Opt. Soc. Am., 60:1110 (1970).
12. J.B. Gruber, E.F. Westrum, Jr., and R.D. Chirico, J. Chem. Phys.,
    (in press).
13. N.C. Chang, J.B. Gruber, R.P. Leavitt, and C.A. Morrison, J. Chem.
    Phys., (in press).
14. G.H. Dieke, Spectra and Energy Levels of Rare Earth Ions in
    Crystals, Interscience, New York (1968).
15. B.G. Wybourne, Spectroscopic Properties of Rare Earths, Inter-
    science, New York (1965).

ENTHALPY OF FORMATION OF BARIUM LANTHANIDE(IV) OXIDES: $BaCeO_3$,

$BaPrO_3$, AND $BaTbO_3$

Lester R. Morss, Chemistry Division, Argonne National
Laboratory, Argonne, Illinois 60439; Nahla Mensi,
Department of Chemistry, Rutgers University, New
Brunswick, New Jersey 08903

ABSTRACT

The perovskites $BaCeO_3$, $BaPrO_3$, and $BaTbO_3$ have been prepared
and the stabilization of the lanthanide(IV) ions in these oxides
has been determined as a function of their structures.

INTRODUCTION

Jacobson, Tofield, and Fender (1-3) studied the crystallographic and magnetic properties of the ternary oxides $BaCeO_3$,
$BaPrO_3$, and $BaTbO_3$. They recognized that these distorted perovskites stabilize lanthanide(IV) ions in octahedral sites and offer
opportunities to observe interesting magnetic and covalent effects.
Quite recently, Brauer and Kristen (4,5) identified the extent of
stabilization of Nd(IV) and Dy(IV) in these perovskites. Since we
are interested in the stabilization of lanthanide and actinide(IV)
ions in crystalline hosts, we chose to initiate these studies with
the lanthanide(IV) perovskites.

EXPERIMENTAL

Each compound was prepared twice as shown in Table 1. Sample
F of $BaTbO_3$ was provided by A. J. Jacobson (1). All samples were
found by X-ray powder diffraction to be in agreement with published
crystal structures (1). $BaCeO_3$ and $BaTbO_3$ were analyzed for Ln(IV)
by dissolving in $1\underline{M}$ HCl - $0.1\underline{M}$ KI solution and titrating with thio-
sulfate; $BaCeO_3$ requires 10-60 min to dissolve in this medium.
$BaPrO_3$ reacts too vigorously for iodometric analysis, so it was

Table 1.   Enthalpies of reaction of $BaLnO_3$ at 25.0°C.

Prep.	Solvent	Mass/mg	$\Delta H(kJ\ mol^{-1})$	Prep.	Solvent	Mass/mg	$\Delta H(kJ\ mol^{-1})$
$BaCeO_3$-A	1M HCl-0.01M KI	66.92	-350.4	$BaPrO_3$-C	0.1M $HClO_4$	47.46	-303.6
$BaCeO_3$-A	1M HCl-0.01M KI	70.35	-349.4	$BaPrO_3$-D	0.1M HNO_3	63.14	-302.5
$BaCeO_3$-A	1M HCl-0.01M KI	119.50	-347.7	$BaPrO_3$-D	0.1M HNO_3	72.46	-308.6
$BaCeO_3$-B	0.9M HCl-0.1M KI	58.10	-341.5		Ave. (95% conf.)		-309.4±5.6
$BaCeO_3$-B	0.9M HCl-0.1M KI	71.18	-341.3	$BaTbO_3$-E	0.1M HCl	61.71	-340.8*
$BaCeO_3$-B	0.9M HCl-0.1M KI	59.63	-345.5	$BaTbO_3$-E	0.1M HCl	71.96	-338.2*
$BaCeO_3$-B	0.9M HCl-0.1M KI	71.46	-341.0	$BaTbO_3$-E	0.1M $HClO_4$	62.50	-340.2
	Ave. (95% conf.)		-345.3±3.7	$BaTbO_3$-E	0.1M $HClO_4$	68.55	-340.3
$BaPrO_3$-C	0.1M HNO_3	70.59	-317.2	$BaTbO_3$-E	0.1M $HClO_4$	74.66	-339.5
$BaPrO_3$-C	0.1M HNO_3	68.11	-316.0	$BaTbO_3$-F	0.1M $HClO_4$	48.94	-334.4
$BaPrO_3$-C	0.1M HNO_3	51.60	-316.2	$BaTbO_3$-F	0.1M $HClO_4$	47.86	-336.2
$BaPrO_3$-C	0.1M $HClO_4$	51.61	-312.8	$BaTbO_3$-G	0.1M $HClO_4$	144.99	-338.3
$BaPrO_3$-C	0.1M $HClO_4$	54.73	-298.1		Ave. (95% conf.)		-338.5±1.6

*Corrected for small amount of Cl_2 produced.

Prep. A: $BaCO_3$ + CeO_2 (1060°C, 90 hours with 3 grindings). Anal: $BaCeO_{3.00±0.01}$.

Prep. B: $BaCO_3$ + CeO_2 (1050°C, 65 hours with 3 grindings). Anal: $BaCeO_{2.99±0.01}$.

Prep. C: BaO_2 + $PrO_{1.833}$ (1020°C, 45 hours with 2 grindings). Anal: $BaPrO_{2.99±0.01}$.

Prep. D: $BaCO_3$ + $PrO_{1.833}$ (1050°C, 82 hours with 3 grindings). Anal: $BaPrO_{3.02±0.01}$.

Prep. E: $Ba(CH_3COO)_2 \cdot H_2O$ + $TbO_{1.818}$ (1000°C, 35 hours, 2 grindings). Anal: $BaTbO_{3.007±0.004}$.

Prep. F: $BaCO_3$ + $TbO_{1.5}$ (1000°C, 60 hours with 3 grindings). Anal: $BaTbO_{2.96±0.01}$.

Prep. G: BaO_2 + $TbO_{1.818}$ (1000°C, 48 hours with 2 grindings). Anal: $BaTbO_{2.998±0.004}$.

reduced with H_2 at 800°C and either the mass loss or the mass of water produced was measured. Analyses are shown in Table 1.

Samples of $BaPrO_3$ and $BaTbO_3$ were found by thermogravimetric analysis to be stable at 1200°C in oxygen (the maximum attainable temperature) and to reduce completely to $BaLnO_{2.5}$ at 700-750°C in hydrogen. These ternary oxides are thus much more stable than the binary dioxides.

Samples of all three compounds were reacted in a solution calorimeter (6) to derive their enthalpies of formation. Thorough equilibration before and after each reaction yielded precise enthalpies of solution (95% confidence + 0.5%) despite the long reaction times. $BaCeO_3$ was reacted with $\overline{H}Cl$-KI solution, as used in the titration analyses, because the dissolution was most rapid in this medium and because the product solution contains $Ce^{3+}(aq)$, which is much better characterized than Ce(IV,aq). $BaPrO_3$ and $BaTbO_3$ were reacted with dilute HNO_3 or $HClO_4$ since Pr(IV) and Tb(IV) oxidize water to O_2 rapidly.

RESULTS AND DISCUSSION

The chemical reactions corresponding to the calorimetric data of Table 1 are shown below. Following each reaction is the equation used to calculate the enthalpy of formation. All species not followed by (s) are in solution. Enthalpy data from the literature

(7-10) include corrections for non-standard states whenever available. We note a difference between the enthalpies of reaction of the two samples of $BaCeO_3$, probably due to nonstoichiometry.

$$BaCeO_3(s) + 2HCl(1\underline{M}) + \frac{3}{2}I^- \to BaCl_2 + \frac{1}{2}I_3^- + 3H_2O + Ce^{3+} \qquad (1)$$

$\Delta H_f^\circ(BaCeO_3, s) = \Delta H_f^\circ(BaCl_2) + \frac{1}{2}\Delta H_f^\circ(I_3^-) + 3\Delta H_f^\circ(H_2O) + \Delta H_f^\circ(Ce^{3+})$
$-2\Delta H_f^\circ(HCl, 1\underline{M}) - (3/2)\Delta H_f^\circ(I^-) - \Delta H(\text{reaction } 1) = -864.06\pm1.72 +$
$\frac{1}{2}(-51.9\pm0.1) + 3(-285.85\pm0.04) + (-700.4\pm2.1) - 2(-164.36\pm0.08) -$
$(3/2)(-56.9\pm0.2) - (-345.3\pm3.7) = -1688.6\pm4.6 \text{ kJ mol}^{-1}$.

$$BaPrO_3(s) + 5H^+ \to Ba^{2+} + Pr^{3+} + \frac{1}{4}O_2(g) + \frac{5}{2}H_2O \qquad (2)$$

$\Delta H_f^\circ(BaPrO_3, s) = \Delta H_f^\circ(Ba^{2+}) + \Delta H_f^\circ(Pr^{3+}) + (5/2)\Delta H_f^\circ(H_2O) -$
$\Delta H(\text{reaction } 2) = -533.5\pm1.7 + (-706.2\pm1.6) + (5/2)(-285.83\pm0.04)$
$- (-309.4\pm7.3) = -1644.9\pm8.0 \text{ kJ mol}^{-1}$.

$$BaTbO_3(s) + 5H^+ \to Ba^{2+} + Tb^{3+} + \frac{1}{4}O_2(g) + \frac{5}{2}H_2O \qquad (3)$$

$\Delta H_f^\circ(BaTbO_3, s) = \Delta H_f^\circ(Ba^{2+}) + \Delta H_f^\circ(Tb^{3+}) + (5/2)\Delta H_f^\circ(H_2O) -$
$\Delta H(\text{reaction } 3) = -533.5\pm1.7 + (-698\pm6) + (5/2)(-285.83\pm0.04) -$
$(-338.5\pm1.6) = -1607.6\pm6.4 \text{ kJ mol}^{-1}$.

Table 2 lists all perovskites $BaMO_3$ for which an enthalpy of formation is known, in the order of increasing ionic radius of M(IV) (11). Two other structural parameters, the tolerance factor t (12) and the change in molecular volume for the reaction

$$BaO(s) + MO_2(s) \to BaMO_3(s) \qquad (4)$$

are also tabulated, as well as $\Delta H(4)$, referred to as $\Delta H(\text{complex})$. With the exception of $BaMoO_3$ and $BaPrO_3$, these data show that $\Delta H(\text{complex})$ becomes less favorable as t decreases; that is, as the size of M(IV) increases beyond the ideal value for the octahedral hole created by close-packed (BaO_3) layers. $\Delta H(4)$ may be unusual for M=Mo because MoO_2 is stabilized by Mo-Mo bonds (13). We suspect that thermodynamic data for one or more Pr species is in error.

Brauer and Kristen (4,5) found that Nd(IV) and Dy(IV) are formed to a greater extent in $BaCeO_3$ than in other perovskites. Nd(IV) and Dy(IV) need large octahedral holes, as provided by $BaCeO_3$. However $BaThO_3$ has even larger octahedral holes and does not stabilize Nd(IV) well (4). Preparative evidence indicates that $BaThO_3$ is not very stable (14) and we plan to provide quantitative thermochemical data to support this conclusion.

Table 2. Structural and thermodynamic parameters for $BaMO_3$.

Compound	Ionic radius of M(IV)/Å[a]	t [b]	Change in molec. vol., Eq. (4)/Å³	ΔH(complex) kJ mol⁻¹ [c]
$BaTiO_3$	0.605	0.97	-9.6	-163
$BaMoO_3$	0.650	0.95	-9.0	-92
$BaHfO_3$	0.71	0.921	-5.8	-134
$BaZrO_3$	0.72	0.917	-3.9	-126
$BaTbO_3$	0.76	0.90	+0.8	-88
$BaPrO_3$	0.85	0.864	+0.8	-147
$BaCeO_3$	0.87	0.856	+2.9	-52

[a] Ref. (11).

[b] $t = (r_{Ba^{2+}} + r_{O^{2-}})/[\sqrt{2}(r_{M^{4+}} + 4_{O^{2-}})]$.

[c] ΔH(complex) refers to reaction (4). Data from Ref. (9), this work, and U. S. Nat. Bur. Stand. Tech. Note 270 series.

ACKNOWLEDGEMENTS

    This work was supported by the Rutgers University Research Council and by the U. S. Department of Energy, Office of Basic Energy Sciences. We also acknowledge calorimetric measurements made by C. W. Williams and a $BaTbO_3$ sample provided by A. J. Jacobson.

REFERENCES

1. A.J. Jacobson, B.C. Tofield and B.E.F. Fender, Acta Cryst., B28:956 (1972).
2. A.J. Jacobson, B.C. Tofield and B.E.F. Fender, Proc. Tenth Rare Earth Res. Conf., Carefree, Arizona, 1973.
3. B.C. Tofield, A.J. Jacobson and B.E. Fender, J. Phys. C., 5:2887 (1972).
4. G. Brauer and H. Kristen, Z. Anorg. Allg. Chem., 456:41 (1979).
5. G. Brauer and H. Kristen, Z. Anorg. Allg. Chem., 462:35 (1980).
6. D.G. Nocera, L.R. Morss and J.A. Fahey, J. Inorg. Nucl. Chem., 42:55 (1980).
7. V.B. Parker, D.D. Wagman and D. Garvin, NBSIR 75-968, U.S. National Bureau of Standards, Washington, D.C. (1976).
8. L.R. Morss, Chem. Rev., 76:827 (1976).
9. G.C. Fitzgibbon, E.J. Huber, Jr. and C. E. Holley, Jr., J. Chem. Thermodynamics, 5:577 (1973).
10. L.R. Morss and C. W. Williams, to be published.
11. R.D. Shannon, Acta Cryst., A32:751 (1976).
12. J.B. Goodenough and J.M. Longo, Landolt-Börnstein Tables, Group III, Vol. 4a, Springer-Verlag, Berlin, pp. 126-314 (1970).
13. B.G. Brandt and A.C. Skapski, Acta Chem. Scand., 21:661 (1967).
14. Gmelins Handbuch, 8th Ed., Thorium, Erg-Bd. C2, pp. 14-17.

ON MIXED CRYSTALS OF $Pr_{1-y}Tb_yO_x$ GROWN UNDER HIGH OXYGEN PRESSURES[†]

B. Chang, M. McKelvy and L. Eyring

Department of Chemistry and the Center for
Solid State Science, Arizona State University
Tempe, Arizona 85287

ABSTRACT

Mixed $Pr_{1-y}Tb_yO_x$ crystals were prepared by oxidizing, hydrothermal synthesis. X-ray powder and electron diffraction were used
to characterize these and their reduction products. Two fluorite
phases of small composition range were observed for $.1 \leq y \leq .8$
rather than the single phase expected from previous work on this
system. The ordered intermediate phases present in the original
preparations at high terbium content are further developed in the
electron beam irradiated materials.

INTRODUCTION

The phase analysis and crystal chemistry of the binary (1) and
ternary (2) rare earth oxides have recently been reviewed. The binary phases of higher oxides of Ce, Pr and Tb are members of a homologous series (R_nO_{2n-2}, $4 \leq n \leq \infty$) of fluorite-related oxides. Although these series are compositionally the same and structurally
very similar, only the structural prototype, R_7O_{12}, appears to be
isostructural across the series. Which system would the mixed oxides
resemble? The ternary $Pr_{1-y}Tb_yO_x$ is particularly interesting since
the kinetic inhibition due to low metal atom mobility (2) is partially relieved by electron mobility. This has been shown to be very
important in mixed valence systems. In spite of structural variation,
Vegard's law has been shown to have wide applicability in the mixed
oxides as shown for Ce-Pr-O (3), Ce-Tb-O (4) and Pr-Tb-O (5). This

[†] This work was performed with support from the NSF under Grant
DMR 80-06584.

study extends the structural observations to phases in the ternary system, Pr-Tb-O, prepared at high oxygen pressures.

EXPERIMENTAL PART

The preparative apparatus and technique have been previously described (6,7,8). A René 41 cold-seal autoclave, heated by a Kanthal wound tube furnace, was used as the hydrothermal reaction vessel. Pressure and temperature were measured with a Bourdon gauge (±34 bars) and a thermocouple mounted in an exterior thermocouple well on the autoclave (±25°C when calibrated against internal temperatures), respectively.

A gas driven water intensifier provided the initial specimen pressure. Thereafter, the reaction vessel pressure was controlled by the heating and cooling of a second interconnected cold-seal autoclave. The second autoclave was heated by a Kanthal wound furnace, controlled by a Thermac Series 6000 temperature controller coupled with a Data Trak Model FGE 5110 programmer. Consequently, all pressure changes for crystal growth have continuous rates.

The terbium oxide and praseodymium oxide used were 99.999% pure provided by Research Chemicals and the Lindsay Division of American Potash and Chemical Corp., respectively. The concentrated nitric acid was reagent grade from the J.T. Baker Chemical Co.

The oxides were heated at 1000°C for 24 hours and slow cooled to 200°C to give $PrO_{1.833}$ and $TbO_{1.75}$. Mixed oxide specimens were then prepared for $TbO_{1.75}/PrO_{1.833} + TbO_{1.75}$ molar ratios varying from 0 to 1 in 0.1 increments. These samples were contained in capsules made from thin walled gold tubing (4.5 mm ID, 5.0 mm OD). One end of a 35 mm length of tubing was crimped and welded shut. 2.937 x 10^{-4} moles of $TbO_{1.75} + PrO_{1.833}$ and 50 mg of conc. HNO_3 were added to the capsule. After the second capsule end was crimped, the capsules were leak checked by weight measurement upon heating at 110°C for two hours.

In the hydrothermal procedure specimens were brought to a pressure of 1650 bar, isobarically heated to 810°C and allowed to equilibrate for 24 hours, followed by depressurization. The depressurization was carried out isothermally in four steps over a 72 hour period to a final pressure of 1150 bar. Each step included a slow uniform pressure decrease, followed by a smaller relatively rapid pressure increase to an isobaric plateau, which was held for several hours. After equilibration at 1150 bar, the reactor was cooled isobarically to 400°C and then smoothly brought to atmospheric pressure on cooling to room temperature. Any hydroxide formed was removed by soaking the samples in 0.2N HNO_3 for 24 hours.

The X-ray data were taken using a CDC-700 Guinier camera with $CuK_{\alpha1}$ radiation (~10% thoria being used as an internal standard). The transmission electron microscopy specimens were prepared by grinding crystals under liquid nitrogen to facilitate formation of thin crystal edges (9). The crystals were then suspended in acetone and dried onto thin holey carbon film grids. Observation of selected

area electron diffraction were made on a JEM 100B electron microscope. Analyses of the acid soluble residues $(R(OH)_3)$ were made
using the ICP method.

RESULTS

Guinier X-ray powder diffraction patterns were taken of
$Pr_{1-y}Tb_yO_x$ for $0 \leq y \leq 1$ at intervals of 0.1 in y. The diagrams consisted of lines from at least two and, in some cases, three phases
except for $y = 0$ or 1. These were all indexed as a cubic fluorite
phase including TbO_x-rich fluorite-related compounds of lower symmetry. The cubic phases are marked with open triangles and squares
while the pseudo-cubic phases are represented by open circles in
Figure 1. The phases with compositions in the range $y = 0.3 - 0.6$
exhibited rather broad diffraction lines and considerable diffuse
scatter.

The rare earth material not crystallized as oxide was precipitated on cooling as the hydroxide. X-ray diffraction patterns were
taken of the crystals in the reactor at the end of the run as well
as of the oxides remaining after the dissolution of the hydroxides
in 0.2N HNO_3. ICP analysis of the solution established that the relative rare earth concentrations from the hydroxide were the same as
the original overall composition. Furthermore, it was shown that 32,
28 and 27% of the original material remained in solution at the end
of crystal growth and precipitated on cooling from runs with $y =$
0.8, 0.6 and 0.4, respectively.

Selected area electron diffraction patterns were taken from
thin edges of fractured crystals. These areas all received appreciable electron irradiation during alignment and some crystals were
deliberately irradiated further to cause oxygen loss in the vacuum
of the microscope. Two compositions were observed by this method:
one with $y = 0.3$ and one with $y = 0.8$. Some of the patterns had
been seen before whereas others appeared to be new. These films often had a multitude of irregular spots and diffuse scatter such as
is characteristic of a material with small domains of different
structures.

DISCUSSION

Crystal growth of the $Pr_{1-y}Tb_yO_x$ occurs during depressurization at constant temperature, followed by isobaric cooling and the
crystallization of $R(OH)_3$. The cycling of the pressure would cause
dissolution of small crystals, then regrowth on larger nuclei under
the varying pressure. Leaching with 0.2N HNO_3 to remove the hydroxide could also cause a disproportionation of any intermediate oxides with the dissolution of a reduced phase (10).

In previous studies of similar fluorite mixed crystals (3,5,
11) compositional sequences usually follow Vegard's law closely.
Where they do not there are uncertainties in the O/M or M/M' ratio.
In these high oxygen pressure runs the O/M ratio is probably either

2.00 (F) or 1.82 (δ)(8). Vegard's law lines for these two types of phases are drawn in Figure 1. All cubic phases are assumed fluorite type RO_2 with a composition determined by Vegard's law as indicated by the arrows. The non-cubic phases with high TbO_x content may have an uncertain oxygen composition as well. There is a qualitatively correct compensation in the Tb composition as the phases separate based on their lattice parameters and relative amounts as determined by relative X-ray intensities.

In every hydrothermal preparation of mixed oxides there were two fluorite-type phases present except for the specimens where y = 0.9. In this case there was a fluorite-type and a δ-phase (12). All of the original preparations richer in TbO_x than y = 0.6 contained a phase of lower symmetry as indicated by the splitting of the main fluorite lines. In the mixtures where y = 0.7 and 0.8 this intermediate phase was not the δ-type. Subsequent preparations marked with squares in Figure 1, including one at y = 0.8, had only two fluorite phases. Small but significant differences in the final oxygen pressure in this critical region probably determines whether an intermediate phase will remain unoxidized.

The appearance of two phases in each case is contrary to expectations (3,5). If it is assumed that Vegard's law assigns true compositions to these phases there are two obvious possible explanations. 1) The pressure dependence of the equilibrium condition may be great enough that the miscibility gap observed reflects a true thermodynamic requirement. 2) The separation may be a fortuitous event in which a phase at one end of the compositon range nucleates and grows until the composition falls below that required to sustain further crystallization of this composition and then nucleation and growth of the other phase proceeds. In this case the separation is non-equilibrium and temporal. Although the first suggestion is unexpected there is an observation which strongly supports it.

After the oxide crystallization is complete at 810°C the system is cooled and brought to atmospheric pressure. The remaining cations in solution crystallize during cooling as $R(OH)_3$. These crystals have lattice parameters consistent with their analysis which is the same composition as the starting mixtures in each case. It is difficult to see how this could occur unless the oxide phases separate essentially at equilibrium. The relative amounts of each phase present as determined by diffraction intensity is in qualitative agreement with either interpretation. Needless to say, this is a surprising result and suggests that many systems which form continuous mixed crystals following Vegard's law at low pressures may show a miscibility gap at high oxygen pressures.

In the composition range below y = 0.4 both of the fluorite phases appear to be enriched in Pr. This would require that the solution remaining be richer in Tb but these residues were not analyzed.

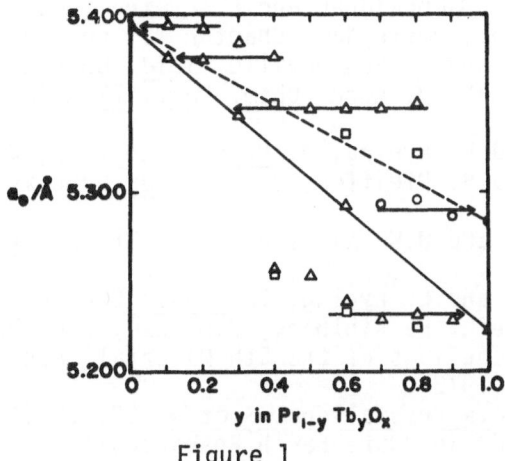

Figure 1

Diagram of the products of hydrothermal growth of mixed oxides of $Pr_{1-y}Tb_yO_x$. Lattice parameters are shown for phases resulting from hydrothermal treatment of mixed praseodymium and terbium oxides in the molar fractions shown as the absissa. Triangles and squares mark the lattice parameters of fluorite phases. Circles mark phases of lower symmetry indexed as pseudocubic phases. The solid line represents the Vegard's law relationship expected for the fluorite-type mixed oxides. Arrows indicate the approximate composition the phases would have if Vegard's law applied to both the fluorite and the δ-phase type oxides.

The electron diffraction patterns suggest that crystals sub-jected to beam irradiation suffer oxygen loss and that they order, at least in patches, to intermediate phases. This study has not been full enough to record a large number of these intermediate phases, but the fact that they form and that there are a wide variety of them is demonstrated. Some patterns had been previously observed in the binary systems. some had not.

REFERENCES

1. L. Eyring, in "Handbook on the Physics and Chemistry of Rare Earths", K.A. Gschneidner and L. Eyring (eds.), North-Holland Publishing Co., Amsterdam, Chapter 27, pp. 337-399 (1979).
2. D.J.M. Bevan and E. Summerville, ibid, Chapter 28, pp. 401-524.
3. J.D. McCullough, J. Amer. Chem. Soc., 72, 1386 (1950); 74, 5225 (1952).
4. J. Kordis and L. Eyring, J. Phys. Chem., 72, 2030 (1968).
5. G. Brauer and B. Pfeiffer, Journal for Praktische Chemie, 4, 23 (1966).
6. R.A. Laudise and J.W. Nielsen, Solid State Physics, 12, 149 (1961).
7. J.M. Haschke and L. Eyring, Inorganic Chemistry, 10, 2267 (1971).
8. M.Z. Lowenstein, L. Kihlborg, K.H. Lau, J.M. Haschke and L. Eyring, "Proceedings of the 5th Materials Research Symposium", pp. 343-351 (1972).
9. S. Iijima, Acta Crystallogr. Sect A, 29, 18 (1973).
10. A.F. Clifford, in "Rare Earth Research II", K.S. Vorres (ed.), Gordon and Breach, New York, pp. 45-50 (1964).
11. J. Kordis and L. Eyring, J. Phys. Chem., 72, 2044 (1968).
12. J. Sawyer, B.G. Hyde and L. Eyring, Bull. Soc. Chim. France, 1190 (1965).

A KINETIC STUDY OF THE OXIDATION OF ZETA PHASE PRASEODYMIUM OXIDE:

$$\frac{10}{9} Pr_9O_{16} + \frac{1}{9} O_2 \rightarrow Pr_{10}O_{18}\,^\dagger$$

Tadashi Sugihara, Sheng H. Lin and LeRoy Eyring

Department of Chemistry and the Center for
Solid State Science, Arizona State University
Tempe, Arizona 85287

ABSTRACT

Thermodynamic studies of phase transformations between the in-termediate phases (Pr_nO_{2n-2}) reveal reproducible hysteresis loops. In this report, kinetic studies between the zeta (n = 9) and the epsilon (n = 10) phases have been carried out by measuring the weight change of the sample as a function of time after a sudden increase of oxygen pressure. The experiments covered temperatures of 464, 477, 491 and 503°C and oxygen pressure varied from 0.01 to 50 Torr.

Since the initial stage of a solid state reaction involves nucleation and growth, this mechanism is considered first. The in-duction period required by the Avrami equation (1) is missing in the oxidation of zeta phase as is clear from the typical experimen-tal curve shown in Figure 1. This indicates that the nucleation step occurs virtually instantaneously, covering the surface of each reaction particle with a layer of product. This process could lead to a moving boundary controlled mechanism in which the reaction in-terface, a proportionately smaller replica of the initial exterior surface, shrinks toward the center of the crystal. In this case (1), the expression for f, weight fraction, is represented by

$$Kt = R \left[1 - (1 - f)^{1/3}\right] + \frac{1}{2} - \frac{1}{2}(1 - f)^{2/3} - \frac{1}{3}f \qquad (1)$$

where $R = kd/kpr_o$ and $K = kd(C^* - Ceq)/C_or_o^2$. Here C_o represents the concentration of the reactant and C^*, Ceq are concentrations of the diffusing species in the product layer, and kd, kp are the diffusion constant and the reaction constant respectively.

---

† This work was supported by the NSF through Grant DMR 78-05722 and will be reported fully elsewhere.

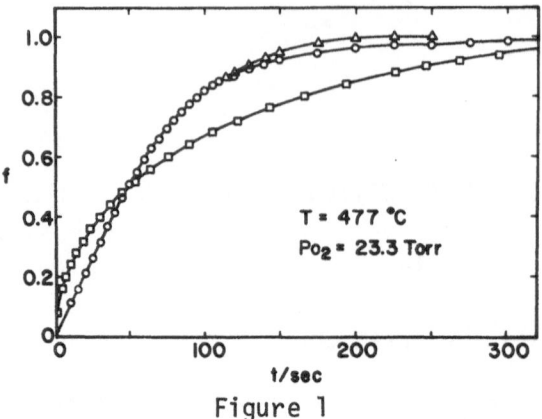

Figure 1

A typical kinetic run at 477°C and 23.3 Torr and the fitting of various models. The fraction of reaction (f) is plotted against time: O, observed; △, phase boundary controlled mechanism; ☐, diffusion mechanism.

Two limiting cases of Eq. 1 are distinguished. Case (1): when R is very large, the so-called phase boundary reaction controlled model is obtained with the equation

$$f = 1 - (1 - K't)^3 \tag{2}$$

where $K' = K/R = kp (C^* - Ceq)/C_o r_o^2$. Case (2): if $kp \gg kd$, one obtains diffusion model, by letting $R \to 0$.

$$Kt = \frac{1}{2} - \frac{1}{2} (1 - f)^{2/3} - \frac{1}{3} f \tag{3}$$

Theoretical curves are depicted for diffusion (Eq. 3) and for phase boundary control (Eq. 2) with the experimental one in Figure 1. The theoretical curves are chosen to fit the experimental plot at f = 0.5. As can be seen from the figure the agreement between the phase boundary controlled model and experiment is satisfactory. The reason for a deviation above f ~ 0.9 seems to be attributable to the existence of an epsilon pseudophase. In conclusion, the rate determining step of the oxidation kinetics of this odd-even reaction is revealed to be the reaction speed at the reaction interface and diffusion of the absorbed species (in this case O_2) is faster than the chemical reaction.

REFERENCE

1. H. Eyring, S. H. Lin and S. M. Lin, "Basic Chemical Kinetics," John Wiley and Sons, New York (1980) Chapter 9.

PREDICTED STABILITIES OF RARE EARTH DIHALIDES

Steven Bratsch* and Herbert B. Silber (1)

Division of Earth and Physical Sciences
The University of Texas at San Antonio (UTSA)
San Antonio, Texas 78285

Using a modified Born-Haber cycle we have established a linear relationship between the enthalpies of formation and the sum of the ionic radii, given by:

$$(\Delta H_f - Q_n) = A/(r_+ + r_-) + B \qquad (1)$$

Q_n is the sum of the sublimation energy and the ionization energies (2). Equation (1) is obeyed for the Ln_2O_3 (c), Ln^{3+} (aq), LnI_3 (c), LnO (c), LnF_2 (c) and LnF_3 (c) systems (2). For the plus-two ions, Ba(II), Eu(II) and Ca(II) can be utilized to establish the constants for equation (1), since the ionic radii and thermodynamic properties of Ca(II) and Yb(II) are nearly equal. The enthalpies of disproportionation, ΔH_{disp}, can be calculated for the following reaction:

$$LnX_2 \text{ (c)} = 1/3 \text{ Ln (c)} + 2/3 \text{ LnX}_3 \text{ (c)} \qquad (2)$$

Table 1. Calculated constants for Equation (1).

System	r_- A°	A KJ-A°/mol	B KJ/mol
$LnCl_2$	1.81	−5,630	−725
$LnCl_3$ (hexagonal)			
La–Tb	1.81	−12,460	−622
$LnCl_3$ (monoclinic)			
Tb–Lu	1.81	−14,610	185
$LnBr_2$	1.96	−5,860	−635
$LnBr_3$	1.96	−13,600	−295

Table 2.   Enthalpy data in KJ/mol.

Ln	$\Delta H_f(LnCl_2)$	$\Delta H_f(LnCl_3)$	ΔH_{disp}^{Cl}	$\Delta H_f(LnBr_2)$	$\Delta H_f(LnBr_3)$	ΔH_{disp}^{Br}
Ba(3)	−859	−	−	−757	−	−
La	−493	−1075	−224	−391	−910	−216
Ce	−555	−1058	−150	−451	−893	−144
Pr	−670	−1051	−31	−566	−886	−25
Nd	−693	−1039	0	−588	−873	+6
Pm	−695	−1032	+ 7	−589	−865	+12
Sm	−796	−1037	+105	−689	−870	+109
Eu(3)	−819	−923	+204	−711	−756	+207
Gd	−486	−1007	−185	−378	−840	−182
Tb	−594	−991	−67	−484	−833	−71
Dy	−685	−1002	+17	−575	−849	+9
Ho	−667	−993	+5	−555	−836	−2
Er	−643	−992	−18	−531	−831	−23
Tm	−719	−992	+58	−606	−826	+55
Yb	−788	−942	+160	−675	−773	+160
Lu	−	−973	−	−	−800	−

The results are summarized in the Tables. Based upon the enthalpies only; Pm, Sm, Eu, Dy, Tm and Yb should form stable dihalides with Cl^- or Br^-; $NdBr_2$ should be stable; and $NdCl_2$ and $HoCl_2$ are borderline. These conclusions are in agreement with those of Kim, except for the marginally stable dihalides (4). The tribromides are within 1% of recently reported values (5).

REFERENCES

1. Supported by a grant from the Robert A. Welch Foundation of Houston, Texas.
2. S. Bratsch and H. B. Silber, submitted.
3. "Selected Values of Chemical Thermodynamic Properties", NBS TN 270-7, 1968-73.
4. Y.-C. Kim and J. Oishi, J. Less-Common Metals, 65: 199 (1979).
5. C. Hurtgen, D. Brown and J. Fuger, J. C. S. Dalton, 70 (1980).

VAPORIZATION AND THERMODYNAMIC PROPERTIES OF SAMARIUM DICARBIDE AND NONSTOICHIOMETRIC DISAMARIUM TRICARBIDE

John M. Haschke, Rockwell International, P.O. Box 464, Golden, CO 80401; Thomas A. Deline, University of Michigan, Ann Arbor, MI 48109

INTRODUCTION

As noted in our recent report on the phase equilibria in the samarium + oxygen + carbon system (1), an interest in the physico-chemical properties of the lanthanide carbides at high temperature is promoted by their importance in nuclear reactor technology. Initial workers reported the existence of three carbide phases (LnC_2, Ln_2C_3, Ln_3C) for several lanthanides (2), and the objective of this study was the determination of thermodynamic values for the samarium carbides from equilibrium vapor pressure data for the SmC_2+C, $Sm_2C_3 + SmC_2$ and $Sm_3C + Sm_2C_3$ two-phase regions. Our reinvestigation of the Sm + O + C system (1) yielded two major findings: (a) The existence of the cubic Sm_3C phase could not be confirmed; however, an anion-deficient fcc phase with identical lattice parameters was found at the $SmO_{0.5}C_{0.4}$ composition. (b) The cubic Pu_2C_3-type phase, designated as SmC_y in this report, was found to be substoichiometric, to exist over the range $1.36<y<1.45$ and to have a carbon-rich phase boundary which decreased in carbon content with increasing temperature. Consequently, this study has been limited to the SmC_2+C and $SmC_y + SmC_2$ regions.

EXPERIMENTAL

Carbide samples were prepared by reaction of the elements in sealed Ta vessels and were analyzed by chemical and X-ray diffraction methods. Equilibrium pressures were measured by a target-collection effusion technique.

RESULTS

Both $SmC_2(s)$ and $SmC_y(s)$ vaporize incongruently with the formation of $Sm(g)$ as the only significant vapor species (1). The dicarbide reaction has been thoroughly characterized in several studies reviewed by Seiver and Eick (3). Our results for SmC_2 are in excellent agreement with earlier values.

The vaporization reaction for SmC_y is described by equation 1.

$$2/(2-y) \ SmC_y(s) \rightarrow y/(2-y) \ SmC_2(s) + Sm(g). \tag{1}$$

The complex coefficients result from the retrograde dependence of the carbon-rich phase boundary of SmC_y on temperature. Equilibrium studies with the Pu_2C_3-type phase show that $y = 1.42$ and $a = 0.8434$ nm at 1600 K, and $y = 1.44$ and $a = 0.8466$ nm at 1400 K (1). The equilibrium pressure of log P is defined by equation 2.

$$\log P \ (Sm,g, \ atm, \ 1372 \ K < T \ 1636 \ K) = [(12.009 - 28849T^{-1} +$$

$$7018500T^{-2}) \pm 0.038]. \tag{2}$$

Since a different vaporization reaction occurs at each temperature, the results have been evaluated only by the third-law method. For $y = 1.43$ (T = 1500 K), $\Delta H°$ for equation 1 at 298 K is 282 kJ mol^{-1}. Enthalpies of formation of SmC_2 and $SmC_{1.43}$ at 298 K have been derived from the equilibrium results. For SmC_2, $\Delta H_f°$ from the second- and third-law results are (-96 ± 8) and (-104 ± 3)kJ mol^{-1}, respectively. For $SmC_{1.43}$, $\Delta H_f°$ is (-90 ± 8)kJ mol^{-1}. These values are in excellent agreement with calorimetric results for CeC_2 and $CeC_{1.5}$ (4).

ACKNOWLEDGMENT

Support of the U. S. Department of Energy, Contract DE-ACO4-76DPO3533, during preparation of the manuscript is gratefully acknowledged.

REFERENCES

1. J.M. Haschke and T.A. Deline, Inorg. Chem., 19:527 (1980).
2. F.H. Spedding, K.A. Gschneidner, Jr., and A.H. Daane, J. Am. Chem. Soc., 80:4499 (1958).
3. R.L. Seiver and H.A. Eick, High Temp. Sci., 3:292 (1971).
4. F.B. Baker, E.J. Huber, Jr., C.E. Holley, Jr., and N.H. Krikorian, J. Chem. Thermod., 3:77 (1971).

COMPLEXITY OF PHASE EQUILIBRIA IN THE MM-Co-Fe SYSTEM BETWEEN 2:7 AND 1:5 STOICHIOMETRIES

E.M.T. Velu, S. Laha, E.C. Subbarao, K.P. Gupta,
A.K. Majumdar, T.A. Padmavathi Sankar, S. Pandian and
U. Ramakrishna
Indian Institute of Technology, Kanpur 208016
India

INTRODUCTION

Phase relations in the MM-Co-Fe system have been reported previously (1-2) based primarily on X-ray and microstructural results. The studies have been supplemented in the present work by studying additional alloys over a narrow composition interval between 2:7 and 1:5 stoichiometries using thermomagnetic analysis (TMA).

EXPERIMENTAL

Indian Mischmetal (MM) and 99.9% pure cobalt and iron were used. The mischmetal contains 52% Ce, 20.1% La, 15.7% Nd, 4-8% Pr, 6% Fe (all in w/o) and the balance unidentified impurities. The melting, annealing and phase analysis procedures are the same as those reported previously (1-2).

RESULTS AND DISCUSSION

The compositions of alloys used for phase study are given in Table 1 together with their TMA results and microstructural data. Alloys 0-3 and 6-5 are of single phase 1:5. The T_c of the 1:5 phase decreases slightly from 504°C at low Fe concentration and then increases steadily with increasing concentration of Fe up to 576°C at 12.4 a/o Fe (Fig. 1). The $\underline{a}$ and $\underline{c}$ values for the Fe free 0-3 alloy agree with the values reported by Khan (3) for the $CeMMCo_5$ phase. With increasing Fe content there is a decrease in $\underline{a}$ and an increase in $\underline{c}$ up to 12 a/o Fe.

Table 1. Compositions Selected in the MM-Co-Fe System and the
          Phases Identified

Alloy	Composition (a/o, analysed)			No. of Phases	TMA[a] Results, Magnetic Transition Temperatures (°C)					
	RE	Co	Fe		X_1	2:7	5:19	X_2	1:5	X_3
0-2[b]	10.2	80.8	0	3			227	302	490	
0-3[b]	16.6	83.4	0	1					504	
B-5	20.6	75.8	3.6	3			250	320	495	
6-1	22.3	72.3	5.4	2		127	266			
6-2	20.5	74.4	5.1	3			270	328	537	
6-4	18.0	76.7	5.3	2			261		525	
6-5	16.6	77.0	6.4	1					532	
12-1	21.2	67.3	11.5	3		184	283			
12-3	19.4	68.3	12.3	2	170		288			
12-5	15.7	71.9	12.4	2					576	
15-1	21.5	63.3	15.2	2		202				
15-4	18.0	66.7	15.3	2	164		273			
L-2	21.1	73.1	5.8	2			295	338		
L-3	21.1	69.6	9.3	3		168	294			
L-6	21.7	73.9	4.4	2		124	266			776[c]
L-8	21.9	70.1	8.0	3		165	290			743[c]
L-10	22.6	67.9	9.5	3		186[c]				762[c]
L-11	22.5	65.3	12.2	2		150[c]	198[c]			724[c]
4	17.1	73.6	9.3	3			270		537	
9	19.0	78.2	2.8	3			224	290	484	
10	20.0	71.5	8.5	3			273	320	528	
12	21.8	64.6	13.6	3		175	260			734[c]
27	19.5	77.7	2.8	3			224	294	480	
36	21.8	75.4	2.8	3		65	222			

a T_c reported here is the mid point of the cooling curves unless
   otherwise stated.
b Alloys prepared with synthetic mischmetal.
c T_c measured by plotting permeability vs. temperature.

    Alloy 0-2 with a composition lying between 5:19 and 1:5
stoichiometries shows three magnetic transitions (Table 1). The
227°C and 490°C transitions correspond to 5:19 and 1:5 phase, respectively. The 302°C transition was also observed for a number of
other Fe containing alloys like 6-2, 9, 10, 27 and B-5 (Table 1).
This is absent for the alloys like 6-4 and 4 which lie closer to
the 1:5 composition. This transition does not occur in the as cast
alloy and disappears for alloys annealed at 1000°C. Low temperature
annealing at 800°C and 700°C gave the same results as annealing at
900°C. Further work is in progress to establish the conditions of
stability of this phase.

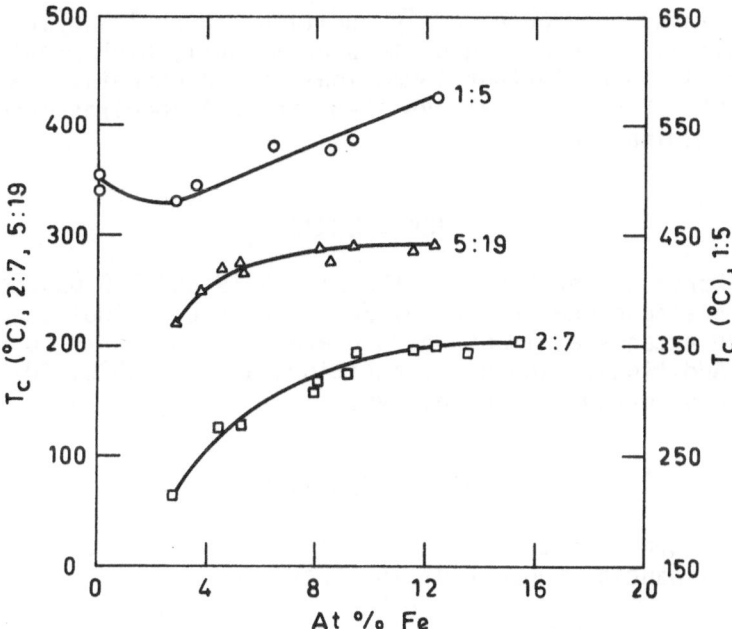

Fig. 1. Variation of Curie temperature (T_C) with Iron Content for 2:7, 5:19 and 1:5 Phases in the MM-Co-Fe System.

On the RE rich side of 5:19 stoichiometry the 5:19 phase is in equilibrium with 2:7 phase. Alloy 36 for example has two magnetic transitions (65°C and 222°C) corresponding to 2:7 and 5:19. In this region alloys with increasing concentration of Fe show their T_Cs markedly raised (Fig. 1). The T_C of the 2:7 phase increases from 65°C (alloy 36) to 202°C (alloy 15-1) as the Fe varies from 2:8 to 15.2 a/o. However, no significant change in lattice parameters appears to occur with Fe content. Similarly the T_C of 5:19 phase also increases steadily up to 6 a/o Fe from 222°C to 295°C and assumes a constant value above 8 a/o Fe. The magnetic transition corresponding to 5:19 is present only as a trace in 12-3 and 15-4. The additional information obtained by TMA allows a refinement of the earlier ternary phase diagram in this region by indicating a larger extension of the 5:19 phase than observed earlier (2).

For many alloys lying along the 2:7 stoichiometric line such as L-6, L-8, L-10, L-11 and 12, a magnetic transition temperature

$\geq$700°C is invariably observed in addition to the low temperature ones (Table 1). The origin of this anamolously high magnetic transition is under further investigation. A magnetic transition around 170°C observed in alloys 12-3 and 15-4 requires more detailed examination.

CONCLUSIONS

The Curie temperatures of the 2:7, 5:19 and 1:5 phases increase with increasing concentration of Fe. TMA results indicate that the 5:19 phase extends at least up to 6 a/o Fe into the MM-Co-Fe ternary system. Additional magnetic transitions around 750°C, 300°C and 170°C were observed in some alloys.

ACKNOWLEDGMENTS

We are grateful to Department of Electronics, Government of India for financial support of this work.

REFERENCES

1. E.M.T. Velu, E.C. Subbarao, H.O. Gupta, K.P. Gupta, S.N. Kaul, A.K. Majumdar, R.C. Mittal, T.A. Padmavathi Sankar, G. Sarkar, M.V. Satyanarayana, K. Shankarprasad and J. Subramanyam, J. Less-Common Metals, 71:219 (1980).
2. R.C. Mittal, M.V. Satyanarayana, K.P. Gupta, H.O. Gupta, S.N. Kaul, A.K. Majumdar, K. Shankaraprasad, T.A. Padmavathi Sankar, E.C. Subbarao and E.M.T. Velu, J. Less-Common Metals, 78:245 (1981).
3. Y. Khan, Acta Cryst., B30:1533 (1974).

LOW TEMPERATURE-HIGH MAGNETIC FIELD HEAT CAPACITY STUDIES OF WEAKLY AND NEARLY FERROMAGNETIC AND MIXED VALENCE RARE EARTH MATERIALS

K. A. Gschneidner, Jr., K. Ikeda and O. D. McMasters

Ames Laboratory[*] and Department of Materials Science and Engineering, Iowa State University, Ames, IA 50011

Low temperature heat capacity measurements from 1 to 20 K as a function of magnetic field from 0 to 10 T (100 kOe) have been made on materials which exhibit unusual properties due to spin fluctuations. The spin fluctuating materials studied to date include the itinerant ferromagnet Sc_3In (1), the mixed valent compound $CeSn_3$, and the nearly ferromagnetic materials: RCo_2 [R = Sc, Y and Lu (2)], Sc and $Pd_{1-x}Ni_x$ ($x = 0.005$ and 0.01) alloys. In all three types of materials the high magnetic fields quench spin fluctuations. In the nearly ferromagnetic materials and the mixed valent compound this quenching leads to a depression of the electronic specific heat constant of $\sim10\%$ for the former and 23% for the latter — this change is over an order of magnitude larger than has been observed heretofore.

The situation for the weakly itinerant ferromagnet Sc_3In is more complex. Above about twice the Curie temperature, $T_c = 6.0$ K, the spin fluctuation contribution to the heat capacity is enhanced by the magnetic field as long as the field is <5 T, but above 5 T the spin fluctuations are quenched. At or below T_c the magnetic field always lowers the heat capacity, completely quenching itinerant ferromagnetism in Sc_3In at 12 T.

[*]Operated for the U.S. Department of Energy by Iowa State University under contract No. W-7405-ENG-82. This research was supported by the Director of Energy Research Office of Basic Energy Sciences, WPAS-KC-02-01.

At least four different behaviors of the spin fluctuation contribution to the heat capacity of solids have been identified. These are: 1) the quenching of the enhancement of the electronic specific heat constant, γ, such as found in $LuCo_2$ (2); 2) the quenching of the enhancement of γ and an increase in the T^3 contribution to the heat capacity, which is due to an induced moment on the magnetic atom; 3) a $T^3 \ln(T/T_s)$ contribution to the heat capacity in addition to the effect on γ and the T^3 contribution, as seen in $CeSn_3$; and 4) the quenching of itinerant ferromagnetism by application of an external magnetic field as exhibited by Sc_3In (1). Some of these behaviors are qualitatively understood in terms of current theory, but the others are not. This is not surprising since we have just discovered these behaviors and theorists have not had time to examine them.

REFERENCES

1. K. Ikeda and K. A. Gschneidner, Jr. J. Magnetism Magnetic Mater. __22__, 207 (1981).
2. K. Ikeda and K. A. Gschneidner, Jr., Phys. Rev. Lett. __45__, 1341 (1980).

ANOMALOUS BEHAVIOR OF CERIUM AND EUROPIUM IONS IN TERNARY MOLYBDENUM

CHALCOGENIDES (CHEVREL PHASES)

M.B. Maple* and M.S. Torikachvili,*[†] Institute for Pure and
Applied Physical Sciences, University of California, San
Diego, La Jolla, CA 92093; R.P. Guertin,** Physics Depart-
ment, Tufts University, Medford, MA 02155; S. Foner, Fran-
cis Bitter National Magnet Laboratory,[§] Massachusetts
Institute of Technology, Cambridge, MA 92139

I. INTRODUCTION

A large class of ternary Mo chalcogenides can be formed with the
general formula $M_xMo_6X_8$ where M is a metal, $0 \leq x \leq 4$, and X = S, Se
or Te (1). These compounds have a relatively complex crystal structure with rhombohedral symmetry and a rhombohedral angle that is close
to 90° (2). The crystal structure of the ternary Mo chalcogenides
(Chevrel phases) is shown in Fig. 1 for a compound with the formula
$REMo_6S_8$ (RE = rare earth).

The ternary Mo chalcogenides have been studied extensively largely because of their remarkable superconductive properties. Some of
the compounds have moderately high superconducting transition temperatures (T_c) (3) and extremely high upper critical magnetic fields
(H_{c2}) (4-7). For example, the compound $PbMo_6S_8$ has values of T_c
and $H_{c2}(0)$ of ~15 K and ~60 Tesla, respectively (3,4,5). It has been
suggested that the extraordinary superconducting properties of the
$M_xMo_6X_8$ compounds may originate from Mo_6X_8 molecular units or "clusters" (8) in which the superconducting electrons, primarily associated
with the d-states of Mo, (7,9) are relatively confined.

---

* Research supported by the U.S. Department of Energy under Contract
  No. DE-AT03-76ER70227.
[†] On leave from Instituto de Fisica "Gleb Wataghin," Universidade
  Estadual de Campinas, Campinas-SP, 13100, Brazil. Also supported
  by FAPESP.
**Research supported by the National Science Foundation under Grant
  No. DMR80-10530.
[§] Supported by the National Science Foundation.

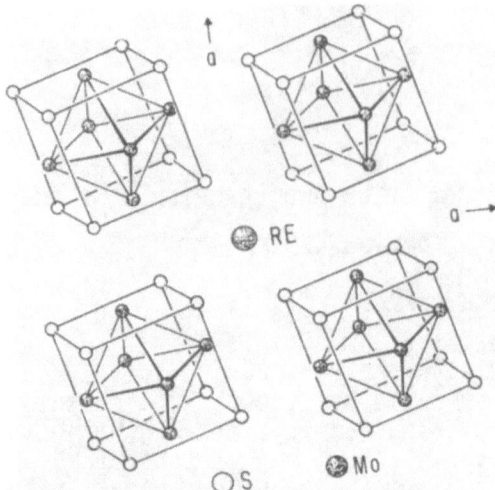

Figure 1.   Crystal structure of the RE Mo chalcogenides. Two rhom-
             bohedral axes (a) are indicated in the figure; the third
             axis projects outward. The RE ion is surrounded by
             eight Mo_6S_8 units. After Marezio et al. (2).

The "cluster character" of the $RE_xMo_6X_8$ compounds may also account for the small exchange interaction between the conduction electrons and the magnetic moments of the RE ions with partially-filled
4f electron shells (7,10). This allows all of the compounds (except
those with RE = Ce and Eu) to remain superconducting in spite of the
moderately large concentration of RE ions. It also enables many of
the compounds to order magnetically via the RKKY interaction with
rather low magnetic ordering temperatures that are comparable to T_c.

Extensive investigations on both series of $RE_xMo_6S_8$ (11) and
$RE_xMo_6Se_8$ (12) compounds have uncovered a number of interesting phenomena that arise from the interplay between superconductivity and
long-range magnetic order. Many $RE_xMo_6S_8$(RE = Gd, Tb, Dy and Er) (11)
and $RE_xMo_6Se_8$ (RE = Gd, Tb and Er) (12) compounds exhibit the coexistence of superconductivity and long-range antiferromagnetic ordering
of the RE magnetic moments. In contrast, experiments on $Ho_xMo_6S_8$ have
shown that long-range ferromagnetic order destroys superconductivity
at a second critical temperature $T_{c2} \sim \theta_f < T_{c1}$, where θ_f is the Curie
temperature and T_{c1} is the upper critical temperature at which the
compound first becomes superconducting (11). Recently, neutron scattering measurements on $Ho_xMo_6S_8$ revealed evidence for a spiral magnetization state with a wavelength $\lambda \sim 10^2$ Å that occurs above T_{c2} in

the superconducting state (13). Similar phenomena have been observed in the series of $RERh_4B_4$ compounds which have been studied in parallel with the $RE_xMo_6X_8$ compounds (12).

Equally interesting is the anomalous behavior of Ce and Eu ions in ternary Mo chalcogenides which we discuss herein. The physical properties of these materials display temperature, magnetic field and pressure dependences that are reminiscent of the Kondo effect or valence (interconfiguration) fluctuations associated with RE ions with unstable valence (10). It is noteworthy that Ce and Eu belong to the class of RE ions (along with Sm, Tm, Yb and, possibly, Pr) that have been found to exhibit valence instabilities in various metallic alloys and compounds (and, in the case of Ce, even the element itself!).

II. EVIDENCE FOR KONDO-LIKE SCATTERING OF CONDUCTION

ELECTRONS BY Ce and Eu IONS IN TERNARY MOLYBDENUM CHALCOGENIDES

A number of the physical properties of the Ce and Eu ternary and pseudoternary Mo chalcogenide compounds suggests an explanation in terms of some type of Kondo scattering of conduction electrons by the Ce and Eu magnetic moments. With the exception of Ce and Eu, all of the $RE_xMo_6S_8$ and $RE_xMo_6Se_8$ compounds are superconducting (14) with T_c's within each series that display a systematic dependence on RE[10] (the T_c of $Yb_xMo_6S_8$ is anomalously high in comparison to the other superconducting $RE_xMo_6S_8$ compounds, since nearly half of the Yb ions are divalent (15), while all the other RE ions are trivalent). A consistent analysis of these data could be made (10) by decomposing T_c within each series of $RE_xMo_6S_8$ and $RE_xMo_6Se_8$ compounds into two contributions; i.e.,

$$T_c = T_{c_o} - \Delta T_c \tag{1}$$

where T_{c_o} is the transition temperature in the absence of magnetic pair breaking interactions, while ΔT_c is the depression of T_c due to the magnetic pair breaking interactions. In the limit of relatively low concentrations n of paramagnetic RE ions, ΔT_c can be expressed as

$$\Delta T_c = nAN(E_F)\mathcal{J}^2(g_J-1)^2 J(J+1) \tag{2}$$

where A is a constant, $N(E_F)$ is the density of states at the Fermi level, $\mathcal{J}$ is the exchange interaction parameter, and g_J and J are, respectively, the Landé g-factor and total angular momentum of the Hund's rule ground state of the RE ion under consideration (16). The exchange interaction parameter $\mathcal{J}$ characterizes the strength and sign of the exchange interaction, which for RE ions is given by

$$\mathcal{K}_{int} = -2\mathcal{J}(g_J-1)\underset{\sim}{J}\cdot\underset{\sim}{s} \tag{3}$$

where s is the conduction electron spin density at the site of the RE ion. The variation of T_{c_o} with RE was estimated by linear interpolation between the nonmagnetic end member La and Lu molybdenum sulfide and selenide compounds. The resultant values of $\Delta T_c = T_{c_o} - T_c$ vs RE can be described qualitatively by Eq. (2) for fixed n and with the assumption that $N(E_F)$ does not vary with RE. A similar analysis based on a different estimate of T_{c_o} has been made by Fischer et al. (17).

From the ΔT_c vs RE data for both series of $RE_xMo_6S_8$ and $RE_xMo_6Se_8$ compounds, a value $|\mathcal{J}| \sim 0.01$ eV was inferred (10). Electron paramagnetic resonance experiments on the compound $Gd_xMo_6Se_8$ yielded the value $\mathcal{J} \sim +0.01$ eV (18) in support of the analysis of the T_c vs RE data in terms of paramagnetic pair breaking effects. The scaling of ΔT_c according to the deGennes' factor $(g_J-1)^2J(J+1)$ implies that the Ce and Eu molybdenum chalcogenides would be superconducting if the Ce and Eu ions interacted with the conduction electrons via the same mechanism as the other RE ions. This suggests that the failure of these compounds to become superconducting down to 50 mK may be due to anomalously strong exchange scattering associated with the Kondo effect (10). The Kondo effect would be generated by a <u>negative</u> exchange interaction between the spins of the conduction electrons and the Ce and Eu spins of the form

$$\mathcal{H}_{int} = -2\mathcal{J}\underset{\sim}{S}.\underset{\sim}{s} \tag{4}$$

where $\underset{\sim}{S}$ is the spin of the RE ion and $\mathcal{J} < 0$. This form for $\mathcal{H}_{int}$ would appear to be appropriate for Ce and Eu molybdenum chalcogenides; magnetic susceptibility measurements indicate that Ce is trivalent and has a crystalline electric field split doublet ground state with an effective spin of 1/2, while Eu is divalent and is therefore an S-state ion with a spin of 7/2 (10). Moreover, Mössbauer isomer shift measurements on $Sn_{1-x}Eu_xMo_6S_8$ compounds reveal a well-defined Eu^{2+} resonance (19). According to the Schrieffer-Wolff transformation (20) the negative exchange interaction parameter $\mathcal{J}$ can be expressed as

$$\mathcal{J} \sim \frac{\langle V_{kf}^2 \rangle}{E_f} < 0 \tag{5}$$

where V_{kf} is the matrix element that mixes conduction electron and localized 4f electron states, and E_f is the energy separating the localized 4f state and the Fermi level.

The basis for the conjecture that Kondo-like scattering of conduction electrons by the Ce and Eu magnetic moments quenches superconductivity in the Ce and Eu molybdenum chalcogenide compounds is the following observation: Certain RE impurities, particularly Ce, which produce a Kondo effect in the normal state, when dissolved in a superconducting matrix generally cause an anomalously rapid depression of the T_c of the matrix, in comparison with the other RE ions. These

latter ions cause smaller depressions of T_c that scale with the
deGennes' factor $(g_J-1)^2 J(J+1)$ [see Eq. (2)] (21). Sometimes even
reentrant superconductive behavior is observed when $T_K \ll T_{c_o}$, where T_K
is the Kondo temperature and T_{c_o} is the T_c of the matrix (21). Moreover, the magnetic state of such RE ions can often be changed with
pressure or by alloying with another element; this occurs, for example, in the system $La_{1-x}Ce_x$ when it is subjected to pressure (22) or
when La is alloyed with Th (21,23). In the regime where the localized
magnetic moment is still well-defined (equivalently, the exchange interaction Hamiltonian is still operative), the change in the magnetic
state of the RE ions can be attributed to an increase in $|\mathcal{J}|$ with
pressure, and, in turn, to an increase of $\langle V_{kf}^2 \rangle$ and/or a decrease in
E_f with pressure (21,24) [see Eq. (5)]. Similar types of behavior may
be expected for Ce and Eu ions in ternary Mo chalcogenides, a point to
which we will return later.

One of the classic earmarks of the Kondo effect is the minimum
in the electrical resistivity (ρ) that occurs at low temperatures in
dilute RE systems, which can evolve into an even more pathological
temperature dependence of ρ in concentrated RE systems (25). Figure 2
shows the electrical resistivity (normalized to its value at 294 K),
$\rho(T)/\rho(294 \text{ K})$, vs temperature for the Ce and Eu molybdenum sulfide and
selenide compounds and for $Lu_{1.2}Mo_6S_8$ as originally reported by Maple
et al. (10). The resistivity of the nonmagnetic reference compound
$Lu_{1.2}Mo_6S_8$ decreases monotonically with decreasing temperature and
then drops abruptly to zero at its T_c of ~2 K. The ρ vs T curve for
both the $CeMo_6S_8$ and $Ce_{1.1}Mo_6Se_8$ compounds indicates that ρ decreases
monotonically with decreasing temperature, with pronounced negative
curvature above a minimum which occurs at 11 K and 6-7 K, respectively, for the two compounds. For the $CeMo_6S_8$ compound there is a
sharp drop in resistivity at 2.5 K, attributed to magnetic ordering.
Magnetization measurements indicate that $CeMo_6S_8$ and $CeMo_6Se_8$ compounds order magnetically (probably antiferromagnetically) at 2.8 K
and ~ 0.4 K, respectively (10,26,27). A magnetic specific heat anomaly at 2.5 K has been observed for $CeMo_6S_8$ (10).

By far the most striking of the ρ vs T curves in Fig. 2 are those
for the Eu molybdenum sulfide and selenide compounds. As the temperature is lowered, ρ for the $Eu_{1.2}Mo_6S_8$ compound increases by a factor
of two between 294 and 100 K; increases more rapidly below 100 K and
then saturates to a value of the order of 7.2 ρ(294 K) as $T \to 0$. The
temperature dependence of ρ for the $Eu_{1.2}Mo_6Se_8$ sample is equally
striking, but qualitatively different. With decreasing temperature,
ρ rises about 10% between 294 and 100 K; decreases rapidly below 100 K
to a value ~ 0.44 ρ(294 K) at a temperature ~ 20 K, and then again
rises slightly as $T \to 0$. Although qualitatively different from each
other, these features in the ρ vs T curves for the two Eu compounds
are similar to those which have been observed in concentrated RE
metallic systems and attributed to Kondo-like or valence fluctuation
phenomena (10,25).

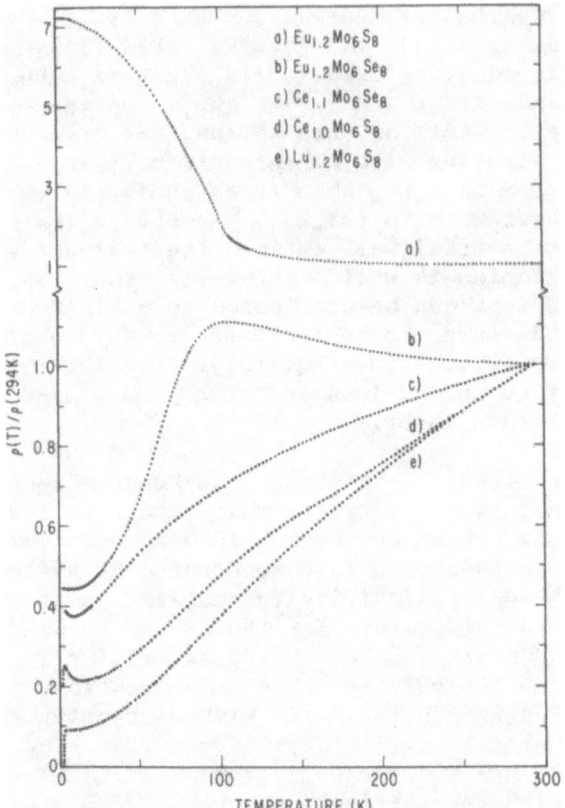

Figure 2.   Electrical resistivity ρ(T) as a function of
temperature for five RE molybdenum chalcogenides (normalized to room temperature
values).   After Maple et al. (10).

     The absence of superconductivity and the electrical resistance
anomalies in the Ce and Eu molybdenum chalcogenide compounds appear
to be associated with Ce and Eu valence instabilities.   The well-defined magnetic moments for Ce^{3+} and Eu^{2+} in these compounds suggests
that an exchange Hamiltonian with a negative exchange interaction generated by hybridization of conduction electron and localized 4f electron states may provide the most appropriate description.   Such a
Kondo lattice approach is probably not unreasonable for Ce, but seems
questionable for Eu in view of the remarkable but qualitatively different behaviors of ρ(T) for the Eu molybdenum sulfide and selenide
compounds.

III. EXCHANGE FIELD ENHANCEMENT OF H_{c2} OF TERNARY

MOLYBDENUM CHALCOGENIDES BY Eu ADDITIONS

Prior to the experiments on ternary $RE_xMo_6S_8$ and $RE_xMo_6Se_8$ compounds reviewed in Section II, Fischer et al. (28) had studied super-conductivity in the pseudoternary systems $Sn_{1.2-x}Eu_xMo_{6.35}S_8$ and $Pb_{1-x}Eu_xMo_{6.35}S_8$ in zero and high magnetic fields. For example, in the $Sn_{1.2-x}Eu_xMo_{6.35}S_8$ system, T_C vs x in zero field initially varied only weakly with x up to x ~ 0.6 and then dropped rapidly, but smoothly towards zero at x ~ 1.0 (28). Measurements of $H_{c2}(T)$ for several values of x revealed an appreciable enhancement of $H_{c2}(0)$ over the 27.5 Tesla upper critical field of $Sn_{1.2}Mo_{6.35}S_8$. The maximum value achieved for $H_{c2}(0)$ was about 40 Tesla at x ~ 0.6, which amounts to a 45% increase in $H_{c2}(0)$. Similar behavior was observed for $Pb_{1-x}Eu_x$-$Mo_{6.35}S_8$.

The explanation proposed for the enhancement of H_{c2} in the $Sn_{1.2-x}Eu_xMo_{6.35}S_8$ and $Pb_{1-x}Eu_xMo_{6.35}S_8$ systems (28) was the exchange field compensation effect (Jaccarino-Peter effect) (29). This involves the compensation of the applied magnetic field by an oppositely oriented, or negative, exchange field, thereby increasing H_{c2} above the paramagnetic limiting field H_p towards the orbital critical field H^*_{c2}. Nuclear magnetic resonance and Mössbauer effect studies by Fradin et al. (30) on $Sn_{0.5}Eu_{0.5}Mo_6S_8$ revealed a negative s-band polarization at the Mo sites, giving additional evidence for this interpretation of the enhancement of H_{c2} in the $Sn_{1.2-x}Eu_xMo_{6.35}S_8$ system. Band structure calculations by Jarlborg and Freeman (31) also support this conclusion.

Since the exchange field is given by an expression of the form

$$H_{ex} = 2 \mathcal{J} S / g \mu_B \tag{6}$$

then negative H_{ex} implies that $\mathcal{J}$ is negative. This is precisely the condition for the occurrence of the Kondo effect, and, in fact, originally motivated some of the investigations described in Section II. Thus, the observation of exchange field enhancement of H_{c2} in ternary Mo chalcogenides is consistent with some type of Kondo scattering in these systems.

Recently, Torikachvili and Maple (32) studied superconductivity in zero and high magnetic fields in the $La_{1.2-x}Eu_xMo_6S_8$ system. Their data for T_c vs x and H_{c2} vs T for various values of x are shown in Figs. 3 and 4, respectively. The shape of the T_c vs x curve is very similar to the one obtained by Fischer et al. (28) for the $Sn_{1.2-x}Eu_x$-$Mo_{6.35}S_8$ system. Moreover, the H_{c2} vs T curves show an enhancement effect similar to that observed in the $Sn_{1.2-x}Eu_xMo_6S_8$ system. However, the analysis of the H_{c2} vs T data for the $La_{1.2-x}Eu_xMo_6S_8$ system is complicated by the fact that the orbital critical field should

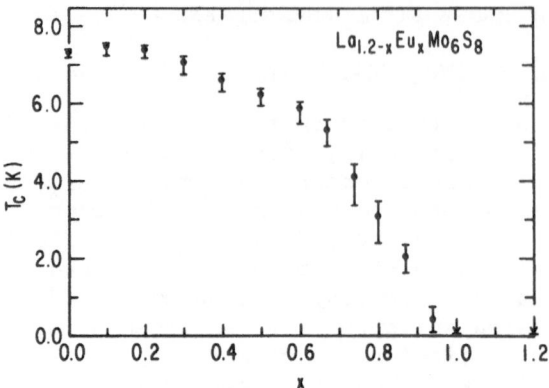

Figure 3.   Superconducting transition temperature T_c vs
Eu concentration x for $La_{1.2-x}Eu_xMo_6S_8$.   After
Torikachvili and Maple (32).

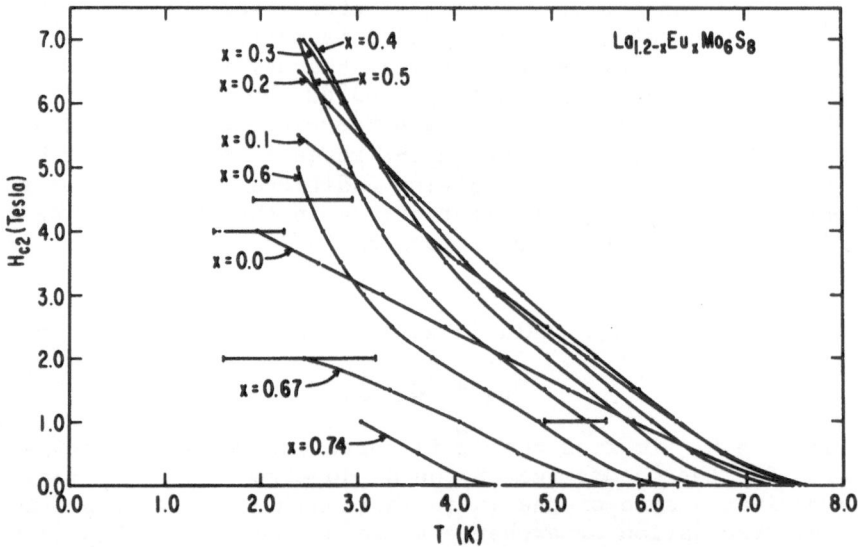

Figure 4.   Upper critical magnetic field H_{c2} vs temperature
T of $La_{1.2-x}Eu_xMo_6S_8$ for different Eu concentrations x.   After Torikachvili and Maple (32).

change with x, since La is trivalent and Eu divalent, the latter fa-
voring higher critical fields.   Assuming the absence of exchange scattering that presumably inhibits superconductivity in the Eu-rich compounds at atmospheric pressure, $H_{c2}(0)$ should vary from 4 Tesla at
x = 0 to ~40 Tesla at x = 1.2 (32), and this change would also contribute to the enhancement of H_{c2}.   The most apparent manifestation of

the exchange field compensation effect is therefore positive curvature of H_{c2} vs T data in Fig. 4, which reflects the increase of the Eu spin polarization, and, in turn, the increase in the magnitude of the negative exchange field as H increases and T decreases; i.e., $|H_{ex}| \propto \langle S_z \rangle = S B_S(\mu H/k_B T)$ where S is the spin (7/2 for Eu) and $B_S(\mu H/k_B T)$ is the corresponding Brillouin function.

The dependence of T_c on x and the enhancement of H_{c2} are similar in the $Sn_{1.2-x}Eu_xMo_{6.35}S_8$, $Pb_{1-x}Eu_xMo_{6.35}S_8$ and $La_{1-x}Eu_xMo_6S_8$ systems. Therefore, the magnetic state of the substituted Eu ion and the interaction of the Eu ion with the conduction electrons appear to be invariant for all three of these systems, in spite of the difference in the metal ion (Sn^{2+}, Pb^{2+}, La^{3+}) constituent of the superconducting matrix. This suggests that the form of T_c vs x and the enhancement of H_{c2} are determined primarily by the features of the electronic structure associated with the Chevrel phase crystal structure and the Eu solute ion, whereas the primary function of the metal M ion is to determine the superconducting properties of the $M_xMo_6S_8$ matrix which are then modified by the Eu substitution.

IV. PRESSURE-INDUCED SUPERCONDUCTIVITY IN Eu-RICH

PSEUDOTERNARY MOLYBDENUM CHALCOGENIDES

An abrupt occurrence of superconductivity in the compound $Eu_{1.2}Mo_6S_8$ under pressure above ~ 7 kbar with an onset temperature ~ 11 K has been reported independently by Chu et al. (33) and Harrison et al. (34). No superconductivity was detected in $Eu_{1.2}Mo_6Se_8$ to temperatures as low as 1.2 K and pressures up to 18 kbar (33). Recently, Torikachvili et al. (35,36) reported measurements of the pressure dependence of T_c of $La_{1.2-x}Eu_xMo_6S_8$ pseudoternary compounds for several values of x between 0 and 1.2. The results they obtained are displayed in Fig. 5. These data reveal the sudden appearance of superconductivity for $Eu_{1.2}Mo_6S_8$ at ~ 7 kbar with a maximum T_c (defined from the 50% value of the ac magnetic susceptibility change between the normal and superconducting states) of ~ 8 K at ~ 15 kbar, confirming the observations of Chu et al. (33) and Harrison et al. (34). Moreover, the data show that the depression of T_c of $La_{1.2}Mo_6S_8$ by Eu additions <u>decreases</u> with pressure, indicating that the strength of the pair breaking by the Eu ions decreases with pressure. This pressure induced decrease in pair breaking strength is most likely associated with the rapid onset of superconductivity above ~ 7 kbar for the $La_{0.2}Eu_{1.0}Mo_6S_8$ and $Eu_{1.2}Mo_6S_8$ compounds.

The data shown in Fig. 5 are also reminiscent of the behavior of $La_{1-x}Ce_x$ alloys under pressure in which the depression of T_c of La by Ce additions initially increases with pressure, attains a maximum value near 15 kbar, and then decreases with pressure (22). A $La_{1-x}Ce_x$ alloy with x = 0.16 was observed to become superconducting abruptly

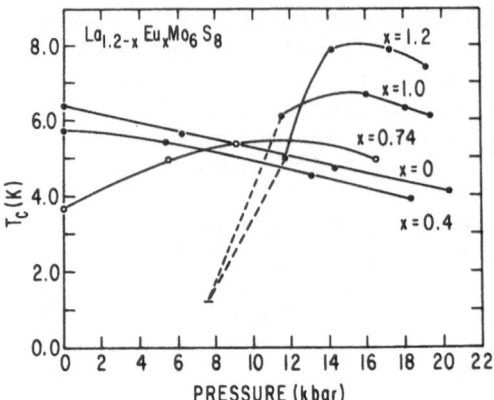

Figure 5.   Superconducting transition temperature T_c vs
pressure of $La_{1.2-x}Eu_xMo_6S_8$ for different Eu
concentrations x.   After Torikachvili et al.
(35,36).

above ~ 27 kbar (22), similar to the behavior of the Eu-rich $La_{1.2-x}$-
$Eu_xMo_6S_8$ compounds. The T_c vs P data for the $La_{1-x}Ce_x$ system under
pressure have been attributed to a pressure-induced demagnetization
of the Ce ions which can be described in terms of an increase of T_K
with pressure at lower pressures and a continuous transition to a non-
magnetic state with intermediate valence at higher pressure (~ 100
kbar) (21,22). Thus, it is tempting to ascribe the pressure
dependence of T_c of the $La_{1.2-x}Eu_xMo_6S_8$ compounds shown in Fig. 5 to
a demagnetization of Eu as a function of pressure, due to a continuous
increase of T_K with pressure, or to a valence transition in which the
admixture of the nonmagnetic $4f^6$ configuration (Eu^{3+}, J = 0) with $4f^7$
(Eu^{2+}) increases with pressure. Both of these alternatives should
result in a decrease of the Eu magnetic moment with pressure.

On the other hand, the sharp onset of superconductivity in the
$Eu_{1.2}Mo_6S_8$ compound could be due to a pressure-induced insulator-metal
transition that occurs near ~ 7 kbar, a viewpoint that has been considered by Chu et al. (33) and Harrison et al. (34). It would not be
unnatural to attribute the low temperature resistivity anomaly in
$Eu_{1.2}Mo_6S_8$ to an insulating ground state, nor the diminuition of this
resistivity anomaly under pressure reported by Chu et al. (33) to a
pressure-induced insulator-metal transition (some intermediate valence
compounds, notably SmB_6, are believed to be narrow gap semiconductors)
(37). Thus, a Eu valence instability may also be implicated here.
Other possibilities would be a crystallographic phase transformation
in the neighborhood of 7 kbar or a pressure induced disproportionation
phenomenon.

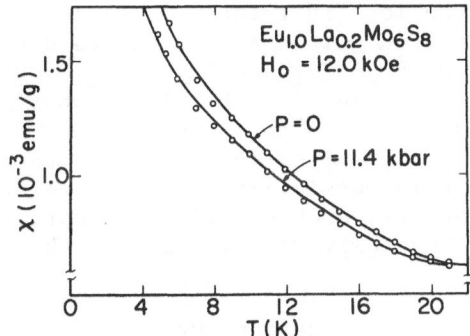

Figure 6.   Magnetic susceptibility χ vs temperature T of
$Eu_{1.0}La_{0.2}Mo_6S_8$ at 0 and 11.4 kbar.   After
Torikachvili et al. (36).

In an attempt to resolve these questions, dc magnetization measurements under pressure on a compound of composition $La_{0.2}Eu_{1.0}Mo_6S_8$
were undertaken (36).   Two important findings emerged: First, the
compound was found to exhibit a sizable Meissner effect, indicating
that bulk superconductivity had indeed been induced under pressure.
Second, the normal state magnetization and susceptibility were found
to decrease under pressure.   The dc magnetic susceptibility χ measured
in a magnetic field of 1.2 Tesla, is plotted vs temperature T at zero
pressure and 11.4 kbar in Fig. 6.   Such a decrease in the magnetic
susceptibility with pressure could be explained by an increase of T_K
with pressure (the s-electron polarization), or a valence transition
in which a greater amount of the nonmagnetic $4f^6$ (Eu^{3+}) configuration
becomes admixed with $4f^7$ (Eu^{2+}).

V.   SUMMARY

Compared to the other RE ions, the behavior of Ce and Eu in ternary Mo chalcogenides is anomalous, as we have summarized in this review.   Many of the remarkable properties that the Ce and Eu substi-
tuents impart to the ternary Mo chalcogenides can be qualitatively
explained in terms of a negative exchange interaction between the conduction electrons and the Ce and Eu magnetic moments.   The negative
exchange interaction is generated by mixing between the conduction
electron and localized 4f electron states.   This is reasonable in view
of the unstable valence of Ce and Eu.   The onset of superconductivity
at ~ 7 kbar and the decrease of the magnetic susceptibility of Eu-rich
$La_{1.2-x}Eu_xMo_6S_8$ compounds may be due to an increase of T_K with pressure, or to a valence transition in which the admixture of the configuration $4f^6$ and $4f^7$ increases with pressure.   Although the Kondo lattice approach seems to describe qualitatively most aspects of the

behavior of Ce and Eu ions in ternary Mo chalcogenides, it fails to account for the difference in the temperature dependence of the electrical resistivity of $Eu_xMo_6S_8$ and $Eu_xMo_6Se_8$, nor does it explain why $Eu_xMo_6S_8$ becomes superconducting at ~ 11 K (onset) near 7 kbar while $Eu_xMo_6Se_8$ remains normal down to 1.2 K up to ~ 18 kbar. Finally, the ground state of $Eu_xMo_6S_8$ may be insulating at zero and low pressures ≲ 7 kbar, although this possibility could be related to a Eu valence instability. Further research on this fascinating problem is clearly warranted and will hopefully resolve many of these questions.

REFERENCES

1. R. Chevrel, M. Sergent and J. Prigent, J. Solid State Chem., 3: 515 (1971).
2. M. Marezio, P.D. Dernier, J.P. Remeika, E. Corenzwit and B.T. Matthias, Mat. Res. Bull., 8:657 (1973).
3. B.T. Matthias, M. Marezio, E. Gorenzwit, A.S. Cooper and H. Barz, Science, 175:1465 (1972).
4. Ø. Fischer, H. Jones, G. Bongi, M. Sergent and R. Chevrel, J. Phys. C, 7:L450 (1974).
5. S. Foner, E.J. McNiff and E.J. Alexander, Phys. Lett., 49A:269 (1974).
6. S. Foner, E.J. McNiff, R.N. Shelton, R.W. McCallum and M.B. Maple, Phys. Lett., 49A:269 (1974).
7. Ø. Fischer, Appl. Phys., 16:1 (1978).
8. J.M. Vandenberg and B.T. Matthias, Proc. Natl. Acad. Sci. USA, 1336 (1977).
9. O.K. Andersen, W. Klose and H. Nohl, Phys. Rev. B, 17:1209 (1978); and references to earlier work quoted therein.
10. M.B. Maple, L.E. DeLong, W.A. Fertig, D.C. Johnston, R.W. McCallum and R.N. Shelton, in "Valence Instabilities and Related Narrow Band Phenomena," R.D. Parks (ed.), Plenum, New York, pp. 17-29 (1977).
11. M. Ishikawa, Ø. Fischer and J. Müller, J. de Physique, 39:C6-1379 (1978); and references cited therein.
12. M.B. Maple, J. de Physique, 39:C6-1374 (1978); also, "Science and Technology of Rare Earth Materials," E.C. Subbarao and W.E. Wallace (eds.), Academic, New York, pp. 167-193 (1980); and references cited therein.
13. J.W. Lynn, G. Shirane, W. Thomlinson and R.N. Shelton, Phys. Rev. Lett., 46:368 (1981).
14. Ø. Fischer, A. Treyvaud, R. Chevrel and M. Sergent, Solid State Commun., 17:721 (1975); R.N. Shelton, R.W. McCallum and H. Adrian, Phys. Lett., 56A:213 (1976).
15. P. Bonville, J.A. Hodges, P. Imbert, G. Jehanno, R. Chevrel and M. Sergent, Revue Phys. Appl., 15:1139 (1980).
16. A.A. Abrikosov and L.P. Gor'kov, Soviet Phys. JETP, 12:1243 (1961).

17. Ø. Fischer, M. Ishikawa, M. Pelizzone and A. Treyvaud, J. de Physique, 40:C5-89 (1979).
18. S. Oseroff, R. Calvo, D.C. Johnston, M.B. Maple, R.W. McCallum and R.N. Shelton, Solid State Commun., 27:20 (1978).
19. J. Bolz, J. Hauck and F. Pobell, Z. Phys. B, 25:351 (1976); J. Bolz, G. Crecelius, H. Maletta and F. Pobell, J. Low Temp. Phys., 28:61 (1977).
20. J.R. Schrieffer and P.A. Wolff, Phys. Rev., 149:491 (1966).
21. M.B. Maple, in "MAGNETISM: A Treatise on Modern Theory and Materials," H. Suhl, (ed), Academic, New York, Vol. V, Ch. 10 (1973); also Appl. Phys., 9:179 (1976); and references cited therein.
22. M.B. Maple, J. Wittig and K.S. Kim, Phys. Rev. Lett., 23:1375 (1969).
23. J.G. Huber, W.A. Fertig and M.B. Maple, Solid State Commun., 15: 453 (1974); C.A. Luengo, J.G. Huber, M.B. Maple and M. Roth, Phys. Rev. Lett., 32:54 (1974).
24. B. Coqblin, M.B. Maple and G. Toulouse, Int. J. Magn., 1:333 (1971).
25. M.B. Maple, L.E. DeLong and B.C. Sales, in "Handbook on the Physics and Chemistry of Rare Earths," K.A. Gschneidner, Jr., and L. Eyring (eds.), North-Holland, Amsterdam, Ch. 11 (1978).
26. D.C. Johnston and R.N. Shelton, J. Low Temp. Phys., 26:561 (1977).
27. M. Pellizone, A. Treyvaud, P. Spitzli and Ø. Fischer, J. Low Temp. Phys., 29:453 (1977).
28. Ø. Fischer, M. Decroux, S. Roth, R. Chevrel and M. Sergent, J. Phys. C, 8:L474 (1975).
29. V. Jaccarino and M. Peter, Phys. Rev. Lett., 9:290 (1962).
30. F.Y. Fradin, G.K. Shenoy, B.D. Dunlap, A.T. Aldred and C.W. Kimball, Phys. Rev. Lett., 38:719 (1977).
31. T. Jarlborg and A.J. Freeman, Phys. Rev. Lett., 44:178 (1980).
32. M.S. Torikachvili and M.B. Maple, to appear in Solid State Commun.
33. C.W. Chu, S.Z. Huang, C.H. Lin, R.L. Meng, M.K. Wu and P.H. Schmidt, Phys. Rev. Lett., 46:276 (1981).
34. D.W. Harrison, K.C. Lim, J.D. Thompson, C.Y. Huang, P.D. Hambourger and H.L. Luo, Phys. Rev. Lett., 46:280 (1981).
35. M.S. Torikachvili and M.B. Maple, to appear in Proc. 16th Int. Conf. Low Temp. Phys., to be held Aug. 19-26, 1981, Los Angeles, CA.
36. M.S. Torikachvili, M.B. Maple, R.P. Guertin and S. Foner, to be published.
37. R.M. Martin and J.W. Allen, in "Valence Fluctuations in Solids," L.M. Falicov, W. Hanke and M.B. Maple, (eds.), North-Holland, Amsterdam, pp. 85-90 (1981); and references cited therein.

NEW FLUORIDES WITH Ce^{IV}, Pr^{IV}, Nd^{IV}, Tb^{IV}, AND Dy^{IV}

Rudolf Hoppe
Institut für Anorganische und Analytische Chemie
Justus-Liebig-Universität Giessen
Heinrich-Buff-Ring 58, 6300 Giessen
West Germany

During recent investigations of fluorides with tetravalent Ce, Pr, Nd, Tb, and Dy, numerous new fluorides, e.g. $CsRbKTbF_7$, Cs_2RbCeF_7, Rb_2KPrF_7, were obtained and studied with x-ray methods. As is the case with Cs_2KNdF_7, Cs_2RbNdF_7, and Cs_3NdF_7 (all orange) and Cs_2KDyF_7, Cs_2RbDyF_7 and Cs_3DyF_7, all fluorides were cubic and isotypic to $(NH_4)_3ZrF_7$.

The preparations were started with fluorination of chlorides such as Cs_2KNdCl_6 (cubic, elpasolite type structure). Intermediates such as Cs_2KNdF_6 were then fluorinated under high pressures (50-150 bar F_2, Monel autoclave) and optimal conditions, primarily 400°C, 2-3 h.

The analytical F-content (e.g. Cs_2RbNdF_7: 21.2% obs., 21.29% calc., Cs_2KDyF_7: 22.1% obs., 22.15% calc.) confirms the synthesis of 'pure' samples.

Raman spectra of fluorides with Nd(IV) or Dy(IV) are quite different from those of analogous fluorides with Nd(III) or Dy(III) (e.g. $Cs_2RbNdF_7 \neq Cs_2RbNdF_6$) and correspond well with those containing Ce(IV), Pr(IV), and Tb(IV), see Table 1.

The synthesis of fluorides such as $BaPrF_6$ (a = 802.4, b = 1158.0, c = 561.6 pm) and $SrTbF_6$ (a = 769.5, b = 1101.5, c = 539.7 pm), both colorless and isotypic to $RbPaF_6$, as well as that of $SrPrF_6$ (a = 709.4, c = 729.8 pm, LaF_3 type structure) is difficult. Colorless samples were obtained only by high pressure fluorination of $BaPrO_3$, $SrTbO_3$, and $SrPrO_3$, respectively.

The first fluoride of Pr(IV) with lithium is Li_2PrF_6 which was

315

Table 1. Lattice Constants and Raman Frequencies
of Rare Earth Fluorides.

	a/pm	Raman Frequencies/cm^{-1}			
CsRbKCeF₇	944.1	228(m)[+]	311(w)	368(w)	532(s)
Rb₃CeF₇	953.7	210(m)	276(w)	325(w)	505(s)
Rb₂CsCeF₇	966.6	204(m)	283(w)	332(w)	506(s)
Cs₂RbCeF₇	979.0	200(m)	276(w)	325(w)	506(s)
Cs₃CeF₇	994.9	199(m)	273(w)	323(w)	506(s)
K₂RbPrF₇	928.8	207(m)	287(w)	329(w)	512(s)
Rb₂KPrF₇	939.2	205(m)	285(w)	329(w)	512(s)
Rb₃PrF₇	951.8	204(m)	282(w)	325(w)	507(s)
CsRbKPrF₇	952.3	201(m)	284(w)	324(w)	508(s)
Cs₂KPrF₇	966.6	197(m)	274(w)	319(w)	502(s)
Rb₂CaPrF₇	964.0	198(m)	280(w)	322(w)	507(s)
Cs₂RbPrF₇	978.3	196(m)	277(w)	317(w)	502(s)
Cs₃PrF₇	992.6	196(m)	275(w)	316(w)	502(s)
Cs₂KNdF₇	960.5	n.o.[++]	n.o.	n.o.	510
Cs₂RbNdF₇	974.8	n.o.	n.o.	n.o.	510
Cs₃NdF₇	987.3	n.o.	n.o.	n.o.	n.o.
K₂TbF₇	909.0	235(m)	307(w)	364(w)	531(s)
K₂RbTbF₇	917.3	230(m)	311(w)	365(w)	534(s)
Rb₂KTbF₇	927.4	225(m)	307(w)	363(w)	529(s)
Rb₃TbF₇	940.8	227(m)	306(w)	366(w)	527(s)
CsRbKTbF₇	939.9	225(m)	308(w)	363(w)	530(s)
Cs₂KTbF₇	952.5	218(m)	301(w)	356(w)	525(s)
Rb₂CsTbF₇	952.9	220(m)	303(w)	360(w)	525(s)
Cs₂RbTbF₇	970.6	216(m)	303(w)	357(w)	524(s)
Cs₃TbF₇	989.0	218(m)	297(w)	354(w)	519(s)
Cs₂KDyF₇	947.4	n.o.	n.o.	n.o.	n.o.
Cs₂RbDyF₇	963.0	n.o.	n.o.	n.o.	520
Cs₃DyF₇	979.5	n.o.	n.o.	n.o.	n.o.

[+] relative intensities are given in parentheses on a strong (s),
medium (m), weak (w) scale.   [++] peaks not observable.

synthesized in two different ways: (a) Appropriate amounts of LiCl
and 'Pr₆O₁₁' were dissolved in concentrated HCl solutions by heating.
The residue obtained after evaporation was dried in a stream of HCl
gas. (b) Solutions obtained likewise with concentrated HNO₃ were
evaporated, the residue was then heated in a stream of O₂ yielding
'Li₂PrO₃'. Fluorination under the usual conditions (p(F₂) = 1 bar,
400°C, 1 d) followed by high pressure fluorination (150 bar F₂, 450°C,
2 d) yielded colorless samples of Li₂PrF₆ isotypic to Li₂ZrF₆
(a = 504.7, c = 484.6 pm). The Raman spectrum (Table 1) corresponds
well with that of Li₂ZrF₆ (one additional peak at 334 cm^{-1} is probably
due to decomposition during the exposure to laser emission. Colorless
"scheelites" CaPr(LiF₄)₂ and CdPr(LiF₄)₂ (a = 519.6, c = 1088.0 and
a = 517.7, c = 1082.2 pm, respectively) were prepared by fluorination
of chlorides such as 'CaPrLiCl₇' and 'CdPrLiCl₇', respectively, in
two steps: (a) F₂:N₂ = 1:5, 12 h, 350°C, (b) p(F₂) = 50 bar, 2 d,
500°C.

TERNARY HALIDES OF THE RARE EARTH ELEMENTS: PHASES AND STRUCTURES

Gerd Meyer
Institut für Anorganische und Analytische Chemie
Justus-Liebig-Universität Giessen
Heinrich-Buff-Ring 58
6300 Giessen, West Germany

PHASES, SYNTHESIS

Many efforts have been made to determine compound formation in rare earth metal trihalide/alkali metal halide systems by thermal analysis (1). X-ray methods were almost never applied. However, there seems to be a limit of four formula types:

ARE_2X_7, $A_3RE_2X_9$, A_2REX_5, A_3REX_6 (A=K,Rb,Cs, X=Cl,Br)

Although the $A_3RE_2X_9$ and A_2REX_5 type compounds melt incongruently there is a straightforward and fast synthetic route for these and for the ARE_2X_7 and A_3REX_6 type phases as well. This starts with aqueous HCl (or HBr) solutions of the appropriate amounts of metal halides (AX, REX_3). With the exception of AX:REX_3=1:2 ratios evaporation to dryness leads to the monohydrates $A_2REX_5 \cdot H_2O$ as the main intermediate product in the residue. The desired product is then obtained by heating the residue in a dry HCl (or HBr) gas stream (2). Not every possible phase is stable in every feasible system. One interest in these systems is therefore to elucidate the stability ranges of the phases and their structures as well as phase transitions where they occur.

A_3REX_6 TYPE COMPOUNDS

The A_3REX_6 type phases are the most stable ones. They exist most likely in all of the systems, and melt congruently at high temperatures. Unfortunately, a densest packing of spheres is not possible because of the relation of A:X = 1:2. Therefore, phase transitions are observed frequently (K_3ScCl_6 is tetramorphic) to enable optimal space filling with increasing temperature. They occur at fairly low

317

temperatures and are chiefly displacive in character. This makes
crystal growth very difficult if not impossible. The room temperature
crystal structures of both K_3ScCl_6 and K_3LuCl_6 were derived from powder data: they crystallize with the K_3MoCl_6 structure type (3) with
the monoclinic ($P2_1/a$) lattice constants: $\underline{a}$=1226.5(6), 1251.5(5) pm
(for K_3ScCl_6 and K_3LuCl_6, respectively), $\underline{b}$=758.0(1), 767.6(2), $\underline{c}$=
1282.8(3), 1301.1(5) pm, β=109.03(2), 109.79(3)°. The structure con-
tains [$RECl_6$] octahedra isolated from one another. Raman spectroscopy
shows that this is also true for all A_3REX_6 type compounds. The
$(NH_4)_3RECl_6$ type compounds apparently play an important role in the
"ammonium chloride-method" for the preparation of anhydrous rare earth
trihalides: a 12-molar "excess" is needed to produce these compounds
according to the equation

$$RE_2O_3 + 12\ NH_4Cl = 2\ (NH_4)_3RECl_6 + 3\ H_2O + 6\ NH_3.$$

In a second step they are decomposed in vacuo at about 300°C to form
$RECl_3$, either in one step or with $NH_4RE_2Cl_7$ as an intermediate.

A_2REX_5 TYPE COMPOUNDS

Octahedral coordination is also present in A_2REX_5 type compounds.
The crystal structure of Cs_2DyCl_5 (4) (orthorhombic, Pbnm, $\underline{a}$=1523.1
(3), $\underline{b}$=954.9(3), $\underline{c}$=749.8(3), Z=4) was determined from single crystal
X-ray data. [$DyCl_6$] octahedra are connected via cis-corners to form
zig-zag-chains with a Dy-Cl-Dy angle of 180°. The chains are arranged
in [001] direction in the manner of a densest packing of rods and are
connected by Cs^+ ions. Figure 1 shows a structure field map of
A_2RECl_5 type compounds (pure ones and solid solutions).

$A_3RE_2X_9$ TYPE COMPOUNDS

All of the enneahalodimetallates(III) of the rare earth elements
$A_3RE_2X_9$, which are observed for A=Rb,Cs and X=Cl,Br,I contain con-
facial bioctahedra [RE_2X_9] as the main structural feature. These are
built when two-dimensionally close-packed layers of composition AX_3
are stacked with the sequences either h_6 (alternative description:
ABABAB..., $Cs_3Tl_2Cl_9$ type (5,6) or $(chc)_2$ (ABACBC..., $Cs_3Cr_2Cl_9$ type
(7,8) with two thirds of the X_6 octahedra occupied by RE. The structure field map is shown in Fig. 2 for $Cs_3RE_2X_9$ type compounds based on
molar volumes, it includes the In(III) halides for better separation
of both structure types. Rubidium compounds $Rb_3RE_2Br_9$ were obtained
with RE=Lu,Er; the only chloride of this type is $Rb_3Sc_2Cl_9$.

ARE_2X_7 TYPE COMPOUNDS

A lesser content of alkali halide does not favor more octahedral

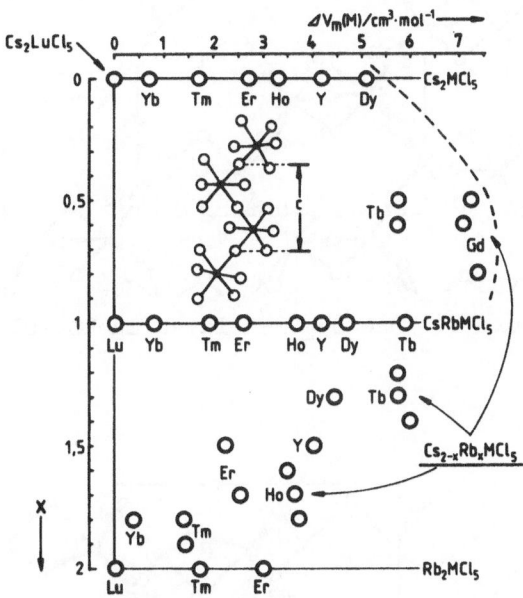

Figure 1. Structure field map for A_2MCl_5 (M=RE) type compounds.

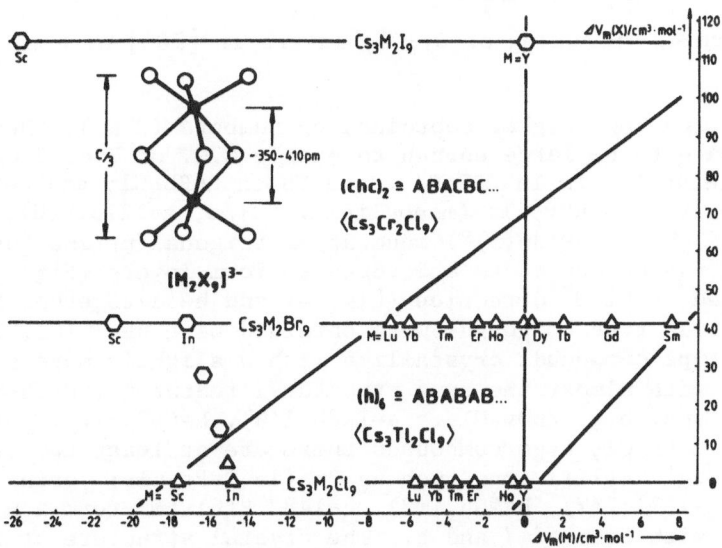

Figure 2. Structure field map for $Cs_3M_2X_9$ (M=RE) type compounds.

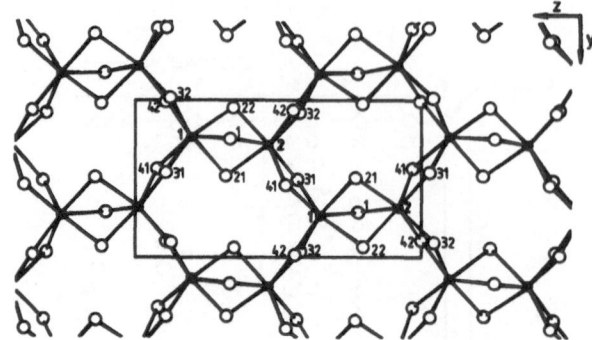

Figure 3.   One Dy_2Cl_7 -layer in KDy_2Cl_7.

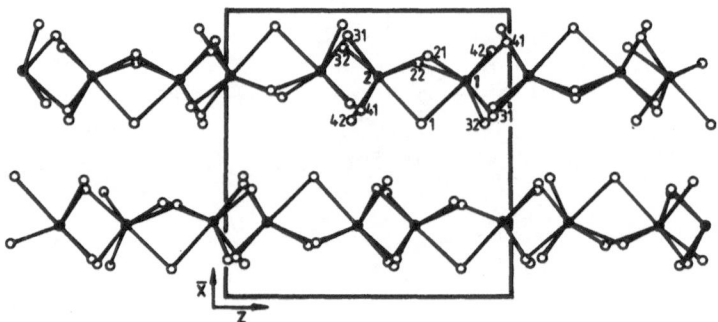

Figure 4.   Stacking of Dy_2Cl_7 layers in [100] direction.

'condensation' but higher coordination numbers (C.N.). Hence, the RE
cations have to be large enough to enable C.N. of 7 or 8 which is the
case at least for Tm in KRE_2Cl_7, and Yb in $RbRE_2Cl_7$ and $CsRE_2Cl_7$,
respectively. In KDy_2Cl_7 (monoclinic, $P2_1/a$, $\underline{a}=1273.9(8)$, $\underline{b}=688.1(5)$,
$\underline{c}=1262.1(6)$ pm, $\alpha=89.36(3)°$) monocapped trigonal prisms $[DyCl_7]$ are
connected via common faces and edges to form layers (Fig. 3) which
are stacked in [100] direction (Fig. 4) and held together by K^+. Isotypic KRE_2Cl_7 type compounds were obtained with RE=Tm-Gd, Y.
$RbRE_2Cl_7$ type compounds crystallize with a slightly more symmetric
structure with almost the same structural features (RE=Yb-Sm,Y, ortho-
rhombic, Pnma, e.g. $RbDy_2Cl_7$: $\underline{a}=1288.1(3)$, $\underline{b}=693.5(2)$, $\underline{c}=1267.2(3)$
pm). For $CsRE_2Cl_7$ type compounds there are at least two structure
types. 'Big' RE(III) ions such as Pr (in $CsPr_2Cl_7$: orthorhombic
$P2_12_12_1$, $\underline{a}=1671.7(7)$, $\underline{b}=966.1(3)$, $\underline{c}=1482.9(6)$ pm) form a network
structure with C.N. of 7 and 8. The crystal structure of the
$CsRE_2Cl_7$ type compounds with RE=Dy-Yb is not yet known.

M \ A	AM₂Cl₇ K	Rb	Cs	A₃M₂Cl₉ K	Rb	Cs	A₂MCl₅ K	Rb	Cs	A₃MCl₆ K	Rb	Cs
La	○		●		◇	○		◇	◇			○
Ce	◇		◇	○		◇	○		◇	○		○
Pr	◇		●	○		◇	○		◇	○		○
Nd	◇	○	◇	○	◇	◇	○	○	◇	○	○	○
Sm	○	●	○	◇	◇	◇	○	○	◇	○	○	○
Eu	○	●		◇			○			○		
Gd	●	●		◇			○			○		
Tb	●	●		◇			○			○		
Dy	●	●	◐	◇		◇	◇		●	○		○
Y	●	●	○	◇	◇	◇	●	◇	◇	◐	○	○
Ho	●	●		◇		●	◇		●	◐		
Er	●	●	○	◇		●	◇		●	◐		○
Tm	●	●				●			●	◐		
Yb	◇	●	○	◇		●	◇		●	◐		○
Lu	◇			◇		●	◇	●	●	●		
In	◇			◇	●	●	○			●	●	●
Sc	◇	◇	◇	◇	●	●	●	○	◇	◇	●	●

Figure 5. Compound formation in the alkali metal chloride/rare earth metal trichloride systems: ● crystal structure known, ○ observed, ◇ not observed in the phase diagram, ◐ phase investigated by X-ray diffraction, structure not known so far, no entry means that the respective phase diagram has not been investigated.

Figure 5 summarizes schematically our present knowledge of compound formation in the alkali chloride/rare earth metal trichloride systems.

REFERENCES

1. R. Blachnik and D. Selle, Z. anorg. allg. Chem., 454:90 (1979) and references therein.
2. G. Meyer, 'Ternary and Quaternary Chlorides and Bromides of the Rare Earth Elements', in Inorganic Syntheses, Vol. 22, S.L. Holt (ed.), to be published.
3. Z. Amilius, B. van Laar, and H.M. Rietveld, Acta Crystallogr., B 25:400 (1969).
4. G. Meyer, Z. anorg. allg. Chem., 469:149 (1980).
5. J.L. Hoard and L. Goldstein, J. Chem. Phys., 3:199 (1935), H.M. Powell and A.F. Wells, J. Chem. Soc., 1008 (1935).
6. G. Meyer, Z. anorg. allg. Chem., 445:140 (1978).
7. G.J. Wessel and D.J.W. IJdo, Acta Crystallogr., 10:466 (1957).
8. D.H. Guthrie, G. Meyer, and J.D. Corbett, Inorg. Chem., 20:1192 (1981).

INVESTIGATION OF THE "DIADOCHIC" INCORPORATION OF THE RARE EARTH ELEMENTS IN CaF_2 AT ELEVATED TEMPERATURE AND PRESSURE

B.A. Bilal, P. Becker, V. Koss and H. Nies

Hahn-Meitner-Institut für Kernforschung Berlin GmbH

D-1000 Berlin 39, FRG

INTRODUCTION

The fractionation of trace elements in minerals such as the rare earth elements (REE) in fluorite is used as an indicator of the geochemical conditions of mineral formation. According to a model suitable for describing this fractionation, two processes are assumed to be responsible for this phenomenon: the different degree of complexation of the trace elements in the mineral bearing solution and the various extent of their "diadochic" incorporation (replacing the cation of the host mineral in lattice) due to their different ionic radii. The coprecipitation of trace elements during crystallization of the mineral may follow a homogeneous distribution law (Henderson-Krack)

$$\frac{n_{Tr}}{n_{Hos}}\bigg|_{solid} = D \frac{[Tr]}{[Hos]}\bigg|_{solution} \qquad [1]$$

or a logarithmic one (Doerner and Hoskins)

$$\ln \frac{{}^{o}[Tr]}{[Tr]} = \lambda \ln \frac{{}^{o}[Hos]}{[Hos]} , \qquad [2]$$

where only the surface layer of the growing mineral is considered to be equilibrated with the mineral bearing solution

D or λ = linear or logarithmic distribution coefficient
n_{Tr} or n_{Hos} = number of moles of the trace element or of the cation of the host mineral

^{o}[Tr] or ^{o}[Hos] = concentration of the trace element or of
                        the host element in the mineral bearing
                        solution at the beginning of the
                        crystallization
[Tr] or [Hos] = concentration of the trace element or of the
                host element in the mineral bearing solution
                at any time during the mineralization

It is generally assumed that a logarithmic distribution controls
the coprecipitation of trace elements if the crystallization takes
place at low temperature, whereas a homogeneous distribution is
obtained at high temperature due to diffusion and recrystallization.

Previous study in our laboratory on the distribution of
different ions between calcite and calcite-bearing solution at
normal p- and T-conditions showed that trace elements are indeed
incorporated only in the surface layer of the crystal. A good
agreement with equation [1] was only obtained when taking n_{Hos} as
the number of mole of Ca at the crystal surface, which was determined by means of ^{47}Ca/Ca-isotope exchange technique (1).

To evaluate a model describing the fractionation of the REE in
fluorite, we ran an experimental program to investigate the incorporation of trace amounts of these elements (labeled with the
corresponding radioactive isotopes) in synthetic fluorite (CaF_2)
which is equilibrated with a fluorite bearing model solution. The
solution contained the complexing agents F^-, Cl^-, SO_4^{2-}, OH^- and had
pH=7. In this chemical milieu the distribution of the REE between
solid CaF_2 and the solution is strongly affected by their complexation with the different ligands present. Taking into account that
the complexation of Ca by means of these ligands is negligible and
only the ligands F^- and OH^- participate significantly to form
fluoro-, hydroxo- and fluorohydroxo complexes of the REE (2-7), the
equilibrium concentration of the REE in equation [1] is expressed
by

$$[REE] = \frac{C_{REE}}{1 + \sum_{ij} \beta_{ij}[F]^i[OH]^j} , \qquad [3]$$

where C_{REE} is the total concentration of REE in the equilibrated
solution and β_{ij} the overall formation constants of the formed
complexes. Substituting [REE] (respectively ^{o}[REE]) from equation
[3] in equations [1] and [2], we obtain

$$\frac{REE}{Ca}\bigg|_{solid} = \frac{D}{1 + \sum_{ij} \beta_{ij}[F]^i[OH]^j} \frac{C_{REE}}{[Ca]}\bigg|_{solution} \qquad [4]$$

$$\frac{REE}{Ca}\bigg|_{solid} = D_{eff} \frac{C_{REE}}{[Ca]}\bigg|_{solution} \qquad [4.1]$$

and

$$\ln \frac{{}^oC_{REE}}{C_{REE}} = \lambda \ln \frac{{}^o[Ca]}{[Ca]} - \ln \frac{1+\sum\limits_{ij} \beta_{ij}[F]^i[OH]^j}{1+\sum\limits_{ij} \beta_{ij} {}^o[F]^i {}^o[OH]^j} \qquad [5]$$

where D_{eff} and λ are obtained from the slope.

To calculate the distribution of REE on coprecipitation in fluorite under hydrothermal condition, the "physical" distribution coefficients D and λ and the stability constants of the resulting complexes have to be determined. The complex formation study of REE in fluorite bearing model systems under normal p-T-conditions has been reported previously (2-8). The investigation of these processes at T = 200 - 300°C and p = 50 - 3000 bar is still in progress in our laboratory. D, however, can be directly determined on carrying out the experiments in acidic solution (pH.< 1) (where complex formation is excluded) at a temperature at which a homogeneous distribution is expected. To compare the incorporation of the different REE, that is to determine their fractionation on co-precipitation with CaF$_2$ owing to a homogeneous distribution law, it is, however, not necessary to know the stability constants of the complexes coming in question, because it is sufficient to compare the effective distribution coefficient D_{eff}, which includes these constants. In this paper some preliminary results of our experimental program are reported.

EXPERIMENTAL

To determine the logarithmic distribution coefficient, 10 samples of a saturated solution of CaF$_2$ (1.2 x 10^{-3}M) which had pH=7 and contained 2.2 x 10^{-7}M ^{141}Ce, 1.1 x 10^{-7}M ^{160}Tb, 5.5 x 10^{-8}M ^{169}Yb and the "background" electrolyte NaCl (1M), Na$_2$SO$_4$ (10^{-2}) were evaporated at 25°C for different times (1-10 days) to obtain various degrees of CaF$_2$ crystallization. The initial concentration of the background electrolyte was adjusted in the different sample to obtain the same final concentration after evaporation. After centrifugation the specific activity of the REE was determined and the corresponding concentration was calculated. The Ca^{2+} concentration was determined by atomic absorption and the free fluoride was measured with a fluoride membrane electrode.

In a second set of samples of the same solution, the experiments were run in an autoclave under pressure of 50 bar and at temperature of 120°C. A fractionated crystallization of CaF_2 was obtained on decreasing the temperature (10°C steps) and waiting for 2 days between every step to establish the distribution equilibria. In a third set the distribution experiments were run at T = 200°C and p = 50 bar in an acidic solution of pH = 1 using [169]Yb to find out which type of distribution is controlling the coprecipitation of the REE at the temperature of hydrothermal solutions. In these experiments solid CaF_2 ($\approx 5 \times 10^{-5}$M) was equilibrated (for 2 days) with 20 ml saturated solution (CaF_2 solubility = 3.5×10^{-2}M) which contained NaCl (1M) and about 10^{-8}M Yb. To compare the "physical" distribution coefficients within the series, the behaviour of the three REE nuclides [144]Ce, [152]Eu and [169]Yb was checked simultaneously.

RESULTS AND DISCUSSION

Fig. 1 shows that the results obtained from the first set fit the logarithmic distribution law. The λ of the first member of series (Ce) is about 4 times more than that of the elements in the middle (Tb) and that at the end of the series (Yb), whereas the λ of Yb is slightly smaller than that of Tb. The ordinate values at the intersection of the lines

$$\ln \frac{{}^{o}C_{REE}}{C_{REE}} = \text{Ce (0.7), Tb (0.5), Yb (0.2)}$$

are in agreement with the values calculated from the (negative) expression

$$\ln \frac{1 + \sum\limits_{ij} \beta_{ij}[F]^i[OH]^j}{1 + \sum\limits_{ij} \beta_{ij} {}^{o}[F]^i {}^{o}[OH]^j}$$

using the β_{ij} determined in (2-8). Fig. 2 illustrates the ratio $C_{REE}/{}^{o}C_{REE}$ as a function of the precipitation degree, μ, of Ca which controls the concentration of the free fluoride in solution. Fig. 3 shows the fractionation of the three REE in crystal as a function of μ.

No reproducable results were obtained in the second set of experiments because of the uncontrolled adsorption of the radio- activity on the walls of the autoclave at high temperature. This problem could not be avoided by coating the walls with different materials. The only successful way was to decrease the pH to values < 3.

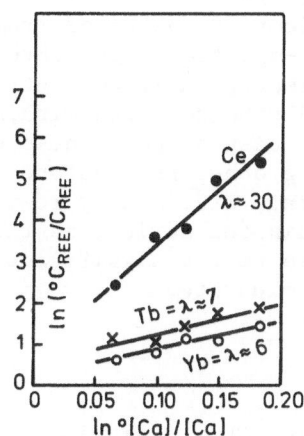

Fig. 1. Logarithmic distribution coefficient of the REE at normal p-, T-conditions.

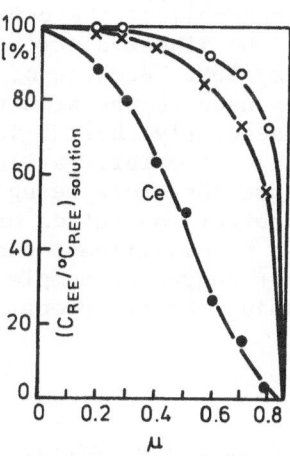

Fig. 2. $^{C}REE/^{o}C_{REE}$ as a function of the precipitation degree μ of CaF_2.

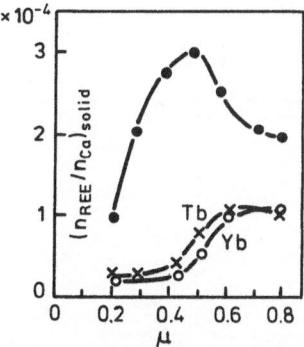

Fig. 3. Fractionation of the REE as a function of the precipitation degree μ of CaF_2.

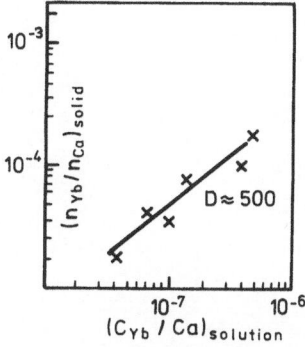

Fig. 4. Linear distribution coefficient of Yb for the 250°C-50 bar experiment.

The result of the third set is shown in Fig. 4. A linear function is obtained on plotting $(Yb/Ca)_{solid}$ versus $(C_{Yb}/[Ca])_{solution}$ which indicates that a homogeneous distribution law is controlling the coprecipitation of the REE at this temperature. The linear distribution coefficient, D, obtained from the slope is ≈ 500. Because of the complex γ-spectra of the three REE isotopes used in the simultaneous experiment, their concentration and consequently their D-values could only be computed within a wide range of error, particularly in the case of Eu. These experiments are therefore being repeated using other REE isotoped to obtain better resolution of the γ-spectra. Preliminary results indicate no significant change of D within the series, so that the different degree of complexation seems to be the significant factor determining the fractionation of REE in fluorite.

REFERENCES

1. V. Koss and P. Möller, Z. Anorg. Allg. Chem., 410:165 (1974).
2. B.A. Bilal, F. Hermann and W. Fleischer, J. Inorg. Nucl. Chem., 41:347 (1979).
3. B.A. Bilal and P. Becker, J. Inorg. Nucl. Chem., 41:1607 (1979).
4. B.A. Bilal and V. Koss, J. Inorg. Nucl. Chem., 42:629 (1980).
5. B.A. Bilal and V. Koss, J. Inorg. Nucl. Chem., 42:1064 (1980).
6. B.A. Bilal and P. Becker, The Rare Earth in Modern Science and Technology, Vol. 2 (1980).
7. V. Koss, Dissertation, Technische Universität Berlin (1980).
8. B.A. Bilal and V. Koss, J. Inorg. Nucl. Chem., to be published.

CRYSTAL STRUCTURES OF EuMgF$_4$, SmMgF$_4$ AND SrMgF$_4$

E. Banks, R. Jenkins and B. Post

Polytechnic Institute of New York
333 Jay Street, Brooklyn, New York   11201

ABSTRACT

Crystal structures of EuMgF$_4$ and SrMgF$_4$ were determined using a computer-controlled x-ray powder diffractometer. Single crystal growth has been unsuccessful to date because the materials melt incongruently. It was initially assumed that these fluorides had structures similar to that of BaMgF$_4$ described by Keve, Abrahams and Bernstein, who reported an orthorhombic structure Ama2. We found these compounds to be centrosymmetric; they were assigned to space group Cmcm.

INTRODUCTION

Several years ago, a number of papers appeared concerning new ternary fluorides of formula type BaMF$_4$ where M represents a divalent transition metal atom, Eibschutz and Guggenheim (1), Eibschutz (2). These materials are typically orthorhombic, and are reported to exhibit ferroelectricity at and below room temperature and antiferromagnetic ordering below room temperature. Another example is the crystal structure of BaMnF$_4$ reported by Keve, et al. (3). There have also been reports describing the non-linear optic behavior of similar compounds such as BaMgF$_4$ and BaZnF$_4$ Bergman and Crane (4). Some years ago, a study was instituted in the Department of Chemical Physics at the Polytechnic Institute of New York, to extend these studies and more recent research in this area has been directed towards the preparation of divalent rare-earth magnesium fluorides.

EXPERIMENTAL

The compounds in question do not grow from the melt because of their incongruent melting behavior. The most successful preparations so far have been obtained by solid state reaction under reducing conditions. As an example, $EuMgF_4$ was prepared by reacting together stoichiometric amounts of EuF_3, MgF_2 and Mg metal, in a sealed graphite crucible at 800-900°C for several hours. (5). Due to the difficulty in preparing the required phase, none of the specimens examined was completely pure. The best specimen examined, consisted of about 85% $EuMgF_4$ and about 5% each of EuF_3, MgF_2 and graphite. The worst specimen examined consisted of about 60% $SmMgF_4$ with the remainder SmF_3, MgF_2 and possibly SmF_2. The presence of the impurities greatly reduced the possible number of reflections usable for the structure analysis, and also prevented the use of profile fitting methods.

Diffractograms were recorded by step-scanning the goniometer between 10-130°2θ using a step size of 0.02 and a count time of 2 sec/step. Peaks were identified in the raw data files using an automated peak hunting algorithm. This algorithm defines the presence and position of a peak by the minima observed in the second derivative of the intensity distribution, in combination with data smoothing and least square fitting.

Preliminary indexing of the $EuMgF_4$ pattern was performed by comparing $Sin^2\theta$ values. Following initial assignment of (hkl) values a cell indexing program was employed to assign indices to all lines and to calculate cell parameters. Examination of the data showed the following systematic absences: (hkl): h=k=2n, (0kl): k=2n, (h0l): =2n, (hk0):h=k=2n. Based on these, plus the evidence of centrosymmetry from second harmonic analysis, the space group Cmcm (D_{2h}^{17}) was assigned to this structure. Using the assumption that all three compounds are isostructural, the patterns for $SmMgF_4$ and $SrMgF_4$ were idexed using as starting indices three similar lines from the $EuMgF_4$ pattern. Table 1 summarizes the cell data for the three materials.

STRUCTURE DETERMINATION

Since the overall scattering function of $EuMgF_4$ is dominated by the europium atoms, the structure was first considered just in terms of this atom type. Following the location of the europium atoms from a Patterson map, a series of electron density maps was prepared to establish the positions of the magnesium atoms. Finally, difference maps were used to locate the fluorines. The final R factor obtained was 0.102.

Table 1. Cell Parameters for $EuMgF_4$, $SmMgF_4$ and $SrMgF_4$

1) **Europium Magnesium Fluoride**   2) **Samarium Magnesium Fluoride**

a = 3.935 ±0.007Å   a = 3.965 ±0.003Å

b = 14.434 ±0.010Å   b = 14.440 ±0.009Å

c = 5.664 ±0.004Å   c = 5.661 ±0.004Å

X-ray volume = 321.7 Å³   X-ray volume = 324.1 Å³

X-ray density = 5.21 g/cm²   X-ray density = 5.13 g/cm²

Measured density = 5.35 g/cm²   Measured density = 4.91 g/cm²

F(N) = 22.9   F(N) = 25.8

3) **Strontium Magnesium Fluoride**

a = 3.917 ±0.002Å

b = 14.451 ±0.008Å

c = 5.637 ±0.003Å

X-ray volume = 319.1 Å³

X-ray density = 3.90 g/cm²

Measured density = 3.89 g/cm²

F(N) = 23.2

Fig. 1 shows the atomic arrangement of $EuMgF_4$ at the levels
x=0 and x=1/2. It will be seen that each europium atom is surrounded by eight fluorines and each magnesium is surrounded by
six fluorines. The Eu-F distances vary between 2.439Å and 2.925Å
with the Mg-F average distance at about 2Å. Table 2 lists the
final atomic coordinates.

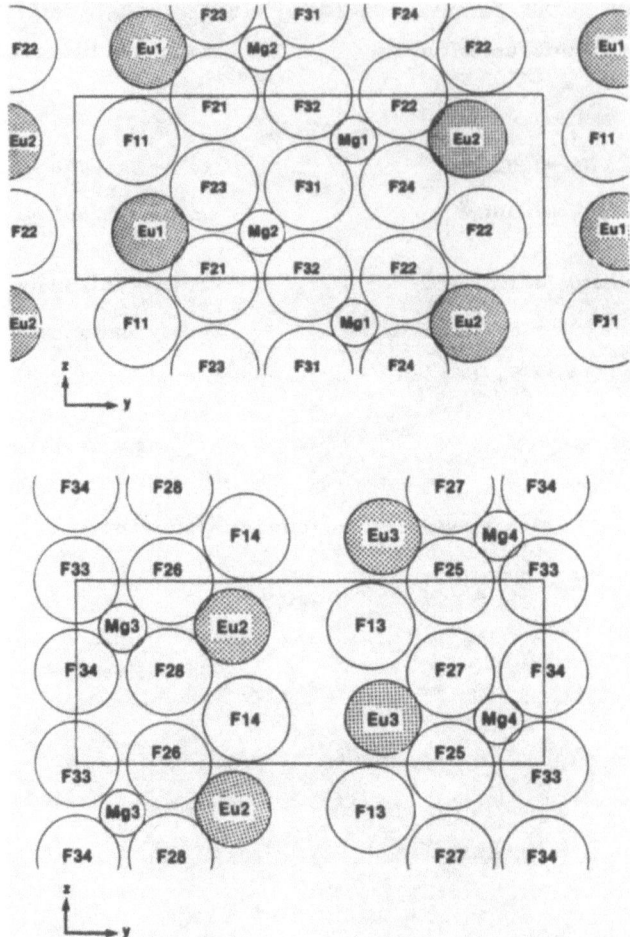

Figure 1.   Atomic arrangement for EuMgF$_4$ at the levels 0 and 1/2

Table 2.   Final Atomic Coordinates

Position	Atom	M=Eu		M=Sm		M=Sr	
		y	z	y	z	y	z
4c	M	0.659	1/4	0.655	1/4	0.656	1/4
4c	Mg	0.070	3/4	0.092	3/4	0.067	3/4
4c	F1	0.608	3/4	0.584	3/4	0.606	3/4
8f	F2	0.835	0.007	0.829	0.014	0.825	0.052
4a	F3	0	1/2	0	1/2	0	1/2

x coordinate = 0

Fig. 2 shows the coordination polyhedron for Eu in EuMgF$_4$ along with the various bond lengths. The asymmetric distribution of the fluorine atoms around the europium is clear and can be compared with the arrangement around barium in BaMnF$_4$, see Fig. 3 We believe, however, that a low temperature form of EuMgF$_4$ possibly exists which might be isostructural with BaMnF$_4$.

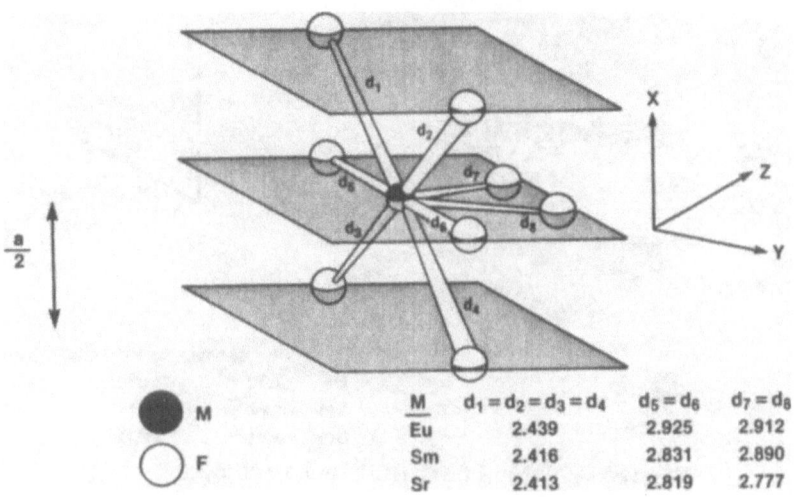

M	$d_1=d_2=d_3=d_4$	$d_5=d_6$	$d_7=d_8$
Eu	2.439	2.925	2.912
Sm	2.416	2.831	2.890
Sr	2.413	2.819	2.777

Figure 2.   Eu in EuMgF$_4$

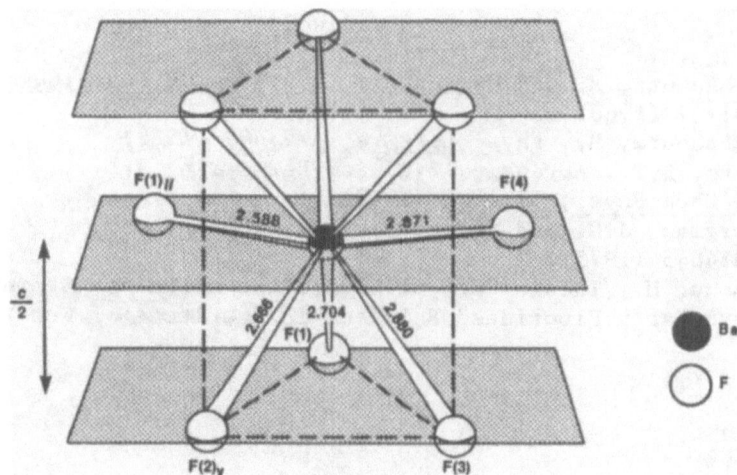

Figure 3.   Ba in BaMnF$_4$

Fig. 4 shows the octahedral arrangement of fluorine atoms around the magnesium along with the respective bond lengths. This arrangement is similar to MnF_6 octahedra in $BaMnF_4$.

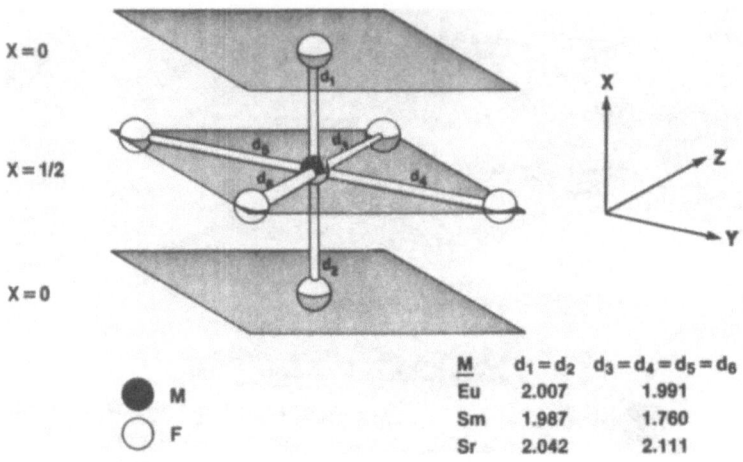

M	$d_1 = d_2$	$d_3 = d_4 = d_5 = d_6$
Eu	2.007	1.991
Sm	1.987	1.760
Sr	2.042	2.111

Figure 4. Mg coordination in $EuMgF_4$

REFERENCES

1. Eibschutz, M. and Guggenheim, M.J., _Solid State Comm._, 6:737 (1968)
2. Eibschutz, M., _Phys. Letters_, A29:409 (1969)
3. Keve, E.T., Abrahams, S.C. and Bernstein, J.L., _J. Chem Phys._, 51:4928 (1969)
4. Bergman, J.G. and Crane, C.R., _J. Appl. Phys._, 46:4645 (1975)
5. Shone, M., Thesis-"New Divalent Rare Earth and Strontium Rare Earth Fluorides" Polytechnic Institute of New York (1980)

STRUCTURAL PHASE TRANSITIONS IN $Cs_2NaLnCl_6$

G.P. Knudsen, F.W. Voss and R. Nevald, Dept. of Electro-
physics, The Technical University of Denmark, DK-2800
Lyngby, Denmark; H.-D. Amberger, Inst. für Anorg. und
Angew. Chemie, Universität Hamburg, D-2 Hamburg 13,
Germany

ABSTRACT

The crystal structure of the elpasolites $Cs_2NaLnCl_6$, where Ln is a
rare-earth-ion, has been investigated by NMR-measurements and high
resolution elastic neutron scattering. Our measurements show that
the compounds with the light rare-earths, contrary to those with
the heaviest rare-earths, undergo a cubic to tetragonal phase transition. The transition for the compounds in this series takes place
at still lower temperatures with increasing atomic number of the
rare-earth and does not occur at all for the last part of the series.

INTRODUCTION

Materials of the cubic elpasolite structure constitute an ideal model
system for studying and testing many basic magneto-physical effects.
Magnetization and NMR-measurements performed by Nevald et al. (1)
showed that the $Cs_2NaLnCl_6$ elpasolites with the light rare-earth-
ions go through a phase transition and are tetragonal distorted at
low temperatures. To determine the type of transition and the dimensions of the tetragonal unit cell, NMR-measurements have been extended and elastic neutron scattering carried out at Risø National
Laboratory. Single crystals of $Cs_2NaLnCl_6$, Ln = Ce, Pr, Nd and Yb,
and powders with Ln = Pr, Tm and Yb were investigated. In the face
centered cubic phase, space group O_h^5, the Ln-ions are surrounded
octahedrally by six Cl- and Na-ions along the cubic axes ±x and ±½
away respectively. The Cs-ions are located in positions ±1/4, ±1/4,
±1/4 of all sign combinations. The Ln-Cl distance x is the only free
parameter in the cubic phase. The elpasolite structure is closely
related to the perovskite and antifluorite structures, which have

respectively the same ions at the Na- and Ln-sites and no ion at the Na-site at all. Until now at least three different space groups for the tetragonal phase are reported: D_{4h}^3 by Meyer and Gaebell (2), C_{4h}^5 or D_{4h}^{17} by Aleksandrov et al. (3). Applying inelastic neutron scattering in the cubic phase we have found that a soft optical mode is associated with the second order phase transition in $Cs_2NaNdCl_6$. The symmetry of this mode determines the low temperature space group to be C_{4h}^5.

RESULTS

The results of the NMR and neutron diffraction experiments are collected in table 1 and figures 1 and 2. NMR gives information concerning structural details via the quadrupole interaction as described by Voss et al. (4). A quadrupole splitting of the Na and Cs resonances is observed at lower temperatures in the compounds with light rare-earths indicating a phase transition to tetragonal symmetry. No splitting is found in the Ho or heavier compounds. By elastic neutron scattering the cell dimensions are determined, and are in agreement with the results for the cubic phase of Ce-, Pr-, Nd-, Sm- and Tm-compounds by Morss et al. (5) and for Ho- and Yb-compounds by Amberger et al. (6). The low temperature unit cell parameters are found by separating the close lying diffraction peaks coming from domains of different orientation. The transition temperature has been determined, by analyzing carefully the merging of two diffraction peaks, when the structural phase transition is approached from below as described in fig. 3. It is found that the transition temperature decreases with increasing atomic number of the Ln-ion in $Cs_2NaLnCl_6$. The same behavior is observed by Meyer and Gaebell (2) in the series $Cs_2NaLnBr_6$.

Table 1. Structure data for $Cs_2NaLnCl_6$. Cell dimensions, a(Å) from Morss et al. (Ce-Sm and Tm) and Amberger et al. (Ho and Yb). α is the linear expansion coefficient. x, in fractions of a(Å), is the Ln-Cl distance at 300 K.

Ln	a(Å) 10K	c(Å) 10K	a(Å) 300K	$\alpha(10^{-4}Å/K)$	x	pha. tr.	T_c(K)
Ce	10.785	10.993	10.946	3.1	0.245	yes	178
Pr	10.763	10.944	10.912	2.3	0.246	yes	159
Nd	10.756	10.904	10.889	2.9	0.245	yes	137
Sm			10.834			yes	~100
Ho			10.726			no	
Tm	10.61	10.61	10.686	2.6		no	
Yb	10.59	10.59	10.672	2.8	0.245	no	

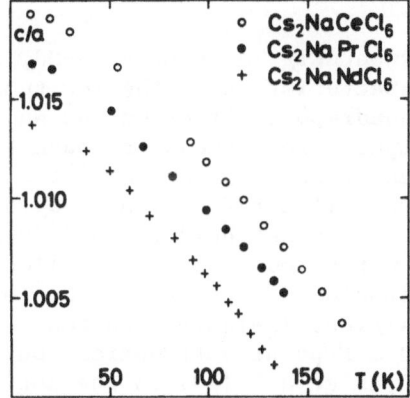

Figure 1. The temperature dependence of the c/a-ratio in the three crystals investigated by neutron diffraction.

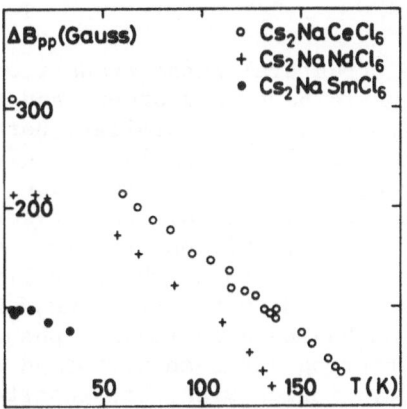

Figure 2. Quadrupole splitting of the Na resonance versus temperature. ΔB_{pp} is the distance between the two satellite NMR-lines.

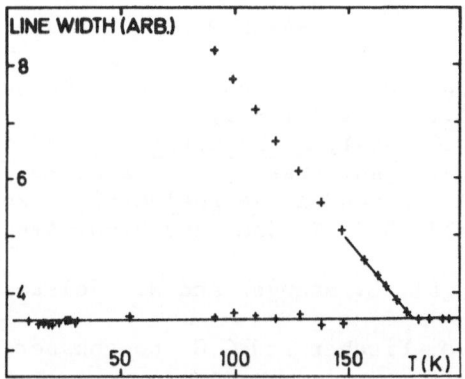

Figure 3. Determination of the transition temperature from the merging of two unresolved diffraction peaks. The line widths resulting from fits to one or two gaussians are shown as the sloping and horizontal lines, respectively. The crossing of the two lines is taken as the transition temperature.

DISCUSSION

The structural phase transition in $Cs_2NaLnCl_6$, with Ln = Ce-Nd, is found to be second order, and the characteristics of the low temperature phase are similar. Both the quadrupole effect on the Na site and the c/a-ratio at a given temperature decrease through the series of compounds. The observed transition temperatures for the compounds in the Cl- and Br-series are a linear function of Gold- et al. (8). This observation, however, does not hold if one includes all the more than 300 elpasolites investigated up to now. Following the limited systematics, $Cs_2NaDyCl_6$ should be the heaviest Cl-compound having a tetragonal phase transition. Inelastic neutron scattering has been performed (9) and a Γ-point soft optical phonon mode, similar with that reported by Lynn et al. (10) in the anti-fluorite K_2ReCl_6, is observed in $Cs_2NaNdCl_6$. The condensation of this rotary mode of symmetry F_{1g} (see for instance Lentz (11)) determines the space group of the tetragonal phase. The space group is found to be C_{4h}^5, where the Cl-octahedra are rotated around the four-fold axis. From this follows that the angle of rotation is the primary order parameter and the c/a-ratio is a secondary order parameter.

ACKNOWLEDGEMENTS

The authors are grateful to H. Bjerrum-Moeller and J. Kjems for good advice and interesting discussions. We would like to thank V. Frank for valuable conversations.

REFERENCES

R. Nevald, F.W. Voss, O.V. Nielsen, H.-D. Amberger and R.D. Fischer, Solid State Commun., 32:1223 (1979).

G. Meyer and H.-C. Gaebell, Z. Naturforsch., 33b:1476 (1978).

K.S. Aleksandrov, A.T. Anistratov, V.I. Zinenko, I.M. Iskornov, S.V. Misul and L.A. Shabanova, Ferroelectrics, 26:653 (1980).

F.W. Voss, R. Nevald, G.P. Knudsen and H.-D. Amberger, "paper of this conf.".

L.R. Morss, M. Siegal, L. Stenger and N. Edelstein, Inorg. Chem., 9:1771 (1970).

H.-D. Amberger, R.D. Fischer and G.G. Rosenbauer, Trans. met. Chem., 1:242 (1976).

V.M. Goldschmidt, Naturwissenschaften, 14:477 (1926).

D. Babel, R. Haegele, G. Pausewang and F. Wall, Mat. Res. Bull., 8:1371 (1973).

To be published.

J.W. Lynn, H.H. Patterson, G. Shirane and R.G. Wheeler, Solid State Commun., 27:859 (1978).

A. Lentz, J. Phys. Chem. Solids, 35:827 (1974).

SYNTHESIS AND CRYSTAL DATA FOR ALKALINE EARTH-LANTHANIDE PHOSPHATES WITH THE EULYTITE STRUCTURE

G.J. McCarthy, D. Krabbenhoft, R.G. Garvey and C. Roob

Departments of Chemistry and Geology, North Dakota State

University, Fargo, ND 58105

Eulytite, $Bi_4(SiO_4)_3$, is a cubic mineral, S.G. = $I\bar{4}3d$, with irregular (BiO_6) polyhedra and (SiO_4) tetrahedra. With the substitution of P for Si, the charge balancing $(3A^{2+} + Ln^{3+})$ for $4Bi$ is possible. Eulytite structure phosphates, $A_3Ln(PO_4)_3$, are one of the ternary phases that form in the $AO-Ln_2O_3-P_2O_5$ systems currently under investigation in our laboratories. Synthesis and crystal data for phosphate eulytites where A = Ca, Sr, Ba, Pb and B = La $\rightarrow$ Lu, Y, Sc, In are included here.

Synthesis was by air firing of batches prepared either from mixed carbonate, oxide and phosphate powders or from solutions of A and Ln nitrates and ammonium hydrogen phosphates. The latter batches were more reactive and gave complete reaction with shorter firing times and lower temperatures. Firing temperature was an important parameter. In $Ca_3Ln(PO_4)_3$ preparations, temperatures above 1300°C were required. At 900-1200°C, the stable assemblage is $[\beta-Ca_3(PO_4)_2 + LnPO_4]$ (1). $Sr_3Ln(PO_4)_3$ phases with small Ln ionic radii dissociate into $Sr_3(PO_4)_2 + LnPO_4$ during long-term firings at 800°C. Dissociation has not been observed in the Ba analogs.

Unit cell parameter data for each phase are given in Table 1. When the data are plotted against Ln^{3+} radii, the expected curvilinear relationship is observed for the larger Ln's, but, starting in the middle of the Ln series, there is an inflection and distinct decrease in slope in spite of the steadily decreasing size of the Ln. Engel, (3) who first observed this anomaly for the Pb eulytites, has proposed that the smaller Ln's do not have close contact with surrounding oxygens. This explanation is currently under investigation by structure refinement from x-ray and neutron powder data.

Table 1. Cell parameters[a] of eulytite structure phases.

Ln	$Ca_3Ln(PO_4)_3$	$Sr_3Ln(PO_4)_3$	$Pb_3Ln(PO_4)_3$	$Ba_3Ln(PO_4)_3$
La	9.941	10.192	10.347	10.521
Ce	9.921	10.166	10.347	10.506
Pr	9.905	10.154	10.328	10.501
Nd	9.890	10.148	10.315	10.492
Sm	9.875	10.129	10.305	10.473
Eu	9.867	10.124	10.304	10.466
Gd	9.857	10.114	10.299	10.465
Tb	–	10.109	10.293	10.464
Dy	–	10.109	10.293	10.460
Ho	–	10.098	10.291	10.460
Y	–	10.096	10.297	10.464
Er	–	10.095	10.301	10.460
Tm	–	10.091	10.306	10.464
Yb	–	10.092	10.289	10.461
Lu	–	10.091	10.293	10.455
(In)	–	10.072[b]	10.228[b]	10.408[b]
(Sc)	–	10.066	10.220	10.395
(Bi)	9.978	10.200	10.372	10.517

[a] Estimated standard deviation is ±0.003.
[b] After G. Engel (3).

REFERENCES

1. G.J. McCarthy and D.E. Pfoertsch, J. Solid State Chem., 38: 128-129 (1981).
2. R.D. Shannon, Acta Cryst., A32:751 (1976).
3. G. Engel, Zeit. Anorg. Allg. Chem., 387:22 (1972).

THE CRYSTAL STRUCTURE AND STOICHIOMETRY OF THE $Ca_{2+x}Nd_{8-x}(SiO_4)_6O_{2-1/2x}$ SYSTEM*

J. A. Fahey, Bronx Community College, Bronx, NY 10453;
W. J. Weber, Pacific Northwest Laboratory, Richland,
WA 99352

INTRODUCTION

A systematic study of $Ca_{2+x}Nd_{8-x}(SiO_4)_6O_{2-1/2x}$, one of the mineral-like phases formed by the rare earths in the "supercalcine-ceramic" nuclear waste forms, has been undertaken. Until now, the structure and stoichiometry of this apatite phase has only been inferred from the hexagonal symmetry revealed by its powder diffraction data (1). The goal of this study was to obtain a complete set of atomic coordinates and the temperature factor and occupation factor for each atom in this structure by applying the Rietveld profile analysis technique (2,3) to powder x-ray diffraction data. Samples of several bulk compositions in the range $0 \leq x \leq 4$ were prepared in order to evaluate the solid solution limits of the apatite phase.

SAMPLE PREPARATION

Samples of three different bulk compositions, $x = 0$, 2 and 4, were prepared by dissolving appropriate amounts of Nd_2O_3 (99.99%) and CaO (99.9%) in 35% nitric acid and adding the resulting solution to an ammonia stabilized silicate solution containing the desired amount of SiO_2. The solutions were evaporated to dryness, heated at 500°C for 2 h to decompose nitrates to oxides, pressed into pellets and fired at 1250°C for 6 h. A second set of pellets were refired at 1600°C for 5 h. After cooling, at 250°C/h, the samples were ground to 325 mesh and examined by powder x-ray diffraction.

*Work supported by the U.S. Department of Energy under Contract DE-AC06-76RLO 1830 and by the Northwest College and University Association for Science.

Analysis of the x-ray data indicated the presence of a single
phase with a hexagonal (apatite) structure in each of the samples.

REFINEMENT

X-ray diffraction measurements were made at room temperature
on a Philips XRG-3100 diffractometer with a fine-focus copper
target and equipped with a theta-compensating slit. A graphite
monochromator was mounted on the detector and tuned for $CuK\beta$ radiation.
The diffraction data were collected using the Angle Mode Programmer
step-scanning system over a two theta range of 8.0° to 100.0° with
a step width of 0.02° and a counting time of 20 seconds per step.

Structural refinement of $Ca_2Nd_8(SiO_4)_6O_2$ appeared to represent
a highly favorable opportunity to use powder x-ray diffraction
data with a slightly modified version of the Rietveld-Hewat profile
analysis program (4). The peak shape was approximated by a modified
Lorentzian function fitted over a range of 2.0 half-widths (FWHM)
on each side of the peak. The data reduction program was modified
to allow for the theta compensating slit of the diffractometer.
Scattering factors appropriate to $Ca(+2)$, $Nd(+3)$, $Si(+4)$ and $O(-2)$
were taken and corrected for the real part of the dispersion term
(5). No provisions exist in the present version of the program
for the imaginary components.

The data taken on $Ca_2Nd_8(SiO_4)_6O_2$ was readily refined assuming
the apatite structure and coordinates of $Ca_{10}(PO_4)_6F_2$, with two
$Nd(+3)$ and two $Ca(+2)$ in the positions at the 4f site. The results
shown in Table 1 are for a model which allowed site exchange
between the $Ca(+2)$ and $Nd(+3)$ ions on the 4f and 6h sites and
individual isotropic temperature factors for each atom, except in
the event that two different atoms shared the same site, in which
case they were assigned the same temperature factor. The refined
occupation parameters show a distribution of two $Nd(+3)$ and two
$Ca(+2)$ ions in the 4f site and six $Nd(+3)$ ions in the 6h site.
The R factor ($R_I=12.6\%$) is defined by Khattak and Cox (6). An
attempt to refine a preferred orientation parameter did not produce
any significant improvement.

The refinement of the data set taken on the sample prepared at
1250°C with bulk composition of x = 2 and based on a model consistent with the bulk composition of the sample was not satisfactory.
A reasonably good refinement was obtained on the basis of the
formula $Ca_3Nd_7(SiO_4)_6O_{1.5}$. The results of this refinement are
shown in Table 2.

DISCUSSION

The structure of $Ca_2Nd_8(SiO_4)_6O_2$ has been shown to be isostruc-

Table 1.   Structural Parameters in Room-Temperature $Ca_2Nd_8(SiO_4)_6O_2$

Crystal System: Hexagonal   Class: 6/m
Space Group: $P6_3/m$ Z = 2
Unit Cell Parameters: a = 9.5297(4)   c = 7.0184(2)
Preferred Orientation Parameter: (G_{001}) = 0.02(1)

Atom	Site	Occupation	X	Y	Z	B (Å^2)
Ca(1)	4f	1.90(6)	0.333	0.667	0.016(2)	1.6(1)
Nd(1)		2.10(6)				
Ca(2)	6h	0.08(5)	0.233(1)	-0.010(1)	0.25	1.8(1)
Nd(2)		5.92(6)				
Si	6h	6	0.398(3)	0.372(3)	0.25	3.3(2)
O(1)	6h	6	0.322(5)	0.479(5)	0.25	3.0(5)
O(2)	6h	6	0.599(6)	0.471(6)	0.25	6.0(6)
O(3)	12i	12	0.343(3)	0.252(3)	0.063(3)	1.0(3)
O(4)	2a	2	0.00	0.00	0.25	4.6(8)

Table 2.   Structural Parameters in Room-Temperature $Ca_3Nd_7(SiO_4)_6O_{1.5}$

Crystal System: Hexagonal   Class: 6/m
Space Group: $P6_3/m$ Z = 2
Unit Cell Parameters: a = 9.5406(5)   C = 7.0135(2)
Preferred Orientation Parameter: (G_{001}) = 0.02(1)

Atom	Site	Occupation	X	Y	Z	B($\text{Å})^2$
Ca(1)	4f	1.82(5)	0.333	0.667	0.011(2)	2.1(1)
Nd(1)		2.18(5)				
Ca(2)	6h	1.18(6)	0.234(1)	-0.013(1)	0.25	2.0(1)
Nd(2)		4.82(5)				
Si	6h	6	0.401(3)	0.367(3)	0.25	3.0(3)
O(1)	6h	6	0.308(4)	0.477(5)	0.25	2.8(6)
O(2)	6h	6	0.601(6)	0.475(5)	0.25	4.9(6)
O(3)	12i	12	0.350(3)	0.260(3)	0.061(3)	1.2(3)
O(4)	2a	1.5	0.00	0.00	0.25	4.8(8)

tural with $Ca_{10}(PO_4)_6F_2$. The metal ions occupying the 4f site are coordinated by six silicate oxygens at an average distance of 2.56 Å. The Nd(+3) ions occupying the 6h site are coordinated by six silicate oxygens at an average distance of 2.49 Å and by one anionic oxygen at 2.27 Å. The anionic oxygen in 2a is coordinated by three Nd(+3) ions. The structure of $Ca_3Nd_7(SiO_4)_6O_{1.5}$ is similar except that five Nd(+3) and one Ca(+2) ions occupy the 6h site and 1/4 of the anionic oxygens are missing in a random fashion.

The calculated Nd-O and Si-O distances are consistent with those of similar lanthanide compounds (7). The structural data for the silicate ion provides an internal standard for evaluating the refinement. The average Si-O distance (1.62 ± 0.03 Å) in Ca_2Nd_8 $(SiO_4)_6O_2$ is in good agreement with the 1.63 Å value reported for several silicates (8). The average O-Si-O angle (109° ± 4) is close to the ideal tetrahedral value.

Comparison of the O(4)-Ca(2) and O(4)-Nd(2) distances in the two structures shows a displacement of 0.03 Å. This displacement is caused by the increased metal ion repulsion around the partially occupied 2a site and represents the average of a 0.12 Å displacement about one vacant site and a zero displacement for three fully occupied sites. This displacement is directed along the x and y axis and is reflected in the increased a parameter of the x = 1 structure over the x = 0 structure. The general expansion of the lattice is somewhat offset by the change in average metal ion radii of 1.096 Å in the x = 0 structure to 1.083 Å in the x = 1 structure and is reflected in the 0.0049 Å contraction in the c axis.

The observed variations in cell volume of the apatite phase reported in this paper with bulk compositions, x, are: 550.4 ± 0.5 $Å^3$ for x = 0 and 552.4 ± 1.3 $Å^3$ for x = 2 and x = 4 in the samples prepared at 1250°C, and 550.8 ± 1.0 $Å^3$ for x = 2 and x = 4 in the samples prepared at 1600°C. This is consistent with a composition limit of x ~ 1 at 1250°C and x ~ 0 at 1600°C for this system. These conclusions were confirmed by SEM-EDX analysis, which revealed the presence of a glass-like phase in addition to the apatite phase in samples with bulk compositions of x = 2 and x = 4. Since the glass-like phase is nondiffracting, its presence was not detected in the x-ray diffraction data. Additional work is under way to establish the exact shape of the solubility region.

REFERENCES

1.   G. J. McCarthy, Proceedings of the 12th Rare Earth Research Conference, Vol. II, 665-676 (1976).
2.   H. M. Rietveld, Acta. Cryst., 22:151-152 (1967).
3.   H. M. Rietveld, J. Appl. Cryst., 2:65-71 (1969).
4.   A. W. Hewat, UK Atomic Energy Authority Research Group Report, RRLL 73/897, AERE Harwell, Oxfordshire (1973).
5.   D. T. Cromer and J. T. Waber, Acta. Cryst., 18:104-109 (1965).
6.   G. P. Khattak and D. E. Cox, J. Appl. Cryst., 10:405-411 (1977).
7.   J. Felsch, J. Solid State Chem., 5:266-275 (1972).
8.   D. W. J. Cruickshank, J. Chem. Soc., 5486 (1961).

ELECTRON BEAM-INDUCED REDUCTION OF $TbO_{2-\delta}$: A HIGH RESOLUTION

ELECTRON MICROSCOPE STUDY[†]

L. Eyring and R. T. Tuenge

Department of Chemistry and the Center for
Solid State Science, Arizona State University
Tempe, Arizona 85287

ABSTRACT

Terbium forms a homologous series of oxides Tb_nO_{2n-2} where n is an integer between 4 (Tb_4O_6 = Tb_2O_3) and ∞ (TbO_2). The structures of phases for n = 4, 7, 11, 12 and ∞ have been reported (1,2,3). In addition, and in common with Ce and Pr, Tb, a disordered fluorite α-phase is formed at higher temperatures and pressures, in the composition range TbO_{2-x}, where $0 \leq x \leq 0.28$. A study of α-TbO_{2-x} and the sequential stages of its reduction to form ordered intermediate phases has been carried out.

Crystals of highly oxidized terbium oxide approximately 0.5 mm across were grown hydrothermally using a technique described elsewhere in this volume (4). The techniques of high-resolution transmission electron microscopy (HRTEM) were applied to thin crystals (5) of the α-phase, heated in the electron beam.

Figure 1 illustrates the progressive change resulting from electron beam heating. On the right, diffraction patterns mark the growing dominance of the β phases. The images on the left are of an identical region of the crystal taken between the diffraction patterns. The earliest order appearing is frequently fringes almost along [131] with about 5.7 Å spacing marked "A" in Fig. 1a and "A'" in Fig. 1d, where they are displaced by the ordered phases of the homologous series. Developing regions of ι (n = 7), δ (n = 11), β(2) and β(3) appear in this region of the crystal in Fig. 1b and c.

[†] This work was supported by the NSF through Grants DMR 77-08473 and DMR 80-06584 and will be reported fully elsewhere.

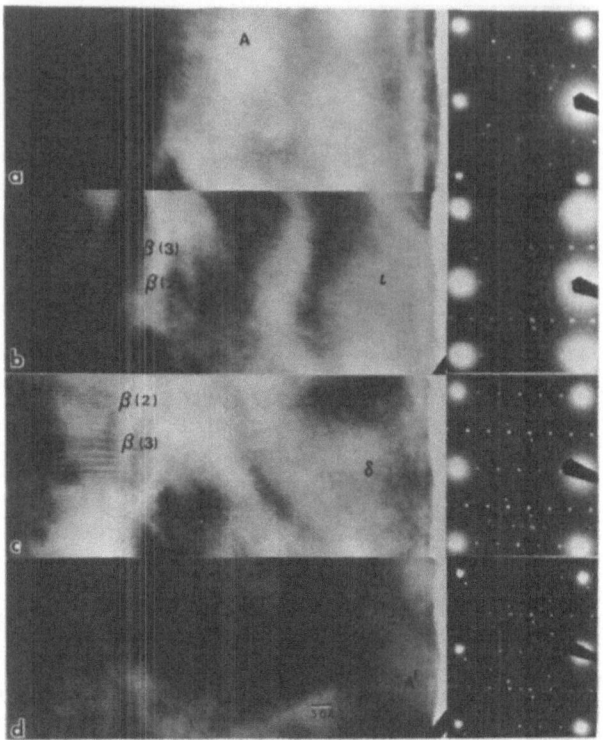

Figure 1.   A sequence of electron diffraction patterns and
            concomitant images of an area of a crystal of
            α-TbO$_{2-x}$ as it orders when reduced in the electron
            beam of the microscope.

The growing and almost complete encroachment of $\beta(2)$ and $\beta(3)$ into
the disordered regions or those consisting of mixed ordered domains
is the most striking feature of the representative event recorded
in Fig. 1.

REFERENCES

1. P. Kunzmann and L. Eyring, J. Solid State Chem., 14, 229 (1975).
2. R. T. Tuenge and L. Eyring, unpublished work.
3. L. Eyring in "Handbook on the Physics and Chemistry of Rare
   Earths," K. A. Gschneidner and L. Eyring (eds.) North-Holland
   Amsterdam (1979) Chapter 27, pp 337-399.
4. B. Chang, M. McKelvy and L. Eyring, this volume.
5. S. Iijima, Acta Crystallogr., Sect A, 29, 18 (1973).

CRYSTAL STRUCTURE AND PROPERTIES OF (LaO)CuS and (LaO)AgS

M. Palazzi, C. Carcaly, P. Laruelle et J. Flahaut
Laboratoire de Chimie Minérale Structurale Associé au
C.N.R.S. n° 040200 - FACULTE DE PHARMACIE - 4, avenue
de l'Observatoire - 75270 Paris - Cedex 06, France

ABSTRACT

The title compounds belong to a new class of materials having alternative sheets of oxide and sulfide. Electrical property measurements and electron microscope characterization are reported.

PREPARATION

(LaO)CuS is prepared by slow oxidation of $LaCuS_2$ at 800°C in a nitrogen flow bubbled through water at 30°C. (LaO)AgS is obtained by firing a mixture of $(LaO)_2S$ and Ag_2S at about 630°C in an evacuated sealed silica ampule.

(LaO)CuS is red and (LaO)AgS is green.

CRYSTALLOGRAPHIC DATA AND CRYSTAL STRUCTURE

The two phases are isostructural. They crystallize in the tetragonal system, space group P4/nmm, Z = 2. Crystal data are given in Table 1.

Each kind of atom is in a plane perpendicular to the four-fold axis. The (LaO) sheets alternate with the sulfide sheets, "CuS" or "AgS". The oxygen atoms are bonded only to the lanthanum atoms. The sulfur atoms are simultaneously bonded to the lanthanum atoms and the second cation (Cu or Ag). The Cu or Ag atoms are in the center of sulfur tetrahedra. The lanthanum atoms are bonded to the O and S

347

Table 1. Crystal data for (LaO)CuS and (LaO)AgS.

(LaO)CuS			(LaO)AgS	

a = 4.000 Å ; c = 8.53 Å   a = 4.050 Å ; c = 9.039 Å

	x	y	z		z
La	1/4	1/4	0.1484	La	0.1357
Cu	1/4	3/4	1/2	Ag	1/2
S	1/4	1/4	0.6633	S	0.6931
O	1/4	3/4	0	O	0

R = 0.029 for 182 reflections   R = 0.055 for 154 reflections

Cu − S = 2.437 Å   Ag − S = 2.67 Å
La − S = 3.253 Å   La − S = 3.25 Å
La − O = 2.367 Å   La − O = 2.37 Å

atoms, and are 7-coordinated. In the La-O sheets, the La-O distances
are relatively short, as expected for a mainly covalent bonding.

IONIC MOBILITY

Four measures of the mobility of the Cu and Ag mobility in the
two phases have been investigated.

Electron Microscopy. In the electron microscope beam, (LaO)AgS
crystals extrude Ag even when the beam intensity is kept low. The
metal, readily identified by electron diffraction, appears on the
crystals as fibres or droplets as shown in Fig. 1. Apparently, Ag^+
ions are driven out by the local electric field. No corresponding
migration of Cu^+ from (LaO)CuS was observed.

Electrolysis. The electrolysis was performed in the following
cell under constant current: +M/(LaO)MX/Pt (M = Cu or Ag). For
(LaO)CuS, copper dendrites are slowly formed at 300°C (Fig. 2). For
(LaO)AgS, a silver precipitate (Fig. 3) appears at the cathode, even
at 25°C. The silver movement is accompanied by formation of sulfur
and lanthanum oxysulfide at the anode.

EMF Measurements. The potentials of the following cells were
measured in open circuit: M/(LaO)MS/S,C/Pt (M = Cu or Ag). The results were: (LaO)CuS E = some mV
             (LaO)AgS E = 0.200±0.005 V.

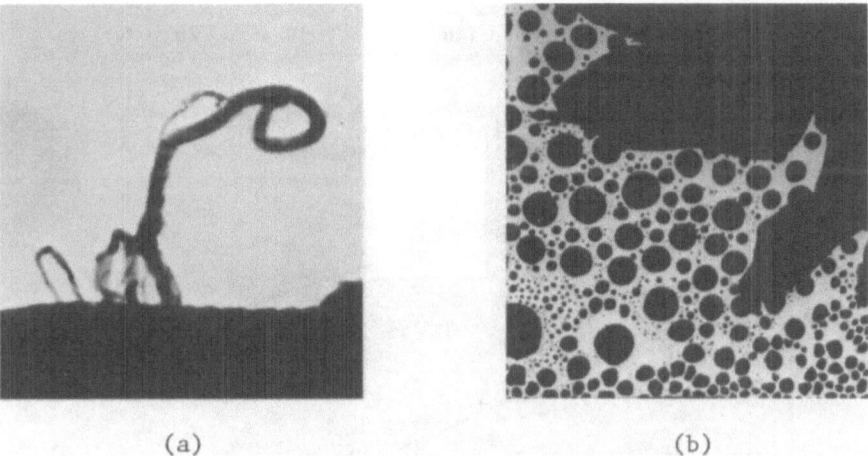

Figure 1.   Fibres or droplets of Ag formed on the surface of
            (LaO)AgS in the electron beam.

Conductivity Measurement. Measurements of the electrical resistivity were performed using the complex impedance method with polycrystalline samples, and four-probe technique with single crystals.
The results are shown in Fig. 4.

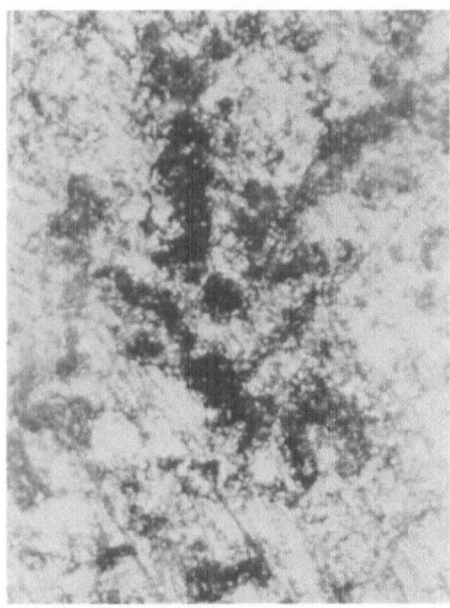

Figure 2. Cu dendrites formed
during electrolysis of (LaO)CuS.

Figure 3. Ag precipitate formed
during electrolysis of (LaO)AgS.

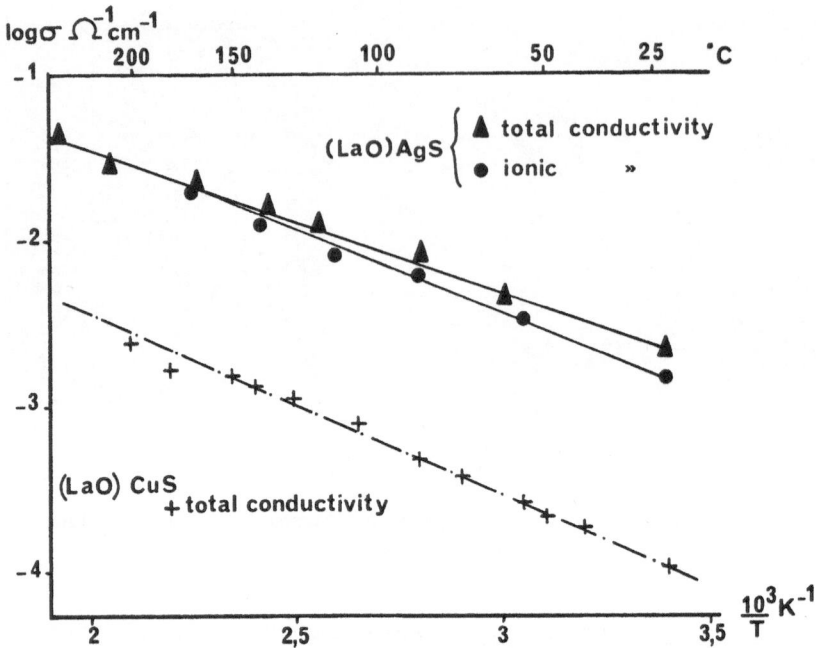

Figure 4.   Electrical conductivity of (LaO)MS phases.

CONCLUSIONS

The electrical conductivity has a bidimensional character in relation to the insulator behavior of the (LaO) sheet. The conductivity
of (LaO)AgS is purely ionic, with a large mobility of the silver cations. The conductivity of (LaO)CuS has a dominant metallic character
and has mainly an electronic origin.

REFERENCES

1. M. Palazzi, C. Carcaly and J. Flahaut, J. Solid State Chem., 35:
    150-5 (1980).
2. M. Palazzi et S. Jaulmes, Acta Cryst. (in press, 1981).

THE CRYSTAL CHEMISTRY OF THE EUROPIUM ARSENIDES

Forrest L. Carter, Chemistry Division
Naval Research Laboratory, Washington, DC 20375

L. D. Calvert, Chemistry Division, National Research
Council of Canada, Ottawa Canada K1A OR9

Recently seven well-defined crystal structures of the europium arsenides, Eu_5As_3, Eu_4As_3, Eu_5As_4, Eu_2As_2, Eu_3As_4, Eu_2As_3, and $EuAs_3$ (1-6), and two closely related oxides, Eu_4As_2O and Eu_3As_3TaO (7,8) have been reported from the same laboratory. In order to better understand the crystal chemistry of these materials and the nature of the elements, Eu and As, the crystal structure results are studied here via the calculation of polyhedral atomic volumes (PAVs), coordination numbers (CN), partial coordination number coefficients, f, Pauling metallic valences, and harmonic analysis (9,10). These approaches have been useful in the past in understanding and quantifying the crystal chemistry of rare earth intermetallics, halides and hydrides (10-12).

Europium PAVs range from 35.0 to $42.5A^3$ while arsenic PAVs vary from 15.7 to $19.7A^3$. Thus neither Eu nor As approach their respective elemental volumes of 48.1 and $21.5A^3$. Europium in the pure arsenides is primarily divalent (exception Eu_4As_3) with some tendency toward trivalency but Eu PAVs show little order with the coefficient, f, in contrast to trivalent rare earths (11). However the divalent Eu PAVs does increase with increasing asphericity. This behavior is consistent with a rather soft large atom. In Eu_4As_3, Eu is clearly trivalent with a strongly contracted trivalent radius while Eu_1 in Eu_4As_2O is normal trivalent with Eu_2 divalent. (PAVs: Eu_1, 31.1 and Eu_2, $41.8A^3$).

Arsenic PAVs correlate reasonably well with PCN coefficient f for these compounds with elemental As included. As volume also increases with increasing asphericity if Eu_4As_3 and elemental As

is excluded. It is particularly interesting that As PAV plotted
against the partial CN based on bond order segregates into five
groups corresponding to PCN = 0, 1, 2, 3 and 4. This suggests
that arsenic character is more traditionally tetravalent than
metallic.

The bonding of both Eu and As in Eu_4As_3 (anti-Th_3P_4
structure) is interesting. Europium, as indicated, is clearly
trivalent with enhanced bonding d character. Here arsenic expands
its formal valence to four (corresponding to As^{+1}) however the
effective charge is reduced to $\sim$ -1/3 since bond polarization trans-
fers about 1-1/3 electrons to arsenic.

REFERENCES

1. Y. Wang, L. D. Calvert, E. J. Gabe and J. B. Taylor, Acta
   Cryst., B34:2281 (1978).
2. Y. Wang, L. D. Calvert, and J. B. Taylor, Acta Cryst., B36:221
   (1980).
3. Y. Wang, L. D. Calvert, E. J. Gabe and J. B. Taylor, Acta
   Cryst. B34:1962 (1978).
4. Y. Wang, E. J. Gabe, L. D. Calvert, and J. B. Taylor, Acta
   Cryst., B33:131 (1977).
5. Y. Wang, L. D. Calvert, M. L. Smart, J. B. Taylor, and E. J.
   Gabe, Acta Cryst., B35:2186 (1979).
6. S. Ono, F. L. Hui, J. G. Despault, L. D. Calvert, and J. B.
   Taylor, J. Less-Common Metals, 25:287 (1971).
7. Y. Wang, L. D. Calvert, E. J. Gabe, and J. B. Taylor, Acta
   Cryst., B33:3122 (1977).
8. Y. Wang, L. D. Calvert, M. L. Smart, J. B. Taylor, Acta
   Cryst., B36:131 (1980).
9. F. L. Carter, Acta Cryst., B34:2962 (1978).
10. F. L. Carter "Polyhedral Atomic Volumes and Generalized
    Coordination Numbers for the Rare Earth Chlorides and Bromides"
    in The Rare Earths in Modern Science and Technology, G. J.
    McCarthy and J. J. Rhyne, Plenum Press, New York, p. 225-231
    (1978).
11. F. L. Carter, "Atomic Volume Contraction in Rare Earth Nickel
    Intermetallics; A Function of Partial Coordination Number
    Coefficient," in The Rare Earths in Modern Science and
    Technology, Vol. 2, G. J. McCarthy, J. J. Rhyne, and H. B.
    Silver (eds.), Plenum Press, New York, p. 299-304 (1980).
12. F. L. Carter, J. Less-Common Metals, 74:245 (1980).

CRYSTAL STRUCTURES AND PHASE RELATIONSHIPS WITHIN TERNARY SYSTEMS:

RARE EARTH METAL-NOBLE METAL-BORON

Peter Rogl, Institute of Physical Chemistry,
University of Vienna, A-1090 Wien, Austria and
Hans Nowotny, Institute of Materials, Science
University of Connecticut, Storrs, Connecticut, 06268

INTRODUCTION

Borides with a ratio boron/metal≈2 are classified as nettype borides, due to the characteristic formation of two dimensional rigid boron networks (1,2). Whereas most boron rich regions of ternary transition metal-boron systems have been characterized, there is still considerable interest in the combinations: rare earth metal-noble metal-boron.

EXPERIMENTAL

Impurity level of starting materials, details of sample preparation (arc melting) of ternary alloys: RE-T-B (RE = rare earth metal, T = noble metal), metallographic as well as X-ray techniques applied, can be found in earlier publications on ternary rare earth borides (3,4).

RESULTS AND DISCUSSION

A study of phase equilibria of the systems [Y,RE] - [Ru,Os] -B at 1400°C revealed the existence of new ternary boron rich compounds.

Single phase powder-(Guinier) X-ray patterns with sharp reflections are easily obtained from arc melted and subsequently heat treated (1400°C, 24 hrs; 10^{-4}Pa) samples of composition (in at %): RE(11) T(22) B(67). The powder patterns of the new phases can be indexed completely on the basis of a primitive orthorhombic unit cell. Only minor variations of lattice parameters were observed with heterogeneous alloys, thus narrow homogeneity regions are attributed to the new phases of a formula: RE_2TB_6 (RE = Gd,Tb,Dy, Ho,Er,Tm,Lu,Y; T = Ru,Os) (see Table 1).

Lattice parameters, composition, intensities as well as the observed extinctions: (hoℓ), h ≠ 2n; (okℓ), k ≠ 2n are compatible

Table 1. Crystallographic data for ternary borides RE_2TB_6 (T = Ru, Os). Y_2ReB_6-type structure; Pbam, Z = 4.

Compound	a(Å)	b(Å)	c(Å)	v(Å³)
Y_2RuB_6	9.1498(16)	11.5139(32)	3.6501(2)	384.5
Gd_2RuB_6	9.2323(30)	11.5843(36)	3.6902(3)	394.9
Tb_2RuB_6	9.1796(20)	11.5407(19)	3.6650(2)	388.3
Dy_2RuB_6	9.1433(23)	11.5067(18)	3.6489(2)	383.9
Ho_2RuB_6	9.1083(24)	11.4666(19)	3.6407(3)	380.2
Er_2RuB_6	9.0822(31)	11.4439(56)	3.6266(4)	376.9
Tm_2RuB_6	9.0605(28)	11.4203(31)	3.6091(3)	373.4
Lu_2RuB_6	9.0138(39)	11.3880(17)	3.5826(8)	367.8
Y_2RuB_6	9.1592(10)	11.5311(25)	3.6473(2)	385.2
Gd_2OsB_6	9.2425(36)	11.5912(41)	3.6814(3)	394.6
Tb_2OsB_6	9.1889(19)	11.5612(43)	3.6567(2)	388.5
Dy_2OsB_6	9.1542(26)	11.5211(51)	3.6435(2)	384.3
Ho_2OsB_6	9.1229(15)	11.4909(35)	3.6320(2)	380.7
Er_2OsB_6	9.0973(21)	11.4644(21)	3.6169(3)	377.2
Tm_2OsB_6	9.0677(29)	11.4336(36)	3.6037(3)	373.6
Lu_2OsB_6	9.0331(35)	11.4113(29)	3.5842(5)	369.5

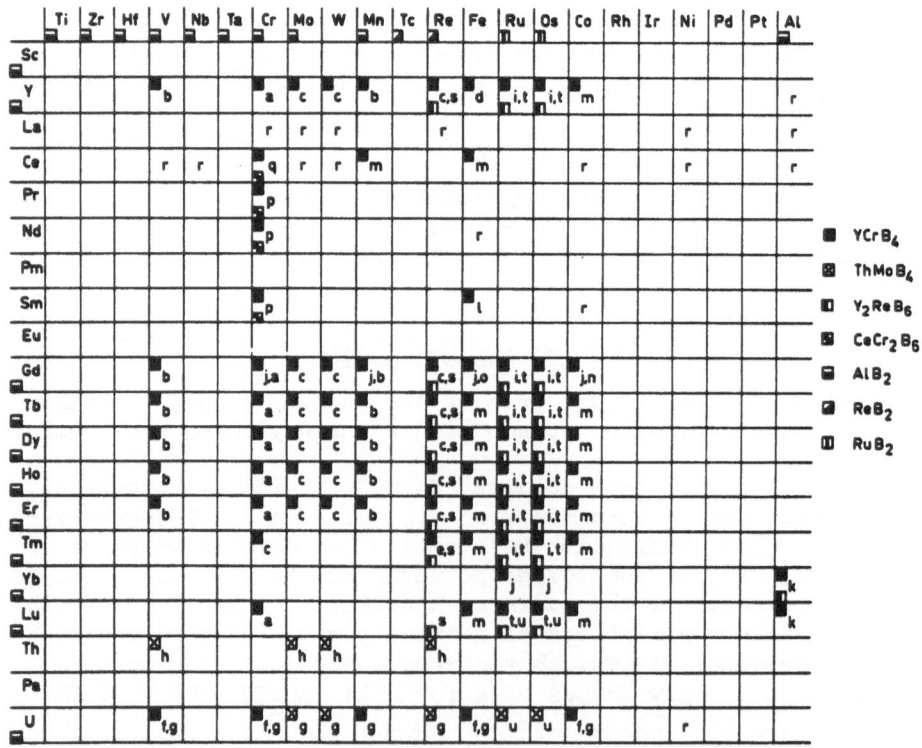

Fig. 1. Formation of $(RE,T)B_2$-borides and different structure types.
References: a(6), b(7), c(8), d(9), e(10), f(11), g(12), h(13), i(14), j(15), k(16), l(17), m(18), n(19), o(20), p(21), q(22), r(23), s(5), t(this work), u(24).

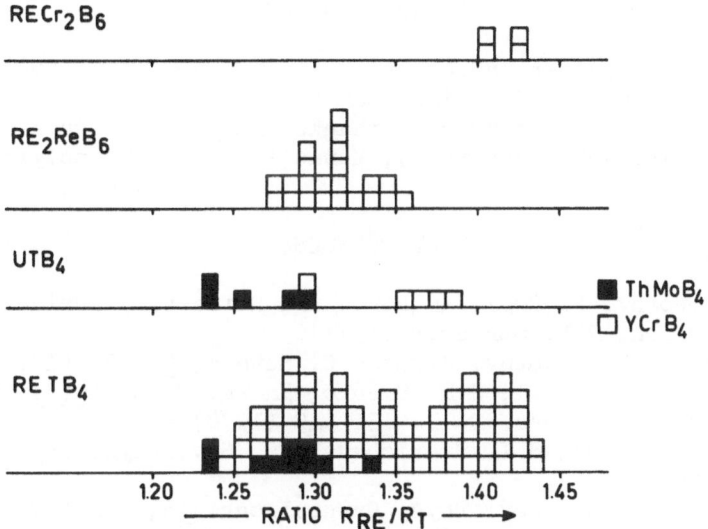

Fig. 2.   Histogram for the occurrence of phases with the $YCrB_4$,
$ThMoB_4$, Y_2ReB_6 and $CeCr_2B_6$ structures.   Metallic radii
after (25).

with symmetry Pbam and prove structural analogy with the structure
type of Y_2ReB_6 (5).   Using the atom parameters as derived from a
single crystal study of Y_2ReB_6 (5), observed and calculated powder
intensities are in excellent agreement.   Intensity calculations can
be obtained on request.   Graphs of the lattice parameters and volumes versus the corresponding values of the trivalent ionic radii
R_{RE}^{3+} reveal linear dependency; the values for the yttrium compound
are ranging between those for Tb and Dy respectively.

With five-, six- and seven-membered boron rings in a typical
planar boron net type arrangement, the crystal structure of Y_2ReB_6
is intermediate between the structure types of AlB_2 and $YCrB_4$
according to the topochemical mode: $Y_2RuB_6 = YB_2 + YRuB_4$.   As far
as the formation of AlB_2-type RE-diborides is concerned (see Fig.1)
the size factor R_{RE}/R_B was found to be restrictive: GdB_2 is a
high-temperature phase and for a still larger R_{Sm}/R_B, SmB_2 can be
obtained by high temperature - high pressure - synthesis only.
From the close structural resemblance of REB_2 and RE_2TB_6, $RETB_4$-
phases it is thus conceivable that similar restrictions in size are
valid, which in turn explain the experimental observation, that the
larger rare earth metals La → Eu will not form with these structure
types (Fig. 2).

ACKNOWLEDGEMENT

This work was sponsored by the Austrian Science Foundation
(Fonds zur Förderung der Wissenschaftlichen Forschung in Osterreich)
under grant 3620. The authors thanks are due to Degussa (Deutsche
Gold und Silberscheideanstalt), Hanau for kindly supplying us the
noble metals.

REFERENCES

1. H. Nowotny, P. Rogl, in Boron and Refractory Borides, ed. M.V.
    Matkovich, N.Y. Springer (1977).
2. P. Rogl, H. Nowotny, J. Less C. Metals, 61, 39 (1979).
3. K. Hiebl, P. Rogl, J.M. Sienko, Inorg. Chem. 19(2), 3316 (1980).
4. P. Rogl, Mat. Res. Bull. 13, 519 (1978).
5. Yu. B. Kuz'ma, S.I. Svarichevskaya, Sov. Phys. Crystallogr.
    17, 569 (1972).
6. Yu. B. Kuz'ma, Sov. Phys. Crystallogr. 15, 312 (1970).
7. Yu. B. Kuz'ma, Dopov.Akad.Nauk. Ukr.RSR, Ser.A., 8, 756 (1970).
8. Yu. B. Kuz'ma, S.I. Svarichevskaya, Dopov.Akad.Nauk. Ukr.RSR,
    Ser.A., 34, 166 (1972).
9. G.F. Stepanchikova, Yu. B. Kuz'ma, Vestn. L'vov. Univ. Ser.
    Chim., 19, 37 (1977).
10. S.I. Mikhalenko, Vestn. Lvov. Univ.Ser. Chim., 16, 58 (1974).
11. I.P. Valovka, Yu. B. Kuz'ma, Dopov.Aka.Nauk. Ukr.RSR, Ser.A.,
    7, 652 (1975).
12. P. Rogl, H. Nowotny, Monatsh. Chem. 106, 381 (1975).
13. P. Rogl, H. Nowotny, Monatsh. Chem. 105, 1082 (1974).
14. P. Rogl, Mater. Res. Bull., 13, 519 (1978).
15. R. Sobczak, P. Rogl, J. Solid State Chem., 27, 343 (1979).
16. S.I. Mikhalenko, Yu. B. Kuz'ma, M.M. Korsukova, V.N. Gurin,
    Neorgan. Mater. 16, 1941 (1980).
17. N.F. Chaban, Yu. B. Kuz'ma, Neorgan. Mater. 13, 923 (1977).
18. G.F. Stepanchikova, Yu. B. Kuz'ma, Vestn.Lvov. Univ. Ser.Chim.,
    19, 37 (1977).
19. H.F. Braun, K. Yvon, Acta Cryst. 36B, 2400 (1980).
20. N.F. Chaban, Yu. B. Kuz'ma, Neorgan. Mater. 13, 923 (1977).
21. S.I. Mikhalenko, Yu. B. Kuz'ma, Dopov.Akad.Nauk. Ukr.RSR. Ser.
    A., p. 951 (1977).
22. Yu. B. Kuz'ma, N.S. Svarichevskaya, V.I. Fomenko, Neorgan.
    Mater. 9, 1542 (1973).
23. No compound formation observed.
24. P. Rogl, to be published.
25. E. Teatum, K. Geschneidner, J. Waber, LA-2345, U.S. Department
    of Commerce, Washington, D.C. (1960); R_U=1.73Å.

FAST DIFFUSION AND ELECTROTRANSPORT OF COBALT, IRON AND NICKEL IN

α-YTTRIUM

I. C. I. Okafor and O. N. Carlson
Ames Laboratory-DOE and Department of
Materials Science and Engineering
Iowa State University, Ames, Iowa 50011

ABSTRACT

Most rare earth metals including yttrium are commonly purified
by vacuum distillation. However, the vapor pressures of iron,
cobalt and nickel are close to that of yttrium making their separation from yttrium by that method rather difficult. This study was
carried out to determine the feasibility of removing these and
similar impurities from yttrium by the electrotransport process as
well as a continuation of our studies on the fast diffusion
phenomenon.

The yttrium metal used in this study was prepared at the Ames
Laboratory by the metallothermic reduction process as described by
Carlson et al. (1). An yttrium alloy containing 0.05% each of
cobalt, iron and nickel was prepared by arc melting. Portions of
this alloy and of the base yttrium were swaged into 0.25 cm diameter rods and butt-welded in tandem to form a composite specimen
that served as a diffusion couple in these experiments. The technique used in this investigation is similar to that described in a
recent paper by Weins and Carlson (2) on the fast transport behavior
of these same solutes in thorium.

The transport experiments were carried out in α-yttrium at
several temperatures in the range of 1290 to 1620 K for periods
ranging from one to nine hours. The concentration profile of the
specimen following each run was determined by a laser source mass
spectrometry technique similar to that described by Schmidt et al.
(3). All three solutes were found to exhibit fast diffusion and
fast electromigration behavior in α-yttrium. Plots of log D versus
1/T for the diffusivity data gave values for D_0 and the activation

Table I. Diffusion constants, activation energies for diffusion
          and activation energies for electrotransport for cobalt,
          iron and nickel in α-yttrium.

Solute	D_o $(m^2 s^{-1} x10^{-6})$	Q_D $(kJ mol^{-1} K^{-1})$	Q_E $(kJ mol^{-1} K^{-1})$
Cobalt	1.4	83.3	71.6
Iron	4.0	92.1	85.8
Nickel	5.8	96.5	95.0

energy, Q_D, for the diffusion equation $D = D_o \exp - (Q_D/RT)$. Values
of D_o and Q_D for each solute are given in Table I. Similar plots
for the electrotransport data in which log VT/j is plotted as a
function of reciprocal temperature (where V is the migration velocity and j is the current density) gave slightly lower values for the
activation energy for electrotransport, Q_E. These are also shown in
Table I for each solute. The earlier results of Murphy et al. (4)
for iron in α-yttrium are incorporated into the iron results with
good agreement being observed between the results of the two studies.

The effective valence of cobalt was found to range between
−7.5 and −6.8, iron between −4.2 and −2.5 and nickel between −4.5
and −2.2 over the temperature regime investigated. Some temperature
dependence in Z^* is noted for nickel and possibly for iron but not
for cobalt.

ACKNOWLEDGEMENTS

This work was supported by the U.S. Dept. of Energy, Basic
Energy Sciences. Special thanks are extended to R. J. Conzemius
for the analytical services. A paper describing this work in
greater detail has been submitted to the Journal of Less-Common
Metals.

REFERENCES

1. O. N. Carlson, J. A. Haefling, F. A. Schmidt and F. H.
    Spedding, J. Electrochem. Soc., 107:540 (1960).
2. W. N. Weins and O. N. Carlson, J. Less-Common Metals, 66:99
    (1979).
3. F. A. Schmidt, R. J. Conzemius, O. N. Carlson and H. J. Svec,
    Anal. Chem., 46:810 (1974).
4. J. E. Murphy, G. H. Adams and W. N. Cathey, Met. Trans.,
    6A:343 (1975).

TERNARY COMPOUNDS IN RE(Au,Ga)$_2$ AND RE(Ag,Ga)$_2$ ALLOYS

A. E. Dwight

Department of Physics, Northern Illinois University
DeKalb, Illinois   60115

A series of intermetallic compounds was examined, ranging
from R.E.Au$_2$ or R.E.Ag$_2$ to R.E.Ga$_2$, where R.E. includes Gd-Lu
and Sc. Alloys were arc melted under argon and homogenized at
700°C. Debye-Scherrer X-ray patterns were taken with Cu and Cr
radiation, and indexing was verified by comparing d_c, d_o, I_c and I_o.

The sequence of structure types is MoSi$_2$, CeCu$_2$, CaIn$_2$ and AℓB$_2$.
The selection of structure type is determined by electron concentration and relative atomic size. This note is primarily concerned
with the crystal structures of the equiatomic compounds, e.g. GdAuGa.
The diffraction patterns were obviously orthorhombic, and could be
indexed as either CeCu$_2$ or TiNiSi-types. Both structures are built
up from very similar flat layers in an abab sequence. Space groups
are Imma for CeCu$_2$ and Pnma for TiNiSi, which means that reflections
of the class h+k+ℓ≠2n(e.g. 111 and 102) may not be present for the
CeCu$_2$ type but may be visible for the TiNiSi-type. The 111 and 102
reflections are not visible on our patterns although intensities calculated for the TiNiSi-type structure show I_c for 111 in GdAuGa is
341 and for 111 in YAuGa is 567, the highest possible for this series.
It is concluded that the CeCu$_2$-type structure is the more probable.
Pt was substituted for Au to form GdPtGa, and in this compound the
111 reflection is strong, clearly proving the structure to be TiNiSi-
type. In the R.E.AgGa series the 111 line could not be seen, and the
CeCu$_2$-type is assumed. Fig. 1 shows a projection of the layers in
the a_o direction. Unit cell constants are given in Table 1.

ACKNOWLEDGEMENT

A portion of this work was performed at Argonne National
Laboratory, under auspices of the U.S. Department of Energy.

Table 1. Unit cell parameters for R.E.AuGa and R.E.AgGa compounds*

Compound	a_o	b_o	c_o	V/M
GdAuGa	4.502	7.161	7.733	62.3
TbAuGa	4.493	7.104	7.719	61.59
DyAuGa	4.469	7.083	7.693	60.88
YAuGa	4.474	7.075	7.721	61.09
HoAuGa	4.466	7.036	7.683	60.35
ErAuGa	4.455	6.995	7.677	59.8
TmAuGa	4.440	6.954	7.678	59.27
LuAuGa	4.441	6.885	7.656	58.52
GdAgGa	4.569	7.153	7.819	63.88
TbAgGa	4.542	7.119	7.796	63.03
DyAgGa	4.521	7.054	7.798	62.17
YAgGa	4.526	7.066	7.804	62.39
HoAgGa	4.526	7.001	7.802	61.8
ErAgGa	4.496	6.981	7.776	61.02
TmAgGa	4.492	6.914	7.782	60.43
LuAgGa	4.488	6.86	7.762	59.74

*a_o, b_o, c_o in Å, V/M in Å^3. a_o, b_o, c_o ± 0.010

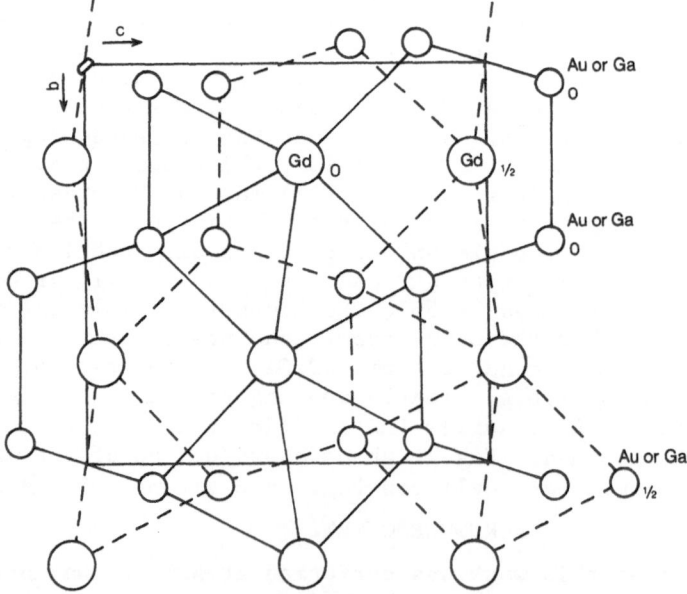

Fig. 1. GdAuGa (CeCu$_2$-type) viewed down [100].

MAGNETIC PROPERTIES OF ErM_3, M = (Ni, Fe, Co)

B. Decrop and J. Deportes, Laboratoire L. Neel, CNRS,
BP 166X, 38042 Grenoble, France; B. Kebe, C. Crowder,
and W. J. James, Graduate Center for Materials Research,
Univ. of Missouri-Rolla, Rolla, MO 65401; W. Yelon,
Univ. of Missouri Research Reactor, Columbia, MO 65201

INTRODUCTION

The RM_3 compounds crystallize in the rhombohedral $PuNi_3$ type structure (1). There are two non-equivalent R.E. sites, R_I (3a) and R_{II} (6c), and three M sites, 3a, 6c, and 18h. Previous studies give evidence of complex magnetic structures for the Er compounds; at 4.2K $ErNi_3$ has a noncollinear magnetic structure (2) while $ErCo_3$ and $ErFe_3$ have collinear $\vec{c}$ axis structures (3,4). At 42K the latter compound exhibits a spin reorientation. The effect of temperature and applied magnetic fields on the magnetic structures of these compounds has not been investigated. Accordingly, we have undertaken a comparative study using single crystal magnetization measurements and powder neutron diffraction experiments.

RESULTS

Single crystals of each compound were prepared using the Czochralsky technique. The polycrystalline samples were prepared by melting the elements in the proper stoichiometric amounts in an inductively coupled water cooled boat. X-rays and neutron diffraction patterns confirmed the absence of the $ErNi_2$ and Er_2Ni_7 phases often observed as impurities. Magnetization measurements were performed at the Service National des Champs Intenses in Grenoble using magnetic fields up to 150 kOe. The temperature was varied from 4.2 to 300K.

$ErNi_3$: Figure 1 shows some magnetization measurements with the magnetic field applied along the $\vec{c}$ axis of the rhombohedral structure and perpendicular to that direction ($\vec{a}$). At 4.2K, a spontaneous

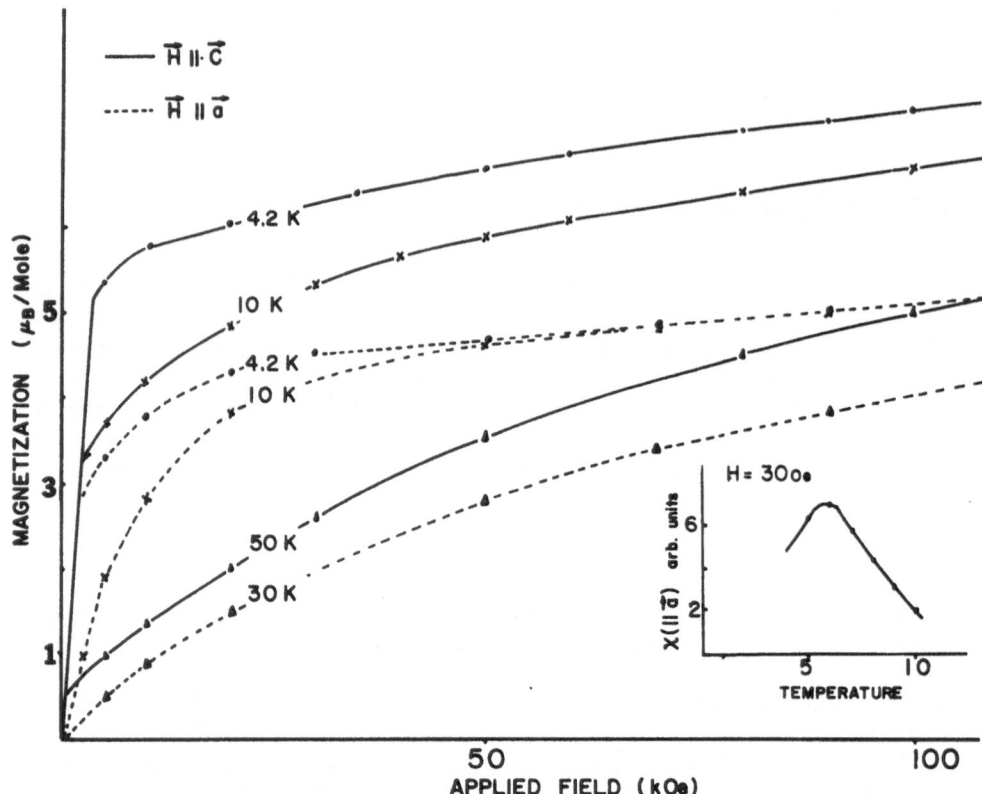

Figure 1.   ErNi$_3$: Magnetization versus H (insert:
             susceptibility measurements).

magnetization is observed along $\vec{c}$ and in the basal plane of the structure.   At zero field the resultant of the moments is therefore in an
intermediate direction between $\vec{c}$ and the basal plane.   The compound
is not saturated even under 150 kOe due to the strong anisotropy of
Er$_3^+$.   At 10K, no spontaneous magnetization is observed in the basal
plane. The resulting magnetization is therefore parallel to $\vec{c}$.   Its
amplitude decreases rapidly with increasing temperature and vanishes
at 60K, the ordering temperature of ErNi$_3$.

The susceptibility perpendicular to $\vec{c}$ measured under 30 Oe decreases sharply at 6.5K (Fig. 1 insert) which corresponds to the temperature where the magnetization direction moves toward $\vec{c}$.

In order to determine the variation of the magnetic structure
with temperature, we have undertaken powder neutron diffraction
experiments at the University of Missouri Research Reactor at 5.5, 10,
25, and 100K.

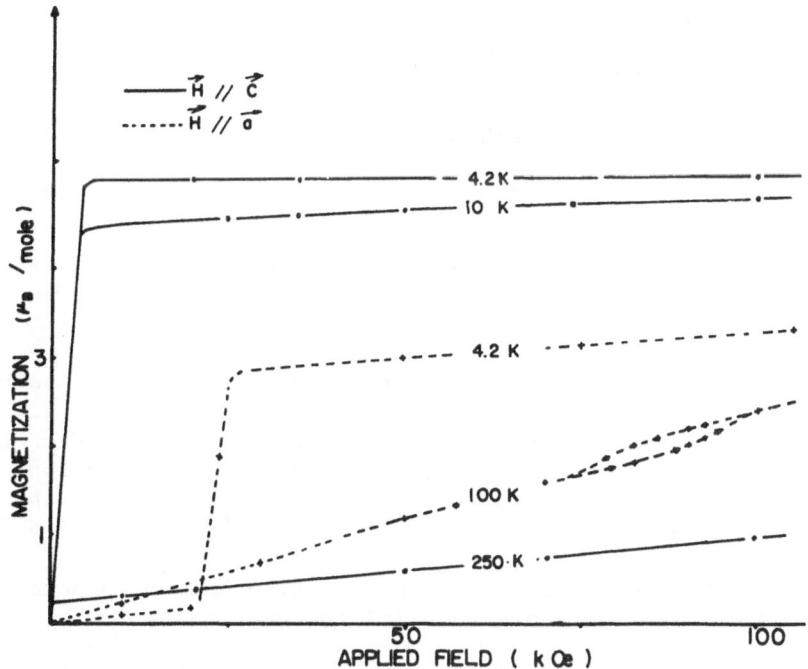

Figure 2.   ErCo₃: Magnetization curves versus H.

At 100K, nuclear peaks characteristic of the PuNi₃ structure are observed. Below T_c magnetic contributions appear. However at 25 and 10K there is no magnetic contribution to reflections such as 001; moments are therefore parallel to $\vec{c}$ and the structure is collinear. At 5.5K a magnetic contribution appears at the 001 positions; the moments are no longer along $\vec{c}$. The refinement of the diffraction pattern leads to a noncollinear structure in agreement with that obtained at 4.2K by Paccard et al. (2). Table I shows the moments resulting from the refinements. Note that the moment on the Er_{II} site decreases more rapidly than that on the Er_I site.

ErFe₃: Magnetization and neutron diffraction measurements (4) show that the compound has a collinear magnetic structure at low temperature (T < 42K); the moments are along $\vec{c}$. A magnetic field applied perpendicular to that direction induces a sudden increase

Table I.   Thermal variation of Er moments.

	$T°$ (K)	*4.2K	5.5	10	25
Moments (μ_β)	Er_I	8.5 ± 0.8	8.4 ± 0.8	6.3 ± 0.5	4.6 ± 0.5
	Er_{II}	7.5 ± 0.5	5.3 ± 0.4	3.4 ± 0.2	1.4 ± 0.2

*from Ref. (2)

of the magnetization at a critical field H_c which is temperature dependent. H_c is 22 kOe at 4.2K and decreases to zero at 42K where the spontaneous magnetization turns suddenly toward an intermediate direction. The magnetic structure at zero field is then noncollinear. The angle θ_1 of the Er_I moment with $\vec{c}$ is smaller than θ_2 of the Er_{II} moments. At 77K $\theta_1 = 38°$, $\theta_2 = 58°$, and $\theta_{Fe} = 57°$.

Above 250K the spontaneous magnetization has no component along $\vec{c}$; the moments are therefore parallel to the b axis.

$ErCo_3$: The magnetization measurements are shown in Fig. 2. Below $T_c = 401K$, the direction of easy magnetization is always along the c-axis. A magnetic field applied in the basal plane induces a transition similar to that observed in $ErFe_3$, but the critical field increases with the temperature and a large hysteresis is associated with the transition.

DISCUSSION

The local symmetry of the two R.E. sites is axial; however, in a point charge model their different surroundings assign opposite signs to the second order terms of their crystal field Hamiltonian. As a result, the easy direction of the Er_I moments is parallel to $\vec{c}$ while that of Er_{II} is perpendicular to $\vec{c}$. This situation favors the stability of a noncollinear configuration.

In $ErNi_3$, Ni atoms are not magnetic; the magnetic order results solely from the interaction between Er moments.

If $K_2 > 0$ and $K_2 < 0$ are assumed to be the second order terms of the local anisotropy of Er_I and Er_{II} sites and w the exchange energy between Er_I and Er_{II} moments, the minimization of the total energy (exchange plus anisotropy energy) leads to the existence of a noncollinear configuration if

$$\left| \frac{w}{2K_1} + \frac{w}{2K_2} \right| < 1 \qquad\qquad (1)$$

Below 6.5K, this condition is satisfied. The decrease of the moments and therefore of the anisotropy with temperature is much more pronounced on Er_{II} sites than on Er_{I} sites; accordingly, at 10K the anisotropy of Er_{I} dominates and the structure is now collinear and along $\vec{c}$. All refinements removing constraints on a Ni moment gave best agreement when the Ni moment was zero. Therefore, it is unlikely that the inability to saturate $ErNi_3$ is caused by polarization of Ni.

In $ErFe_3$ and $ErCo_3$ the exchange field on the Er moments comes mainly from Er-3d interactions which are stronger than Er-Er interactions and thus stabilize a collinear structure at low temperature.

Magnetization measurements for YCo_3 and $GdCo_3$ show a strong c axis anisotropy on Co moments ($H_A \simeq 70$ kOe). This c axis anisotropy combined with that of the Er_I moment always favors an easy direction of magnetization along $\vec{c}$.

In $ErFe_3$, the anisotropy of the Fe moments is small. Since at 4.2K the moments are along $\vec{c}$, the anisotropy of the Er_I moments must be preponderant. Its relative influence decreases with an increase of temperature and above 42K the magnetic structure becomes noncollinear. The sudden reorientation of the moments at 42K which corresponds to a transition between two configurations of the same energy can be induced below that temperature by a magnetic field applied in the basal plane. This transition cannot be explained by a simple model of local anisotropy which would require a progressive rotation of the magnetic moments.

Crystal field calculations are in progress to see if the transition results from changes in the crystal field energy levels.

ACKNOWLEDGEMENTS

The authors wish to thank the Department of the Army and the National Science Foundation for support under Grants DAAG-29-80-C-0084 and NSF INT 7826549.

REFERENCES

(1) J. S. Smith and D. A. Hansen, Acta. Cryst., 19:1819-24 (1965).
(2) D. Paccard, J. Schweizer, and J. Yakinthos, J. de Physique, C1, 32:663 (1971).
(3) J. K. Yakinthos and J. Rossat-Mignot, Phys. Stat. Solidi (b), 50:747 (1972).
(4) B. Kebe, W. J. James, J. Deportes, R. Lemaire, W. Yelon, and R. K. Day, J. Appl. Phys. (in press, 1981).

TRANSPORT AND MAGNETIC PROPERTIES OF THE $Gd_4(Co_{1-x}Ni_x)_3$ SERIES ($o \leqslant x \leqslant 0.2$)

E. Gratz, G. Hilscher, H. Kirchmayr, H. Sassik

Institut für Experimentalphysik, Technical University

Karlsplatz 13, A-1040 Vienna, Austria

INTRODUCTION

Recently we investigated the magnetic properties and the transport properties of the $(Gd,Y)_4Co_3$ pseudobinary system (hexagonal Ho_4Co_3 structure (1)) in order to study the influence of the substitution of magnetic Gd^{3+} by the nonmagnetic Y^{3+}. We found that the variation of most of the physical parameters is due mainly to the change of the 4f moment concentration. We have, for example, found that the magnetic ordering temperature decreases nearly linearly with decreasing Gd content. Furthermore, the induced magnetic moment on the Co sites (about 0.6 μ_B/Co atom for Gd_4Co_3) decreases rapidly and remains constant in the high concentration range (1).

The reason for conducting the present investigation of the isostructural $Gd_4(Co,Ni)_3$ pseudobinary system is threefold:

i) to investigate how much Co can be substituted by Ni before the hexagonal Ho_4Co_3 structure becomes unstable, because it is known that Gd_4Ni_3 does not exist;

ii) to investigate the influence of the variation of the 3d electron concentration due to the Ni substitution on physical parameters such as Curie temperature, magnetization and resistivity;

iii) to discuss why the structure becomes unstable when comparing this system with $Gd(Co_{1-x}Ni_x)$ (2).

RESULTS AND DISCUSSION

The samples were prepared by high frequency induction melting
in a protective argon atmosphere. After an annealing procedure of
200 hours at 550°C, pure phase samples could be formed only over the
concentration range $0 \leqslant x \leqslant 0.2$. For samples with $x \geqslant 0.25$ traces
of the Gd(Co,Ni) phase (orthorhombic CrB-structure) were identified.
Therefore the present investigations were performed on samples with
$0 \leqslant x \leqslant 0.2$. Fig. 1 shows the concentration dependence of the
magnetic ordering temperature for this system together with data
obtained for the system $(Gd,Y)_4Co_3$. In contrast to the T_C vs. x
behaviour of the latter, hardly any change of T_C vs. x were found
for $Gd_4(Co,Ni)_3$. The linear increase for $(Gd,Y)_4Co_3$ on the one
hand and the constancy of T_C for $Gd_4(Co,Ni)_3$ on the other implies
that the large magnetic R-moments are mainly responsible for the
magnetic behaviour in both series. The spontaneous magnetization
for $Gd_4(Co,Ni)_3$ (also shown in Fig. 1) exhibits a small increase
with increasing Ni concentration. Such a behaviour can be explained
under the assumption that with increasing Ni concentration a common
3d band becomes progressively filled, which manifests itself in a
reduction of the induced magnetic moment on the 3d-band. Because
of the antiparallel alignment of the 3d moment with that of Gd, the
bulk magnetization increases with this reduction of the 3d moment.
Temperature dependence of the electrical resistivity is given in
Fig. 2. Although it is not possible to give a detailed explanation
of the shape of the ρ vs. T curve the following comments can be
made:

   i) Both localized magnetic moments and the itinerant magnetic
      moments (due to the unfilled 3d-band) contribute to the
      resistivity. The kink at T_C (T_C is taken from initial
      susceptibility measurements) in the low Ni concentration
      range is caused by the breakdown of order in the arrangement
      of the localized moments at T_C.

  ii) The magnitude of ρ at room temperature in samples with
      $0.05 \leqslant x \leqslant 0.2$ is about 250 $\mu\Omega$cm. This is much higher than
      the limit of about 150 $\mu\Omega$cm which is considered a critical
      value in compounds containing d-elements (3) which, how-
      ever, seems to be correct only for nonmagnetic compounds
      (as especially Nb_3Sn).

 iii) The kink at T_C mentioned in (i) becomes progressively
      suppressed with increasing residual resistivity ρ_0 and
      simultaneously increasing ρ at T_C.

These measurements clearly show that the critical magnitude of the
resistivity in magnetic compounds with 3d elements is in the range
of 250 $\mu\Omega$cm. About the same magnitude is found in other rare earth
compounds with a complex crystal structure containing transition

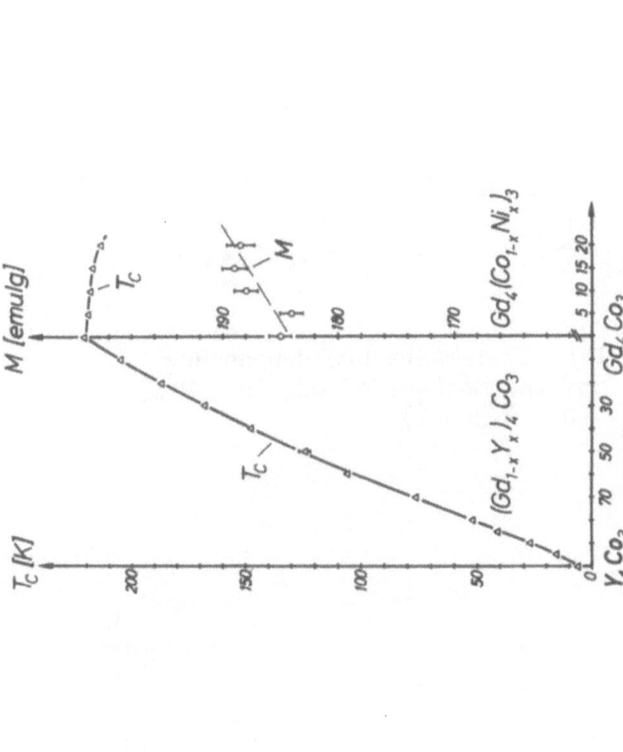

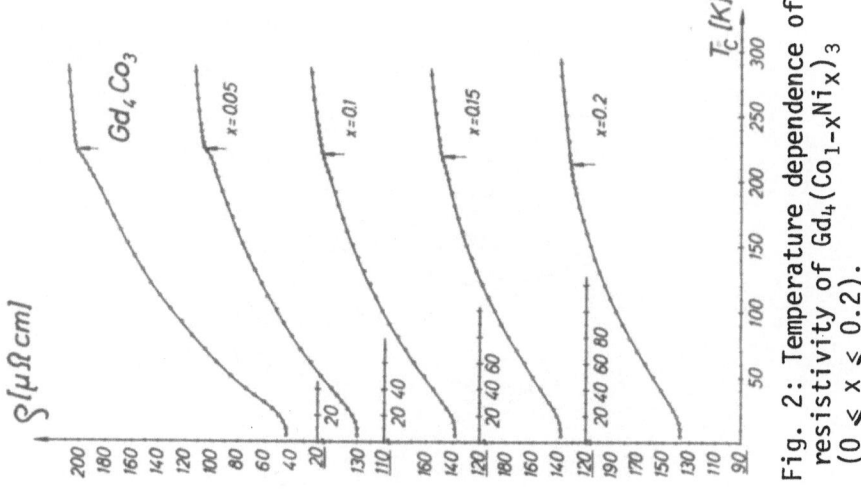

Fig. 2: Temperature dependence of resistivity of $Gd_4(Co_{1-x}Ni_x)_3$ (0 ⩽ x ⩽ 0.2).

Fig. 1: Compositional dependence of T_C in $(Gd,Y)_4Co_3$ and $Gd_4(Co_{1-x}Ni_x)$ (0 ⩽ x ⩽ 0.2), respectively magnetization at 7T and 4.2 K.

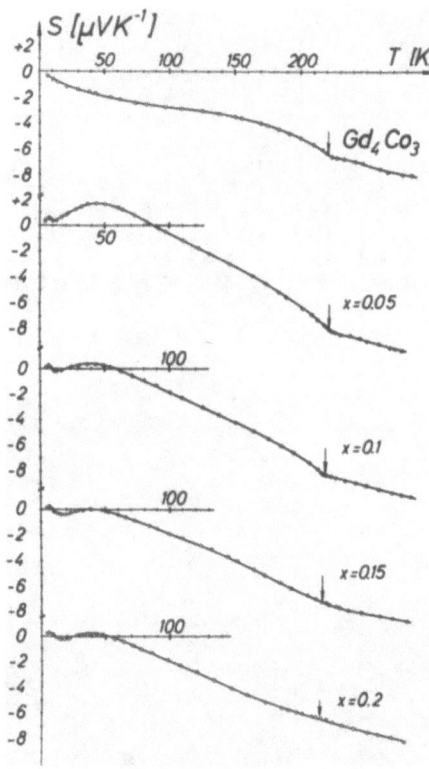

Fig. 3: Temperature dependence
of thermopower of $Gd_4(Co_{1-x}Ni_x)_3$
($0 \leqslant x \leqslant 0.2$).

metals with unfilled 3d shells (see e.g. the resistivity of R$_6$(Fe$_x$Mn$_{1-x}$)$_{23}$ compounds in (4)). The disappearance of the kink at T$_c$ is closely related to the fact that near the "high resistivity limit" any structure (e.g. a kink at T$_c$) in the ρ vs. T curve becomes suppressed.

The temperature dependence of the thermopower (S) is given in Fig. 3. These measurements show that the thermopower is a very sensitive function of the 3d electron concentration. The substitution of only 5% Co by Ni changes the S vs. T curves drastically. No such drastic effect could be found in the isostructural (Gd,Y)$_4$Co$_3$ series where the magnetic Gd^{3+} ion is substituted by the nonmagnetic Y^{3+} ion.

In this and in a previous paper (2) we have studied pseudo-binary systems in which the CrB-structure of Gd(Co$_x$Ni$_{1-x}$) and the Ho$_4$Co$_3$-structure in Gd$_4$(Co$_{1-x}$Ni$_x$)$_3$ becomes unstable with increasing x. In both cases we assume that the instability of the structure will be reached if due to the increase (in Gd$_4$(Co,Ni)$_3$) and the decrease (in Gd(Co,Ni)) of the 3d electron concentration, the Fermi level reaches the d-band edge.

REFERENCES

1. E. Gratz, V. Sechovsky, E.P. Wohlfarth and H.R. Kirchmayr, "The Magnetic and Transport Properties of the Compounds (Gd,Y)$_4$Co$_3$", J. Phys. F 10:2819 (1980).
2. E. Gratz, G. Hilscher, H.R. Kirchmayr and H. Sassik, "Transport and Magnetic Properties of Gd(Co$_x$Ni$_{1-x}$) Series (0 $\leqslant$ x $\leqslant$ 0.3)" in The RARE EARTHS IN MODERN SCIENCE AND TECHNOLOGY" Vol. 2, G.J. McCarthy, J.J. Rhyne and H.B. Silber, (Eds.), Plenum Publishing Corporation p. 327 (1980).
3. P.B. Allen in Superconductivity in d- and f-band Metals, H. Suhl and M.B. Maple (Eds.), Academic Press Ind., New York, p. 291 (1980).
4. E. Gratz and H.R. Kirchmayr, "Temperature Dependence of the Electrical Resistivity of R$_6$(Fe$_{1-x}$Mn$_x$)$_{23}$ Compounds in the Temperature Range 4.2 to 300 K" J. Mag. Mag. Mat. 2:187 (1976).

STRUCTURES AND MAGNETISM OF SOME POLYCOMPONENT 2:17 RARE EARTH-TRANSITION METAL SYSTEMS[*]

W. E. Wallace, M. Merches and R. S. Craig

Department of Chemistry, University of Pittsburgh

Pittsburgh, PA   15260

ABSTRACT

Structures and magnetic properties of the polycomponent systems $Er_{2-x}Pr_xCo_{13}(Fe,Mn)_4$ are presented. Replacement of Er by Pr and Co by Fe and/or Mn both increase the anisotropy fields; the effects are not quite additive.

I.   INTRODUCTION

The present study is a continuation of a series of investigations carried out in recent years in this laboratory dealing with 2:17 stoichiometry materials which are of interest in connection with the fabrication of high energy permanent magnets (1). In two previous studies (2,3) the R_2Co_{17} systems with R = Er, Tm and Yb have been modified by partially replacing Co with Fe and/or Mn to increase the anisotropy field. Also, systems have been studied in which Er, Tm or Yb was partially replaced by Pr, again to enhance H_A. The present work involved the study of alloy systems in which there was replacement in both the rare earth and Co sublattice. One of the objectives of the present work was to ascertain whether the effects of two kinds of replacement are additive. To anticipate what is indicated in greater detail below the effects on H_A of the two kinds of substitution are mutually reinforcing but not quite additive.

---

[*]The present work was supported by a grant from the Army Research Office.

II. EXPERIMENTAL DETAILS

The methods used in preparation of the alloys and in making the magnetic measurements are those that are now standard in this laboratory. They have been described elsewhere (3).

III. RESULTS AND DISCUSSION

The structures and magnetic properties for the systems studied are listed in Table 1.

While the substitution of Pr for Er is essentially without effect on T_c, the same is not true for μ_{sat}, the saturation magnetization. The data in Fig. 1 and in Table 1 show a progressive rise in μ_{sat} with rising Pr content in the system. This is expected since the Er-Co coupling is antiferromagnetic whereas Pr normally couples ferromagnetically with Co (4). The increment in moment expected is 12.2 μ_B per Pr^{3+} (i.e., $\mu_{Pr}-[-\mu_{Er}]$) if Pr and Er are carrying the full moment of the free tripositive ion. The

Table 1. Structures and Magnetic Properties of Some Polycomponent 2:17 Systems[a]

	a (Å)	c (Å)	T_c(°K)	μ_{sat} (μ_B/fu)	H_A(298°K) (kOe)
$Er_2Co_{13}Fe_4$[b]	8.325	8.179	1190	14.75	42
$Er_{1.6}Pr_{0.4}Co_{13}Fe_4$[b]	8.351	8.194	1180	19.97	45
$Er_{1.2}Pr_{0.8}Co_{13}Fe_4$	8.369	8.207	1170	23.88	39
$Er_{0.8}Pr_{1.2}Co_{13}Fe_4$[b]	8.389	8.226	1150	25.03	
$Er_{0.4}Pr_{1.6}Co_{13}Fe_4$[c]	8.424	12.360	1142	30.16	
$Er_{0.4}Pr_{1.6}Co_{13}Fe_3Mn$[c]	---	---	1066	30.99	
$Er_2Co_{13}Mn_4$	8.365	8.235	610	5.92	37
$Er_{1.6}Pr_{0.4}Co_{13}Mn_4$[b]	8.393	8.241	604	8.90	35
$Er_{1.2}Pr_{0.8}Co_{13}Mn_4$[b]	8.408	8.246	598	14.54	
$Er_{0.8}Pr_{1.2}Co_{13}Mn_4$	8.435	12.420	592	18.66	
$Er_{0.4}Pr_{1.6}Co_{13}Mn_4$[c]	8.468	12.418	586	23.60	27
$Er_{0.4}Pr_{1.6}Co_{13}Mn_3Fe$[c]	8.477	12.409	---	---	
$Er_{0.4}Pr_{1.5}Co_{13}Mn_2Fe_2$[c]	8.471	12.364	979	28.90	28

a. All these systems exhibited uniaxial anisotropy.
b. Th_2Ni_{17} structure.
c. Th_2Zn_{17} structure.

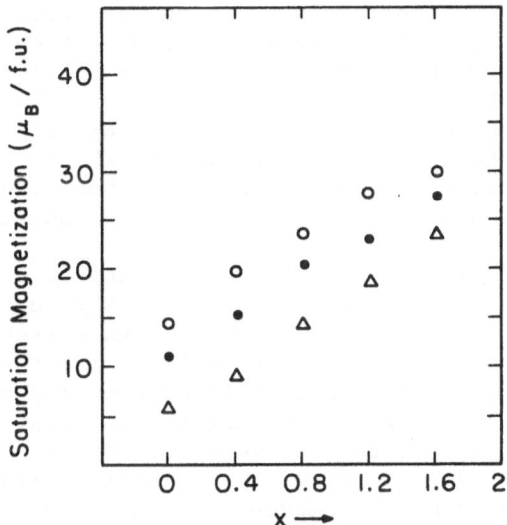

Fig. 1. Saturation magnetization at 4.2 K for several polycomponent systems. ● $Er_{2-x}Pr_xCo_{17}$, o $Er_{2-x}Pr_xCo_{13}Fe_4$, Δ $Er_{2-x}Pr_xCo_{13}Mn_4$.

The increments observed for the three sets of systems shown in Fig. 1 are 10.4, 10.6 and 11.2 μ_B per Pr^{3+}, which are close to the expected value. This makes it clear that Er and Pr are coupling antiparallel, resembling the behavior observed by Wallace, Volkmann and Hopkins (5) for (R,R´)Co$_5$ systems and by Merches, Wallace and Craig (3) for (R,R´)$_2$Co$_{17}$ systems. In those studies R and R´ were observed to couple antiparallel when one rare earth was light and the other was heavy.

The data in Table 1 show that the replacement of Fe by Mn in the polycomponent systems results in a striking decrease in μ_{sat}. The saturation moments of Er_2Co_{17} and $Er_2Co_{13}Fe_4$ are 11.22 and 14.75 μ_B/fu, respectively. From these data Fe appears to be carrying a moment of about 2.6 μ_B in this ternary. If so, replacement of 4 atoms of Fe by 4 atoms of Mn would result in a decline in moment of about 10 μ_B if Mn is not carrying a moment. The difference in moment between $Er_2Co_{13}Fe_4$ and $Er_2Co_{13}Mn_4$ is about 11 μ_B, which suggests that Mn is acting essentially as a non-magnetic diluent in the systems. In the quaternary systems containing Pr the corresponding decline is less, implying that Mn is supplying some moment which is coupled parallel to that of Co.

At room temperature the several samples studied exhibited uni-axial anisotropy. The anisotropy fields are given in Table 1. These are to be compared with H_A values for Er_2Co_{17} and $Er_{1.6}Pr_{0.4}Co_{17}$ obtained earlier (3) - 17 and 24 kOe, respectively. Substitutions on both sublattices increase H_A for reasons which have been described in the earlier studies (2,3). The present work shows that the effects of the two kinds of substitution are mutual-ly reinforcing although they are not exactly additive.

From the data in Table 1 it is found that over half of the systems studied have maximum energy products in excess of 30 MGOe. However, about half of the systems with high energy products have very low H_A values, < 25 kOe. The systems with the largest anisot-ropy fields, i.e., in the range 40 to 45 kOe, may not be suitable systems for permanent magnet fabrication, at least until a way is found to produce in them an intrinsic coercivity which is an appre-ciable fraction (1/3 or more) of its anisotropy field. Perhaps inclusion of Zr as a minor component, as used in Sm_2Co_{17}-based systems (6), will bring this about. Experiments along this line are currently under way.

REFERENCES

1. See W. E. Wallace and K. S. V. L. Narasimhan, in The Science and Technology of Rare Earth Materials, E. C. Subbarao and W. E. Wallace, eds. Academic Press, New York (1980), p. 393, for references to earlier work on 2:17 systems.
2. W. E. Wallace, M. Merches, G. K. Shenoy and P. J. Viccaro, Sol. State Commun., submitted.
3. M. Merches, W. E. Wallace and R. S. Craig, J. Mag. Mag. Mat., submitted.
4. See W. E. Wallace, Rare Earth Intermetallics, Academic Press, Inc., New York (1973), chapter 10.
5. W. E. Wallace, T. V. Volkmann and H. P. Hopkins, Jr., J. Sol. State Chem. $\underline{3}$, 510 (1971).
6. T. Ojima, S. Tomizawa, T. Yoneyama and T. Hori, Jap. J. Appl. Phys. $\underline{16}$, 671 (1977) and IEEE Trans. Mag. $\underline{MAG-13}$, 1317 (1977).

EVIDENCE FOR THE NONCOLLINEARITY OF THE MAGNETIC STRUCTURE OF Er_6Mn_{23}

B. Kebe, C. Crowder, and W. J. James, Department of
Chemistry and GCMR, University of Missouri-Rolla, Rolla,
MO 65401; J. Deportes and R. Lemaire, Laboratoire L.
Neel, BP 166X, 38042 Grenoble, France; W. Yelon,
University of Missouri Research Reactor, Columbia, MO
65201

INTRODUCTION

In the R.E.-3d intermetallic compounds, the coupling of the
spin moments is generally antiparallel. This leads to ferromagnetic configurations in the light rare earths and ferrimagnetic
arrangement in the heavy rare earths as observed for the Fe, Co,
and Ni compounds. The situation appears to be different in Mn compounds. From polycrystalline magnetization measurements of R_6Mn_{23}
compounds, De Savage et al. (1) have proposed ferromagnetic structures for all the compounds excepting Gd_6Mn_{23}. Single crystal magnetization measurements (2) suggest a noncollinear arrangement of
the R.E. moments at low temperatures. To give more evidence to
this noncollinearity, we have studied the magnetic structure of
Er_6Mn_{23} from 1.8K to 300K by neutron diffraction.

The crystallographic structure of R_6Mn_{23} compounds has already
been described elsewhere (2). The Er compound exhibits ferromagnetic behavior (T_C = 486K). Single crystal magnetization measurements have been made along the three principal crystallographic
directions [100], [110], and [111]. At 4.2K [111] is the easy
direction of magnetization, but even at 150 kOe, saturation is not
achieved. Magnetization measurements along [100] and [110] reveal
a large magnetocrystalline anisotropy. Above 100K the anisotropy
disappears, and a large susceptibility is superposed upon the
spontaneous magnetization.

EXPERIMENTAL

Polarized neutron studies on the isotype compound Y_6Mn_{23} (3)

(only Mn is magnetic) give evidence for a ferrimagnetic coupling of Mn moments. Polarized neutron studies on a Er_6Mn_{23} single crystal at 300K have been carried out. The [111] direction was maintained vertical and 20 reflections collected. The nuclear and magnetic contributions to the intensities were separated by measuring the efficiency ratio and intensities. The thermal variation of the magnetic structure was deduced from powder neutron diffraction patterns collected on the multicounter D_{1B} at the Institut Laue-Langevin (ILL). The sample was rotated continuously to avoid preferred orientation. The temperature was varied from 1.8 to 300K. In the meantime a similar study was performed at the University of Missouri Research Reactor at 8.5 and 77K.

RESULTS

The polarized neutron results confirm the collinear structure at 300K. The moments on the four different Mn sites (4b, 24d, $32f_1$, $32f_2$ are given in Table I. The coupling of the moments in the different Mn sites is the same as observed for Y_6Mn_{23}. The Mn moments deduced in Er_6Mn_{23} are smaller than those in the Y compound; the Er moment is 0.5 μ_B at 300K.

All the observed reflections are characteristic of the Th_6Mn_{23} type structure. Only slight variations are observed in the intensities of the different reflections between 300 and 100K. Below 100K a pronounced increase in the variation of the intensities (Fig. 1) is observed resulting mainly from the magnetic contribution of the Er atoms. A collinear model does not give a satisfactory fit to the neutron data. At 4.2K we have attempted to fit

Table I. Thermal Variation of Er and Mn Moments

	Y_6Mn_{23}		Er_6Mn_{23}		
	4.2K*	300K*	300K*	100	4.2
R	--	--	.50 (5)	1.35 (6)	7.5 (1)
Mn_1 (4b)	-2.81 (12)	-2.25 (9)	-1.88 (9)	-2.5 (2)	-2.6 (2)
Mn_2 (24d)	-2.07 (4)	1.72 (4)	-1.44 (4)	-1.6 (1)	-1.7 (1)
Mn_3 ($32f_1$)	1.79 (4)	1.52 (3)	1.21 (4)	-1.6 (1)	1.6 (1)
Mn_4 ($32f_2$)	1.77 (4)	1.27 (3)	1.05 (3)	-1.5 (1)	1.5 (1)

*polarized neutron diffr. () standard deviation.

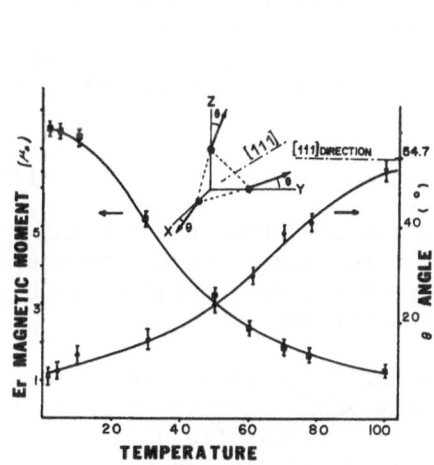

Figure 1. Thermal variation of the intensity of some peaks.

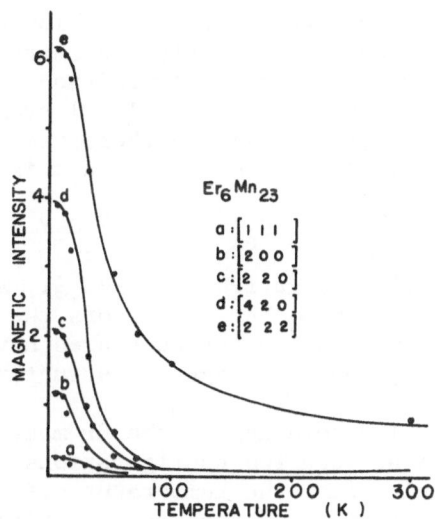

Figure 2. Variation of $|M_{Er}|$ and θ (angle of M_{Er} to [100] with temperature.

our data with a noncollinear model keeping the same Mn sub-lattice found for Y$_6$Mn$_{23}$. The best fit is obtained with the Er moments inclined at $\theta = 9°$ from the cube axis on which the Er lies. The Er moment is 7.5 μ_β, considerably less than the free ion value ($g_J \mu_\beta = 9 \mu_\beta$). However, a better fit is obtained by assigning smaller values to Mn moments (Table I) with no significant change in θ. The thermal variation of the Er moment and its angle with the cube axis is shown in Fig. 2.

DISCUSSION

The Curie temperatures of Er$_6$Mn$_{23}$ and Y$_6$Mn$_{23}$ are of the same order of magnitude. This shows that the Mn-Mn interactions are dominant; they stabilize the strong ferrimagnetic coupling of the Mn sites. In the Mn compounds the coupling between Mn atoms is very sensitive to the distance of separation. Due to the lanthanide contraction the Mn-Mn distances are smaller for the Er compound than for Y$_6$Mn$_{23}$; accordingly, a slight decrease of the Curie temperature and therefore smaller Mn moments are observed in the Er compound. Moreover, in Lu$_6$Mn$_{23}$ where even smaller Mn-Mn distances are expected, the Curie temperature and the spontaneous magnetization are indeed lower ($T_C = 301K$, $\sigma_S = 9.2 \mu_\beta$).

The magnetic configuration of Er atoms is a result of crystal field and exchange field effects on the Er moments. As a consequence of the special position of the Er atoms on the cube axis, there results an axial local symmetry. The second order terms of the crystal field are therefore dominant. In the Stevens notation the Hamiltonian of an Er atom is

$$H = s \, \mu_B \, H_{ex} + \alpha_J \, V_2^0 \, O_2^0 \tag{1}$$

$\alpha_J \, V_2^0$ is negative in Er_6Mn_{23}. At low temperatures, the crystal field term $\alpha_J \, V_2^0 \, O_2^0$ favors the [100] direction. The Er atoms on the three different cube axes have easy directions normal to each other, thus the magnetic structure is noncollinear.

The rotation of the Er moments is not associated with a transition between two configurations of nearly equal energy. Rather, it results from the temperature effect on the crystal field levels in the presence of the exchange field. This result is confirmed by our specific heat measurements which show no anomaly.

We have attempted to fit the variation of the spontaneous magnetization along [111] with the temperature using the exchange field H_{ex} = 30 kOe (4). The best fit is obtained for $\alpha_J \, V_2^0$ = 1.2K; however, the agreement at low temperature is not satisfactory and must result mainly from the negative Er-Er interaction that we have neglected.

ACKNOWLEDGEMENTS

The authors wish to thank the Department of the Army and the National Science Foundation for support under DAAG-29-80-C-0084 and NSF INT 7826549.

REFERENCES

(1) B. F. De Savage, R. M. Bozorth, F. E. Wang, E. R. Callen, J. Appl. Phys., 36:992 (1965).
(2) K. Hardman, W. J. James, J. Deportes, R. Lemaire, and R. Perrier de la Batie, J. de Phys., Coll. C5, 40:204 (1979).
(3) A. Delapalme, J. Deportes, R. Lemaire, K. Hardman, and W. J. James, J. Appl. Phys., 50(3):1987 (1979).
(4) G. Hilscher and H. Rais, Phys. F:Metal Phys., 8 (3):511 (1978).

MÖSSBAUER INVESTIGATION OF THE EFFECT OF ANNEALING ON THE

ORDERING TEMPERATURE OF AMORPHOUS DyFe$_2$

C. Bucci: American Univ., Washington, DC 20016 and Naval
   Surface Weapons Ctr., Silver Springs, MD 20910
E. Bauminger: Hebrew University, Jerusalen, Israel
H. Savage: Naval Surface Weapons Ctr., Silver Spring
   MD 20910

An important feature of several amorphous metallic alloys is the so-called "structural relaxation" (1) which occurs, in general, at relatively high temperatures where structural changes are likely to be thermally activated. As a consequence of such a relaxation, considerable variations in the mechanical, magnetic, and transport properties are observed even though the systems are still considered amorphous.

In an amorphous rare-earth iron alloy, DyFe$_2$, we investigated the magnetic behavior by using Mössbauer spectroscopy in a temperature region between 300K and 700K, where structural relaxation is expected to take place. The hyperfine fields of ^{57}Fe and ^{161}Dy nuclei (values and relative spreads)(2), the magnetic ordering temperature (3), magnetization and coercivity (4) have been reported on this system, particularly at low temperatures. The interpretation of these properties, when related to the crystalline counterparts, is still subject to a great deal of speculation. In the presence of topological and/or chemical disorder one deals with spatial fluctuations of exchange interactions (7) and local anisotropy, with rare-earth iron charge transfer (5) and with an enhanced localization of electron states (6). The comparison between the amorphous and the crystalline state apparently involves a complex admixture of such different effects. It is hoped that small changes "within" the amorphous phase can be more directly correlated to the fundamental interaction in the system.

In this experiment the ^{57}Fe Mössbauer absorption was measured. One can also examine ^{161}Dy spectra, but this nucleus was not observed since the 4f-shell of Dy is well shielded from the influence of the neighborhood and the effects of structural relaxations are expected

to be more evident on the magnetic moment and exchange interactions
of the Fe atoms.

The amorphous bulk-sputtered $DyFe_2$ samples were ground into
powders and compressed in a boron nitride holder which was previously outgassed at 850K. Mössbauer spectra taken between 80K and
300K confirmed that the hyperfine field, its distribution and the
ordering temperature agree quite well with previous measurements
on bulk-sputtered samples: H(hf) = 217 kOe with an average spread
of 35% and T_c = 295K $\pm$ 2K. The calculated spectra used to fit the
data were Zeeman sextets with asymmetric Lorentzian lines.

The samples were subjected to a series of annealings in a
vacuum furnace at increasing temperatures, T_a, for fixed lengths of
time. The time necessary to reach T_a and to return to room temperature (RT) was kept within 60 sec. Between subsequent annealings
Mössbauer spectra were measured both at low temperature in a cryostat and, when necessary, near or above RT in the same furnace used
for the annealings. The ordering temperature, T_c, was determined
both by directly observing the collapse of the magnetically split
spectrum and by measuring the count-rate at zero source velocity as
a function of temperature. In the "rate" method (Fig. 1), the value
of T_c is obtained from the minimum rate. It agrees quite well with
the value deduced from the velocity-scanned spectra.

Figure 2 describes how T_c was found to vary with the annealing
temperature. In region 1 (up to 510K) the amorphous phase is stable
in the same configuration as at low temperatures; even prolonged
annealings in this region do not affect the spectral composition of
the Mössbauer absorption. During the annealings in region 2, T_c

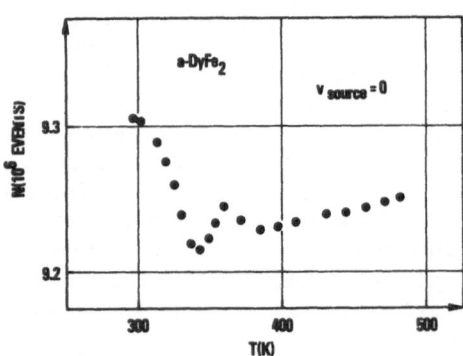

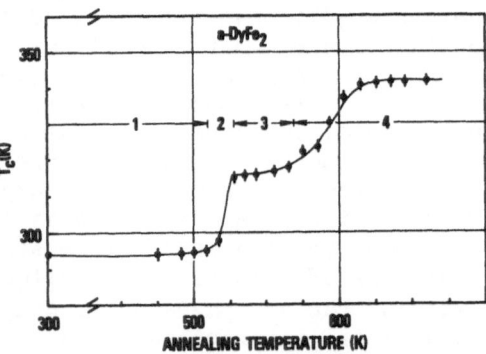

Figure 1. Count rate at zero
velocity measured after annealing at 605K. The minimum transmission occurs at a temperature
which is taken as T_c.

Figure 2. T_c of amorphous $DyFe_2$
vs. annealing temperature. Different regions of interest are
indicated as 1,2,3, and 4. Reference to these regions is indicated
in the spectra shown in Fig. 3.

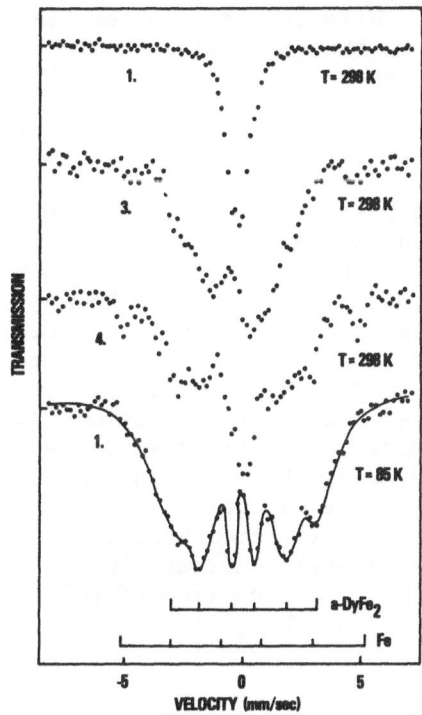

Figure 3. Some Mössbauer absorption spectra measured between subsequent annealings. The numbers on the left side refer to the temperature regions in which the annealings were performed. The measurement temperature is indicated on the right. The curve overlapping the spectrum at 85K gives the best fit.

increases from 295K to 315K and throughout region 3 this new configuration seems to be stable. Except for the increase in T_c, no other changes were observed. Annealings in region 4 (T > 575K) seem to produce further increases in T_c, but simultaneously a central peak and "free-iron" lines appear in the spectra (see Fig. 3, spectrum with label 4). We assume that this indicates a progressive degeneration of the sample caused by the residual oxygen and water in the vacuum furnace, which operated at 5×10^{-6} torr at RT and 6×10^{-5} torr above 550K. This conclusion is consistent with the measured evolution of the spectra of type 4 up to 500K. Relaxation broadening was observed for the "free-iron" lines as if they were due to very small iron particles. We intend to investigate region 4 more thoroughly.

The sharp increase in T_c that takes place when annealing is done near 525K is not reversible, i.e., one cannot return to T_c = 295K (or near it) upon subsequent annealings in region 1.

It is generally assumed that the spatial fluctuations of the exchange interactions, in the present case the Fe-Fe exchange, are responsible for the reduction of the Curie temperature from the crystalline state to the amorphous one. With this assumption, the ratio between the ordering temperature in the two states is simply related to the mean square fluctuation in the exchange, j^2 , (7):

$$T_c^{am} = T_c^{cry} \left[1 - (2/Z) \, j^2/J_o^2 \right] \tag{1}$$

where Z is the average number iron neighbors in the amorphous state and J_o is the exchange for the crystalline system. We assume that Eq. (1) allows us to relate the observed fractional change in T_c to the relative modifications of the spatial fluctuations of the exchange and to changes in Z which have taken place during the structural relaxation in the amorphous phase. Our experiment does not differentiate between changes in Z and in j^2. However, there is considerable evidence that the Fe-Fe distance and coordination number do not change appreciably upon going from the crystalline to the amorphous state in both $DyFe_2$ (8) and $GdFe_2$ (9). Between the two amorphous states found in $DyFe_2$ we speculate that there should be even less change than between crystalline and amorphous state. This seems to be an indication that the observed increase in T_c^{am} from region 1 to region 3 is probably due to a reduction of the spatial fluctuations in the exchange.

ACKNOWLEDGMENTS

The authors gratefully acknowledge several discussions with Drs. R. Abbundi, E. Callen, and J. Cullen.

REFERENCES

1. C. D. Graham, Jr. and T. Egami, J. Magn. and Mag. Mat. 15, p. 1325 (1980).
2. D. W. Forester, R. Abbundi, R. Segnan, and D. Sweger, A.I.P. Conf. Proc. Series 24, 115 (1975).
3. J. J. Rhyne, A.I.P. Conf. Proc. Series 29, 182 (1976)
    D. W. Forester, A.I.P. Conf. Proc. Series 31, 384 (1976)
    N. Heiman, K. Lee and R. I. Potter, J. Appl. Phys. 47, 2634 (1976).
4. A. E. Clark, Appl. Phys. Lett. 23, 11, p. 642 (1973).
5. L. J. Tao, S. Kirkpatrick, R. J. Gambino and J. J. Cuomo, Solid State Comm. 13, 1491 (1973).
6. J. Chappert, J. de Physique, C2, Supp. 3, 40, 107 (1979).
7. C. G. Montgomery, J. I. Krugler and R. M. Stubbs, Phys. Rev. Lett. 25, 669 (1970).
8. E. A. Stern, S. Rinaldi, E. Callen, S. Heald and B. Bunker, J. Magn. and Mag. Mat. 7, p. 188 (1978).
9. G. S. Cargill, A.I.P. Conf. Proc. Series 18, p. 631 (1974).

MAGNETIZATION AND ^{55}Mn HYPERFINE FIELD IN RMn$_2$ (R = RARE EARTH) INTERMETALLIC COMPOUNDS

S. K. Malik, S. K. Dhar and R. Vijayaraghavan, Tata
Institute of Fundamental Research, Bombay 400 005, India;
K. Shimizu, Toyama University, 3190 Gofuku, Japan; W. E.
Wallace, Department of Chemistry, University of Pitts-
burgh, Pittsburgh, PA 15260

INTRODUCTION

The RMn$_2$ (R = rare earth) compounds show some interesting
crystallographic features. The compounds with R = Gd, Tb and Dy
have the cubic Laves phase (C15 type) structure (1), those with Nd
(2), Er, Tm (3) and Lu crystallize in hexagonal Laves phase (C14
type) structure, while PrMn$_2$ has β-Mn type structure (4). The compounds SmMn$_2$ (5) and HoMn$_2$ (6) occur in either the C14 or C15 type
structure. The compounds for R = La, Ce, Eu and Yb are not known
to exist. As noted earlier (3,7), the change in lattice parameter
over the RMn$_2$ series is more rapid than that in the corresponding
Fe, Co and Ni series. This may be taken to imply that the Mn valence (and hence its magnetic moment) changes across the series.
We have carried out magnetization studies on GdMn$_2$, DyMn$_2$ and HoMn$_2$
and ^{55}Mn hyperfine field measurements on several RMn$_2$ compounds to
obtain information about the Mn ion moment. We find that the Mn
moment is non-zero in cubic Laves phase RMn$_2$ compounds, decreases
in going from GdMn$_2$ to DyMn$_2$ and becomes zero or nearly so in
cubic HoMn$_2$ as well as in hexagonal Laves phase compounds.

EXPERIMENTAL

The RMn$_2$ compounds with R = Gd, Tb, Dy, Ho, Er and Tm were
mostly prepared by melting together stoichiometric amounts of the
rare earth (99.9% pure) and premelted Mn (99.99% pure) in a water-
cooled copper boat using rf induction heating. Some of the compounds were prepared by arc melting in argon atmosphere. Powder
x-ray diffraction patterns of the materials were recorded using
CuK$_\alpha$ radiation. All the compounds were found to be single phase

materials crystallizing in appropriate structures. Magnetization
measurements were made at 4.2 K and zero-field NMR studies were
carried out at temperatures close to 1.4 K using spin echo technique.

RESULTS AND DISCUSSION

A. Magnetization

Figure 1 shows a plot of magnetization at 4.2 K versus applied
field for $GdMn_2$, $DyMn_2$ and $HoMn_2$. In each case the magnetization
tends to a saturation value, indicating that these compounds are
magnetically ordered at 4.2 K. The magnetic moment values extrapolated to (1/H → 0) are given in Table 1. It is now well established that in compounds with transition metals, the rare earth
and the transition element moments are coupled antiferromagnetically (8). Further, the Gd^{3+} ion is an S-state ion and, therefore,
its magnetic moment is not influenced by crystal field effects.
From the considerably reduced moment values, particularly for
$GdMn_2$, it is inferred that Mn has a non-zero moment in these compounds. Assuming a moment of 7 μ_B for Gd^{3+} and a simple antiparal-
el coupling between Gd and Mn moments, the Mn moment is estimated
to be about 2 μ_B in $GdMn_2$. Because of crystal field effects and
deviation from simple antiparallel coupling, it is not possible to
get a quantitative estimate of the Mn moment in $DyMn_2$ and $HoMn_2$
from magnetization data alone. Therefore, we have used the nuclear
hyperfine field values to estimate Mn ion moments in the remaining
RMn_2 compounds. The temperature dependence of the magnetization
in $GdMn_2$, $DyMn_2$ and $HoMn_2$ is reported elsewhere (7).

Table 1. Summary of crystallographic, magnetization and ^{55}Mn NMR
measurements. The symbols have their usual meaning.

Compound	Structure Type	μ_{sat} (μ_B)	H_{Mn} (kOe)	Estimated μ_{Mn} (μ_B)
$GdMn_2$	C15	2.9	130	1.5-2[a]
$TbMn_2$	C15	8.0	104	1.2
$DyMn_2$	C15	7.2	70	0.9
$HoMn_2$	C15	6.4	24	0.4
$ErMn_2$	C14	7.88	19	-
$TmMn_2$	C14	5.28	12	-

(a) Obtained from magnetization data alone.

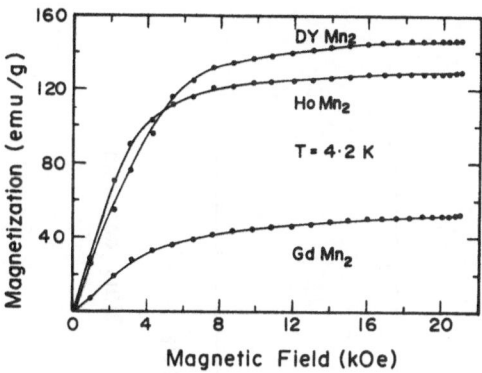

Fig. 1. Magnetization vs
applied field at
4.2 K for GdMn$_2$,
DyMn$_2$ and HoMn$_2$.

B.   Zero-field ^{55}Mn NMR

Fig. 2a,b shows the ^{55}Mn NMR spectra in various RMn$_2$ compounds.
In the cubic Laves phase there is a unique Mn site, while in the
hexagonal phase there are two inequivalent Mn sites. The observation of a single ^{55}Mn NMR line even in hexagonal phase compounds
suggests that the hyperfine field is nearly the same at the two Mn
sites. Further, the hyperfine field at the Mn site is larger in
cubic compounds than in hexagonal compounds, the only exception
being HoMn$_2$. Among the cubic RMn$_2$ compounds, the field is the
largest in GdMn$_2$ and decreases successively in TbMn$_2$, DyMn$_2$ and
HoMn$_2$. If the field at the Mn site arises mainly from its own
local moment (vide infra) the NMR measurements suggest that the Mn
moment decreases in going from GdMn$_2$ to TmMn$_2$.

The hyperfine field at the Mn site mainly consists of three
terms, (i) that arising from the polarization of the Mn core electrons due to its local moment, (ii) the contribution from the
neighboring Mn moments and (iii) the contribution from the neighboring rare earths. The dipolar contribution is usually small and
hence neglected. The rare earth contribution to the Mn field in
GdMn$_2$ may be taken to be the same as the corresponding field in
GdCo$_2$ (9) suitably scaled by the hyperfine interaction constants.
This turns out to be ∿ 25 kOe. The contributions (i) and (ii) are
proportional to the Mn moment and may be combined and written as
$\alpha\mu_{Mn}$. To estimate α we use the experimentally determined μ_{Mn} and
H_{Mn} in GdMn$_2$ and obtain α = 103 kOe/μ_B. This value of α is then
used in conjunction with the observed hyperfine field to get an
estimate of the Mn ion moment in the remaining RMn$_2$ compounds
(Table 1). The hyperfine field at Mn in hexagonal compounds
ErMn$_2$ and TmMn$_2$ is of the same order as that arising from neighboring rare earth ions, implying nearly zero Mn moment in hexagonal
compounds in agreement with neutron diffraction studies (10). The

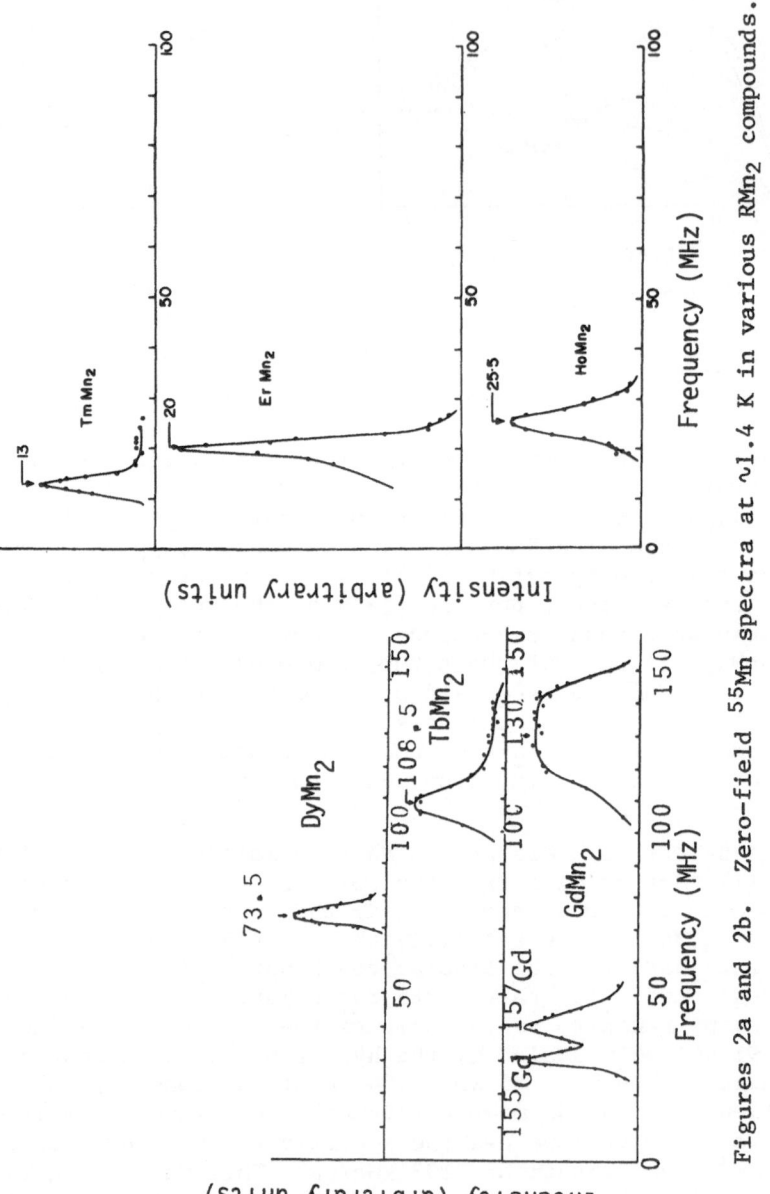

Figures 2a and 2b. Zero-field ^{55}Mn spectra at ~ 1.4 K in various RMn_2 compounds.

low Mn field or small Mn moment in $HoMn_2$ is consistent with its dimorphic nature. The Mn moment is non-zero in other cubic RMn_2 compounds.

REFERENCES

1. K. Nassau, L. V. Cherry and W. E. Wallace, J. Phys. & Chem. Solids 16, 123 (1960).
2. H. R. Kirchmayr, Z. Angew. Phys. 27, 18 (1969).
3. J. H. Wernick and S. E. Haszko, J. Phys. Chem. Solids 18, 207 (1961).
4. H. Oesterreicher, J. Less-Common Metals 23, 7 (1971).
5. Quoted by H. R. Kirchmayr and Carl A. Poldy in "Handbook of Physics and Chemistry of Rare Earths," Karl A. Gschneidner and LeRoy Eyring, eds., Vol. No. 2, p. 55 (1979).
6. S. K. Malik and W. E. Wallace, unpublished work.
7. S. K. Malik and W. E. Wallace, J. of Mag. and Mag. Materials, accepted for publication.
8. See, for example, W. E. Wallace, Rare Earth Intermetallics, Academic Press, Inc., New York, 1973, chapters 10 and 11.
9. S. Hirosawa, T. Tsuchida and Y. Nakamura, J. Phys. Soc. Jap. 47, 804 (1979).
10. G. P. Felcher, L. M. Corliss and J. M. Hastings, J. Appl. Phys. 36, 1001 (1965).

MAGNETIC PROPERTIES OF CUBIC AND HEXAGONAL HoMn$_2$

J. Rhyne and K. Hardman, National Bureau of Standards
Washington, DC 20234

S. Malik and W. Wallace, University of Pittsburgh
Pittsburgh, PA 15260

HoMn$_2$ crystallizes in the cubic Laves phase C-15 (Fd3m) structure which has two inequivalent sites, one for the Ho (a) and one
for the Mn (d). The structure can be transformed to the hexagonal
Laves phase C-14 (P6$_3$/mmc) structure by a careful annealing procedure. In this structure, the Mn occupies two sites (a and h), and
the Ho occupies an f site.

High resolution neutron magnetic profile refinement studies
have been performed on both the cubic and hexagonal phases at 295 K
and 4 K. Bulk magnetization measurements gave a Curie temperature
of 26 K for both phases and essentially equal saturation magnetizations, the latter values being affected by the large crystal field
anisotropy of the Ho, particularly in the hexagonal structure.

The neutron profile refinement produced 4 K sublattice magnetization values for both the Ho and the Mn sites as given in Table 1.
R-factors for the profile fits are also given and indicate excellent
agreement between the experimental data and the nuclear and magnetic
model. Ho and Mn spins are oriented antiparallel resulting in a
ferrimagnetic structure. The spins are aligned in the basal plane
of the hexagonal HoMn$_2$ as expected from the symmetry of the Ho 4f
electron charge distribution. The magnetic neutron results are also
compared to those obtained from bulk magnetization in Table 1. The
bulk moment data were taken in fields up to 20 kOe and extrapolated
to H = ∞ from a 1/H plot of the magnetization. The bulk data, which
give only the vector sum of Ho and Mn moments, are in good agreement
with the neutron results for the cubic phase, but significantly lower
for the hexagonal material in which the large Ho uniaxial anisotropy
more strongly affects the bulk magnetization values. The Ho moment

for the hexagonal phase (9.4 μ_B) is close to the free ion value of 10.0 μ_B while that for the cubic phase (8.12 μ_B) is significantly reduced. Since the c/a ratio of the hexagonal structure is near the ideal value (1.632), the second order crystal field term B_2^0 is zero for both hexagonal and cubic phases. Assuming that the same Curie temperature (26 K) implies nearly equal exchange energies, then the reduced Ho moment in the cubic phase suggests a stronger influence of the B_4^0 crystal field term on the Ho moment in the cubic than in the hexagonal structure.

Table 1.   Cubic and hexagonal HoMn$_2$ structural and magnetic para-
           meters obtained from neutron profile refinement (Neu)
           and magnetization (Mag) data.

	Latt. Param. (A)	$\mu_{Ho}^{4K}(\mu_B)$	$\mu_{Mn}^{4K}(\mu_B)$	$\mu_{HoMn_2}^{4K}$	T_c (Mag)
CUBIC	7.518	8.12	−0.84	6.45 (Neu) 6.4 (Mag)	26 K
HEX	5.282 (a) 8.622 (c)	9.40	−0.64 (a) −1.03 (h)	7.5 (Neu) 6.4 (Mag)	26 K

R factors of fit above and below Curie temperatures ($R_{w.p.}$ = weighted profile R factor, $R_{espd.}$ = calculated statistical "perfect fit" R factor, $R_{mag.}$ = magnetic model R factor only):

	Cubic	Hexagonal
$R_{w.p.}/R_{expd.}$ (295 K)	8.66/4.76	5.01/3.32
$R_{w.p.}/R_{expd.}$ (4 K)	7.73/3.05	6.31/3.08
$R_{mag.}$ (4 K)	4.3	3.55

SUSCEPTIBILITY DENSITIES IN THE PAULI PARAMAGNETS YNi$_5$ AND CeNi$_5$

D. Gignoux, D. Givord, F. Givord, R. Lemaire,
Laboratoire Louis Néel, C.N.R.S., 166X,
F. Tasset, Institut Laue-Langevin, 156X,
38042 Grenoble-cédex, France

INTRODUCTION

The induced magnetic densities in the enhanced Pauli paramagnets YNi$_5$ and CeNi$_5$ have been measured by polarized neutron
diffraction at 100 K in an applied field of 48 kOe. The magnetizations induced on the Ni sites appear more localized in CeNi$_5$ because
of the superposition of a non uniform diffuse and negative polarization to the 3d contribution. These results are discussed from
the band structure of these compounds.

The alloys of Y and Ni form a large number of compounds, the
magnetism of which is associated with the onset of 3d magnetism.
In Y$_2$Ni$_{17}$ (T_c = 160 K), part of Ni atoms exhibit collective electron metamagnetism. YNi$_5$ is a strong Pauli paramagnet. For Y$_2$Ni$_7$
and YNi$_3$, a resurgence of ferromagnetism is observed. Such
behaviors result from a modification of the band structure across
the composition range. The 2 constitutents have a large difference
in electronegativity which gives rise to a strong hybridization of
the 3d and 4d states leading to a zone of the density of states
with a positive curvature (1). Magnetic properties of these various
compounds can be interpreted by considering the position of the
Fermi level in this particular zone (2). A study of the field
induced magnetic densities in YNi$_5$ was undertaken in order to get
more information on the respective contributions of the 3d and 4d
electrons to the susceptibility. CeNi$_5$ is also paramagnetic at any
temperature. Moreover, the peculiar thermal variation of its
susceptibility has been interpreted by Buschow (3) as resulting
from a progressive change from a Ce^{4+} state at 0 K to an intermediate valence state at room temperature, arising through thermal
excitation to the 3+ magnetic state. However, the study of the

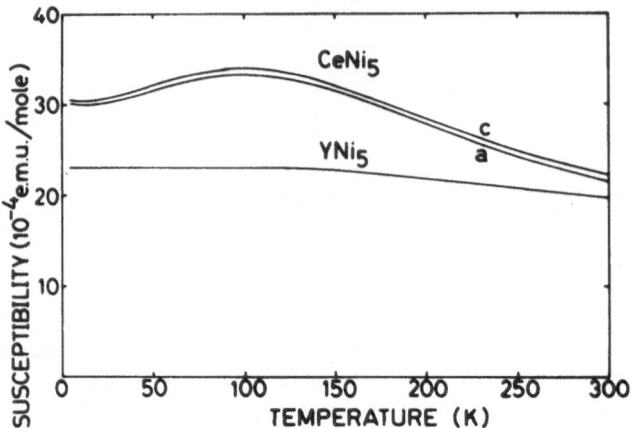

Figure 1. Thermal variations of
the susceptibilities of
YNi₅ and CeNi₅.

L_{III} X-ray absorption edges have shown that in this compound Ce is in a valence state close to 4+ at room temperature (4). Besides, the thermal dilatation in CeNi₅ is identical to that in LaNi₅. A polarized neutron study, similar to that of YNi₅, has brought us information on the origin of the CeNi₅ susceptibility.

Both compounds crystallize in the CaCu₅ hexagonal phase (P6/mmm) where Ca atoms lie in 1a (0,0,0) and Cu atoms in 2 different sites : 2c in the plane z = 0 and 3g in the plane z = 1/2. The two single crystals were prepared by the Czochralski method.

EXPERIMENTAL RESULTS

The magnetic susceptibilities were measured along the $\vec{a}$ and $\vec{c}$ directions from 4.2 K to room temperature (figure 1). The anisotropy is null in YNi₅ and very weak in CeNi₅. As previously measured (3) the CeNi₅ susceptibility presents a broad maximum around 100 K, similar to those observed in YCo₂ (5) or LuCo₂ (6). Moreover, at any temperature, the value of the CeNi₅ susceptibility is close to that of YNi₅.

Measurements with polarized neutrons of reflections (0 k 1) out to sinθ/λ = 0.6 Å⁻¹ for YNi₅ and (h k 0) out to 0.75 Å⁻¹ for CeNi₅ were performed at I.L.L. (Grenoble). A field of 48 kOe was applied parallel to $\vec{a}$ in YNi₅ and $\vec{c}$ in CeNi₅. The experiments were done at 100 K in order to get rid of the influence at low temperature of possible paramagnetic impurities. The polarized beam technique yields the polarization ration $R = [(1+\gamma)/(1-\gamma)]^2$ with $\gamma = F_M/F_N$, F_M and F_N being the magnetic and nuclear structure factors respectively. This relation has been corrected taking account of all corrections due to instrument and crystal imperfections. Nuclear structure factors F_N and extinction parameters were precisely determined from the diffraction at various wave lengths of unpolarized neutrons on 4 circle diffractometers. The structure refinement has led to introduce atomic random substitutions : in YNi_5, 2 % of Ni_{2c} are replaced by Y atoms. In $CeNi_5$, 0.3 % of Ce atoms are replaced by Ni dumbells. The corresponding compositions are $YNi_{4.82}$ and $CeNi_{5.02}$. The extinction parameters are $t = 0.5 \pm 0.2$ μ and $g = 300 \pm 50$ ($\eta = 3.2 \pm 0.5'$) for YNi_5 and $t = 6.7 \pm 0.3$ μ and $g = 568 \pm 29$ ($\eta = 1.7 \pm 0.1'$) for $CeNi_5$.

The measured magnetic structure factors F_M have been used to make projections of the moment densities (figure 2). For the (000) reflections, we have used the magnetic structure factor associated with the bulk magnetization measured in the same experimental conditions. The projections evidence strong positive densities localized on Ni sites, those of the 3g sites being higher than those of the 2c sites. In YNi_5, these densities correspond to a form factor of 3d electrons at the top of the band whereas in $CeNi_5$ they are observed much more localized. However, in this later compound the same type of 3d electrons should also mainly contribute to these

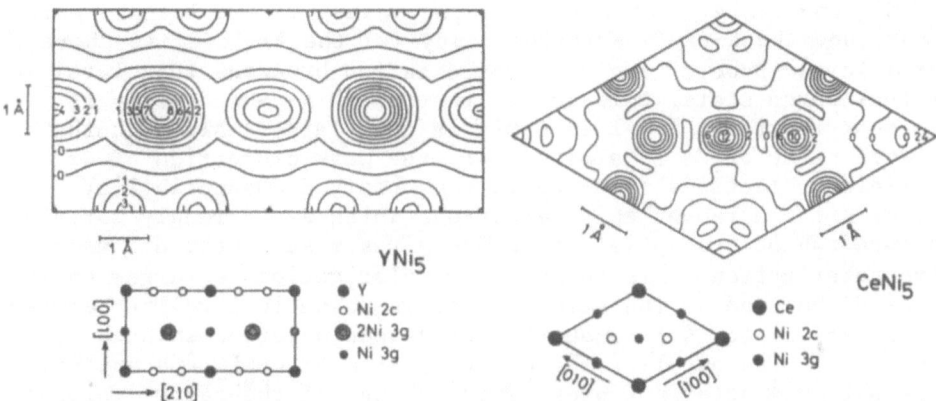

Figure 2. Projections of the magnetic densities in YNi_5 and $CeNi_5$ averaged on a square of 0.4 Å of side (contours are in 10^{-3} $\mu_B/Å^2$).

Table 1. Field (48 kOe) induced magnetization in 10^{-3} μ_B for YNi_5 and $CeNi_5$.

	μ/RNi_5	μ_{3d}/Ni		μ_{4f}/R	μ/RNi_5	μ/RNi_5
	bulk	Ni_{2c}	Ni_{3g}	R	localized	diffuse
YNi_5	19.3(9)	2.4(4)	4.1(6)	–	17.1(1.2)	2.2(2.1)
$CeNi_5$	29.2(9)	7.0(7)	8.2(8)	2.4(6)	41.0(1.8)	−11.8(2.7)

densities. The observed localization results then from the super-position of a contribution of 3d type, similar to that in YNi_5 and of a strong antiparallel polarization, more diffuse and non uniform. Whereas no density is localized on Y, a weak one is found on Ce with a 4f type form factor, within the experimental accuracy. The values of the magnetic moments induced by a 48 kOe applied field are given in the table. More details on the way to get these values are given in references (7) and (8). The difference between the total moment obtained from the neutron study and the bulk one reveals in $CeNi_5$ a large diffuse polarization of the 5d electrons antiparallel to the 3d magnetism.

DISCUSSION

In both compounds the bulk susceptibility originates mainly from the Ni 3d electrons as evidenced by the Fourier projections. On the Ni sites the 3d band is almost filled up, so the 3d magnetic states have a localized antibonding character as in Ni metal. The magnitude of the diffuse magnetic polarization is quite different in YNi_5 and $CeNi_5$. In $CeNi_5$, from the values of the lattice para-meters and the X-ray absorption study (4) the 4f level is above the Fermi level. However a 4f–5d hybridization broadens this level in a virtual bound state. Compared to that in YNi_5, the density of state at the Fermi level in $CeNi_5$ is then larger. Although the valence state of Ce is close to 4+, the weak proportion of 4f hybridized electrons can lead to the observed susceptibility on the Ce site. Moreover these electrons which are strongly correlated enhance the 5d susceptibility which gives rise to the diffuse nega-tive polarization. This non uniform polarization is larger on the Ni sites because of the strong negative magnetic coupling between the 3d and 5d (or 4d) conduction electrons observed in such com-pounds (9). The peculiar thermal variation of $CeNi_5$ susceptibility does not originate in a progressive change of the cerium valence state through thermal excitations. It rather results from the position of the Fermi level in a zone of density of states with a positive curvature due to the 4f-5d and 3d-5d hybridizations and from different thermal variations of the antiparallel 3d and 5d contributions to this susceptibility.

REFERENCES

1. M. Cyrot and M. Lavagna, J. Phys., 40:763 (1979).
2. D. Gignoux, D. Givord, J. Laforest, R. Lemaire and P. Molho, Physics of Transition Metals, 1980, Institute of.Physics Conference Series 65, London 1981, 287.
3. K.H.J. Buschow, M. Brouha, H.J. Van Daal and H.R. Miedema, in "Valence instabilities and related narrow bands phenomena" Conf. of Rochester, Ed. by. R.D. Parks, Plenum Press (New York) 125 (1977).
4. D. Gignoux, F. Givord, R. Lemaire, H. Launois and F. Sayetat, J. Phys. submitted.
5. R. Lemaire, Cobalt, 33:201 (1966).
6. D. Bloch, F. Chaissé, F. Givord, J. Voiron and E. Burzo, J. Phys., C32:659 (1971).
7. D. Gignoux, D. Givord, R. Lemaire and A. Naït Saada, J. Magn. Magn. Mat., 23 (1981) in press.
8. D. Gignoux, F. Givord, R. Lemaire and F. Tasset, to be published.
9. B. Barbara, D. Gignoux, D. Givord, F. Givord and R. Lemaire, Int. J. Magnetism, 4:77 (1973).

THERMAL VARIATION OF ANISOTROPIES OF COBALT IN YCo$_5$ AND NdCo$_5$ UP TO 450 K

J.M. Alameda[+], D. Givord, R. Lemaire and Q. Lu[++]

Laboratoire Louis Néel, C.N.R.S., 166X

38042 Grenoble-cédex, France

INTRODUCTION

In hexagonal rare earth (R) - cobalt RCo$_5$ compounds, the contribution from Co atoms to the anisotropy is very large as shown by the magnetic properties of YCo$_5$. The origin of such a large value in a puzzling question.

The anisotropy energy and the anisotropy of the magnetization in RCo$_5$ are studied, between 4.2 K and 450 K, by high field measurements in YCo$_5$ and by polarized neutron diffraction in NdCo$_5$. The anisotropy energy shows a more rapid decrease with increasing temperature than the one predicted from the temperature variation of the magnetization due to random orientation of magnetic moments. At 4.2 K the anisotropy of the magnetization on Co atoms reaches 4 % of the spontaneous magnetization. The contribution of the Co atoms lying in the axial symmetry site is twice that of Co atoms lying on another site of orthorhombic symmetry.

MAGNETIC MEASUREMENTS

The variation of the magnetization in YCo$_5$ at 4.2 K with fields up to 200 kOe along and perpendicular to c-axis is shown in figure 1. The results give a direct evidence for a large anisotropy of the magnetization which is independent of the direction in the basal plane. Similar measurements have been performed at several temperatures up to 450 K.

Permanent address :
[+]Laboratorio de Magnetismo, Universidad Complutense, Madrid, Spain
[++]Department of Physics, University of Nanjing, China

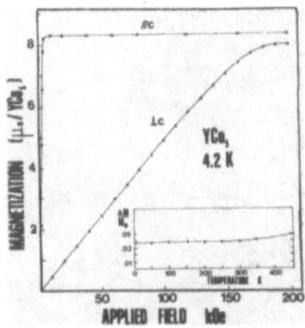

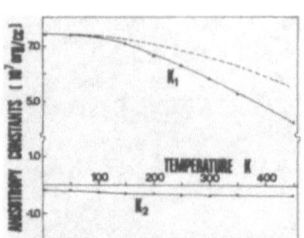

Figure 1. Field variation of the
magnetization in YCo5 at 4.2 K.
Inset : temperature dependence
of the magnetization anisotropy.

Figure 2. Temperature dependences
of K1 and K2 in YCo5. Broken line
- calculated variation according
to Callen and Callen (7).

In order to analyse the results, the anisotropy energy is
written in terms of the anisotropy constants K_1 and K_2, relevant to
the hexagonal symmetry. The anisotropy of the magnetization is
written in the first order as $M(\Theta) = M_s(1 - p \sin^2\Theta)$, where
$p = \Delta M/M_s = (M//^c - M^c)/M_s$. By generalizing the classical formula
of Sucksmith-Thompson (1) K_1, K_2 and p can be determined at each
temperature (2). Their temperature variations are shown in figure
2 and figure 1 (inset).

POLARIZED NEUTRON STUDY

In the CaCu5 structure (P6/mmm) of RCo5, Co atoms occupy two
sites (2c and 3g) with different symmetry. In order to determine
the contribution of each site to the anisotropy of the magnetization of cobalt, a polarized neutron study must be performed during
which the magnetization is successively aligned along the easy
direction and perpendicular to it. In YCo5, this is not possible
in normal laboratory fields due to the large anisotropy. In NdCo5,
the individual anisotropies of Nd and Co atoms are opposite and
their magnitudes are nearly equal at 260 K. At this temperature, a
field of 46 kOe, available on the D3 diffractometer at I.L.L., can
saturate the magnetization both along c and a axes. The details of
the experiment performed will be published elsewhere (3). The
results lead to the determination of magnetic moments and form
factors for both Nd and Co atoms.

Table 1. Analysis of the Co magnetism from the polarized neutron
study of $NdCo_5$.

	M parallel to a-axis		M parallel to c-axis	
	magnetic moment $\mu_T(\mu_B)$	orbital contribution μ_L/μ_T	magnetic moment $\mu_T(\mu_B)$	orbital contribution μ_L/μ_T
Co_{2c}	1.84(3)	.19(4)	1.95(3)	.33(5)
Co_{3g}	1.84(2)	.12(2)	1.90(3)	.20(3)

At 260 K, when the magnetization is along a-axis, μ_{Nd} =
2.12(2) μ_B, and when it is along c-axis, μ_{Nd} = 1.78(2) μ_B ; the
magnetic moment of Nd is larger when it is aligned along a-axis as
the crystal field acting on Nd^{3+} ion favors this axis. The magnetic
moments of Co were analysed in terms of spin and orbital parts. As
shown in table 1, a large orbital contribution to the magnetic
moment is obtained, specially for the site 2c of Co. The anisotropy
of magnetization for the site 2c is about twice that of the site
3g.

DISCUSSION

At 4.2 K, K_1 in YCo_5 reaches 7.4×10^7 erg/cm^3. This very large
value is close to those previously published (4,5). It must be
related to the large orbital magnetization of Co atoms which is
revealed by the polarized neutron study on $NdCo_5$ (table 1) as previously shown in YCo_5 (6). The decrease of the anisotropy when
temperature increases has been considered, by Callen and Callen (7)
in particular, in a model of localized magnetism. The experimental
variation of the anisotropy of YCo_5 is about twice as fast as
deduced from this model (figure 1). In fact, 3d electrons form a
band. Aubert and Michelutti (8) gave a good account for the very
fast decrease with increasing temperature of the anisotropy in
nickel by considering the smearing out of the density of states at
finite temperature. In YCo_5, where the magnetic correlations are
larger than in nickel, the temperature variation of the anisotropy
is intermediate between that observed in nickel and that calculated
according to Callen and Callen (7).

The anisotropy of the magnetization at 4.2 K reaches 0.31 μ_B/
YCo_5. This property, observed at low temperature has an intrinsic
character. The value of p as that of the anisotropy energy is
about ten times that in cobalt (9). Between 4.2 K and 450 K, ΔM
increases from 0.31 to 0.38 μ_B. Callen and Callen (10) have shown

that the anisotropy of the magnetization is temperature dependent
since at finite temperature the disorder of moments is larger
along a hard than along an easy direction. In YCo_5, the anisotropy
of the magnetization calculated in such a model increases by 0.020
μ_B in the temperature range mentioned above. It gives only partial
account for the total thermal variation of the magnetization anisotropy. The anisotropy of the magnetization of Co_{2c} atoms is about
twice that of Co_{3g} atoms. This property must be related to the
different local symmetries of the two Co sites, uniaxial for the
2c site, orthorhombic for the 3g one. Moreover, the orbital moment
is also larger for Co_{2c} atoms. This suggests that the anisotropy
of the magnetization results from the anisotropy of the moment
induced on cobalt atoms by the spin-orbit coupling. The differences
between the orbital contributions when the magnetization is respectively along a-axis or along c-axis (table 1), although inaccurate,
are in agreement with this conclusion.

ACKNOWLEDGEMENTS

We have benefit stimulating discussions with Prof. M. Shimizu.

REFERENCES

1. W. Sucksmith and F.R.S. Thompson, Proc. Roy. Soc. (London),
   225:362 (1954).

2. J.M. Alameda, J. Déportes, D. Givord, R. Lemaire and Q. Lu,
   J. Magn. Magn. Mat., 15:1257 (1980).

3. J.M. Alameda, D. Givord, R. Lemaire, Q. Lu, S.B. Palmer and
   F. Tasset (to be published).

4. E. Tatsumoto, T. Okamoto, H. Fujii and C. Inoue, J. Physique,
   32, Suppl. 1:550 (1971).

5. H.P. Klein, A. Menth and R.S. Perkins, Physica, 80B:153 (1975).

6. J. Schweizer and F. Tasset, J. Phys. F. : Metal Phys., 10:2799
   (1980).

7. H.B. Callen and E. Callen, J. Phys. Chem. Solids, 27:1271
   (1966).

8. G. Aubert and B. Michelutti, Physica, 86-88B:295 (1977).

9. J.P. Rebouillat, Thesis, University of Grenoble, France (1972).

10. E.R. Callen and H.B. Callen, J. Phys. Chem. Solids, 16:310
    (1960).

THE MAGNETIC STRUCTURE OF Y(Mn$_{1-x}$Fe$_x$)$_{12}$

Y. C. Yang (a), Gary J. Long, B. Kebe, and W. J. James,
Department of Chemistry and Graduate Center for Materials
Research, University of Missouri-Rolla, Rolla, MO 65401;
J. Deportes, CNRS, Laboratoire Louis-Néel, F-38042
Grenoble, France

INTRODUCTION

The Y(Mn$_{1-x}$Fe)$_{12}$ compounds with $x \leq 0.67$ crystallize in the
ThMn$_{12}$ structure. The yttrium atoms occupy the $2a$ sites while manganese and iron are distributed on the three non-equivalent sites,
$8i$, $8j$, and $8f$.

The yttrium-manganese system is particularly interesting
because of the magnetic behavior of the resulting ternary compounds
formed upon the addition of iron. There are three yttrium-manganese
compounds: YMn$_2$, Y$_6$Mn$_{23}$, and YMn$_{12}$. The Curie temperature and the
magnetization are lowered by the addition of manganese to YFe$_2$ and
Y$_6$Fe$_{23}$ and by the addition of iron to Y$_6$Mn$_{23}$. There are compositional regions where no long-range ordering occurs in the
Y(Mn$_{1-x}$Fe$_x$)$_2$ and Y$_6$(Mn$_{1-x}$Fe$_x$)$_{23}$ systems [1,2].

Previous studies on Y(Mn$_{1-x}$Fe$_x$)$_{12}$ have shown that its magnetic
behavior is different from the other yttrium-manganese systems [3].
To study this difference we have measured the magnetic, neutron, and
Mössbauer spectral properties of the Y(Mn$_{1-x}$Fe$_x$)$_{12}$ system.

EXPERIMENTAL

Ingots of Y(Mn$_{1-x}$Fe$_x$)$_{12}$ with x = 0.1, 0.2, 0.3, 0.4, 0.5, and
0.6 were prepared by induction melting of the elements in a water-
cooled copper boat under an argon atmosphere. X-ray diffraction
photographs of the samples showed no evidence of other phases.
Neutron diffraction data were collected at room temperature and at

liquid nitrogen temperature by using a two-axis neutron diffractometer located at the University of Missouri Research Reactor. The
refinements were carried out by using a Rietveld line-profile
analysis technique (4). The magnetic susceptibility measurements
were made on a Faraday balance described earlier (5). The Mössbauer
spectral measurements were obtained on a Ranger Instruments spectrometer which utilized a room temperature $^{57}Co(Rh)$ source and was
calibrated with natural α-iron foil (6).

RESULTS AND DISCUSSION

The magnetic susceptibility of $Y(Mn_{1-x}Fe_x)_{12}$ with $x = 0.0$,
0.1, and 0.2 is constant between 80 and 300K as shown in Fig. 1.
This behavior is often associated with Pauli paramagnetism such as
is observed in YMn_2 (1). However, it may correspond to weak band
antiferromagnetism similar to that found in chromium and manganese
(7). Neutron diffraction data discriminate between these two cases
and in the former, no magnetic neutron scattering is observed. In
the latter, new scattering peaks, characteristic of antiferromagnetic ordering, should appear below the Néel temperature. The
results at 300K display peaks characteristic of the $I4/mmm$ space
group and can be indexed with $h + k + l = 2n$. At 78K, extra peaks
which can be indexed with $h + k + l = 2n + 1$ are observed, indicating an antiferromagnetic structure. The thermal variation of
the additional peaks determine the Néel temperature as 120K for
YMn_{12} (8).

The temperature dependence of the magnetic susceptibilities
of $Y(Mn_{1-x}Fe_x)_{12}$ with $x = 0.3$, 0.4, and 0.5, as illustrated in
Fig. 1, is typical of antiferromagnetic materials with Néel temperatures of 230, 253, and 250K respectively. The neutron diffraction results and the magnetic measurements show $Y(Mn_{0.4}Fe_{0.6})_{12}$ is
antiferromagnetic with a ferromagnetic component. The magnetic
transition temperature is approximately 200K.

The average magnetic moments on the $8i$, $8j$, and $8f$ sites, as
refined from the 78K neutron diffraction data are presented in
Table I. In contrast to the $Y(Mn_{1-x}Fe_x)_2$ and $Y_6(Mn_{1-x}Fe_x)_{23}$ compounds, the variation of the moment on the $8i$ and $8j$ sites in
$Y(Mn_{1-x}Fe_x)_{12}$ as a function of composition shows a maximum at
$x = 0.3$. The compositional dependence of the Néel temperature
shows a similar behavior. The absence of the ordering of the iron
atoms on the $8f$ site, when $x < 0.6$, is similar to the behavior
observed in the other systems (2).

The room temperature Mössbauer spectra of $Y(Mn_{1-x}Fe_x)_{12}$ with
$x = 0.1$, 0.3, and 0.6 indicate no long-range magnetic ordering but
do reveal different quadrupole interactions for the different sites.

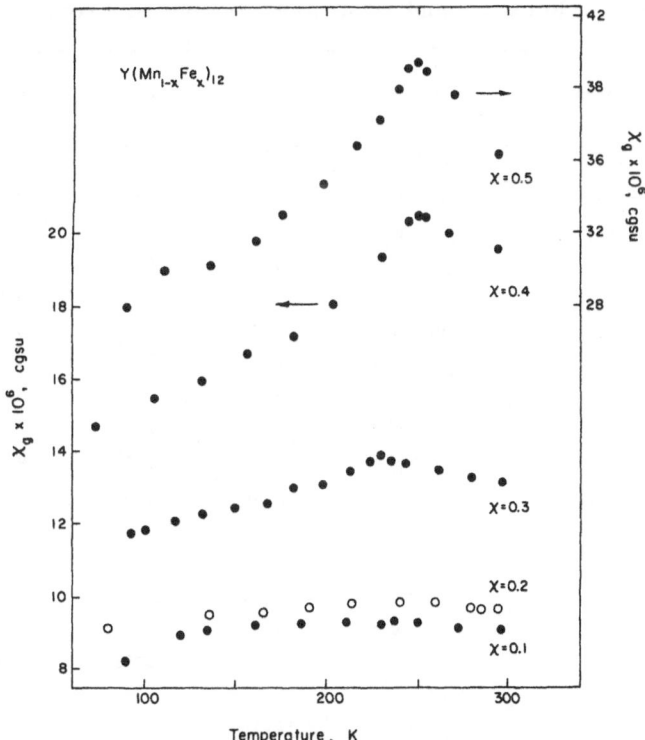

Figure 1. The magnetic susceptibility as a function of temperature for several compositions of $Y(Mn_{1-x}Fe_x)_{12}$.

The 78K Mössbauer spectra are broad and only partially resolved. They do, however, indicate the presence of long-range magnetic ordering of the iron moments on the $8j$ and $8f$ sites with the internal hyperfine fields, moments, and approximate relative areas given in Table I. By using both the average moment on a given site (derived from the neutron data) and the moment at the iron atom on a given site (derived from the Mössbauer spectra) along with the relative fraction of manganese and iron on the site (see Table I), it is possible to estimate the range of moments at the manganese atoms on the site. The limiting moments are presented in Table I and are consistent with the antiferromagnetic behavior observed in $Y(Mn_{0.7}Fe_{0.3})_{12}$ and the ferromagnetic behavior observed in $Y(Mn_{0.4}Fe_{0.6})_{12}$ at 78K.

ACKNOWLEDGEMENT

The authors wish to thank the National Science Foundation for support under grant NSF INT 7826549.

Table I. Neutron and Mössbauer Magnetic Properties.

Compound	Property	$8i$	Site $8j$	$8f$
YMn_{12}	μ^a_{ave}, μ_β	0.4	0.4	0
$Y(Mn_{0.9}Fe_{0.1})_{12}$	μ^a_{ave}, μ_β	0.86	0.89	0
$Y(Mn_{0.7}Fe_{0.3})_{12}$	μ^a_{ave}, μ_β	1.30	1.36	0
	f^a_{Mn}	0.93	0.68	0.46
	f^a_{Fe}	0.07	0.32	0.54
	% of Fe^a	7.2	34.8	58.0
	H^b_{int}, kOe	<5	37	58
	μ^b_{Fe}, μ_β	<0.04	0.25	0.39
	% of Fe^b	13	40	47
	μ_{Mn}, μ_β	.≥1.39	1.89	0.46
$Y(Mn_{0.4}Fe_{0.6})_{12}$	μ^a_{ave}, μ_β	0.98	1.07	0.6
	f^a_{Mn}	0.72	0.29	0.14
	f^a_{Fe}	0.28	0.71	0.86
	% of Fe^a	15.3	38.1	46.6
	H^b_{int}, kOe	<5	49	81
	μ^b_{Fe}, μ_β	<0.04	0.33	0.54
	% of Fe^b	4	23	73
	μ_{Mn}, μ_β	≥1.37	2.86	0.94

[a]Derived from neutron diffraction data. [b]Derived from Mössbauer effect spectral data.

REFERENCES

(a) Permanent Address: Department of Physics, Beijing University, Beijing, China.
(1) H. R. Kirchmayr, J. Appl. Phys., 39:1088 (1968).
(2) G. J. Long, K. Hardman, and W. J. James, Solid State Comm., 34:253 (1980); K. Hardman, W. J. James, G. J. Long, W. B. Yelon, and B. Kebe, in The Rare Earths in Modern Science and Technology, Vol. 2, G. J. McCarthy, J. J. Rhyne, and H. B. Silber (eds.), Plenum Press, New York, pp 315-320 (1980).
(3) Y. C. Yang, B. Kebe, W. J. James, J. Deportes, and W. B. Yelon, J. Appl. Phys. (in press, 1981).
(4) H. M. Rietveld, J. Appl. Cryst., 2:65 (1969).
(5) G. J. Long, G. Longworth, P. Battle, A. K. Cheetham, R. V. Thundathil, and D. Beveridge, Inorg. Chem., 18:624 (1979).
(6) K. H. Pannell, C. C. Wu, and G. J. Long, J. Organomet. Chem., 186:85 (1980).
(7) C. G. Shull and M. K. Wilkinson, Rev. Mod. Phys., 25:100 (1953).
(8) J. Deportes, D. Givord, R. Lemaire, and H. Nagai, Physica, 69B:86 (1977).

MAGNETIC CHARACTERISTICS OF THE INTERMETALLIC COMPOUNDS

$R_2Ni_{17-x}Al_x$ (R = Gd,Tb)

Marin Coldea and Iuliu Pop

Faculty of Physics, Cluj-Napoca University

3400 Cluj-Napoca, Romania

The intermetallic compounds R_2Ni_{17} and R_2Al_{17} are isostructural and form a continuous series of solid solutions with the general formula $R_2Ni_{17-x}Al_x$ (1). They crystallize in a hexagonal structure of the Th_2Ni_{17} type and the unit cell contains two formula units (space group $P6_3/mmc$) (2).

Earlier magnetic data (3-6) point out the ferrimagnetic behavior of the Gd_2Ni_{17} and Tb_2Ni_{17} compounds. However, the results obtained in the ordered state, namely, the temperature dependence of the magnetization, the values of the magnetic moments and the Curie temperatures, are contradictory and the interpretation of the results is different. Laforest et al. (3,4) have shown that the formation of these compounds is always accompanied by free Ni. Consequently, they have corrected their results for free Ni, which amounted to as much as 2%, but they do not give any information about their procedure. The loss of magnetization in two stages observed by Carfagna and Wallace (5,6), and especially the high values of Curie temperatures close to T_C for Ni, arouse suspicion that their samples were not single phases. This discrepancy led us to investigate in more detail the magnetic behavior of the intermetallic compounds Gd_2Ni_{17} and Tb_2Ni_{17} and also of the pseudobinary compounds $R_2Ni_{17-x}Al_x$ (R = Gd, Tb and 0 < x < 2).

Our results are interpreted on the basis of magnetic interactions between Ni-Ni, Ni-R and R-R, taking into account the positions of the magnetic ions in the lattice determined from x-ray data. In the compounds R_2Ni_{17} there are four crystallographically inequivalent sites for Ni atoms (4f,6g,12j,12k) and two for R atoms (2b,2d) (2). The Ni atoms on 6g, 12k and 12j sites give rise to identical

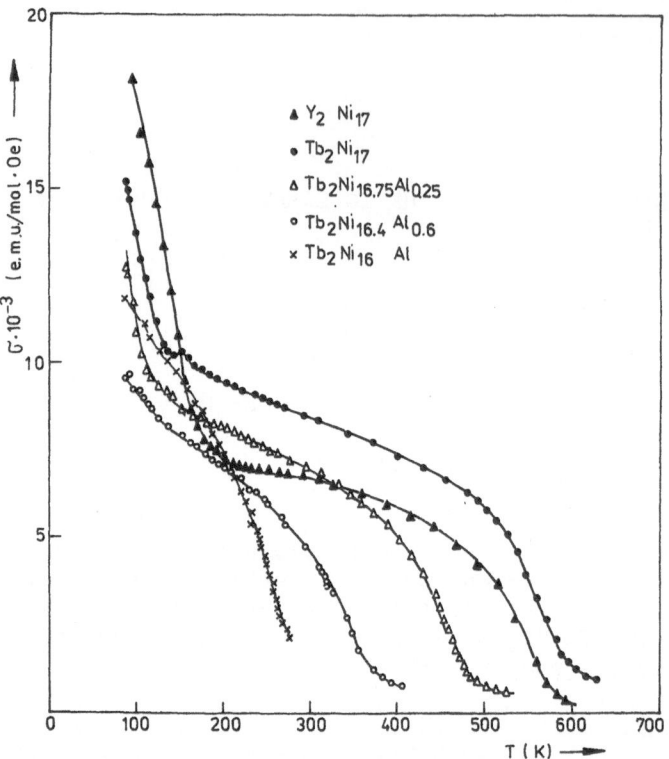

Figure 1.   Magnetization vs temperature for
Y_2Ni_{17} and $Tb_2Ni_{17-x}Al_x$.

regular hexagons. The Ni atoms on 4f sites are located between
the planes formed by the three Ni sites mentioned above.

In order to understand the magnetic behavior of these compounds, we first examine the magnetic properties of the isostructural compound Y_2Ni_{17}. Our results show that the magnetization is
indeed lost in two stages (Fig. 1), as observed by Carfagna and
Wallace (5). However, a discrepancy appears in the Curie temperature obtained by us (T_c = 564 K) and that obtained by Carfagna and
Wallace (621 K). Further, the thermal variation of the reciprocal
magnetic susceptibility (Fig. 2) shows clearly that the compound
Y_2Ni_{17} is ferrimagnetic and not ferromagnetic as was considered by
the above authors. The magnetic susceptibility does not obey the
usual law of Néel-type, which suggests that below T_c the ferrimagnetic structure is not of a collinear type. The ferrimagnetic behavior of Y_2Ni_{17} may be explained, taking into account the different
sites of Ni atoms which form several spin sublattices. The analysis
of the magnetic interactions between these sublattices shows that in
the ground state at 0 K the spins on 4f sites (sublattice A) are
antiparallel to those on 6g, 12j and 12k sites (sublattices B). All

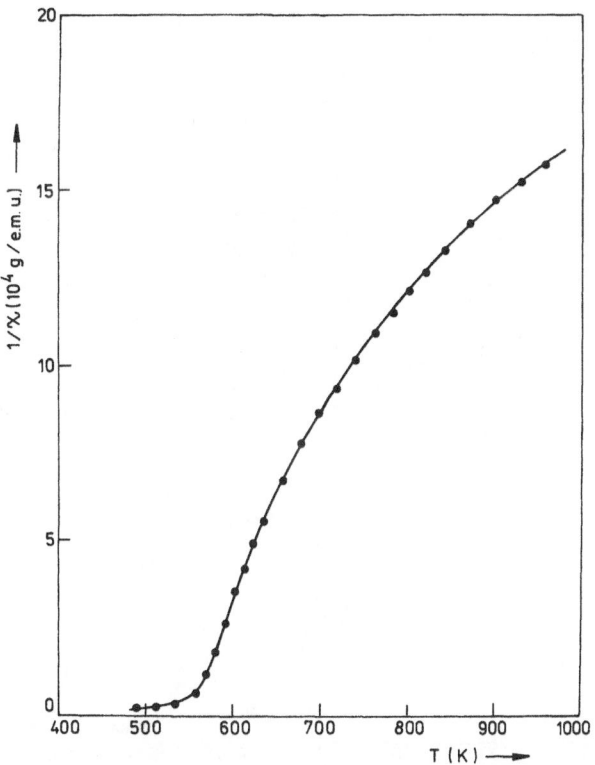

Figure 2. Reciprocal magnetic susceptibility
vs temperature for Y_2Ni_{17}.

B sublattices are equivalent and the nearest-neighbor interactions
between any two B sublattices are equal. The anomaly in magnetization at lower temperatures may be interpreted as a second order
phase transition of order-order type, i.e., from collinear to noncollinear ferrimagnetic structure, according to the Yafet and Kittel
theory (8). Since the separation between Ni atoms on B sublattices
d_{BB} is smaller than d_{AB}, the B-B interaction is dominant. At the
temperature when the condition $\vec{M}_A + \gamma_2\vec{M}_B = 0$ is satisfied, the spins
on B sites fall into two similar sublattices, whose magnetizations
are at an angle to each other, and their resultant is antiparallel
to the magnetization of the spins on the A sublattice. In our case
the coefficient $\gamma_2 = M_A/M_B = 4/30 = 0.13$, which is very close to
that predicted by Yafet and Kittel for achieving the triangular
arrangement in the B sublattice.

In the case of Tb_2Ni_{17} and Gd_2Ni_{17} compounds also, the double
magnetization has been observed (Fig. 3 and 4). This may be explained in the same manner as for Y_2Ni_{17}. The R ions are located

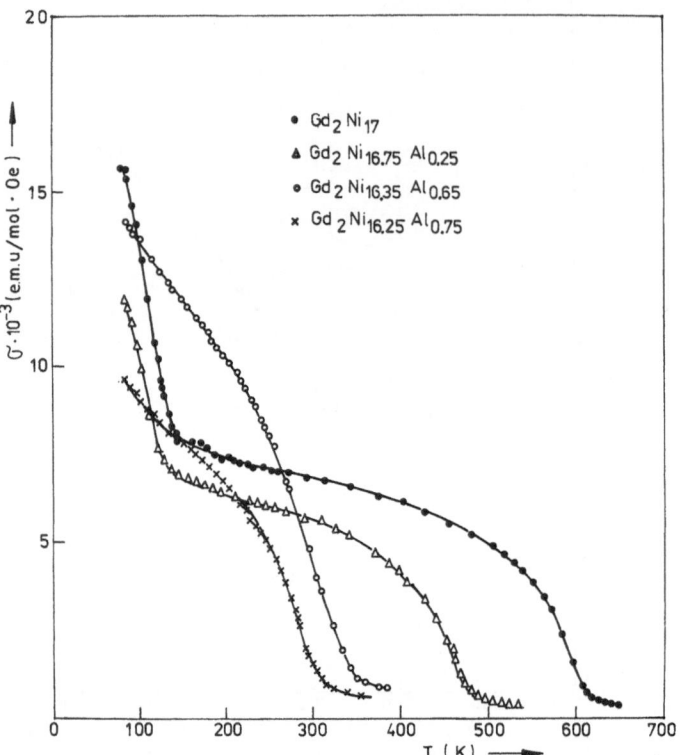

Figure 3.   Magnetization vs temperature
            for $Gd_2Ni_{17-x}Al_x$.

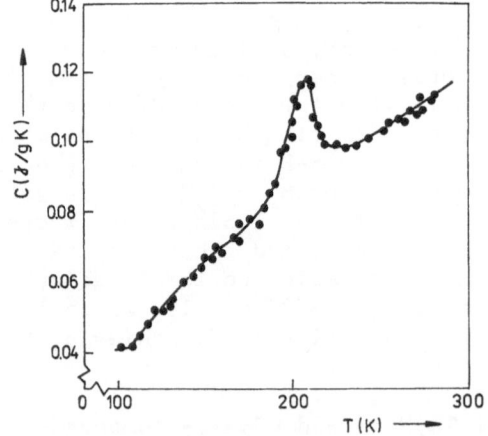

Figure 4.   Specific heat vs temperature
            for Tb_2Ni_{17}.

in the center of the hexagons formed by Ni atoms on 12j sites. The
ground state at 0 K is formed now from three collinear sublattices,
A, B and C, where the sublattice C comprises the R atoms on 2b and
2d sites. Because the B-C interaction is dominant now, the sublattices B and C will be magnetized in opposite directions, so that
the sublattice C is parallel to A. When the triangular arrangement
appears in the B sublattice the spins on the C sites fall into two
similar sublattices whose magnetizations also form a triangular
arrangement due to the strong B-C interaction, their resultant
being antiparallel to that of the B sublattice and consequently
parallel to A. This transition of order-order type is also supported by the anomaly in the specific heat (Fig. 4).

Analysing the symmetry operations for the equivalent atoms,
one may suppose that in the pseudobinary compounds $R_2Ni_{17-x}Al_x$
investigated, the Al atoms occupy the 4f sites, because in this
situation the symmetry operations are least affected. The double
demagnetization is gradually attenuated as Al content increases and
vanishes for $x > 0.65$ (Figs. 1 and 3), when the A-B interaction is
negligible in comparison with the B-C interaction. Above this concentration the compounds are collinear ferrimagnets.

In all the compounds investigated the Ni-Ni distances are comparable with those in pure Ni metal ($d_{Ni-Ni} = 2.72$ Å in Gd_2Ni_{17} and
2.48 Å in pure Ni metal). This shows that the Ni-Ni interaction is
the dominant one, which also explains the high values of T_c and the
monotonic variation of Curie temperatures with Al concentration in
the pseudobinary compounds $R_2Ni_{17-x}Al_x$.

Summarizing the results, one can say that the double demagnetization is indeed a characteristic feature for R_2Ni_{17} compounds, as
was stated by Carfagna and Wallace, and that this behavior may be
well explained by the Yafet and Kittel theory.

REFERENCES

1. I. Pop, N. Dihoiu, M. Coldea and C. Hägen, J. Less-Common
    Metals 64, 63 (1979).
2. W. B. Pearson, Handbook of Lattice Spacings and Structure of
    Metals and Alloys. Pergamon Press, Vol. 2, p. 266 (1967).
3. J. Laforest, R. Lemaire, D. Paccard and P. Pauthenet, C.R.
    Acad. Sci., Paris 264, 676 (1967).
4. E. Burzo and J. Laforest, C.R. Acad. Sci., Paris 274, 114
    (1972).
5. P. D. Carfagna and W. E. Wallace, J. Appl. Phys. 39, 5259
    (1968).
6. W. E. Wallace, Rare Earth Intermetallics. Academic Press, Inc.,
    New York, pp. 140-141 (1973).

7. N. M. Henry and K. Lonsdale, <u>International</u> <u>Tables</u> <u>for</u> <u>X-Ray</u>
    <u>Crystallography</u>. The Kynach Press, Birmingham, England, Vol.
    1, pp. 304-305 (1952).
8. Y. Yafet and C. Kittel, Phys. Rev. <u>87</u>, 290 (1952).

OBSERVATION OF SPIRAL SPIN ANTIFERROMAGNETIC DOMAINS IN SINGLE CRYSTAL TERBIUM

S.B. Palmer, Dept. of Applied Physics, University of
Hull, Hull, U.K.
J. Baruchel, Laboratoire Louis Neel du CNRS, 166X,
38042 Grenoble, France.
S. Farrant and D. Jones, Centre for Materials
Science, University of Birmingham, Birmingham, U.K.
M. Schlenker, Laboratoire Louis Neel du CNRS and
Inst. Laue Langevin, 38041 Grenoble, France

We have carried out the first observation of spiral-spin antiferromagnetic domains by means of polarized neutron diffraction topography of a single crystal of Tb. The antiferromagnetic (AF) phase can be reached by cooling the sample through the Néel temperature T_N or by warming it from below the Curie temperature T_C. Markedly different domain structures are observed for the two cases. The effects of repeated thermal cycling and induced strain have been studied for the two kinds of domain structures.

INTRODUCTION AND EXPERIMENTAL DETAILS

The heavy rare earth element Tb has three magnetic phases. It is paramagnetic at high temperatures and orders in a spiral spin AF state below $T_N \sim 219K$ into a ferromagnetic (F) phase. In the spiral spin phase the magnetic moments lie normal to the hexagonal, $\vec{c}$, axis which is also the helix axis. In a particular basal plane sheet the moments are aligned ferromagnetically but the direction of magnetization rotates between successive Tb sheets by an angle θ. The turn angle θ decreases slowly with temperature from 20° at T_N to 18° at T_C (1).

Measurements of the temperature dependence of the magnetic susceptibility (2), the elastic constant C_{33} and its ultrasonic attenuation α_{33} (3) have revealed differences in the AF phase depending on whether the sample is cooled from above T_N or warmed from below T_C. These anomalies have been attributed to spiral

spin AF domains where the different domains are characterized by
having spirals of opposite sense (4). The present note describes
the application of neutron diffraction topography to visualize the
domains (5) and to study, in a preliminary manner, their properties.

The neutron diffraction pattern of a helimagnet is characterized by pairs of magnetic satellites $hk\ell_{\pm}$ which accompany the
nuclear reflection (1).   If the incident neutron beam is polarized
and the polarization direction, $\vec{p}$, is parallel to both the scattering vector $\vec{h}$ and the propagation vector $\vec{k}$ of the magnetic spiral
then the entire contribution to one of the magnetic satellites
comes from one kind of spiral spin domain while the other kind of
domain diffracts only on the other satellite peak (6).

To record the domain structure using polarized neutron
diffraction topography the single crystal sample was set for a
chosen Bragg reflection with scattering geometry as described above.
A neutron sensitive photographic detector (Kodak Periapical Ultra
Rapide, backed by a 10μm thick ^{167}Gd screen acting as (n–β)
converter) was placed in the diffracted beam. Local variations in
the recorded scattering intensity are associated with defects or
domains in the crystal.   The present work was performed on the
diffractometer S.20 of the Institut Laue-Langevin in Grenoble, at a
wavelength of $\sim 1.5\mathring{A}$ (7).   The Tb sample used was grown by electron
beam melting followed by zone refining. The resulting single
crystal was then placed in a SSE furnace and held just below the
melting point for 1000 hours (8).   After spark machining and etch-
ing the final thickness of the sample was 0.3mm and its diameter
5mm.   The $\vec{c}$ and $\vec{a}$ axes were contained in the plane of the disc.
The mosaic spread of the 00ℓ lattice planes was measured by neutron
diffraction to be less than 2' indicating a high quality single
crystal.

RESULTS AND DISCUSSION

Figure 1 shows topographs of both the 002 reflection and the
associated satellites $002_{\pm}$.   They were carried out at a temperature about two degrees above T_C (T_0) after the sample had been
cooled from the paramagnetic phase.   The central 002 nuclear
reflection topograph (Fig. 1a) shows no images associated with
domains or domain walls and the defect images are very similar to
those obtained with a room temperature topograph (7).   The topograph made with the 002_+ satellite (Fig. 1b) shows irregularly
shaped domains which appear alternately black and white.   The
white regions are those scattering neutrons into this particular
satellite. The contrast of these regions is reversed in Fig. 1c,
the topograph of the 002_- satellite, as would be expected since it
is the domains of opposite chirality that are now scattering.

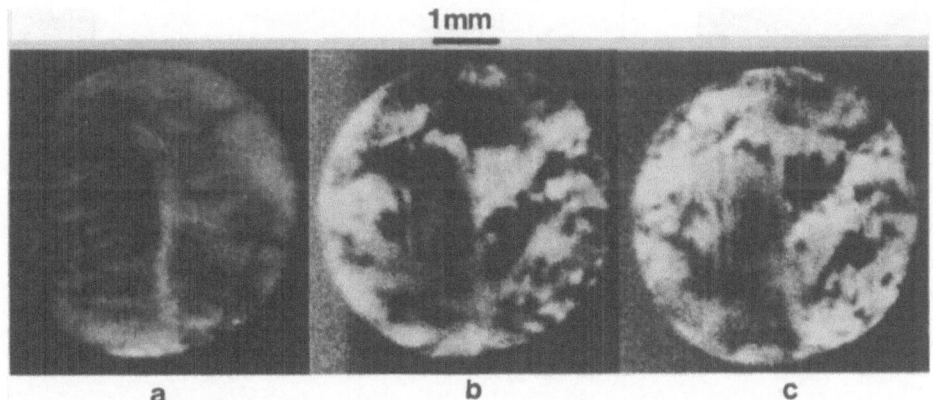

Fig. 1 Topographs of the 002 reflection in Tb at $T_0 \sim T_C + 2K$.
Sample cooled from above T_N. (a) 002 central peak, (b) 002_+
satellite, (c) 002_- satellite. (In all Figs. except 2c $\vec{p}$ is nearly
along [001] and parallel to $\vec{h}$ and $\vec{k}$).

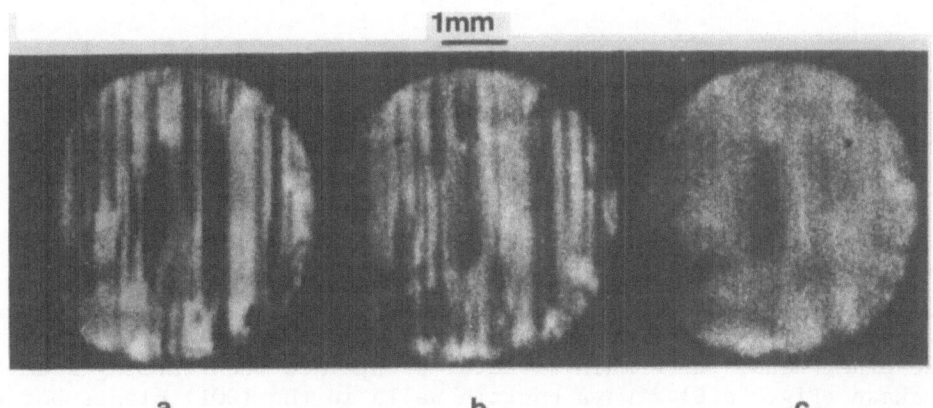

Fig. 2 Topographs of the 002_- satellite in Tb at T_0. Sample warmed
from below T_C. (a) $\vec{p}$ along [001], (b) $\vec{p}$ along [00$\bar{1}$], (c) $\vec{p}$ along
the 'a' sample direction.

    When the sample is cooled into the F phase and then warmed to
T_0, the domain structure is very different (Fig. 2).   Alternate
black and white bands now run in a direction normal to the $\vec{c}$ axis.
The two topographs show the same domain structure with the contrast
being reversed in Fig. 2b with respect to Fig. 2a.   Figure 2c
shows a topograph of the 002_- satellite when $\vec{p}$ has been turned
through 90^0 destroying the symmetry required to visualize the
domains.

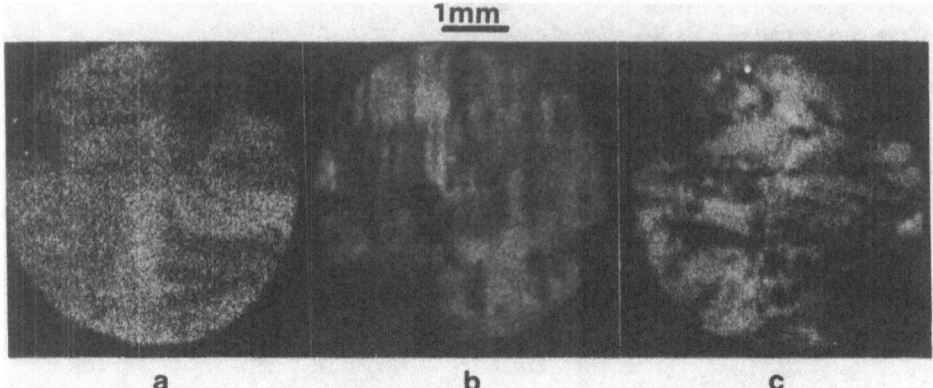

Fig. 3 Topographs of the 002 reflection in Tb at T_0 after mechan-
ically straining the sample. (a) 002 central peak, (b) 002_
satellite after cooling from above T_N, (c) 002_ satellite after
warming from below T_C.

        Repeated sample cooling from above T_N indicates that the
domain structures of Fig. 1b,c are nearly reproducible and some
correlation can be observed when comparing Figs. 1a and 1b,c. On
the other hand, the domain structures obtained when warming from the
F phase (Fig. 2a,b) always contain walls in the (001) planes but the
positions of these walls varies from run to run. This suggests that
the former domain structure could be related to crystal defects
while the latter, more regular arrangement, is likely to be associated with the parent F domain structure. The same sample was
studied after it had been slightly strained (Fig. 3). The nuclear
peak (Fig. 3a) shows enhanced distortion with respect to Fig. 1a
and the domain structure observed on cooling from above T_N (Fig. 3b)
is now totally unlike the original topograph (Fig. 1b,c) whereas
the structure observed when warming from below T_C still contains
walls in the (001) planes (Fig. 3c).

        The authors would like to thank A. del Morel, S. Hosoya and
B.K. Tanner for independently suggesting the relevance of neutron
topography to this problem.

REFERENCES

(1) W.C. Koehler, Magnetic Properties of Rare Earth Metals, ed. R.J. Elliott, Plenum Press, London (1972)

(2) T.J. Mckenna, S.J. Campbell, D.H. Chaplin, G.H.J. Wantenaar and G.V.H. Wilson, J. de Phys., 40: C5-22 (1979)

(3) D.C. Jiles, G.N. Blackie and S.B. Palmer (to be published)

(4) S.B. Palmer, J. Phys. F: Metal Physics, 5:2370 (1975)

(5) M. Schlenker and J. Baruchel, J. Appl. Phys., 49:1996 (1978)

(6) Yu. A. Izyumov, Sov. Phys. JETP, 15:1162 (1962)

(7) J. Baruchel, S.B. Palmer and M. Schlenker, J. de Phys., (to be published)

(8) R.G. Jordan, D.W. Jones and V. Hems, J. Less Comm. Metals, 42:101 (1975)

CRYSTAL FIELD AND QUADRUPOLAR EFFECTS ON THE

THIRD–ORDER MAGNETIC SUSCEPTIBILITY

Pierre Morin and Denys Schmitt

Laboratoire Louis Néel
CNRS, Associé à l' Université
166 X, 38042 Grenoble Cedex - France

The magnetic susceptibility has been used for a long time for studying magnetic interactions as well as crystalline electric field (CEF) effects in rare earth compounds. Less attention has been turned to the third-order magnetic susceptibility which characterizes the initial curvature of the paramagnetic magnetization curves. This third-order susceptibility is very sensitive to both the CEF and quadrupolar interactions ; in addition, it is strongly anisotropic according to the investigated crystallographic direction (1). Such an experimental method then constitutes a new tool for studying CEF effects as well as quadrupolar - magneto-elastic and two-ions - interactions in cubic rare earth intermetallic compounds where they have been shown to be large (2).

The starting Hamiltonian includes the fourth- and sixth-order cubic CEF terms, the Zeeman coupling, and the bilinear (Heisenberg) and quadrupolar interactions written in the molecular field approximation (3) :

$$\mathcal{H} = Wx\, O_4/F_4 \;+\; W\,(1-|x|\,)O_6/F_6 \;-\; g_J\mu_B\,\vec{H}\cdot\vec{J}$$

$$-\left[3\,\theta^*/J(J+1)\right]\,\langle\vec{J}\rangle\cdot\vec{J} \;-\; G_1\!\left[\langle O_2^0\rangle\,O_2^0 \;+\; 3\langle O_2^2\rangle O_2^2\right] \tag{1}$$

$$-\,G_2\!\left[\langle P_{xy}\rangle\,P_{xy} \;+\; \text{cycl.}\right]$$

In this expression, $G_1(G_2)$ is the quadrupolar parameter associated with the tetragonal (trigonal) symmetry lowering mode; both coefficients receive a contribution from the magnetoelastic coupling and from the two-ions indirect quadrupolar interaction(2) θ^* is the bilinear exchange parameter.

Due to symmetry considerations, the magnetization M may be written in the paramagnetic range, as:

$$M = g_J \mu_B \mid <\vec{J}> \mid \quad = \chi_M^{(1)} \ H + \chi_M^{(3)} \ H^3 + \ldots \tag{2}$$

Applying perturbation theory leads to the following expression for the first- and third-order magnetic susceptibilities (1,4) :

$$\chi_M^{(1)} = \chi_0^{(1)} / \ (1 - n\chi_0^{(1)} \ ) \tag{3}$$

$$\chi_M^{(3)} = \left[\ \chi_0^{(3)} + 2 \ G_1 (\chi_2^{(2)})^2 / \ (1 - G_1 \chi_2) \right] / \ (1 - n\chi_0^{(1)})^4 \tag{4}$$

with a magnetic field applied along a fourfold axis ; note that in the case of a threefold axis, G_1 should be replaced by $G_2/12$. Four CEF susceptibilities are involved in these expressions, and may be analytically calculated from the cubic CEF level scheme ; $\chi_0^{(1)}$ and $\chi_0^{(3)}$ are the first- and third-order magnetic susceptibilities without any interactions ; χ_2 is the strain susceptibility appearing in the elastic constants experiments (5) ; $\chi_2^{(2)}$ is the quadrupolar susceptibility occuring in the parastriction (3).

In the Eqs. (3) and (4), $\chi_M^{(1)}$ is reinforced only by the bilinear exchange interactions $n = 3\theta^*/g^2 \mu_B^2 \ J(J+1)$ while $\chi_M^{(3)}$ receives a contribution from n and from the quadrupolar interactions G_1 (or G_2). This quadrupolar contribution may either enhance or reduce the pure CEF contribution $\chi_0^{(3)}$ according to the relative sign of G_1 (or G_2) (see the example below).

The highly anisotropic behaviour of $\chi_0^{(3)}$ may be emphasized by taking some examples with the Tm^{3+} ion. Figure 1 shows the temperature variation of $\chi_0^{(3)}$, when H lies along the $[001]$ or the $[111]$ axis, for 3 different ground states. At low temperature Curie and/or Van Vleck behaviours can be found according to the ground state and the field direction. This variety of situations may then give precise informations on the CEF level scheme.

An experimental method for obtaining $\chi_M^{(3)}$ consists in a detailed analysis of the magnetization curves, reporting M/H as a fonction of H^2. Figure 2 shows such plots for TmCd : the behaviour is linear till the higher order terms (in H^5 ...) become large in Eq. 2. The isotropic null field value is $\chi_M^{(1)}$ whilst the anisotropic $\chi_M^{(3)}$ values are given by the different slopes.

Examples of the temperature behaviour of $\chi_M^{(3)}$ may be provided by the compounds TmCd and $Tm_{0.3}Y_{0.7}Cd$: this system has been intensively studied in recent years because of large quadrupolar interactions which drive TmCd to be quadrupolarly ordered below $T_Q = 3.2$ K (6).

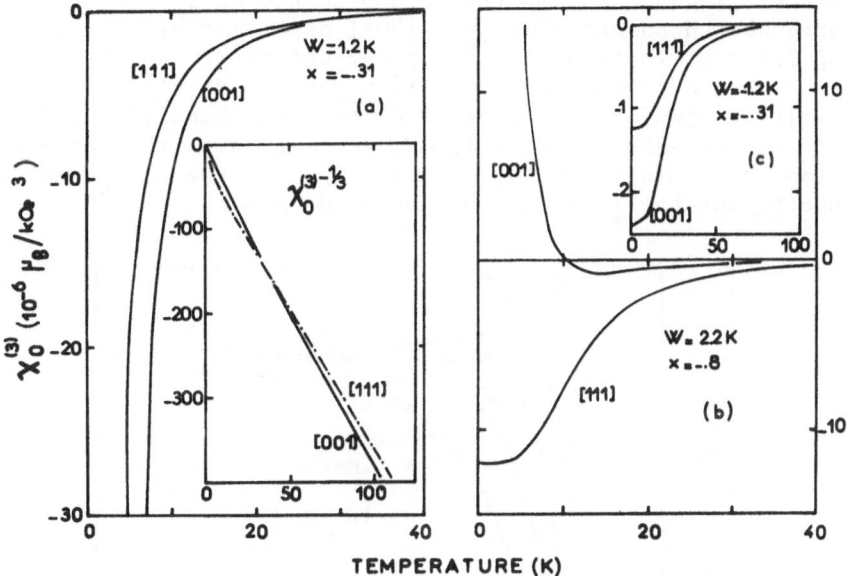

Figure 1 - Calculated variation of $\chi_0^{(3)}$ with a triplet (a),
a non-magnetic but quadrupolar doublet (b) and
a singlet (c) as ground state (J = 6).

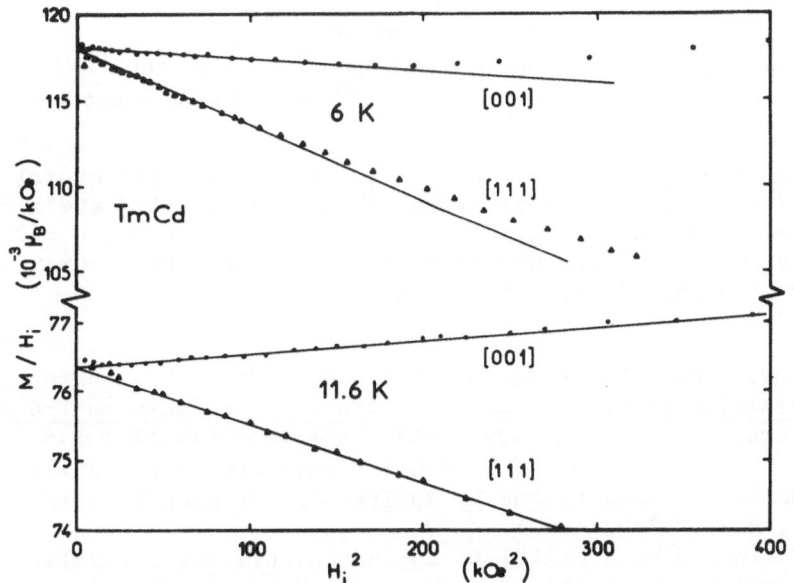

Figure 2 - Experimental values of M/H versus H^2 in TmCd.

Figure 3 shows the temperature dependence of $\bar{\chi}_M^{(3)}$ for both compounds when H points along the [001] axis. The effect of the quadrupolar contribution to $\chi_M^{(3)}$ is obvious here: in TmCd, it is large enough and of opposite sign with respect to $\chi_0^{(3)}$ to change the resulting sign of $\chi_M^{(3)}$ above 9 K. Note the sensitivity for determining the value of G_1 = 13 mK, that is in perfect agreement with other determinations (6). In $Tm_{0.3}Y_{0.7}Cd$, the tetragonal quadrupolar interactions have vanished (G_1 = 0).

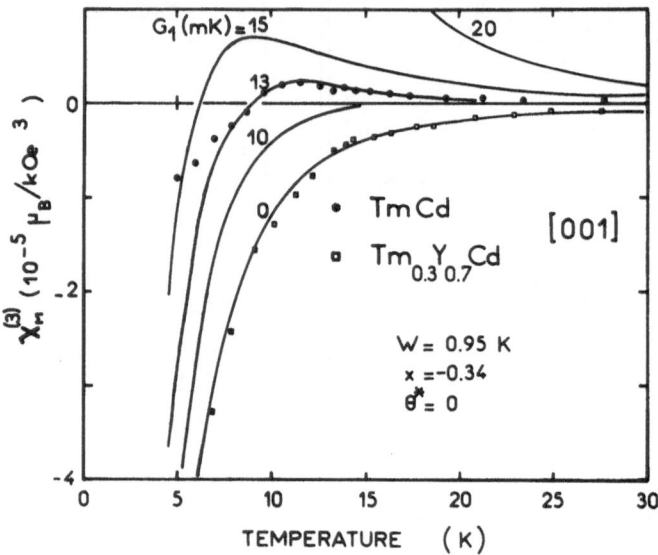

Figure 3 – Experimental and calculated temperature variations of $\chi_M^{(3)}$ along the [001] axis in TmCd system

When H is parallel to the [111] axis, both experimental curves for TmCd and $Tm_{0.3}Y_{0.7}Cd$ are very close to each other and may be fitted with the same trigonal quadrupolar parameter G_2 = – 50 mK. Note that G_1 and G_2 are opposite in sign as in other isomorphous thulium compounds (1,4).

REFERENCES

1. P. Morin, D. Schmitt and E. du Tremolet de Lacheisserie, in Crystalline Electric Field and Structural effects in f-Electron Systems, J.E. Crow (ed.),Plenum Press,New York,pp 61-74 (1980).
2. P.M. Levy, P. Morin and D. Schmitt,Phys.Rev.Lett.,42,1417 (1979).
3. P. Morin, D. Schmitt and E. du Tremolet de Lacheisserie, Phys. Rev., B 21, 1742 (1980).
4. P. Morin and D. Schmitt, to appear in Phys.Rev., (1981).
5. M.E. Mullen, B. Lüthi, P.S. Wang, E. Bucher, L.D. Longinotti, J.P. Maita and H.R. Ott, Phys.Rev., B 10, 186 (1974).
6. R. Aléonard and P. Morin, Phys. Rev., B 19, 3868 (1979).

MAGNETIC PROPERTIES OF SOME SINGLE CRYSTAL RARE EARTH TETRABORIDES

J. Etourneau, B. Chevalier, Chimie du solide, Université
de Bordeaux I, 33405, Talence, France; R. Georges,
J.C. Gianduzzo, Physique du Solide, Université de Bor-
deaux I, 33405, Talence, France; G. Will, W. Schäffer,
Mineralogisches Institut der Universität, 5300 Bonn
Fed. Rep. Germany

INTRODUCTION

The rare earths tetraborides show the tetragonal (P4/mbm) crystal structure (1). The boron skeleton displays two kinds of large
channels along the c-axis (c-channels) and along the (110) directions
(u-channels), at the intersection of which the cations are located.
Since the boron framework stability only requires a double electron
transfer from each rare earth (2), these compounds are metallic conductors and the localized moments of the 3+ rare earths cations are
coupled by indirect exchange via the conduction electrons. Highly
anistropic magnetic properties are expected. Some of them, concerning TbB_4 and HoB_4 are described here, however, the final discussion
will concern PrB_4, DyB_4 and ErB_4, too. The single crystals preparation has been described elsewhere (3). A flux extraction magnetometer has been used for magnetic measurements; neutron diffraction
experiments have been performed on the Katinka diffractometer of the
University of Bonn, using samples enriched with the low absorbing [11]B
isotope.

EXPERIMENTAL RESULTS: TbB_4

The thermal variation of the principal susceptibilities X_a and
X_c indicate a very anisotropic behavior (θ_{pa}=-18K, θ_{pc}=-48K) with
antiferromagnetic ordering below 45K. Both Curie constants are consistent with the presence of Tb^{3+} cations. Neither X_a nor X_c vanishes at absolute zero, and the moments should lie in the (a,b)
plane, with some kind of domains structure. Indeed, the neutrons
diffraction patterns obtained at 4.2K and 50K suggest ferromagnetic
(100) layers with an antiferromagnetic AABBAABB... sequence (Fig. 1).

423

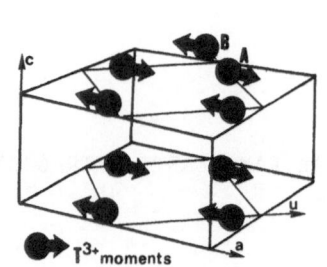

Figure 1. Magnetic struc-
ture of TbB₄ at helium
temperature.

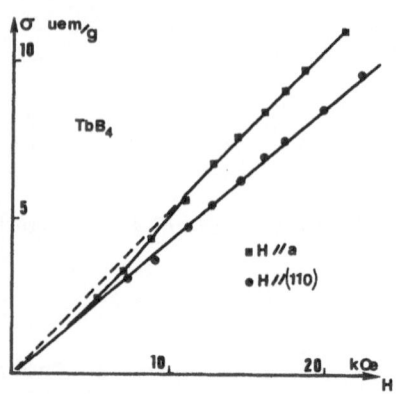

Figure 2. Spin flopping in
TbB₄.

The moments are within the layers and normal to the c-axis. A clearcut spin flopping effect is observed when the magnetic field is
applied along the a-axis, in agreement with the expected domains
structure (Fig. 2).

HoB₄

The thermal variations of X_a and X_c for HoB₄ also show a very
anistropic antiferromagnetic behavior (θ_{pa}=-13.4K; θ_{pc}=-9.8K). The
Néel temperature (6.7K) is lower than previous estimates (4). Again
both Curie constants are consistent with 3+ cations. Since X_c vanishes in the low temperature limit, the moments should lie along the
c-axis. At 2.25K two metamagnetic-like transitions are observed as
the field is applied along this direction (Fig. 3). The situation is
similar to that met with ErB₄ (5), except that a further transition
is expected in higher fields in order to reach the full magnetization
of the rare earth. Another phase transition is observed when the
field is applied along the a-axis (Fig. 4). Its nature is not yet
elucidate, but it should be an exchange anistropy effect. Further
measurements including electrical conductivity as a phase-sensitive
property (6), and neutron diffraction experiments are on course, in
order to determine the complete phase diagram.

DISCUSSION

A quantitative approach performed on ErB₄ (7) has revealed that
the coupling between A-B nearest neighbors (Fig. 1) mainly determines
the magnetic ordering. A point-charge calculation shows that the
crystal field strongly stabilizes the $5d_{c^2-u^2}$ orbitals. Their spatial configuration suggests that the A-B pairs form double chains

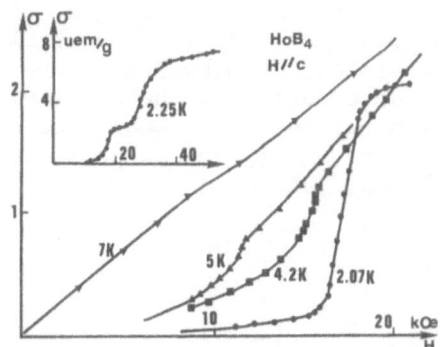

Figure 3. Metamagnetic behavior of HoB$_4$.

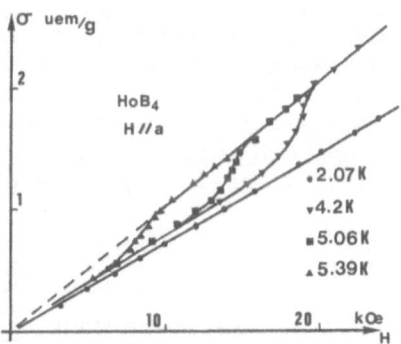

Figure 4. Field variation of the magnetization in HoB$_4$.

along c. The corresponding cation-cation distance is the c parameter ($\approx$4.0Å), whereas it is only 3.7Å within the pairs. Let t and T be the related transfer integrals, then t should be significantly smaller than T. As a result, the $5d_{c^2-u^2}$ orbitals should first mix into bonding and antibonding combinations split by 2T. These in turn give rise to almost one dimensional bands exhibiting the width 2t (Fig. 5). The metallic behavior indicates a band overlap. Thus, 2t should be smaller than T. Now, the overall conduction electron energy variation due to the onset of ferromagnetic or antiferromagnetic ordering may be expressed in terms of t, T, the 4f-5d exchange coupling J, and the Fermi wave number (in the lower band) k_F:

$$\Delta E_F = -(J^2 S^2/4)(dN/dE)_{k_F} \; ; \qquad \Delta E_{A_F} = -N(J^2 S^2/4T)(1-2k_F c/\pi)$$

where dN/dE is the density of states. The relative stability of these configurations depends on the ratio t/T, the critical value being 0.54. As results from the relative orders of magnitude, the

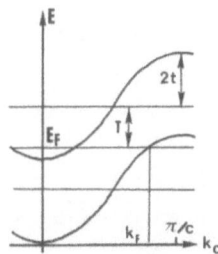

Figure 5. Band structure in the rare earths tetraborides.

actual value of the ratio should be close to the critical one. And it may cross it when going along the rare earths series, accounting for the different behaviors exhibited by PrB_4 and the other rare earths tetraborides. Since the k_F value relative to the critical ratio is slightly lower than π/c, one predicts a very weak band overlap, and a highly magnetic field sensitive electrical properties, as was indeed observed by some of us in PrB_4, TbB_4, and ErB_4 (6).

REFERENCES

1. P. Blum and F. Bertaut, Acta Cryst., 7:81 (1954). G. Will, P. Pfeiffer, W. Schäfer, J. Etourneau and R. Georges, Rev. Chim. Miner., 17:541 (1980).
2. W.N. Lipscomb and D. Britton, J. Chem. Phys., 33:275 (1960).
3. B. Bressel, B. Chevalier, J. Etourneau and P. Hagenmuller, J. of Crystal Growth, 47:429 (1979).
4. A. Berrada, J.P. Mercurio, B. Chevalier, J. Etourneau, P. Hagenmuller, M. Lalanne, J.C. Gianduzzo and R. Georges, Mat. Res. Bull., 11:1519 (1976) and references therein.
5. F. Pfeiffer, W. Schäfer, G. Will, J. Etourneau and R. Georges, J. of Magn. and Magn. Mat., 14:306 (1979).
6. J. Etourneau, J.P. Mercurio, P. Hagenmuller, A. Berrada, R. Georges, R. Bourezg and J.C. Gianduzzo, 6th Int. Symposium on Boron and Borides, J. Less Common Metals, 67:257 (1979).
7. R. Bourezg, Thesis, University of Bordeaux I, France.

TRANSFERRED HYPERFINE INTERACTIONS AND QUADRUPOLE EFFECTS FOR THE DIAMAGNETIC IONS IN RARE-EARTH ELPASOLITES, $Cs_2NaLnCl_6$

F.W. Voss, R. Nevald and G.P. Knudsen, Dept. of Electro-
physics, The Technical University of Denmark, DK-2800
Lyngby, Denmark; H.-D. Amberger, Inst. für Anorg. und
Angew. Chemie, Universität Hamburg, D-2 Hamburg 13
Germany

ABSTRACT

Various experimental investigations of the magnetic properties of
the rare-earth elpasolites $Cs_2NaLnCl_6$ (Ln = Ce-Yb) have been performed, in order to elucidate the weak covalent couplings of the
Ln's with their surroundings. By these experiments, made on single
crystals, four parameters describing the covalency are determined:
the electric field gradient, the isotropic and anisotropic magnetic
shifts and the orbital reduction factor.

INTRODUCTION

The group $Cs_2NaLnCl_6$ crystallizes in the cubic structure O_h^5 (Fm3m),
but whereas the compounds with the heaviest lanthanides stay cubic
to low temperature, those with light lanthanides undergo a phase
transition to tetragonal symmetry. The crystals which remain cubic
are very attractive because of the low number of parameters characterizing various physical effects. The tetragonally distorted crystals are more complicated, but when the temperature dependence of
all the symmetrically free parameters of this case are determined,
we have the opportunity to follow the variation in the covalency as
a function of these parameters. In the study of covalency the Yb-
compound is of special interest because the octahedral crystal
field predicts the ground level to be a Γ_6 doublet. The two Γ_6 wave
functions are combinations of one electron wave functions, whose orbital parts all transform like t_{1u}, relating ground state properties to the covalency of t_{1u} orbitals alone.

THEORY

The Hamiltonian of a nucleus in a crystal with an applied field H_a is

$$H = -\gamma \hbar H_{loc} I_{z'} + Ah(3I_z^2 - I^2 + \eta(I_x^2 - I_y^2)) \qquad (1)$$

where H_{loc} is the field at the nucleus and $6A = 6e^2qQ/(4I(2I-1)h) = \nu_Q$ is the strength of the quadrupole interactions. For a crystalline site with a four-fold axis $\eta = 0$, and for a cubic site $A = 0$ too. In the following we consider $\eta = 0$. This is true for all nuclei except the Cl's in the basal plane of the tetragonally distorted crystals which are not fully analyzed yet. The relation between H_a and H_{loc} is shown by Nevald and Hansen (1) to be

$$\bar{H}_{loc} = \bar{H}_a + (\bar{\bar{\alpha}} + \zeta\bar{\bar{1}})\bar{M}/\rho = \bar{H}_a + (\bar{\bar{\delta}} + \bar{\bar{\epsilon}} + \zeta\bar{\bar{1}})\bar{M}/\rho \qquad (2)$$

where $\bar{M}$ is the magnetization and ρ the mass density. $\bar{\bar{\delta}}$ is the dipole contribution to the field, whereas $\bar{\bar{\epsilon}}$ and ζ are the anisotropic (ATH) and isotropic (ITH) transferred hyperfine interactions respectively. At cubic sites $\bar{\bar{\delta}}$ and $\bar{\bar{\epsilon}}$ equal zero. The ground level of Yb in an octahedral field is of Γ_6 symmetry. The wave functions spanning Γ_6 are given by Thornley (2) and lead to the g value of (-2-2/3) where the first term is the orbital- and the second the spin-contribution. Construction of molecular orbitals Yb-Cl gives a covalent reduction in the g value which can now be written $g(\Gamma_6) = -2k - 2/3$.

EXPERIMENTS

In ^{23}Na- and ^{133}Cs-NMR on the distorted crystals, where the sites are non-cubic, the quadrupole interaction is found very much smaller than the Zeeman coupling. This means that the central transition is only shifted by the magnetic moments of the lanthanides. The quadrupole interaction is deduced from the satellite transitions. In contrary the ^{35}Cl-nuclei are always affected by a much stronger quadrupole interaction. In this case full numerical diagonalization of eq. (1) has been performed. From this it turns out that for the central transition, treating the quadrupole term as a second order perturbation is an excellent approximation. The result is

$$\nu_m = \nu_L - \nu_Q^2(I(I+1) - 3/4)(1-\cos^2\theta)(9\cos^2\theta-1)/16\nu_L \qquad (3)$$

where I equals 3/2 for Cl, θ is the angle between the field gradient axis and H_{loc}, and $\nu_L = \gamma H_{loc}/2\pi$. In the present experiment the NMR frequency is kept fixed and H_a swept through the resonances. The experimental value of ν_Q is found from the distance between the central and satellite transitions with H_a along a four-fold axis. First, however, the transitions have been corrected for the magnetic shift. The local field H_{loc} which is calculated using eq. (3)

or diagonalizing eq. (1) numerically, is inserted in eq. (2) and thereby $\bar{\bar{\alpha}}$ and ζ are determined. For this analysis single crystal magnetization curves have been measured by us (3). With known Cl-position it is possible to calculate the dipole contribution to $\bar{\bar{\alpha}}$ and extract the ATH parameter $\bar{\bar{\varepsilon}}$.

DISCUSSION

The experimental results are collected in Tables 1 and 2. The low temperature Na- and Cs-NMR spectra in Figs. 1 and 2 clearly show the distortion from cubic symmetry of the Ce- and Nd-compounds. The Na spectra further demonstrate that three types of domains are present in the Ce-elpasolite, each domain having the tetragonal axis along one of the four-fold axis of the cubic phase. The strength of the low temperature quadrupole interaction at the Na- and Cs-sites is decreasing uniformly through the series. The shift in sign of ζ from light to heavy Ln-compounds is in agreement with earlier results on other systems. Hitherto doubt has existed, whether the space group symmetry of the tetragonal phase was C_{4h}^5(I4/m) or D_{4h}^{17}(I4/mmm) involving distortion and rotation or only distortion of the Ln-Cl₆ octa-hedra. The Cl-spectra have too many lines for D_{4h}^{17} symmetry whereas qualitative agreement exists with the C_{4h}^5 space group. It has also been confirmed from soft mode study in neutron diffraction by Knudsen et al. (4) that the Nd-compound has the C_{4h}^5 symmetry. In C_{4h}^5 six types of domains are possible because the Cl octahedron around a Ln may rotate in two directions. This is a complication in the unscramb-ling of the part of the spectrum belonging to the Cl-nuclei in the basal planes. From Table 2 and Fig. 3 is seen that the strength of the quadrupole interaction at the Cl is some MHz, which is less than

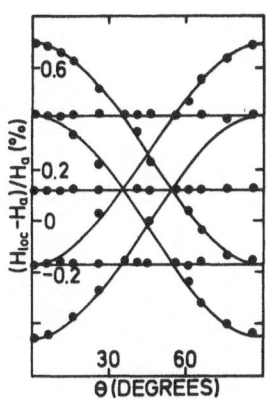
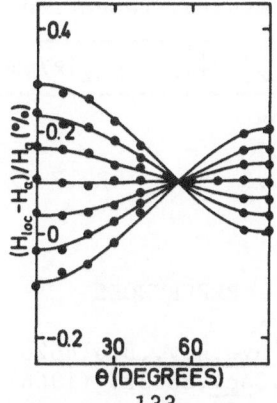
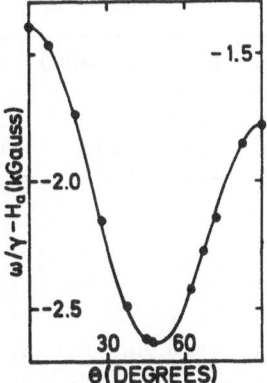

Figure 1. ^{23}Na-NMR in Cs₂NaCeCl₆ for H_a in (100) at 4.2K. ν_m = 30.5 MHz.

Figure 2. ^{133}Cs-NMR in Cs₂NaNdCl₆ for H_a in (100) at 22K. ν_m = 15.1 MHz.

Figure 3. The central ^{35}Cl-NMR in Cs₂NaYbCl₆ for H_a in (110) at 4.2K ν_m = 15.98 MHz.

Dots are experimental results and lines are calculated from Tables 1&2

Table 1. Electric field gradient q, isotropic shift ζ and α_{zz} = $-2\alpha_{xx} = -2\alpha_{yy}$ of the diagonal anisotropic magnetic shift tensor $\bar{\alpha}$. (a) is corrected results from the work of Nevald et al. (5)

	T(K)	$|q|$ (Å^{-3})		ζ (g/cm^3)		α_{zz} (g/cm^3)		
		Na	Cs	Na	Cs	Na	Cs	
Ce	4.2	0.019	0.024	18.3	–	~ 0	–	
Nd	4.2	0.013	0.013	23.2	15.3	-12.4	-6.4	(a)
Sm	4.2	~ 0.004	–	–	–	–	–	
Ho	4.2	0	0	-6.3	–	0	0	(a)
Yb	4.2	0	0	0	0	0	0	

in most other compounds proving covalency to be weak. The isotropic shift at the Cl site is ten to twenty times larger than at F in compounds with similar Ha-Ln distance, which is reasonable. Furthermore the Cl-shift is also positive for the light and negative for the heavy elpasolites, indicating that the filled outer electronic shells of the rare-earth ions must be taken into consideration in constructing molecular orbitals. The ESR line measured on the dense Yb elpasolite is rather narrow, and the g value is in excellent agreement with the value found by Devaney and Stapleton (6) on Yb in Cs_2NaYCl_6. To understand quantitatively the covalency, a detailed experimental investigation of the whole series of elpasolites is in progress. The authors are very grateful to Prof. V. Frank for his assistance and advice.

Table 2. The quadrupole interaction, isotropic magnetic shift ζ and anisotropic shift α_{zz} at the Cl site. For Yb the dipole contribution δ_{zz} is calculated and ε_{zz} extracted. For Yb the g value and orbital reduction factor is shown.

	T(K)	ν_q(MHz)	ζ	α_{zz}	δ_{zz}	ε_{zz}(g/cm^3)	g	1-k
Nd	150	2.4	151	138	–	–	–	–
Nd	4.2	2.33	64	57	–	–	–	–
Yb	4.2	4.013	-250	35.5	58.0	-22.7	2.585±0.004	0.041±0.002

REFERENCES

R.Nevald and P.E.Hansen,Phys.Rev.,B18:4626 (1978).
J.H.M.Thornley,Proc.Phys.Soc.,88:325 (1966).
To be published.
G.Knudsen,F.W.Voss,R.Nevald and H.-D.Amberger,"paper at this conf.".
R.Nevald,F.W.Voss,O.V.Nielsen,H.-D.Amberger,R.D.Fischer,Solid State Commun., 32:1223 (1979).
P.Devaney and H.J.Stapleton,J.Chem.Phys.,63:5459 (1975).

STRUCTURE-PROPERTY INTERPLAY IN THE SYSTEM $RTiO_3$:

R = LANTHANIDE (III)

J.E. Greedan

Institute for Materials Research, McMaster University
Hamilton, Canada L8S 4M1

Few systems illustrate the subtle interplay between structure and physical properties as dramatically as the series $RTiO_3$, R = lanthanide(III), $4f^n$, and Ti(III), $3d^1$. To amplify this statement the crystal structure of this series will be described and the feature or features most crucial in determining physical properties will be identified. We will then examine selected physical properties to see how they vary as a function of the crucial structural parameters.

STRUCTURE

Recently, we determined the structure of five members of the series, R = La, Nd, Sm, Gd, and Y, using single-crystal data[1]. All compounds were found to crystallize in Pbnm (#62, non-standard setting of Pnma). This is the structure originally found by Geller[2] for $GdFeO_3$ and is shown in figure 1. The primitive perovskite cell in the $GdFeO_3$ structure is monoclinic but only slightly distorted from cubic. This gives rise to a high degree of psuedo symmetry which complicates the structural solution, particularly for large R, such as La, where the monoclinic cell edges are of nearly identical length. This has lead previous authors, working from powder X-ray data, to characterize $LaTiO_3$ as cubic and indeed it is possible to index powder patterns of $LaTiO_3$ (Cu radiation, $2\theta < 80°$) on a cubic cell. A further complicating feature was the presence of twinning in the available crystals of $SmTiO_3$, $NdTiO_3$, and $LaTiO_3$. All of these contained the $1\bar{1}0$ twin plane found in $LaFeO_3$, but $LaTiO_3$ exhibited a further rotation twin component about one of the psuedo three-fold axes.

Let us consider some features of the GdFeO$_3$ structure-type and its relationship to the cubic perovskite structure in more detail (Fig. 1). For cubic perovskite the coordination polyhedron of the large cation, R, is a 12-fold cube octahedron and that for the small cation, M, is a regular octahedron. The octahedra share corners in three dimensions and the M-O-M bond angles are 180° as the oxygens are in positions of the type ($\frac{1}{2}$00). The GdFeO$_3$ structure is found where R is too small for 12-fold coordination. A lower energy structure is formed by a cooperative distortion involving a buckling or twisting of the octahedra about each of the cubic axes to give an eight-fold coordination about R. The new cell is now orthorhombic with $a_O \approx b_O \approx \sqrt{2}a_c$ and $c_O \approx 2a_c$ and contains 4 formula units. A primitive cell, outlined by M-atoms in Fig.1, is now not strictly cubic but monoclinic, as mentioned previously. Note the 0 atoms are shifted significantly from their positions in the cubic cell. Among other things this means that M-O-M angles will be significantly less than 180°. As the size of R decreases relative to M, the degree of distortion in the primitive cell increases and the M-O-M angles should decrease.

Goodenough has considered bonding and physical properties in RMO$_3$ perovskites(3). The physical properties, electrical, magnetic, etc., will be determined largely by interactions between the transition metal ions. It is to be noted that there are no close, direct M-M contacts in the perovskite structure and that any inter-action must be of the M-O-M type, through the intervening oxygen.

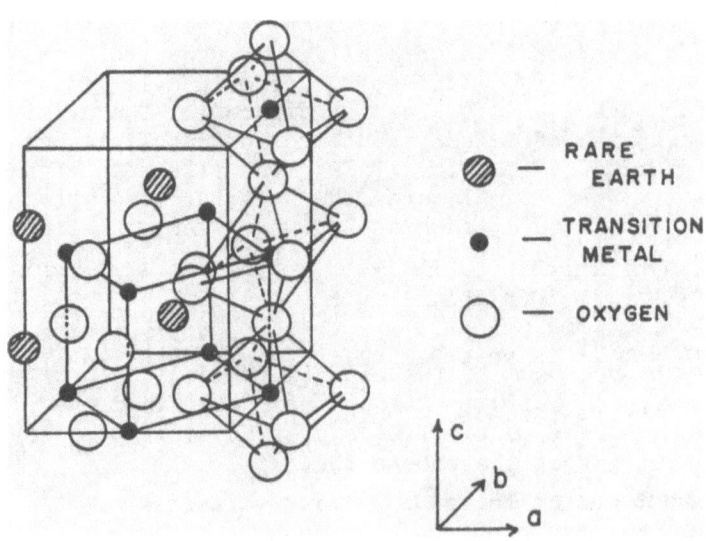

Figure 1. Partial contents of the GdFeO$_3$-type cell.

Goodenough treats the interaction qualitatively using the tight-binding or LCAO-MO approach. The parameter of the theory which describes this interaction is the transfer integral, b. It is approximately proportional to the overlap integral d-p-d. If b is large the d-electrons will be delocalized or collective moving in relatively broad, partially filled bands giving rise to metallic behaviour and Pauli paramagnetism. For small b one anticipates localized d-electrons and insulating or semiconducting behaviour coupled with Curie-Weiss magnetic behaviour. In the case of b intermediate the d-electrons move in filled, highly correlated Hubbard bands which also give rise to semiconducting behaviour and Curie-Weiss magnetic properties. Goodenough envisages critical values of b at which more or less abrupt transitions occur between the various regimes.

Clearly, the structural parameter most closely related to b is the M-O-M angle, as the d-p-d overlap integral should depend strongly on this angle. Here the R atom plays an important role as the M-O-M angle decreases with decreasing R radius leading to a decreasing b. If for a given RMO₃ series the range of M-O-M angles scanned included critical values of b, dramatic changes of physical properties should occur as a function of R. Further, for any given compound in the system a critical value of b may be scanned as a function of temperature or pressure.

It has been argued that the series RTiO₃ fits the above description (4,5,6). In the following we illustrate the remarkable features of this series of compounds often using data obtained from single crystal samples.

ELECTRICAL PROPERTIES

Mean Ti-O-Ti angles as a function of R have been measured for selected RTiO₃ phases(1). The angles vary smoothly from 157° (La) to 142° (Y). Note the considerable distortion from 180°, even for LaTiO₃. In figure 2 we plot resistivity values for single crystals (200K) as a function of Ti-O-Ti angle. A dramatic increase of almost six orders of magnitude occurs from high angles (large b) to low angles (small b). The two lowest resistivity samples, LaTiO₃ and CeTiO₃, are in fact metals, albeit poor metals(6). The others are semi-conductors with activation energies ranging from 0.03-0.04 ev (PrTiO₃ and NdTiO₃) to > 0.20 eV (GdTiO₃ and YTiO₃). Thus, a metal semiconductor transition occurs between Ti-O-Ti angles of 153° (CeTiO₃) and 152° (PrTiO₃).

We mention now another apparent relation between electrical properties and structure. Upon examining the results of our structure solutions for RTiO₃ phases we noted an interesting distortion of the Ti-O₆ octahedron with decreasing R-radius. For R = La the Ti-O bond lengths are not significantly different, while for R = Gd and Y a psuedo-tetragonal distortion occurs. This

distortion is noted in plotting the difference between the axial and
equatorial Ti-O bond lengths as a function of R(III) radius as in
figure 3. Also in figure 4, we note a good correlation between the
observed activation energies for conductivity and this distortion
index. Such is perhaps not unreasonable if conduction in these
materials occurs by the hopping or small polaron process. The depth
of the potential well by which the electrons are trapped should
increase as the environment about the trapping site, Ti(III), becomes
more distorted.

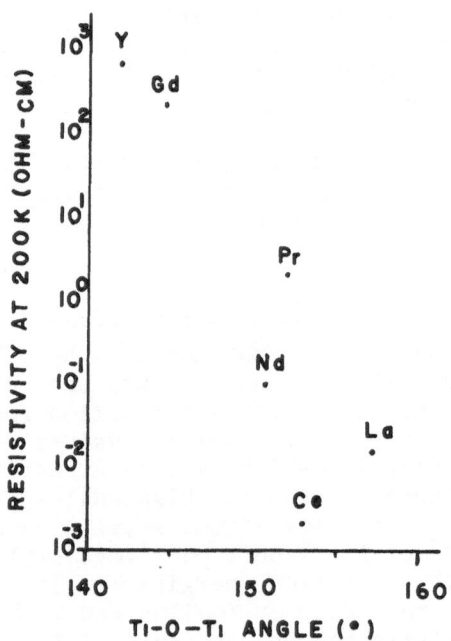

Figure 2. Resistivity at 200K for single crystals of selected
            RTiO₃ as a function of Ti-O-Ti angle.

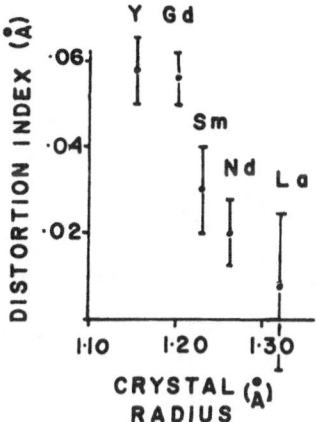

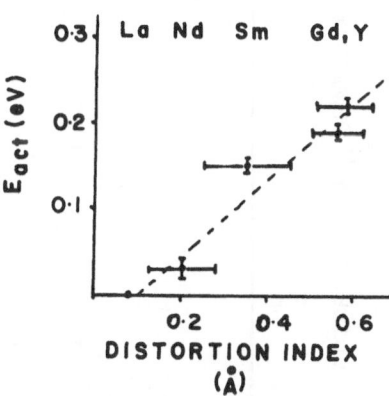

Figure 3. Distortion index as a function of R(III) radius. The distortion index is the difference between the axial Ti-0 and equatorial Ti-0 bond lengths.

Figure 4. Activation energies versus distortion index for semiconducting RTiO₃ phases.

MAGNETIC PROPERTIES

Corresponding changes occur in magnetic susceptibilities as illustrated in figure 5 where data for LaTiO₃, PrTiO₃, and NdTiO₃ are presented. LaTiO₃ shows only (a very high) Pauli paramagnetism, while PrTiO₃ and NdTiO₃ are Curie-Weiss (C-W) paramagnets with no sign of a temperature independent paramagnetic (T.I.P.) term. Data for CeTiO₃ (not shown) can be analyzed in terms of a C-W component from Ce(III) and an T.I.P. contribution of the magnitude found in LaTiO₃ from Ti(III)(6). The Ti(III) contribution to the susceptibilities of these compounds can be estimated by comparing the measured susceptibilities of the isostructural PrScO₃ and NdScO₃ where diamagnetic Sc(III) replaces Ti(III). The derived C-W constants from such measurements are shown in Table 1.

Table 1. Curie-Weiss Constants for PrTiO₃, PrScO₃, NdTiO₃ and NdScO₃.

COMPOUND	C_m(mole-cm^3-K^{-1})	θ_c (K)
PrTiO₃	1.73(3)	-49(3)
PrScO₃	1.34(3)	- 8(2)
Ti(III) in PrTiO₃	0.39(6)	
NdTiO₃	1.77(3)	-35(3)
NdScO₃	1.50(3)	-32(3)
Ti(III) in NdTiO₃	0.27(6)	

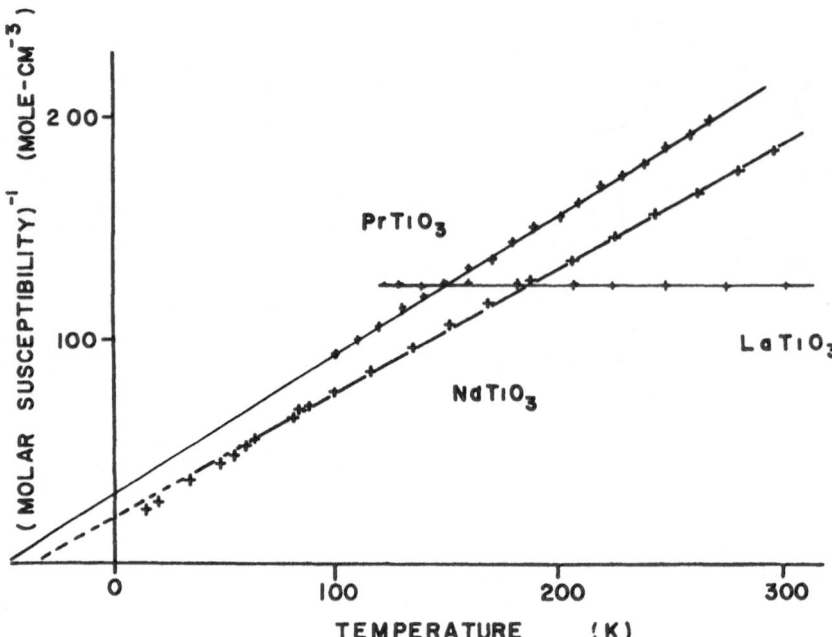

Figure 5. Magnetic susceptibilities for PrTiO₃, NdTiO₃, and LaTiO₃.

From these results we see that Ti(III) makes a contribution of the
C-W type with a C_m near that for a spin-only d^1 ion, 0.37. Note
also that the Curie constants for Pr(III) and Nd(III) in the scandium
phases are significantly smaller than the free ion values, 1.60 and
1.62 respectively, indicating the influence of crystal-field effects.

   All of the above indicates that for LaTiO₃ and CeTiO₃ the
transfer integral, b, may be close to a critical value. Evidence for
this can be found in recently published electrical resistivity data
(6). At high temperatures, T > 125K for LaTiO₃ and T > 60K for
CeTiO₃, the temperature dependence of the resistivity is metallic,
$d\rho/dT > 0$. Below these temperatures $d\rho/dT$ changes sign and the data
can be fitted with some success to a semiconductor model with very
low activation energies, 10^{-3} to 10^{-2} eV. Although a detailed
understanding is not currently available the simplest interpretation
is that a metal-semiconductor transition occurs with temperature
in these two materials.

   The magnetic structures of all of these phases are of interest
and are strongly influenced by the chemical structure. Most of the
RTiO₃ show some form of magnetic order at low temperature. Table 2
lists the critical temperatures and type of order, if known.

   Consider first the phases R = Gd to Lu plus Y. Here one finds
either ferromagnetism (when R is non-magnetic) or ferrimagnetism

Table 2.   Critical Temperatures and Type of Order
for $RTiO_3$.

R	T_c (K)	Order*	R	T_c (K)	Order*
La	125	WFo	Dy	64	Fi, NON-COL.
Ce	116	?	Ho	56	Fi, NON-COL.
Pr	96	?	Er	41	Fi, COL.
Nd	< 4.2K		Tm	58	Fi, COL.
Sm	?	?	Yb	39	Fi, COL.?
Gd	34	Fi, COL.	Y	29	Fo
Tb	49	Fi, NON-COL.	Lu	31	Fo

* Fi ≡ ferrimagnetic, COL. ≡ colinear, NON-COL. ≡ non-colinear
Fo ≡ ferromagnetic, WFo ≡ weak ferromagnetic.

(where R is magnetic)(7,8). Although neutron diffraction data are
not available, evidence from bulk magnetic measurements that $YTiO_3$
is ferromagnetic (and semiconducting) is convincing. The saturation
moment (4.2K) measured on a single crystal is 0.84 μ_B, close to the
spin-only value of 1.0 μ_B(9). We have determined the magnetic
structures of the remaining compounds (except $GdTiO_3$) from powder
neutron diffraction(10,11). A common feature of all the structures
is a ferromagnetic Ti(III) sublattice. Although a detailed theory
is lacking, the fact that the Ti-O-Ti angles are intermediate between
90° and 180° may play an important role in determining this unusual
type of superexchange coupling between d^1 ions. The ferromagnetic
Ti(III) sublattice is in turn responsible for the observation of
ferrimagnetism in these materials. The corresponding RMO_3 compounds
where M = V, Cr, Mn, and Fe have antiferromagnetic M(III) sublattices
(12). Due to the CsCl-type symmetry of the R(III)-M(III) lattice
the result is a near cancellation of the R-M interaction(13).

For the $RTiO_3$ phases two types of ferrimagnetic structures were
found. The simplest structure, R = Er, Tm, and possibly Yb, has
colinear and antiparallel R and Ti moments along [001]. The other
type, R = Tb, Dy, Ho, has the R-moments in the ab plane. The R-
moment at approximately (00¼) makes a small angle, θ, with [010]
([100] for Tb) and the R-moment at approximately (½ ½ ¼) makes an
angle,-θ, with the same direction. This results in an antiferromagnetic component along [100]and a ferromagnetic component along
[010]. Although the Ti(III) moment is difficult to locate the best
refinement places it antiparallel to the ferromagnetic R(III)
component. Although the magnetic structures of the $RTiO_3$ and RMO_3
compounds are quite different there is a common feature. This the
angle which the R moment makes with respect to the crystal axes as
shown in Table 3.

Table 3.   Rare Earth Moment Direction in RMO_3;
          R = Fe, Cr, Ti, Al.

RMO_3	R(III) DIRECTION	RMO_3	R(III) DIRECTION
$TbAlO_3$	$34°$ to [100]		
$TbFeO_3$	$38°$ "	$ErFeO_3$	[001]
$TbCrO_3$	--	$ErCrO_3$	"
$TbTiO_3$	$36°$ "	$ErTiO_3$	"
$DyAlO_3$	$33°$ to [010]	$TmFeO_3$	[001]
$DyFeO_3$	$30°$ "	$TmCrO_3$	"
$DyCrO_3$	$29°$ "	$TmTiO_3$	"
$DyTiO_3$	$31°$ "		
$HoFeO_3$	$27°$ to [010]		
$HoCrO_3$	$26°$ "		
$HoTiO_3$	$24°$ "		

This indicates that the angle is determined entirely by the identity of the R(III) and is most likely a crystal field effect. This hypothesis is currently under investigation.

The magnetic structures of the remaining $RTiO_3$ are unknown. Weak ferromagnetism was found for $LaTiO_3$ (6). The saturation moment (4.2K) is only 7×10^{-3} μ_B. T_C for $LaTiO_3$ occurs at the same temperature as the apparent metal-semiconductor transition, suggesting a connection between the two. There is a smooth decrease in T_C for R = Ce and Pr, but it is remarkable that no obvious sign of magnetic order can be detected in $NdTiO_3$ above 4.2K. The magnetic structures of these interesting materials will probably yield only to investigation with polarized-beam neutron diffraction techniques.

In conclusion the system of isostructural compounds, $RTiO_3$, represents a case where physical properties are extremely sensitive to changes in crystal geometry. A metal-semiconductor transition occurs as a function of R between R = Ce and Pr. This corresponds to a difference in the Ti-O-Ti angle of at most one or two degrees. Further, both $LaTiO_3$ and $CeTiO_3$ show apparent metal to semiconductor transitions with temperature. For the phases where R = Gd to Lu plus Y the Ti-O-Ti superexchange interactions are such that ferromagnetic coupling occurs. This gives rise to magnetic structures which are ferrimagnetic with relatively strong R-Ti coupling in contrast to the related RMO_3 phases where the R-M (M = Fe, Cr, Mn V) interaction is weak.

ACKNOWLEDGMENTS

Financial support from the National Science and Engineering Research Council of Canada and the Research Board of McMaster University is acknowledged. I would like to thank the students, D.A. MacLean, K. Seto, and C.W. Turner, technical associates J.D. Garrett, H.F. Gibbs, and R. Faggiani, and colleagues H.N. Ng, C.V. Stager, M.F. Collins, and W.R. Datars who have contributed to this work.

REFERENCES

1. David A. MacLean, Hok-Nam Ng, and J.E. Greedan, J. Solid State Chem., 30:35 (1979).
2. S. Geller, J. Chem. Phys., 24:1236 (1956).
3. J.B. Goodenough, Prog. Solid State Chem., 5:145 (1971).
4. P. Ganguly, O. Parkash, and C.N.R. Rao, Phys. Stat. Solidi, A36:669 (1976).
5. G.V. Bazuev, G.P. Shveikin, Fiz. Tverd. Tela., 17:3453 (1975).
6. David A. MacLean and J.E. Greedan, Inorg. Chem., 20:1025 (1981).
7. D. Johnston, Ph.D. Thesis, Univ. of California, San Diego, 1975.
8. Carl W. Turner and J.E. Greedan, J. Solid State Chem., 34:207 (1980).
9. J.D. Garrett, David A. MacLean, and J.E. Greedan, Mat. Res. Bull. 16:145 (1981).
10. Carl W. Turner, Malcolm F. Collins, and J.E. Greedan, J. Magn. and Magn. Mater., 20:165 (1980).
11. Carl W. Turner, Malcolm F. Collins and J.E. Greedan, J. Magn. and Magn. Mater. (1981) in press.
12. J.B. Goodenough and J.M. Longo, "Crystallographic and Magnetic Properties of Perovskite and Perovskite-related Compounds", Landolt-Bornstein Tabellen Neue Serie 111/4a (Springer-Verlag, Berlin, 1970).
13. M.A. Gilleo, J. Chem. Phys. 24:1239 (1956).

SOME MAGNETIC PROPERTIES OF THE SYSTEM $La_xGd_{1-x}TiO_3$

J.P. Goral and J.E. Greedan

Department of Chemistry and Institute for Materials
Research, McMaster University, Hamilton, Ontario
L8S 4M1

The solid solutions $La_xGd_{1-x}TiO_3$ (x=0-1) form an isostructural series crystallizing in the $GdFeO_3$ structure. The susceptibility of $LaTiO_3$ is temperature independent down to 130K, at which point it increases dramatically. Hysteresis, consistent with weak ferromagnetism, has also been reported at 4.2K.[1] $GdTiO_3$ appears to be ferrimagnetic (Tc=33K) with a saturation moment of $3.3x10^4$ emu mole^{-1} at 4.2K.[2] In this work, magnetization data as a function of temperature are reported for powder samples of inter-mediate composition (Fig.1).

As $GdTiO_3$ is doped with La, a single critical temperature is observed for compositions up to 40% La. Although it becomes less clearly defined with increasing La content, Tc drops steeply to a value of 12K for the composition $La_{.3}Gd_{.7}TiO_3$.

As $LaTiO_3$ is doped with Gd, Tc decreases smoothly and a low temperature anomaly, LTA, appears below 20K. This behaviour persists to the composition $La_{.5}Gd_{.5}TiO_3$. The high temperature Tc is no longer observable in the 40% La phase. The inflections at the LTA and Tc are both highly field dependent and are greatly supressed on application of a field larger than $\sim$ 0.2T. It should be noted that in $La_7Gd_{.3}TiO_3$, hysteresis is observed below Tc but not above it.

The dependence of Tc on composition in the La rich compounds, along with the existence of hysteresis my be indicative of the same type of magnetic ordering of the Ti^{3+} sublattice as in $LaTiO_3$. Similar behaviour has been observed in the compounds $La_xY_{1-x}TiO_3$ (x=0-0.4) where both rare earth ions are closed shell.

441

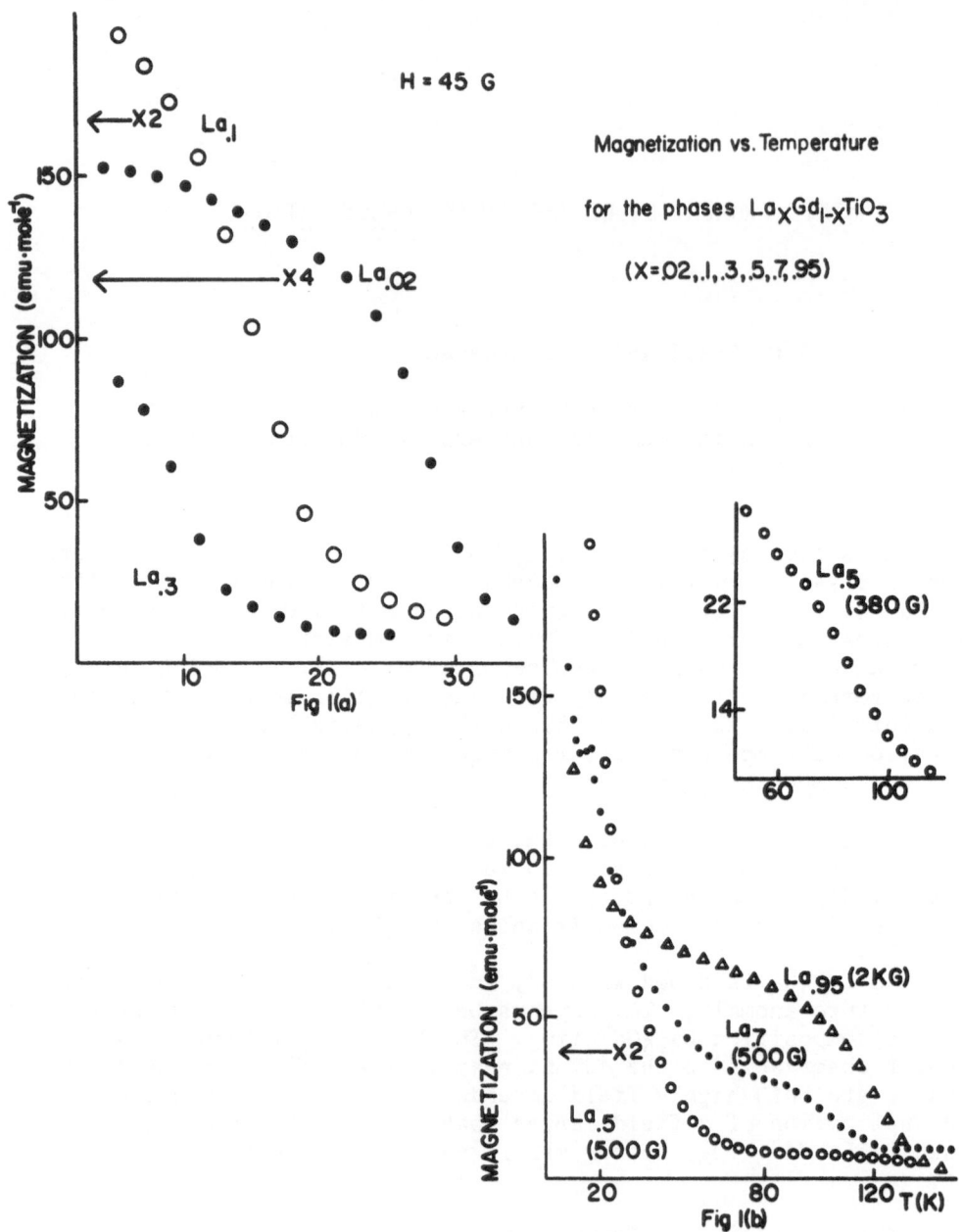

REFERENCES

1. D.A. MacLean and J.E. Greedan, Inorg. Chem. 20:1025 (1981).
2. C.W. Turner and J.E. Greedan, J. Solid State Chem. 34:207 (1980).

ZERO FIELD TEMPERATURE DEPENCENCE OF THE RARE EARTH SUBLATTICE

MAGNETIZATION IN $RTiO_3$; R = Tb, Dy, Ho, Er and Tm

Carl W. Turner[a], Malcolm F. Collins[b], J.E. Greedan[a]

Department of Chemistry[a] and Physics[b], McMaster
University, Hamilton, Ontario, L8S 4M1 Canada

The compounds $RTiO_3$, R = Tb, Dy, Ho, Er and Tm, form an isomorphous series having the $GdFeO_3$ type structure which is an orthorhombic distortion of the cubic perovskite structure. The magnetic properties of these materials differ from those found in the isostructural compounds RMO_3, M = Cr or Fe. This difference in the magnetic behaviour can be understood by considering the isotropic R^{3+}-M^{3+} coupling. In the case of Fe^{3+} or Cr^{3+}, the transition metal moments couple antiferromagnetically so the R^{3+}-M^{3+} isotropic coupling nearly vanishes by symmetry. This coupling would be rigourously zero if the structure was the ideal cubic perovskite.

Since the Ti^{3+} moments couple ferromagnetically in $RTiO_3$[1], the R^{3+}-Ti^{3+} isotropic coupling does not cancel by symmetry. A consequence of this is that the rare earth and titanium moments order at the same temperature with T_c ranging from 40K to 65K[2]. This is to be contrasted with the orthferrites and orthochromites where there are two ordering temperatures; the upper one corresponding to ordering of the transition metal moments and the lower one around 4K due to ordering of the rare earth moments[3].

The temperature dependence of the rare earth magnetization has been determined using neutron diffraction techniques. Figure 1 shows the experimental results for $TbTiO_3$ and $ErTiO_3$. It is evident that the temperature dependence of the rare earth magnetization is better represented by a spin ½ Brillouin function than the appropriate free ion Brillouin function. A possible explanation of this is that the temperature dependence of the rare earth magnetization is governed by the temperature dependence of the titanium sublattice magnetization. As the titanium magnetization arises from one d

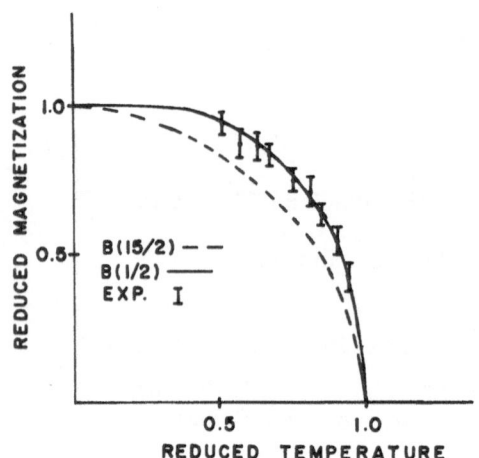

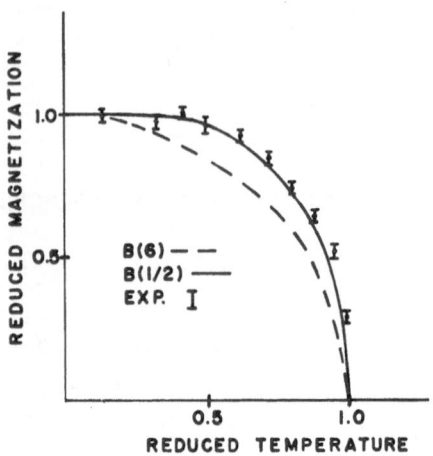

Figure 1(a). Reduced magnetization versus reduced temperature for ErTiO$_3$. Error bars correspond to ± one standard deviation. Also shown are Brillouin functions calculated for J = 15/2 and J = 1/2.

Figure 1(b). Reduced magnetization versus reduced temperature for TbTiO$_3$. Error bars correspond to ± one standard deviation. Also shown are Brillouin functions calculated for J = 6 and J = 1/2.

electron, it will be described by a spin ½ Brillouin function. The two are linked through the R^{3+}-Ti^{3+} isotropic coupling.

It is this coupling which causes the rare earth moments to order with the titanium moments at critical temperatures of 40K for ErTiO$_3$2 and 49K for TbTiO$_3$2. Consequently the rare earth magnetization will have the same temperature dependence as the titanium magnetization.

REFERENCES

1. J.E. Greedan and D.A. MacLean, Inst. Phys. Conf. Sec. no. 37 (1978) 249.
2. C.W. Turner and J.E. Greedan, J. Solid State Chem., 34:207 (1980).
3. J.B. Goodenough and J.M. Longo, Crystallographic and Magnetic Properties of perovskite and Perovskite-related Compounds, Landolt-Bornstein Tabellen Neure Serie 111/4a (Springer-Verlag, Berlin, 1970).

MAGNETIC AND STRUCTURAL PROPERTIES OF CeSb$_{1-x}$Te$_x$ MIXED COMPOUNDS

D. Ravot and J.C. Achard, Equipe de Chimie Métallurgique
des Terres Rares, C.N.R.S., 1, Place Aristide Briand
92190 Meudon-Bellevue; J. Rossat-Mignod, Laboratoire de
diffraction neutronique, département de recherche
fondamentale, Centre d'Etude Nucléaire de Grenoble, 85 X
38041 Grenoble Cedex, France

INTRODUCTION

Cerium monoantimonide CeSb and cerium monotelluride CeTe have
the NaCl structure. These two compounds order antiferromagnetically
at low temperatures (about 16 K and 2 K respectively). While CeTe
is an antiferromagnet of type II (1), CeSb is a more complex antiferromagnet, which shows, depending on temperature and magnetic
field, numerous magnetic phases below T_N. These phases can be
described as a stacking of "up" and "down" (100) ferromagnetic
planes with paramagnetic (100) planes inserted (2). CeTe has an easy
axis along the [111] direction and CeSb along the [100] direction.
The Neel temperature is, in CeTe, smaller than the crystal field
splitting between the two levels Γ_7 and Γ_8 derived from the $^2F_{5/2}$
state of the free Ce^{3+} ion. In CeSb these two energies have the
same order of magnitude.

To study the complex magnetic behavior of CeSb, some authors
(3,4) have investigated the physical properties of diluted solutions
(Ce$_x$La$_{1-x}$Sb, Ce$_x$Y$_{1-x}$Sb, Ce$_x$(La,Y)$_{1-x}$Sb) or substituted compounds
(CeSb$_{1-x}$As$_x$, CeSb$_{1-x}$Bi$_x$). At Ce ~0.6 in Ce$_x$La$_{1-x}$Sb and 0.3 in
Ce$_x$Y$_{1-x}$Sb, the easy direction rotates from [100] as in pure CeSb
to [111]. In the same way the $1/\chi$ behavior becomes more normal for
lowered Ce concentration (4). In the similar compounds Ce$_x$La$_{1-x}$Bi,
the Neel temperature varies linearly with the Ce concentration
for x>0.5. No antiferromagnetic ordering has been observed for
x=0.35 above 2.5 K (5). In CeAs$_x$Sb$_{1-x}$ and CeBi$_x$Sb$_{1-x}$, when x
increases the complex magnetic diagram of CeSb tends to vanish.
Although the parameter varies linearly with the substitution, this
is not the case for T_N (6). These different results have incited

445

us to study the physical properties of the $CeSb_{1-x}Te_x$ mixed compounds in which mixing of two antiferromagnets of different kinds is realized.

SAMPLES AND CHARACTERIZATION

The samples are synthesized by direct reaction between the elements in a molybdenum crucible under an argon atmosphere by heating to 2000°C and slow cooling (4 C/h or 10 C/h). We obtained polycrystalline samples with some single crystals cleaved from the mass. Samples were characterized by metallographic studies, x-ray diffraction, electron microprobe analysis and, in some cases, spectrometric analysis (Table 1). In some samples, small and rare inclusions were detected which have not yet been previously identified. In the $CeSb_{1-x}Te_x$ series, the transition from the black color of CeSb to the blue color of CeTe takes place between $CeSb_{0.70}Te_{0.30}$ and $CeSb_{0.50}Te_{0.50}$. All the compounds are single phase and have the rocksalt structure. The lattice parameters of $CeSb_{1-x}Te_x$ vary linearly with x (Fig. 1). However, it seems that an anomaly was detected in the vicinity of CeSb, producing a small maximum, but the amplitude of this anomaly is of the same order as the precision of the measurement, so more precise measurements will be needed to explore this.

Table 1. Characterization of $CeSb_{1-x}Te_x$ Compounds.

	Composition			
Nominal	Microprobe (+=Inclusion)	Actual	Lattice Constant	T_N
CeSb	CeSb	CeSb	6.425(2)	15.5
$CeSb_{.98}Te_{.02}$	$Ce_{.506}Sb_{.487}Te_{.007}+$	$CeSb_{.96}Te_{.01}$	6.427(2)	12.2
$CeSb_{.95}Te_{.05}$	$Ce_{.507}Sb_{.467}Te_{.026}$	$CeSb_{.92}Te_{.05}$	6.426(3)	3.75
$CeSb_{.90}Te_{.10}$	$Ce_{.508}Sb_{.443}Te_{.049}+$	$CeSb_{.87}Te_{.10}$	6.421(2)	3.10
$CeSb_{.70}Te_{.30}$	$Ce_{.505}Sb_{.355}Te_{.140}+$	$CeSb_{.70}Te_{.28}$	6.410(2)	2.00
$CeSb_{.50}Te_{.50}$	$Ce_{.505}Sb_{.257}Te_{.237}+$	$CeSb_{.51}Te_{.47}$	6.399(3)	1.50
$CeSb_{.30}Te_{.70}$	$Ce_{.499}Sb_{.147}Te_{.354}+$	$CeSb_{.29}Te_{.71}$	6.380(4)	1.30
$CeSb_{.10}Te_{.90}$	$Ce_{.505}Sb_{.071}Te_{.424}+$	$CeSb_{.14}Te_{.84}$	6.370(2)	2.60
$CeSb_{.05}Te_{.95}$	$Ce_{.500}Sb_{.030}Te_{.471}+$	$CeSb_{.06}Te_{.94}$	6.363(5)	2.40
CeTe	CeTe	CeTe	6.360(5)	2.00

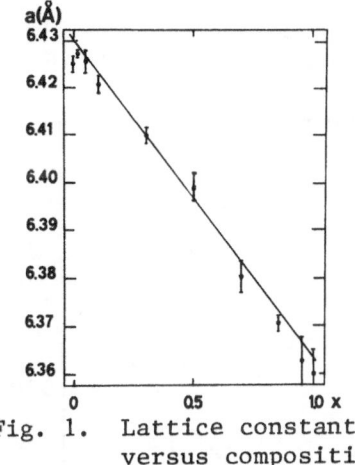

Fig. 1. Lattice constant
         versus composition

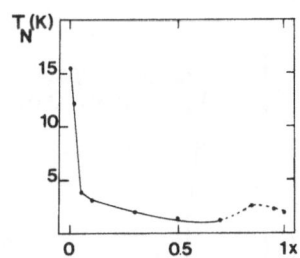

Fig. 2. Neel temperature
         versus composition

SUSCEPTIBILITY MEASUREMENTS

We have measured the susceptibility of our compounds between
1.2 K and 4.2 K in a magnetic field of about 10 G. by the mutual
inductance method and between 4.2 K and 300 K in a magnetic field
of about 4kG in a Faraday balance. All the $CeSb_{1-x}Te_x$ compounds
order antiferromagnetically at low temperatures. The variation of
the Neel temperature with x is shown in Fig. 2. From CeSb, T_N
decreases abruptly with small additions of Te and goes through a
minimum at $CeSb_{0.30}Te_{0.70}$. In the vicinity of CeTe, T_N shows a
local maximum. This variation differs from that in $CeAs_xSb_{1-x}$ and
$CeBi_xSb_{1-x}$ (6). At low temperatures in the antiferromagnetic state,
the Sb rich compound $CeSb_{0.98}Te_{0.02}$ shows an abnormal thermal
variation of its inverse susceptibility with a maximum at T=10.4 K
followed by a minimum at T=8.4 K. This behavior is the trace, as
in CeSb, of the complex magnetic behavior of these compounds as
revealed by transport measurements (7). Contrary to CeSb, in
$CeSb_{0.98}Te_{0.02}$ a deviation from the Curie-Weiss law is evidenced in
the paramagnetic range. For $CeSb_{0.95}Te_{0.05}$, which orders at
T_N=3.8 K, no anomaly has been detected between 1.2 K and T_N. However, in the paramagnetic region, a deviation from the Curie-Weiss
law is evidenced (more pronounced than in $CeSb_{0.98}Te_{0.02}$) which can
be attributed as in CeP and CeAs to the crystal field interaction.
Following Jones (8) we can apply to $CeSb_{0.95}Te_{0.05}$ crystal field
theory to calculate the thermal variation of the inverse susceptibility. We deduce a surprisingly high value of the crystal field
splitting Δ=50 K.

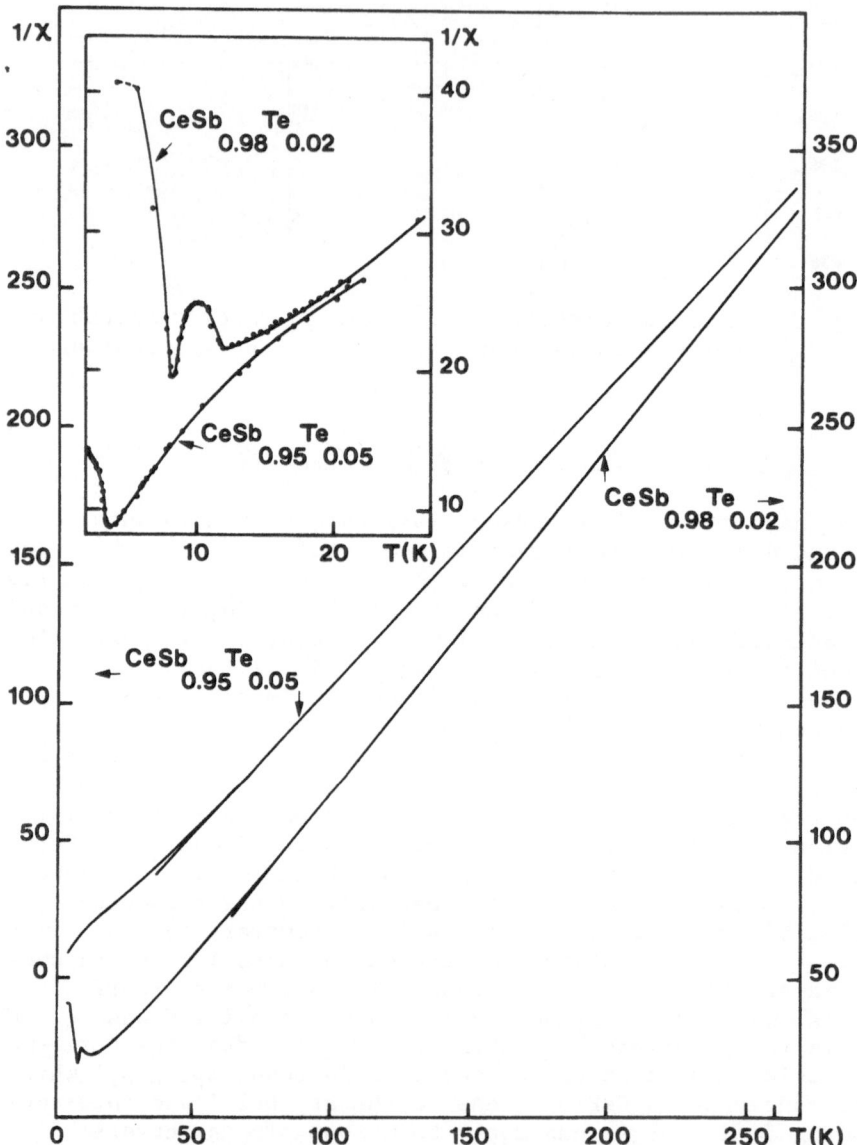

Fig. 3.   Inverse susceptibility versus temperature for
          $CeSb_{0.98}Te_{0.02}$ and $CeSb_{0.95}Te_{0.05}$.

ACKNOWLEDGEMENTS

The authors wish to thank Dr. J.P. Renard and B. Lecuyer of the Orsay University for providing necessary facilities used in the very low temperature susceptibility measurements.

REFERENCES

1. H.R. Ott, J.K. Kjems, F. Hulliger, Phys. Rev. Lett., 42:1378 (1979); D. Ravot, P. Burlet, J. Rossat-Mignod, and J.L. Tholence, J. Phys., 41:1117 (1980).
2. J. Rossat-Mignod, P. Burlet, J. Villain, H. Bartholin, Si-Tcheng Want, D. Florence and O. Vogt, Phys. Rev., B16:440 (1977).
3. B.R. Cooper, A. Furrer, W. Buhrer, O. Vogt, Sol. State Comm., 11:21 (1972); H. Bartholin and O. Vogt, Phys. Stat. Solid. (a), 52:325 (1979).
4. B.R. Cooper and O. Vogt, J. Phys. Coll. C_1, Suppl. 2 and 3, 32 (1971), C_1 1026.
5. T. Tsuchida, A. Hashimoto and Y. Nakamura, J. Phys. Soc. Japan, 36:3:685 (1974).
6. H. Bartholin, O. Vogt, and J.P. Senateur, J.M.M.M., 15-18:1247 (1980).
7. M. Escorne, A. Mauger, D. Ravot, J.C. Achard, to be published and in publication in J. Phys. C.
8. E.D. Jones, Phys. Lett., 22:266 (1966).

EFFECTS OF THE NON-STOICHIOMETRY ON THE TRANSPORT PROPERTIES OF CeSb

D. Ravot and J.C. Achard, Equipe de Chimie Metallurgique
des Terres Rares, C.N.R.S., 1, Place Aristide Briand
92190 Meudon-Bellevue, M. Escorne and A. Mauger
Laboratoire de Physique des Solides, C.N.R.S., 1
Place Aristide Briand 92190 Meudon-Bellevue, France

ABSTRACT

Transport properties of stoichiometric and a non-stoichiometric
CeSb have been investigated in the range 2.8 K < T < 100 K. It is
shown that the main scattering mechanism of the carriers is by spin
fluctuations. There is evidence of magnetic phase transitions.
The existence of various phases at low temperatures strongly depends
on the stoichiometry and crystal quality. At higher temperatures
(T > 40 K) a resistivity minimum, characteristic of a Kondo effect,
is not observed.

INTRODUCTION

Among cerium monopnictides compounds, interest has been
focused on rocksalt structure CeSb and CeBi. In these two compounds, the crystal field splitting of the $^2F_{5/2}$ ground state of
Ce^{3+} into two levels, Γ_7 and Γ_8, and the exchange energy are of the
same order of magnitude. This is responsible for the complex
phase diagram of CeSb suggested by neutron [1] and specific heat
measurements [2]. Recently we have shown that transport measurements [3] are also a powerful means to investigate the magnetic
properties of CeSb. Our measurements, however, were limited to
one non-stoichiometric sample. There we extend our previous
studies to samples closer to stoichiometry.

EXPERIMENTAL

The compounds were obtained by direct reaction between the

elements in a molybdenum crucible under an argon atmosphere. The
maximum temperature reached was 2000°C and the cooling rate was
4 C/h. Typical single crystals grown by this method were 1x1x2 mm^3.

Two different compositions have been studied: Ce/Sb=1 (sample
1) and Ce/Sb = 0.95 (sample 2). In the latter compound, as shown
by spectrometric analysis, initial and final composition were both
Ce$_{0.95}$Sb (3). The samples were characterized by metallographic,
x-ray and spectrometric analysis. The observed lattice constants
were 6.425(2) (sample 1) and 6.424(2) Å (sample 2), which are in
the range of previous published data. Electrical resistance was
measured by a comparator bridge method described elsewhere (4).
The experimental data are expressed in a resistivity scale; however, the shape factor is not precisely defined.

RESULTS AND DISCUSSION

The Neel temperature (T_N) defined by the maximum of the magnetic susceptibility was 16±0.5 K for the two samples. The resistivity (ρ) curve as a function of temperature is shown in Fig. 1
for sample 1 and in ref. (3) for sample 2.

In the paramagnetic configuration, a monotonic increase of ρ
with T is observed in contrast to previous results (5). In particular no Kondo minimum was observed. However, we cannot definitely
conclude that a Kondo effect does not occur in CeSb. At T > 40 K,
the contribution of phonons to the resistivity becomes significant
and increases sharply with temperature, so that it could mask a
minimum of the spin dependent resistivity. Below 40 K, the resistivity is mainly due to the elastic plus inelastic magnetic scattering of the carriers by spin fluctuations. We performed the
calculation for the case of fully incoherent scattering, following
the procedure described in (3). The best fit, shown in Fig. 1,
was obtained with $\Delta = E(\Gamma_8) - E(\Gamma_7) = +28.2$ K for sample 1. Within the
experimental uncertainty this value was the same as for sample 2
($\Delta = +28.5$ K).

Around T_N, the resistivity strongly decreased upon cooling.
The magnitude of this decrease is the result of two additional
effects: (i) a decrease of inelastic scattering processes leading
to a decrease of calculated resistivity in Fig. 1; (ii) the onset
of a magnetic order implying that the scattering is not entirely
incoherent and which is responsible for the difference between the
theoretical and experimental curves.

When cooling below T_N, sharp decreases in the resistivity
curves were observed at definite temperatures associated with magnetic phase transitions (Fig. 2). The various phases observed,
together with the transition temperatures are listed in Table 1.

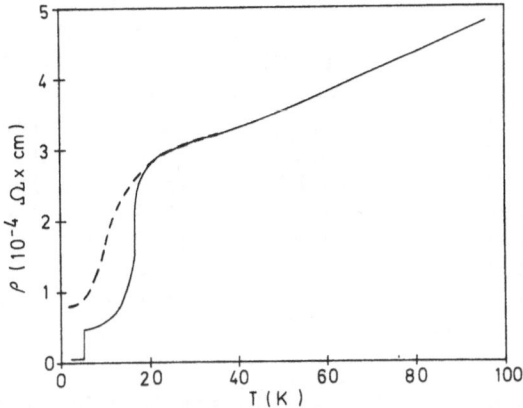

Figure 1. Resistivity as a function of temperature for sample 1.
The broken curve is the theoretical spin dependent resistivity
(see text).

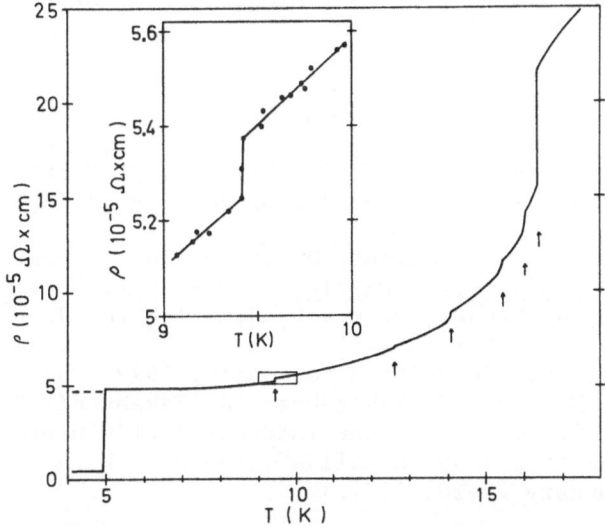

Figure 2. Resistivity as a function of temperature for sample 1.
Arrows point out the phase transitions. The insert is an enlargement illustrating the variations of ρ in the vicinity of one transition. Full circles are experimentals points. At T < 4.9 K, the
broken curve is the experimental curve after a first thermal cycle
(see text).

Table 1.   Temperatures of Discontinuities in the Resistivity Curves
           due to Phase Transitions.

| Phases | Temperatures (K) | | |
Transitions	Ref. (2)	Sample 1	Sample 2
P→AFP1	16.8	16.4	16.0
AFP1→AFP2	16.3	16.0	15.3
AFP2→AFP3	15.8	15.4	15.0
AFP3→AFP4	14.1	13.9	–
AFP4→AFP5	12.8	12.6	–
AFP5→AFP6	9.5	9.4	–
AFP6→AF	7.9	4.95?	–

The phases AFP4,5,6 were observed in sample 1 but not in sample 2.
Moreover, when successive thermal cycles between 4.2 K and 300 K
were performed, a strong transition, at 4.9 K, was observed only
on the first cooling and not in the subsequent cycles. Due to the
large amplitude of the jump, we suspect that it is associated with
the transition from the AFP 6 phase to the AF phase. In the AFP
phases, the paramagnetic planes strongly scatter the free carriers
so that the disappearance of such planes in the AF phase induces
a large drop in resistivity.

REFERENCES

1.   J. Rossat-Mignod, P. Burlet, J. Villain, H. Bartholin,
     Tcheng-Si Wang, D. Florence and O. Vogt, Phys. Rev. B16:440
     (1977).
2.   J. Rossat-Mignod, P. Burlet, H. Bartholin, O. Vogt, R. Lagnier,
     J. Phys. C: Solid State Physic, 13:6391-9 (1980).
3.   M. Escorne, A. Mauger, D. Ravot, J.C. Achard, J. Phys. C: Solid
     State Phys., 14:1821-38 (1981).
4.   M. Escorne, These d'Universite, Paris, 1979.
5.   T. Suzuki, M. Sera, H. Takegahara, H. Takahashi, Y. Yanase,
     T. Kasuya, Proceeding of the International Conference on
     "Valence Fluctuations in Solids", Santa-Barbara, California,
     U.S.A., January 27-30, 1981.

LOW-TEMPERATURE BEHAVIOR OF DyS, DySe, HoS and HoSe

F. Hulliger, M. Landolt and R. Schmelczer*

Lab. f. Festkörperphysik ETH, CH-8093 Zürich

*now with Inst. f. Angew. Physik ETH, CH-8093 Zürich

INTRODUCTION

In the last years the interest in rocksalt-type rare-earth chalcogenides was concentrated on divalent and mixed-valence representatives, i.e. on the Sm, Eu and Tm compounds. For most of the isostructural monopnictides the magnetic structure and the crystal-field splitting of the J ground state has been investigated (1). To our knowledge these data are still unknown for the Dy and Ho monochalcogenides.

Whereas the monopnictides are semimetallic valence compounds the excess valence electrons in the monochalcogenides convert them to colored metals. In both, however, the cation is "magnetically" trivalent, i.e. it has the same f-electron configuration. The additional electron in the chalcogenides modifies the chemical bonds, the crystal electric field, as well as the magnetic exchange interactions. Our experimental results suggest that there is no drastic change in properties so that an extrapolation from the monopnictides is justified. Obviously, the antiferromagnetic exchange interactions are strengthened in the chalcogenides. DyP and DyAs show a HoP-type magnetic order in low fields and then switch to ferromagnetic order in higher fields applied along the [001] direction. In DySb (2-5) and DyBi (6) the ranges of the ferrimagnetic HoP-type order and the ferromagnetic order are preceded by a type II antiferromagnetic order stable in low magnetic fields. The holmium pnictides behave in a similar way with the difference that the transition fields are lower and the range of the HoP-type magnetic structure is smaller (see Fig.8). Thus, HoAs is antiferromagnetic in zero field.

EXPERIMENTS AND RESULTS

On single crystals of DyS, DySe, HoS and HoSe we have investigated
the low-temperature lattice distortion as well as the magnetization
in steady fields up to 100 kOe and in pulsed fields up to ~200 kOe.

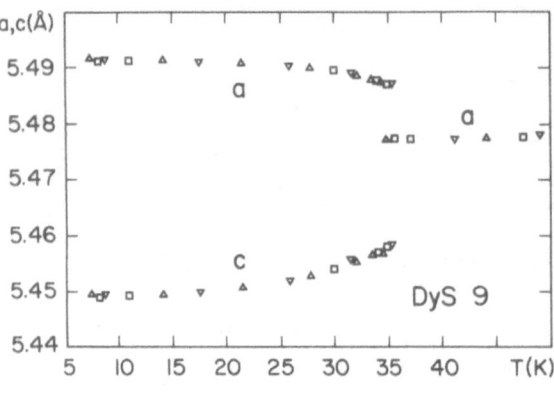

Fig. 1 Pseudo-tetragonal distortion of the NaCl-type cell of DyS, as derived from X-ray diffraction measurements on a (001) single crystal. At 8 K $(a-c)/a_O = 0.0076$, where a_O is the lattice constant of the cubic phase, immediately above the Néel temperature.

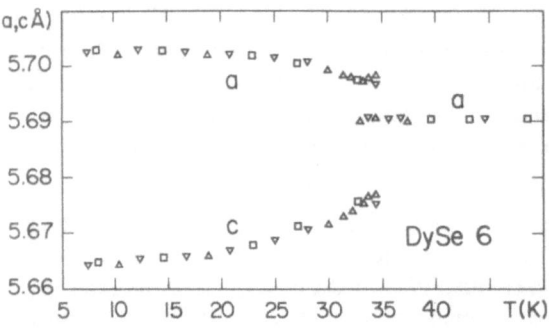

Fig. 2 Pseudo-tetragonal distortion of the cubic unit cell of DySe below the Néel point. At 8 K $(a-c)/a_O = 0.0067$.

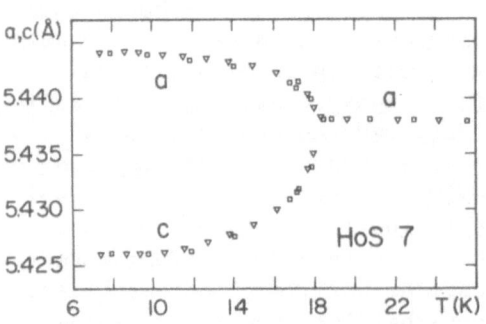

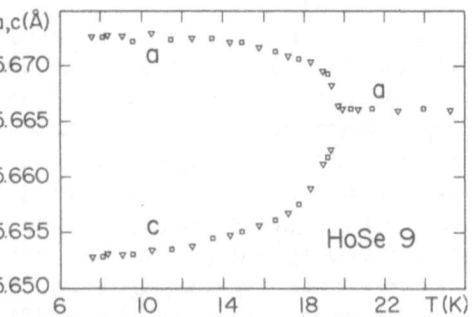

Fig. 3 Pseudo-tetragonal distortion of the rocksalt structure of HoS. At 8 K $(a-c)/a_O = 0.0033$.

Fig. 4 Pseudo-tetragonal distortion of the cubic structure of HoSe. At 8 K $(a-c)/a_O = 0.0035$.

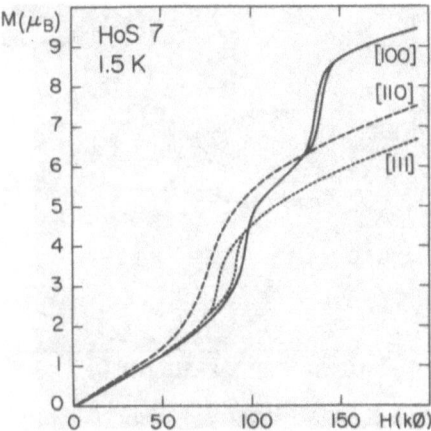

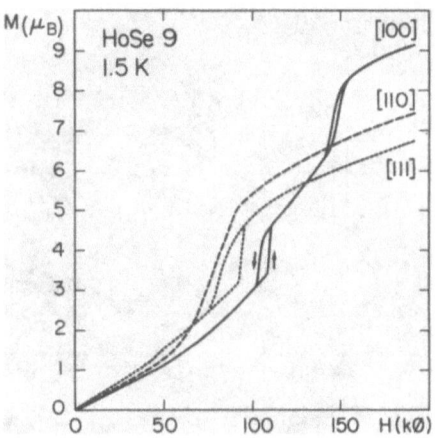

Fig. 5 Magnetization vs effective field at 1.5 K in HoS crystals oriented with applied field along [100], [110] and [111].

Fig. 6 Field dependence of the magnetization of HoSe single crystals at 1.5 K.

At the Néel point DyS and DySe undergo a first–order transition to a pseudo–tetragonal (probably monoclinic) structure (Figs.1 and 2). The feature of the a,c(T) curves is reminiscent of DySb (7) and DyBi (6). In HoS and HoSe (Figs.3 and 4) the phase transition at the Néel temperature appears to be of second order and the a,c(T) curves closely resemble those of HoSb (7) and HoBi (8). The different character of the phase transition is a consequence of the crystalfield splitting of the J ground state in Dy^{3+} (Γ_6 and $\Gamma_8^{(\text{II})}$ lowest) and Ho^{3+} ($\Gamma_3^{(2)}$, $\Gamma_4^{(1)}$ and Γ_1 lowest, very close to each other). At 8 K the distortions $(a-c)/a_0$ reach values of 0.76, 0.67, 0.33 and 0.35% in DyS, DySe, HoS and HoSe, respectively.

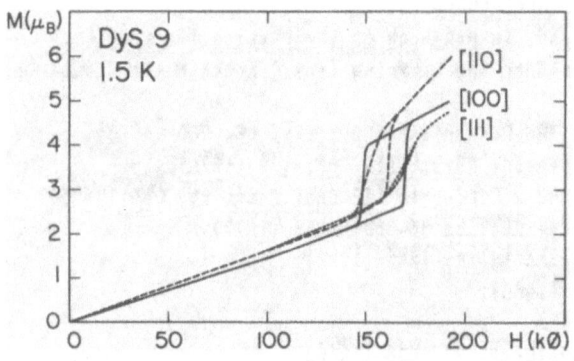

Fig. 7 Magnetization curves for DyS at 1.5 K. The smeared [111] transition may be due to crystal imperfections and the non–ellipsoidal shape of the sample.

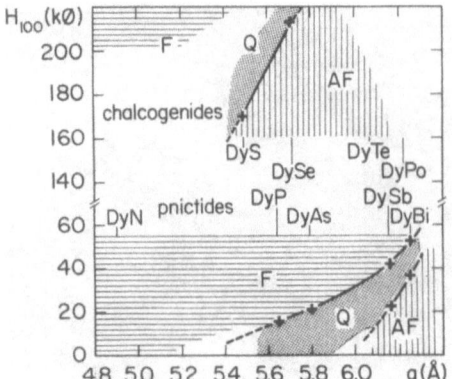 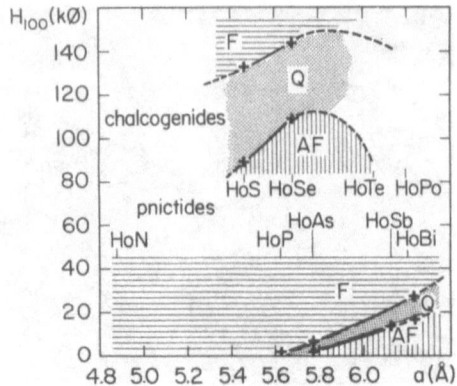

Fig. 8 Magnetic order at 1.5 K with magnetic fields applied along [111] vs. lattice constants for the Dy (left) and the Ho mono-chalcogenides (right) as compared with the monopnictides.

The magnetization behavior of HoS and HoSe (Figs.5 and 6) also corresponds to that of the antimonide (9) and bismuthide (1.8). In the case of DyS (Fig.7) and DySe the second transition lies beyond the limit of our pulsed-field apparatus, but the character of the magnetization curves is certainly identical with that of DySb and DyBi.

From these similarities we deduce that DyS, DySe, HoS and HoSe exhibit a type II antiferromagnetic order below the Néel tempera-tures of 36, 35, 19 and 20 K, respectively. At 1.5 K in magnetic fields applied along [001] a transition to a HoP-type ferrimagnetic order occurs at effective fields of about 170, 213, 89 and 119 kOe, respectively; and at fields >195, >220, 133 and 144 kOe, respectively, the ferrimagnetic alignment switches to ferromagnetic order. We have tried to sketch the magnetic behavior of the monochalcogenides as compared to that of the monopnictides in some kind of phase diagrams (Fig.8) involving the cation—cation distances.

REFERENCES

1. F. Hulliger, "Rare-Earth Pnictides", in Handbook on the Physics and Chemistry of Rare-Earths, Vol.4, K.A.Gschneidner and L.Eyring (eds.),North Holland Publ.Co, Amsterdam, pp. 153-236 (1979).
2. G.Felcher, T.O.Brun, R.J.Gambino and M.Kuznietz, Phys.Rev. B8, 260 (1973).
3. P.Streit, G.E. Everett and A.W. Lawson, Phys.Lett. 50A, 199 (1974).
4. T.O.Brun, G.H.Lander, F.W.Korty and J.S.Kouvel, AIP Conf.Proc. 24, 244 (1975).
5. J.S.Kouvel, T.O.Brun and F.W.Korty, Physica 86-88B, 1043 (1977).
6. F. Hulliger, J.Magnetism Magn.Mater. 15-18, 1243 (1980).
7. F.Lévy, Phys.kondens.Mat. 10, 85 (1969).
8. F.Hulliger, unpublished.
9. G.Busch, P.Schwob and O.Vogt, Phys.Lett. 23, 636 (1966).

ELECTRICAL RESISTIVITY AND MAGNETIC FIELD EFFECTS OF $NdS_{3-x}V_xS_4$

Syed M.A. Taher , Department of Physics, Wichita State
University, Wichita, KS 67208
John B. Gruber, Department of Physics and Chemistry
Portland State University, Portland, OR 97207

Experimental Details

We wish to report electrical resistivity, ρ , of single crystal
($Nd_{3-x}V_xS_4$) where $0 < x < 0.33$ between 2 and 300K. Samples were
prepared either by direct reaction of the elements in closed
quartz tubes (between 600 and 800C) or by passing hydrogen sulfide
over rare earth oxides (R O) and grown as ingots from the melt
between 1800 and 2100C using Bridgeman techniques. X-ray diffraction
measurements reveal all samples possess the high temperature γ -
phase or Th_3P_4 bcc structure. The density of conduction electrons
in these solid solution samples range from 10^{16} to 10^{21} carriers/cm^3.
The electrical resistivity was measured by a four point DC-technique
with pressure contacts at the gold-coated bars of samples having
dimensions of 1 x 2 x 8mm. Overall accuracy of the resistivity
measurements is typically 2% from 2 to 40K and 1% from 40 to 300K.
The resistivity of all samples was also measured under different magnetic fields up to 7.7kG. For magnetic measurements a magnetometer
was used with magnetic fields up to 30kG at temperatures between 4
and 300K.

Table I summarizes some magnetic properties of the samples
studied. These measurements show that the high temperature magnetic
susceptibility follows a Curie-Weiss law, $\chi = C/T-\theta$. The constants
C are quite close to the values expected for a ground state $^4I_{9/2}$ of
the Nd^{+3} ion, particulary in the case of $NdS_{1.49}$. Agreement between
the values of C becomes less satisfactory with increasing positive
exchange interaction as a result of increasing Nd content. Experimental results demonstrate the dominant role of conduction electrons
in establishing the ferromagnetic ordering in the samples with increasing stoichiometry toward Nd_3S_4. The strength of the exchange

interaction is related to the density of the conduction electrons: the influence of these electrons is strong at high values of n and is less noticeable at low values of n. $NdS_{1.49}$ is a semiconductor with carrier activation energy 3 ev and shows no magnetic ordering down to 3K. Below 50K $1/\chi$ departs from linearity, manifested by a faster rise of χ with decreasing T. At T $\sim$ 10K the rate of increase slows down considerably. This may be associated with magnetic ordering or spin glass short range order phenomena.

Discussion of Data

The measured resistivity values with and without a magnetic field H as a function of 1/T is shown in figure 1. The resistivity in each sample decreases linearly with decreasing temperature and goes through a minumum before the Curie temperature is reached on cooling. (Table II) The variations in ρ and temperatures of ρ are attributed to the different Nd concentrations in the samples. Application of a magnetic field reduces the resistivity in all samples at lower temperatures. However, the effects of a magnetic field on the resistivity of the $NdS_{1.49}$ and $NdS_{1.47}$ samples is much more pronounced than in the other samples with higher electron concentration. Such a behavior in ρ vs T is considered to be due to the electron magnon scattering as seen in other ferromagnetic materials.

If we discuss only the temperature region where ρ_{min} occurs, the transport behavior of all our samples is similar to $Ce_{3-x}V_xS_4$[1,2] or $Gd_{3-x}V_xS_4$[3]. In all our samples we find that low temperature behavior in resistivity is influenced by magnetic effects. At higher temperatures we find ρ vs T is linear. The ordering temperatures increase with increasing Nd content; i.e. increasing number of carriers.

If n is greater than $8 \times 10^{19}/cm^3$ the Cutler and Mott model suggests that E_f lies above E_c and the ρ-data describe a metallic behavior in the absence of any magnetic effects. The activation energies (ΔE) for $NdS_{1.49}$ and $NdS_{1.47}$ are obtained using the equation $\rho(t) = \rho_m(t) \exp(\Delta E/KT)$ where $\rho_m(T)$ is the high temperature behavior. Fig. 2 shows the variation of ΔE as a function of T. The peaks in ΔE occur in the neighborhood of the Curie Temperature. We observe they are reduced by an applied magnetic field. We believe that the change in ΔE in these samples may be described in terms of a localized magnetic polaron.[5] The localization is due to the exchange interaction between the conduction electron and 4f spin. The localized states remain stable over a wide temperature range due to the combined effect of magnetic and Coulombic interactions. The magnetic binding energy will be related to $I_{c-f}(<S>_{cluster} - <S>_{lattice})$, where I_{c-f} denotes the exchange interaction term, $<S>_{cluster}$ is the magnetization due to the cluster of spins in the neighborhood of an electron and $<S>_{lattice}$ is the magnetization due to the spins scattered throughout the lattice. An applied magnetic field will increase the <S>$_{lattice}$ and consequently reduce the binding energy, as seen in our measurements. More specifically, spin-clustering is predominant at low electron concentrations in Neodymium sulfides as is demonstrated by the significant quenching of resitivity under magnetic field.

The magnetic effects can be described by three main mechanisms which dominate at different conduction densisies. In compounds with

TABLE I   SOME MAGNETIC PROPERTIES OF Nd$_3$-$_x$V$_x$S$_4$

Sample Composition	Carrier Conc/cc n, (RT)	Curie Temp. paramagnetic θ (K)	Ferro. TC	Cg	μ$_{eff}$
NdS$_{1.37}$	~10^{21}	36	38	7.8	3.58
NdS$_{1.40}$	~10^{21}	20	22.5	8.92	3.46
NdS$_{1.45}$	~10^{20}	16	17.7	9.05	3.49
NdS$_{1.47}$	~10^{20}	13	16	--	--
NdS$_{1.49}$	~10^{16}	-10	--	8.4	3.61

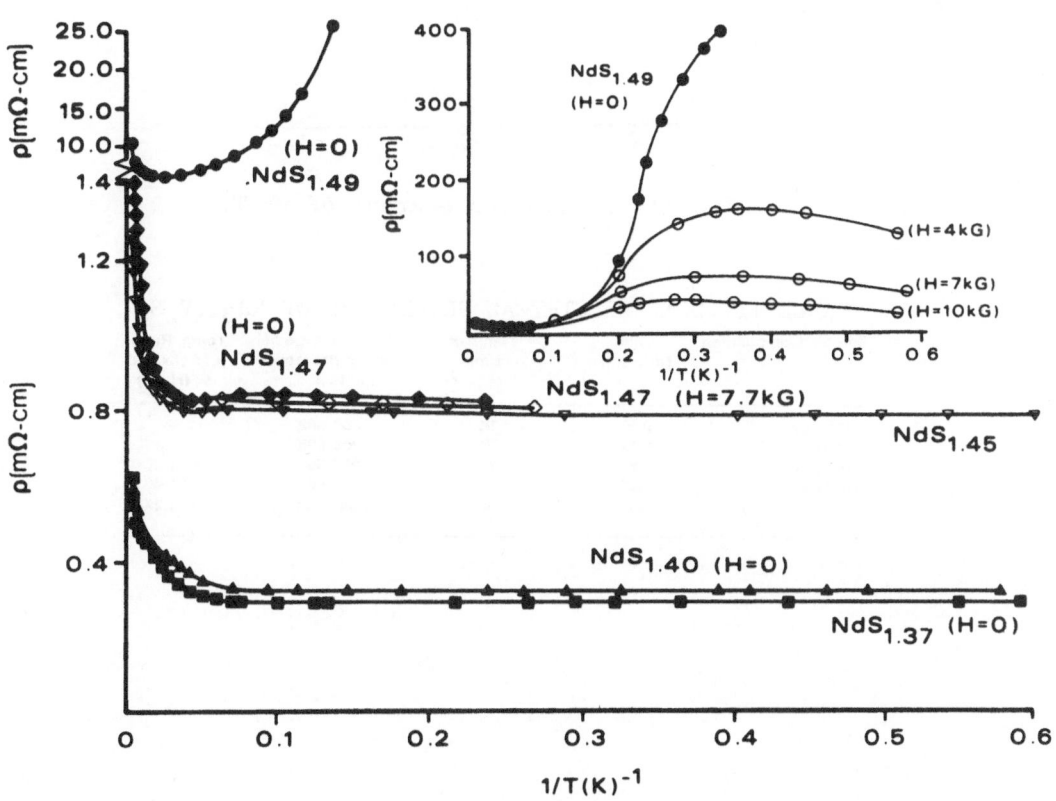

Fig. 1   Electrical Resistivity ρ (m Ω-cm) vs 1/T.

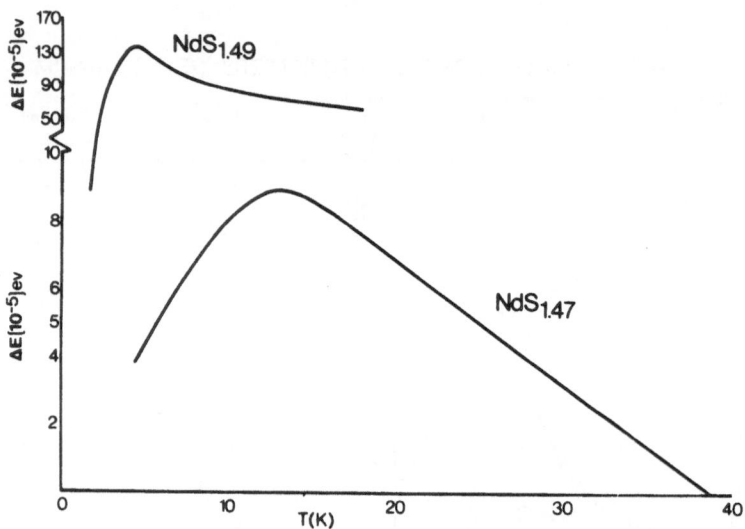

Fig. 2   Activation energy ΔE vs T.

TABLE II   SOME ELECTRICAL PROPERTIES OF $Nd_{3-x}V_xS_4$

Sample Composition	ρ min (mΩ-cm)	Temp. of ρ min	Temp. Range of linear decrease in ρ	Temp. Range of Activated ρ, (H = 0)
$NdS_{1.37}$	0.3	3K	300-60K	--
$NdS_{1.40}$	0.32	3K	300-60K	--
$NdS_{1.45}$	.8	20	300-50	30-14K
$NdS_{1.47}$	.83	25K	300-50K	27-13K
$NdS_{1.49}$	5.1	50	300-50K	50-1.2K

large electron densities the resistivity decreases with temperature
below the Curie point because the spin disorder scattering vanishes
with increasing magnetization. At smaller conduction electron
densities and Fermi energies the electrons are more sensitive to long
range fluctuations of local moments, and a resistivity maximum near
the Curie point results from "critical scattering". In the Nd- compound with the lowest conduction electron concentration (NdS$_{1.49}$)
we get a significant increase of resistivity below 10K. We believe
this results from electron localization combined with thermally activated hopping conductivity at low temperatures.

ACKNOWLEDGMENT
     We thank K. A. Gschneider, Jr., Ames Lab., ISU for helpful
discussions and the opportunity to prepare sulfide crystals. We
thank B. Beaudry, Ames Lab., ISU for preparing the samples. Thanks
are also due to S. Meis and J. Schwartz for the assistance in taking
some of the data. Acknowledgment is made to the Donors of the Petroleum Research Fund, administered by the American Chemical Society,
for the support of this research. (Grant # 11273-G3.5)

REFERENCES
1. M. Cutler, and N.F. Mott, Phys. Rev., 181, 1336, (1969).
2. G. Becher, J. Feldhaus, K. Westerholt and S. Methfessel, J. of
    Mag. and Mag. Mat., 6, 14-16, (1977).
3. T. Penney, F. Holtzberg, L.J. Tai and S. Von Molnar, AIP Conf.,
    Proc. 18, Mag. and Mag. Mat., 1973, 908, (1974).
4. T. Penney, M. W. Shafer, and J. B. Torrance, Phys. Rev., B, 5,
    3669, (1972).
5. J. B. Torrance, M.W. Shafer and T. R. McGuire, Phys. Rev.,
    letters, 29, 1168, (1972).

FARADAY ROTATION OF RARE EARTH ALKALI GERMANATE GLASSES

S.C. Cherukuri and L.D. Pye

N.Y. State College of Ceramics, Alfred University

Alfred, N.Y. 14802

ABSTRACT

The Faraday effect in rare earth alkali germanate glasses has been studied. The variation of the Verdet constant with the concentration of rare earth ions was found to be linear. The compositions that were selected fall in a range that corresponds to a conversion from diamagnetic to paramagnetic character. This result led to the discovery of zero magnetic rotation glasses.

INTRODUCTION

Magnetic field induced circular birefringence is called Faraday rotation. It is a linear magneto-optical effect in which the magnetic field introduces non-zero off-diagonal elements in the dielectric tensor of an isotropic medium and leads to optical rotation (1). The rotation angle of the plane of polarization is given by the equation (2),

$$\Theta = VLH.s \qquad\qquad (1)$$

where L is the length of the medium in meters, H is the applied magnetic field in teslas, s is a unit vector in the direction of propagation and Θ is the angle of rotation in radians. V is a constant of proportionality which is called the Verdet constant and has the units of radians/tesla.meter. By convention, the Verdet constant is considered to be positive for diamagnetic rotation and negative for paramagnetic rotation. This constant has a strong wavelength dependence.

In this work, an alkali germanate base glass, which is known to be diamagnetic in nature was doped with various paramagnetic rare earth ions. A linear variation of the Verdet constant from positive to negative values was anticipated with increasing concentration of rare earth ions.

EXPERIMENTAL TECHNIQUES

The batch compositions of the glasses studied are shown in Table 1. Three different concentrations were used for each rare earth oxide (0.5, 1.5 and 2.5 mole% RE_2O_3 in the batches). All measurements were made at a wavelength of 632.8 nm using a He-Ne gas laser.

Table 1. The mole compositions of the base glass and three other RE glasses.

	Base glass	RE(1)	RE(2)	RE(3)
RE_2O_3	----	0.5	1.5	2.5
GeO_2	76.0	75.5	74.5	73.5
Na_2O	10.0	10.0	10.0	10.0
K_2O	12.5	12.5	12.5	12.5
Al_2O_3	1.5	1.5	1.5	1.5

RESULTS AND DISCUSSION

The Verdet constants are reported in Figs. 1, 2 and 3 for alkali germanate glasses with different concentrations of rare earth ions. A linear relationship was observed between the Verdet constant and the rare earth ion concentration for all investigated rare earths. This result is consistent with theory as well as the previous results obtained by Borrelli (3), Rubinstein (4), and Robinson (5). It has been anticipated that the Verdet constant of rare earth glasses is independent of the base glass particularly at higher concentrations of rare earth ions because the diamagnetic Verdet constant of glasses does not vary much from one base glass to another. This trend is confirmed in Fig. 4, where Verdet constants are plotted against concentrations of Nd^{3+} and Pr^{3+} ions in borate, phosphate, silicate and as determined in this work, germanate glasses. Similar linear plots can be shown for all other rare earth ions.

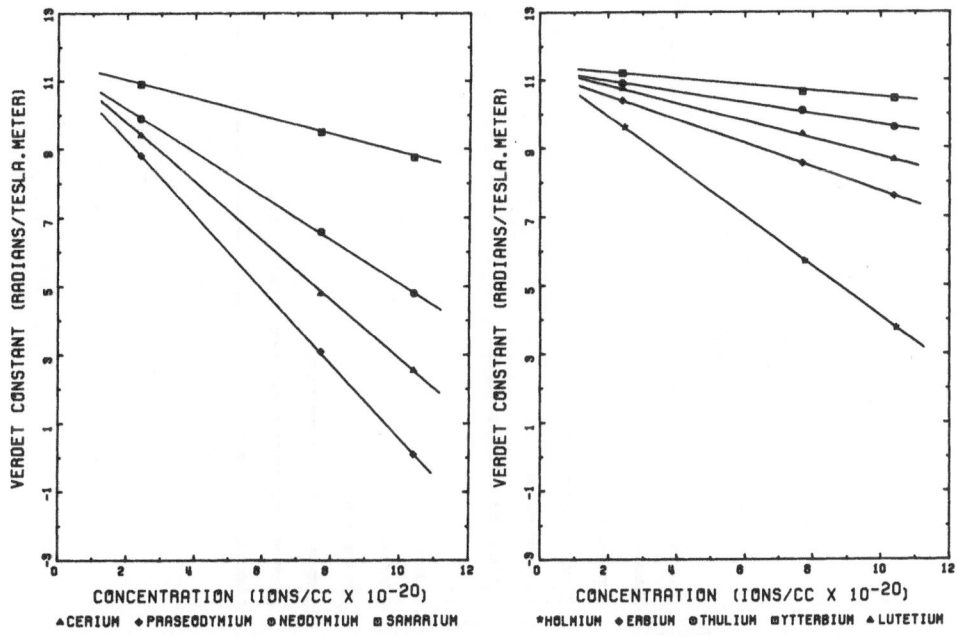

Fig. 1. Verdet constant vs concen-
        tration of rare earth ions.

Fig. 2. Verdet constant vs con-
        centration of rare
        earth ions.

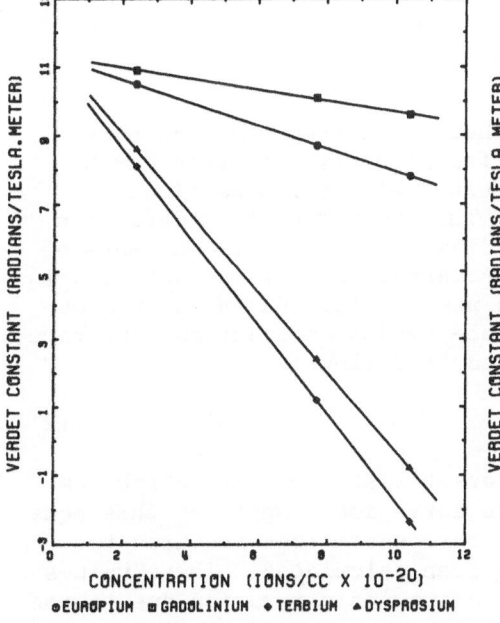
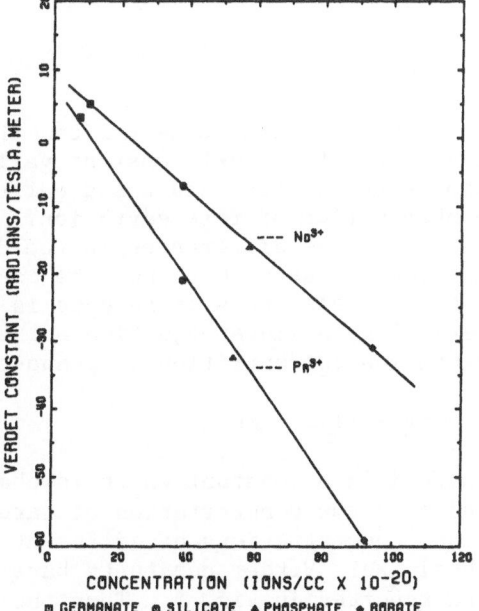
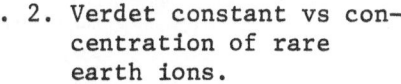

Fig. 3. Verdet constant vs. con-
        centration of rare earth
        ions.

Fig. 4. Verdet constant vs
        concentration of rare
        earth ions.

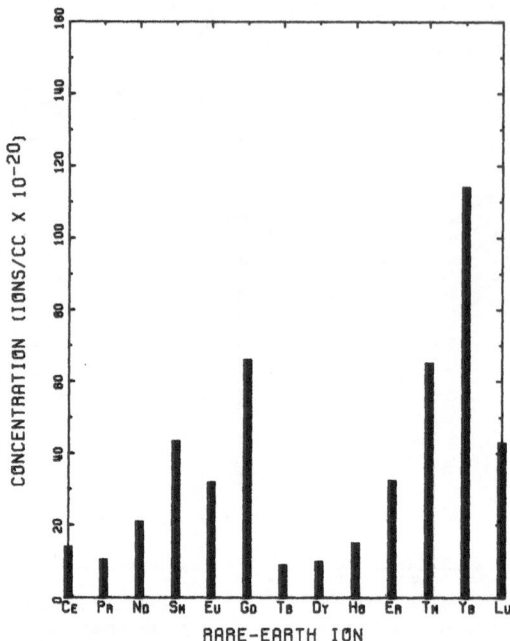

Fig. 5. Concentration of rare earth ions
        for zero rotation.

Thus, it can be genralized that, no matter what the base glass is, the Verdet constant varies linearly with rare earth ion concentration, and has a definite value for a particular concentration of rare earth ion. There might be some deviation due to slight differences in the diamagnetic Verdet constants of the base glasses, but when the concentration of rare earth ion is not too small, this difference is negligible. Based on this observation, a linear equation for the Verdet constant and the rare earth ion concentration is proposed as follows:

$$V = 11.5 - Kc \qquad\qquad (2)$$

where K is a constant which is characteristic of rare earth ion and c is the concentration of rare earth ion. Applying this equation, concentrations of different rare earth ions required to obtain zero Verdet constants have been calculated. These values are reported in Fig. 5. Thus the diamagnetic rotation due to the base glass is exactly nullified by the paramagnetic rotation of added rare earth ions at these concentrations.

REFERENCES

1. A. Nussbaum and R.A. Phillips, <u>Concemporary Optics for Scientists and Engineers</u>, Prentice-Hall, N.Y., pp 370–416 (1975).
2. E. Hecht and A. Zajac, <u>Optics</u>, Addison-Wesley Inc., Reading, Mass., pp 261–262 (1976).
3. N.F. Borrelli, <u>J. Chem. Phys.</u>, 41 (11) 3289–3293 (1964).
4. C.B. Rubinstein et al., <u>J. Appl. Phys.</u>, 35 (8) 2338–2340 (1964).
5. C.C. Robinson and R.E. Graf, <u>J. Appl. Opt.</u>, 3 (10) 1190 (1964).

USING NMR TO STUDY THE PROPERTIES OF RARE-EARTH MATERIALS

CONTAINING HYDROGEN

R. G. Barnes

Ames Laboratory, US-DOE, and Department of Physics

Iowa State University, Ames, IA 50011

Nuclear magnetic resonance provides a variety of approaches to obtaining information about rare-earth materials containing hydrogen.[1,2] Both steady-state and transient (pulsed) techniques may be applied to the NMR of all three hydrogen isotopes (^{1}H, ^{2}D, ^{3}T) as well as to the nuclei of the non-magnetic rare earths (^{45}Sc, ^{89}Y, ^{139}La, ^{175}Lu) and of other metals occurring in compounds and alloys. Valuable information can be derived from measurements made on polycrystalline (powder) samples, a distinct advantage in the case of metallic hydrides which are difficult to obtain in single-crystal form.

NMR has been used to study crystalline, electronic, and magnetic structures as well as their phase transitions in both solid solution and hydride phases of rare-earth metals and inter-metallic compounds, contributing significantly to our understanding of these materials. The resonance second moment which depends on the inter-nuclear dipolar interaction is sensitive to hydrogen locations in the lattice, particularly in the case of the spin - nuclides ^{1}H, ^{3}T, and ^{89}Y. For example, such measurements have confirmed the "premature" occupation by hydrogen of octahedral interstitial sites in the f.c.c. CaF_2 structure of yttrium dihydride.[3] The electric quadrupole interaction of large-spin nuclides (e.g., ^{2}D, ^{139}La) reflects the strength and symmetry of the second-order crystalline electric field (CEF) at the nuclear site. Measurements of the quadrupole interaction parameters are especially useful in studies of structural phase transitions and hydrogen superlattice formation.[4,5] Both the magnetic (Knight) shift of the resonance frequency and spin-lattice relaxation time T_1 depend on the electronic density-of-states at the Fermi level $N(E_F)$, in metallic materials. The dependence of these parameters on temperature and hydrogen concentration shows how hydriding alters $N(E_F)$ and can reveal electronic structure transitions (e.g., metal-insulator).[6] The coupling between magnetic rare-earth ions in hydrides has been studied via its effect on the proton Knight shift and T_1.[7]

NMR also furnishes an especially powerful approach to the study of hydrogen diffusion, and this has been the objective of many NMR investigations.[8] Both T_1 and the spin-spin relaxation time T_2 are influenced by the modulation of the dipolar interaction caused by the atomic motion of diffusion. The electric quadrupole interaction is also modulated by diffusive motion, and the effects of diffusion may be detected in the NMR of both the diffusing hydrogen and stationary metal nuclear species. The hydrogen jump frequency and its activation energy are the parameters most readily determined. Such measurements have shown, for example, that in the dihydride phases of Y and La, hydrogen diffusion increases with increasing hydrogen content and that this trend continues in the trihydride phase of La.[9,10] In intermetallic hydrides as well as in close-packed metal lattices hydrogen may occupy crystallographically inequivalent interstitial sites. NMR studies of such systems indicate that hydrogen diffusion occurs in stages characteristic of different atomic jump mechanisms on or among the several hydrogen sublattices.[11]

REFERENCES

1. R. M. Cotts, Ber. Bunsenges. phys. Chem. <u>76</u>, 760 (1972).
2. R. G. Barnes, in <u>Nuclear and Electron Resonance Spectroscopies Applied to Materials Science</u>, E. N. Kaufmann and G. K. Shenoy, eds., Elsevier North Holland (1981), p. 19.
3. D. L. Anderson, R. G. Barnes, D. T. Peterson, and D. R. Torgeson, Phys. Rev. B <u>21</u>, 2625 (1980).
4. D. G. deGroot, R. G. Barnes, B. J. Beaudry, and D. R. Torgeson, J. Less-Common Metals <u>73</u>, 233 (1980).
5. D. G. deGroot, R. G. Barnes, B. J. Beaudry, and D. R. Torgeson, Z. phys. Chem., N.F. <u>114</u>, 83 (1979).
6. R. B. Creel, R. G. Barnes, B. J. Beaudry, D. R. Torgeson, and D. G. deGroot, Solid State Commun. <u>36</u>, 105 (1980).
7. L. Shen, J. P. Kopp, and D. S. Schreiber, Phys. Letters <u>29A</u>, 438 (1969).
8. R. M. Cotts, in <u>Hydrogen in Metals I. Basic Properties</u>, G. Alefeld and J. Volkl, eds., Springer Verlag (1978), p. 227.
9. D. L. Anderson, R. G. Barnes, T. Y. Hwang, D. T. Peterson, and D. R. Torgeson, J. Less-Common Metals <u>73</u>. 243 (1980).
10. D. S. Schreiber and R. M. Cotts, Phys. Rev. <u>131</u>, 1118 (1963).
11. R. G. Barnes, T. T. Phua, D. R. Torgeson, and D. T. Peterson, Bull. Am. Phys. Soc. <u>26</u>, 335 (1981).

MAGNETIC AND STRUCTURAL PROPERTIES OF $Y_6Mn_{23}D_{23}$

C. Crowder, B. Kebe, and W. J. James, Graduate Center
for Materials Research, University of Missouri-Rolla,
Rolla, MO 65401; W. Yelon, Missouri University Research
Reactor, University of Missouri-Columbia, Columbia, MO
65201

INTRODUCTION

The addition of hydrogen (deuterium) to compounds of the type
R_6Mn_{23} often has a dramatic effect on the magnetic properties of
these compounds. For example, it has been reported (1) that
Th_6Mn_{23} is a Pauli paramagnet whereas $Th_6Mn_{23}H_{23}$ is ferromagnetic
with $T_C = 355°K$. By contrast, Y_6Mn_{23} is ferrimagnetic (2) while
$Y_6Mn_{23}D_{23}$ shows no net spontaneous magnetization, even at 4.2°K (1).
In an attempt to explain the effect of the addition of deuterium,
this study is aimed at ascertaining the crystallographic deuterium
positions and the magnetic structure for $Y_6Mn_{23}D_{23}$. Powder neutron
diffraction data show no evidence of magnetic ordering at 298°K, but
do show evidence of antiferromagnetism at 80°K. A modified Rietveld
profile technique along with magnetic susceptibility measurements
have been used to confirm this.

EXPERIMENTAL

An ingot of Y_6Mn_{23} was prepared by induction melting of 99.9%
pure elements in a water-cooled Cu boat under argon atmosphere.
X-ray diffraction photographs showed a single phase. The ingot was
deuterated with 99.5% deuterium in a stainless steel vessel at
20 atm and 110°C for 2 hours. Pressure-volume measurements indicated ~23 atoms of deuterium absorbed per formula unit. The
resulting compound was finely divided. Neutron diffraction data
were obtained on a two-axis neutron diffractometer at the Missouri
University Research Reactor. A displex cooling unit was used for
data taking at 80°K. Data obtained are shown as "+" signs in
Fig. 1. Refinement of the data was carried out using a Rietveld (3)

473

profile technique as modified by Prince (4). Scattering lengths
and form factors for Y, Mn, and D were taken from Bacon (5).
Calculated profiles are represented by solid lines in Fig. 1.
Magnetic susceptibility was measured by the extraction method at
the Service National des Champs Intense at Grenoble, France. Measurements were performed at temperatures from 2 to 300°K and at
fields up to 150 kOe.

RESULTS AND DISCUSSION

It is known (2) that Y_6Mn_{23} has face-centered cubic symmetry
(Fm3m), isotypic with Th_6Mn_{23}. If the data for the deuteride at
298°K are refined using Fm3m symmetry, a j site is found occupied
by ten deuterium atoms/f.u. This agrees with results obtained by
Commandre, et al. (6). The multiplicity of a j site is 24 atoms/
f.u., but if the symmetry is reduced to I4/mmm, this site degen-
erates to two m sites and one l site, each of multiplicity eight.
Upon refinement using I4/mmm symmetry, it was found that one m site
was fully occupied while the other m site was empty. The l site
contained two deuterium atoms/f.u. The refined cell dimensions
were still nearly cubic with Y and Mn retaining their cubic positions. The X-value (R-profile, weighted/R-expected) for the Fm3m
refinement was 2.1 whereas the I4/mmm refinement gave a X-value of
1.78. Refinement of the deuterium populations gave a total of
23 atoms/f.u., confirming the pressure-volume measurements. Individual site populations and positions are summarized in Table I.
Results obtained by other workers (6,7) using Fm3m symmetry have
been included for comparison.

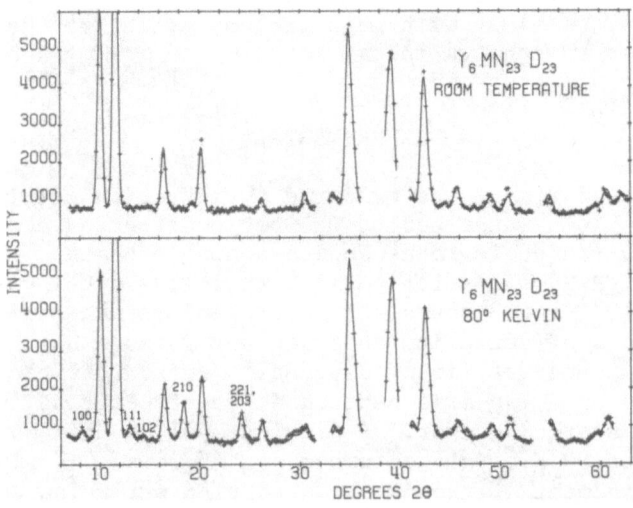

Figure 1. Neutron diffraction patterns for $Y_6Mn_{23}D_{23}$, λ = 1.293 Å.

Table I.   Refinement results for $Y_6Mn_{23}D_{23}$ at 298°K.

| | I4/mmm | | | | Fm3m | | | |
site	position	parameter	value(error)	T.F.*	site	position	parameter	a	b
Y4e	0,0,z	N	2	0.25	Y24e	x,0,0	N	6	6
		z	.209(.001)				x	.235	.208
Y8h	x,0,0	N	4	0.25					
		x	.209(.001)						
Mn2b	0,0,1/2	N	1	2.0	Mn4b	1/2,1/2,1/2	N	1	1
Mn8f	1/4,1/4,1/4	N	4	2.0	Mn24d	0,1/4,1/4	N	6	6
Mn4c	1/2,0,0	N	2	2.0					
Mn16n₁	x,0,z	N	8	2.0	Mn32f₁	x,x,x	N	8	8
		x	.360(.002)				x	.186	.182
		z	.180(.001)						
Mn16n₂	x,0,z	N	8	2.0	Mn32f₂	x,x,x	N	8	8
		x	.251(.002)				x	.377	.370
		z	.375(.001)						
D2a	0,0,0	N	1	5.0	D4a	0,0,0	N	1	1
D16n₃	x,0,z	N	6	5.0	D32f₃	x,x,x	N	8	5.6
		x	.205(.003)				x	.101	.102
		z	.092(.002)						
D16m	x,x,z	N	8	5.0	D96j	0,y,z	N	9.6	9
		x	.366(.002)				y	.167	.167
		z	.152(.003)				z	.370	.373
D16l	x,y,0	N	2	5.0					
		x	.426(.012)						
		y	.174(.008)						
D32o	x,y,z	N	6	5.0	D96k	x,x,z	N	4.3	6.4
		x	.225(.003)				x	.153	.16
		y	.113(.003)				z	.011	.046
		z	.170(.003)						
cell parameters		a	9.031(.004)		cell parameter		a	12.840	12.825
		c	12.787(.009)						
R-profile			4.8%		crystallog. residue			4.8%	
R-profile(weighted)			6.4%		R-profile(weighted)				7.8%
R-expected			3.6%		R-expected				5.2%

*Refined Debye-Waller temperature factor.
(a) M. Commandre, et al.
(b) K. Hardman and J. J. Rhyne

At 80°K, the diffraction pattern shows several new reflections. They are explained by antiferromagnetic ordering where moments are reversed on atoms related by body-centered translations.   This antiferromagnetic model is supported by the magnetic susceptibility measurements.   A plot of susceptibility vs. temperature (Fig. 2) shows $T_N \sim 170°K$.   The magnetization increases linearly with the field for all temperatures between 2.0 and 300°K.   This indicates strong negative interactions between the Mn moments.

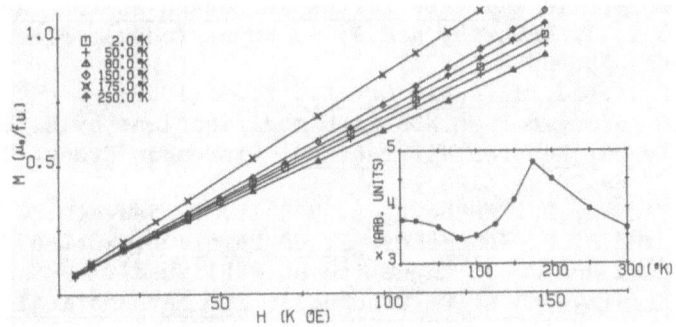

Figure 2.   Magnetic susceptibility of $Y_6Mn_{23}D_{23}$.

Table II. Moments on Mn atoms.

		b site	f site	c site	n_1 site	n_2 site
$Y_6Mn_{23}D_{23}$	$80°K$	-3.8(.25)	0	3.4(.15)	1.3(.10)	1.3(.10)
Y_6Mn_{23} (2)	$4°K$	-4.3	-2.9	-2.9	2.0	2.1

In the diffraction pattern, the 001 and 003 reflections are absent, so a collinear structure with moments parallel to the c-axis is reasonable. Considering the I4/mmm crystallographic symmetry and the collinear structure, there are eight possible Shubnikov magnetic space groups (8). Of these, only the $P_I4/mm'm'$ space group gave a good profile upon refinement. The magnetic R-factor for the other seven space groups varied from 60-95%, while for $P_I4/mm'm'$ it was 16.9%. This is reasonable considering the small size of the magnetic reflections and an expected magnetic R-factor of 18%. Because of the nature of the $P_I4/mm'm'$ space group, there can be no moment along the c-axis for the Mn atoms in the 8f positions and, indeed, none is found. Moment magnitudes for the other Mn atoms compare well with those of the non-deuterated compound (Table II). All other parameters refined to values similar to those at 298°K.

While Y_6Mn_{23} has Fm3m symmetry, it has been found that the deuteride has I4/mmm symmetry. The Mn moments order antiferromagnetically below ∿170°K, parallel to the c-axis. While the moment magnitudes have not changed significantly, it is obvious that the addition of deuterium has affected the Mn-Mn exchange.

ACKNOWLEDGMENT

The authors thank the Department of the Army for support under Grant No. DAAG-29-80-C-0084.

REFERENCES

(1) S. K. Malik, T. Takeshita, and W. E. Wallace, Solid State Comm., 23:599 (1977).
(2) K. Hardman and W. J. James, "Neutron Diffraction of Y_6Fe_{23} and Y_6Mn_{23}," in The Rare Earths in Modern Science and Technology, G. J. McCarthy and J. J. Rhyne (eds.), Plenum Press, New York, pp 408-8 (1977).
(3) H. M. Rietveld, J. Appl. Cryst., 2:65 (1969).
(4) Program received from NBS with modifications by E. Prince.
(5) G. E. Bacon, Neutron Diffraction, Clarendon Press, Oxford, England, pp 38-41 (1967).
(6) M. Commandre, D. Fruchart, A. Roualt, D. Sauvage, C. B. Shoemaker, and D. P. Shoemaker, J. de Physique, 40:L-639 (1979).
(7) K. Hardman and J. J. Rhyne (to be published).
(8) C. J. Bradley and A. P. Cracknell, The Mathematical Theory of Symmetry in Solids, Clarendon Press, Oxford, England, pp 569-681 (1972).

MAGNETIC AND STRUCTURAL PROPERTIES OF $Th_6Mn_{23}D_x$ AND $Y_6Mn_{23}D_x$

K. Hardman, J. J. Rhyne, E. Prince, National Bureau of
Standards, Washington, DC 20234;
H. K. Smith, S. K. Malik, and W. E. Wallace, University
of Pittsburgh, Pittsburgh, PA 15260

The magnetic behavior of the isostructural compounds Y_6Mn_{23} and Th_6Mn_{23} is drastically altered on hydrogenation (or deuteration). Y_6Mn_{23} is a ferrimagnetic compound with $T_c = 486$ K and a bulk magnetization value of 13 μ_B/f.u. (formula unit). The long-range magnetic order vanishes on hydrogen absorption. In contrast, Th_6Mn_{23} is a Pauli paramagnet, while $Th_6Mn_{23}H_{30}$ exhibits long-range magnetic ordering with $T_c = 329$ K and a bulk magnetization value of 18.5 μ_B/f.u.

Th_6Mn_{23} is fcc (space group Fm3m) with 116 atoms per unit cell and four formula units per unit cell. There is one thorium or yttrium site and four manganese sites (b, d, f_1, and f_2). The magnetic structure of Y_6Mn_{23} is collinear with spins along the <111> direction with the b and d sites coupled antiparallel to the two f sites. All magnetically ordered spins within each of the sites are coupled in the same direction. At liquid helium temperatures, the magnetization of the manganese moments is 1.8 μ_B/Mn atom at the f sites and greater than 2 μ_B/Mn atom at the b and d sites.

The compounds $Th_6Mn_{23}D_{16}$ and $Y_6Mn_{23}D_{22}$ were studied at 295 and 4 K by neutron diffraction profile refinement methods. The pressure isotherms of $Th_6Mn_{23}D_x$ and $Y_6Mn_{23}D_x$ were measured at 295 K. $Th_6Mn_{23}D_x$ exhibits a pressure plateau between 17 and 24 atoms of deuterium, while $Y_6Mn_{23}D_x$ has no plateau from 13 to 24 atoms of deuterium.

The site occupancies and atomic positions of the deuterium atoms were determined for both compounds. At 295 K there are 5.67 and 8 deuterium atoms in the tetrahedral f sites of $Y_6Mn_{23}D_{22}$ and

Table 1. $Th_6Mn_{23}D_{16}$–4 K–$P4_2mnm$ Structural Parameters

	x	y	z	N
Th e	0	0	0.213 (0.214)	2
f	0.224 (0.214)	0.224 (0.214)	0	2
g	0.207 (0.214)	0.793 (0.786)	0	2
Mn b	0	0	0.5	1
Mn c	0	0.5	0	2
j_1	0.25 (0.25)	0.25 (0.25)	0.75 (0.75)	4
Mn k_1	0.005 (0)	0.359 (0.358)	0.182 (0.179)	8
Mn k_2	0.017 (0)	0.259 (0.264)	0.363 (0.368)	8
D k_3	0.005 (0)	0.204 (0.204)	0.097 (0.102)	8
D i	0.249 (0.22)	0.500 (0.5)	0	4
j_3	0.353 (0.36)	0.350 (0.36)	0.146 (0.14)	4

$Th_6Mn_{23}D_{16}$ respectively (space group $Fm3m$). However, the octahedral a site (one atom), which is fully occupied in the yttrium compound, is virtually empty in the thorium compound. The trigonal i site (8 atoms) in $Th_6Mn_{23}D_{16}$ is two-thirds full, while the tetrahedral j site (9) was occupied in $Y_6Mn_{23}D_{22}$. The remaining deuterium atoms in $Y_6Mn_{23}D_{22}$ occupied the tetrahedral k site (6.13 atoms).

$Y_6Mn_{23}D_{22}$ remains face-centered cubic ($Fm3m$) at 4 K but undergoes a magnetic phase change around 160 K corresponding to a transition in the magnetic susceptibility. Neutron diffraction data suggest this change to be from the paramagnetic phase to a short-range magnetic cluster phase, in that antiferromagnetic peaks as well as magnetic contribution to nuclear peaks are present.

A structural and magnetic phase change occurs in $Th_6Mn_{23}D_{16}$ below 80 K. At 4 K, $Th_6Mn_{23}D_{16}$ has a tetragonal structure ($P4_2/mmm$). The atom positions and occupancy are given in Table 1. The values in parenthesis are the face-centered cubic equivalent atomic positions. The lattice has been slightly stretched along the c axis and contracted along the a axis from that of an fcc structure with a = 9.077 Å and c = 12.961 Å.

$Th_6Mn_{23}D_{16}$ becomes ferrimagnetic below the ordering temperature. Its magnetic coupling is similar to that of Y_6Mn_{23} except that the d site has broken down into two sites which are c and j. The magnetic moment spins on the b and c sites are coupled antiparallel to the j and two k sites. Although the bulk magnetization values (12.32 μ_B/ f.u. in $Th_6Mn_{23}D_{16}$ and 13.23 μ_B/f.u. in Y_6Mn_{23}) are similar, the magnetic moments are generally less than in Y_6Mn_{23}. The magnitude of the moments in $Th_6Mn_{23}D_{16}$ at 4 K are 0.56, 2.62, 1.67, 0.99, and 0.44 μ_B/f.u. for b, c, j, k_1, and k_2 sites respectively.

VALENCE BONDING, ATOMIC VOLUMES, AND COORDINATION NUMBERS FOR

Y_6Mn_{23}, Th_6Mn_{23}, AND THEIR HYDRIDES

Forrest L. Carter

Chemistry Division, Naval Research Laboratory

Washington, DC 20375

The interesting bonding, magnetic, and hydriding properties (1,2) of Y_6Mn_{23} and Th_6Mn_{23} are examined using Pauling's metallic radii, polyhedral atomic volume (PAV), and partial coordination number coefficient (PCNC) calculations (3,4). The strong contraction that rare earth and early transition metals undergo in their formation of intermetallic compounds with small transition metals (3,5) is reversed upon hydride formation, suggesting that the latter process is partly driven by the compressed state of the large metal atom. This effect has been identified in $LaNi_5$-type compounds (6), in TiFe and TiCu (5), and in Mg_2Ni (7) and is investigated for the subject compounds. PAVs and valences for both selected Mn and H atoms suggest future areas for structure refinements.

In both series the Y and Th PAVs show a substantial volume contraction (10 and 7% respectively) upon compound formation and reexpansion (almost complete for $Th_6Mn_{23}D_{21}$). These contractions are not as large as observed for the rare earth nickel series since Mn is both larger than Ni (12.5 to 14.6 vs $11.1A^3$ for Ni) and multivalent (and hence multi-sized) even in some of its allotropes. The Mn PAVs vary from 12.8 (Mn3) to $13.7A^3$ (Mn4) in Y_6Mn_{23} and from 12.6 (Mn3) to $14.9A^3$ (Mn1) in Th_6Mn_{23}. Upon hydriding the latter compound to $Th_4M_{23}H_{21}$, Mn1 PAV increases to $18.8A^3$ and an unrealistic valence of 1.2. It is likely that future structural refinements will decrease the calculated Mn1 PAV to below $15A^3$.

From the generalization of coordination number (4) the concept of PCNC can be developed which indicates the relative importance of different kind of neighbors. For the Th_6Mn_{23} structure this is indicated in Table 1 where f(A-B) (based on bond order) is

Table 1.   f(A-B) Interaction Matrix

A \ B	Th	Mn_1	Mn_2	Mn_3	Mn_4
Th	.26	.01	.20	.22	.31
Mn1	.07	0	0	0	.93
Mn2	.27	0	0	.47	.25
Mn3	.18	0	.30	.32	.18
Mn4	.36	.10	.21	.25	.06

the PCNC of atom A with its B neighbors. We note especially that
though Mn1 is the center of Th octahedra its main interaction is with
its Mn4 neighbors to the extent of 93%. Previous experience (5,6)
suggests that the H PAV should be between ∿1.5 to 2.5A^3, as is observed for D2 and D3 (notation of refs. 1,2) in the deuterides of
Y_6Mn_{23} and $ThMn_{23}$. However D1 PAV for $Y_6Mn_{23}D_9$ and $Y_6Mn_{23}D_{18}$ is too
large (4.0 and 4.1A^3) respectively and marginal for D1 in $Th_6Mn_{23}D_{21}$
(2.8A^3). In addition we note that the PAV value of D3 in $Y_6Mn_{23}D_{18}$
is much too small (0.65A^3). Accordingly it is anticipated that the
deuteriums in this disordered position, and their neighbors, will
reposition themselves to give a more reasonable D PAV.

Finally we note a correlation between the average Mn valence
and magnetic ordering. Thus $Th_6Mn_{23}D_{23}$ is ferromagnetic and
Y_6Mn_{23} is ferrimagnetic (Mn ave. valence = 3.8 and 4.0 respectively)
while Th_6Mn_{23} is a Pauli paramagnet and $Y_6Mn_{23}D_{23}$ is nonmagnetic to
low temperatures (3.7 and 3.5 respectively). Assuming that bonding
d-character is about 50% this suggests that unoccupied d-orbital
character is required for magnetic ordering.

REFERENCES

1. M. Commandre, D. Fruchart, A. Ronault, D. Sauvage, C. B. Shoe-
    maker, and D. P. Shoemaker, J. Phys. (Paris), 40, L-639 (1979).
2. K. Hardman, J. J. Rhyne, K. Smith, and W. E. Wallace, J. Less-
    Common Metals 74, 97 (1980).
3. F. L. Carter, Atomic Volume Contraction in Rare Earth Nickel
    Intermetallics, in G. J. McCarthy, J. J. Rhyne and H. B. Silver
    (eds.), Rare Earths in Modern Science and Technology II,
    Plenum, New York, 1979, p. 299.
4. F. L. Carter, Acta Cryst. B34, 2962 (1978).
5. F. L. Carter, J. Less-Common Metals 74, 245 (1980).
6. F. L. Carter, J. C. Achard and A. Percheron-Guegan, Hydrogen
    Coordination Number, Volume, and Dissociation Pressure in
    $LaNi_5$ Substituted Hydrides, in G. J. McCarthy, J. J. Rhyne and
    H. B. Silver (eds.), Rare Earths in Modern Science and Technol-
    ogy II, Plenum, New York, 1979, p. 599.
7. G. C. Carter and F. L. Carter, Metal Hydrides for Hydrogen
    Storage, a review in T. N. Veziroglu Proceedings of Inter-
    national Symposium on Metal Hydrogen Systems, Miami Beach, Fla.,
    13-15 April 1981, in press.

STRUCTURE OF Al, Cu and Si SUBSTITUTED LaNi$_5$ AND OF THE CORRESPONDING β-DEUTERIDES FROM POWDER NEUTRON DIFFRACTION. LOCALIZED DIFFUSION MODE OF HYDROGEN IN LaNi$_5$ AND Al and Mn SUBSTITUTED COMPOUNDS FROM QUASIELASTIC NEUTRON SCATTERING

J.C. Achard[+], A.J. Dianoux[o], C. Lartigue[+o]
A. Percheron-Guegan[+], F. Tasset[o]
[+]Chimie Métallurgique des Terres Rares, CNRS, 1, Place
Aristide Briand, 92190 Meudon, France;
[o]Institut Laue-Langevin, 156X, 38042 Grenoble Cedex
France

ABSTRACT

The first part of this paper gives structural parameters obtained from powder neutron diffraction for the following compounds and related deuterides: LaNi$_{5-x}$M$_x$ with M$_x$ = Al$_{0.5}$, Si$_{0.5}$, Cu$_{1.0}$. The second part of the paper is a preliminary report on a neutron time of flight [TOF] spectroscopic measurement which shows for the first time experimental evidence for quasi elastic broadening of hydrogen diffusion peak, both in LaNi$_5$H$_{5.8}$ and LaNi$_4$MnH$_{5.7}$. In the Al substituted phase LaNi$_4$AlH$_{4.3}$ this effect is much reduced.

INTRODUCTION

The effects associated with the partial substitution by Al and Mn for Ni in the compound LaNi$_5$ was studied by neutron powder diffraction at the ILL-Grenoble (1,2). Owing to the continuing practical interest in these substituted systems (3) and the attention shown (4), we have decided to enlarge our neutron study in two ways: (i) a more systematic structural determination of the effect of other substitutions (Cu, Si); (ii) a study of the dynamical behavior of hydrogen in those substituted compounds using neutron spectroscopic methods. We report here on the advancements in both directions.

NEUTRON DIFFRACTION STRUCTURAL STUDY

Experimental

The measurements were made at room temperature using the ILL-DIB multidetector diffractometer, (λ = 1.28Å). The samples were prepared and characterized as described previously (1,5). Activated compounds were deuterated and kept in closed quartz containers under deuterium pressures corresponding to the equilibrium of the β-phases as in (2).

Data Reduction and Results

The structure refinements were made using the Rietveld powder profile least-squares refinement program (6). Refinements were obtained with the same 5 distinct sites model (space-group P6/$_m$ mm), including the possibility of replacing some La atoms by pairs of Ni atoms, as in our previous structural work (2). Results for three intermetallic compounds $LaNi_{5-x}M_x$ (M_x = $Al_{0.5}$, $Si_{0.5}$, $Cu_{1.0}$) are reported in Table 1. The refined parameters for the corresponding deuterides are given in Table 2.

Discussion

Although both Al and Si substitutions sit in the z=1/2 plane, a noticeable difference is in the absence of cell dilatation associated with Si. It seems to indicate a smaller size for Si atoms. Cu atoms are found in both atomic planes with a slight preference for the z=0 plane. This repartition is not surprising since Cu and Ni are known to have similar atomic sizes. A comparison of the deuterium occupation numbers found in $LaNi_{4.5}Al_{0.5}D_{5.4}$ with those previously reported for $LaNi_5D_{6.6}$ and for $LaNi_{4.0}Al_{1.0}D_{4.8}$ (2) shows that the opportunity for the 4h and 12o sites of being occupied by deuterium atoms gradually disappears when Al concentration increases. Concerning the $LaNi_{4.5}Si_{0.5}D_{4.3}$ deuteride it seems to retain some deuterium in the 4h site, a near neighbor of the Si site. This is worth noting since the low deuterium content of this deuteride should have resulted in the only occurrence of 6m, 12n and 3f sites according to a "rule" stated in ref. (2). Conversely this may be considered as an indication of special properties associated with Silicon substitutions. The $LaNi_4CuD_{5.07}$ results have to be compared with $LaNi_5D_{6.6}$ because the Cu atoms go into both Ni sites. Here the most significant fact is probably the disappearance of the 12o site.

NEUTRON TIME OF FLIGHT DYNAMICAL STUDY

Experimental

We used the IN5 TOF spectrometer installed on a neutron guide looking at the ILL high flux reactor cold neutron source. The

Table 1. Atomic Parameters of the Intermetallic Compounds $LaNi_{5-x}M_x$. S_1 and S_2: Numbers of M atoms on (2c) and (3g) Sites Respectively; B: Isotropic Temperature Factors; R: Agreement Value for Integrated Intensities; Standard Deviations are Given in Parentheses.

| | Nominal Composition | | |
	$LaNi_{4.5}Al_{0.5}$	$LaNi_{4.5}Si_{0.5}$	$LaNi_4Cu$
a(Å)	5.040(1)	5.007(1)	5.033(1)
c(Å)	4.023(1)	3.992(1)	4.007(1)
s(La)	0.009(4)	0.029(4)	0.020(3)
s_1(z=0)	0.0	0.0	0.54(4)
s_2(z=1/2)	0.42(4)	0.50(5)	0.44(4)
x (Ni_{III})	0.285	0.287	0.286
z (Ni_{IV})	0.310	0.312	0.311
B_{La}	0.68(7)	0.69(8)	0.72(5)
B_I	0.58(5)	0.63(6)	0.74(5)
B_{II}	0.48(4)	0.68(4)	0.61(3)
Refined Composition			
La	0.991(4)	0.971(4)	0.981(3)
Ni	4.60(5)	4.56(7)	4.06(6)
M	0.42(4)	0.50(5)	0.98(6)
Normalization	$LaNi_{4.64(5)}Al_{0.42(4)}$	$LaNi_{4.69(7)}Si_{0.51(5)}$	$LaNi_{4.14(6)}Cu_{1.00(6)}$
R-factor (%)	3.43	5.44	2.74

incident neutron energy selection is made through the use of a variable speed multichopper device. Activated samples were kept in a suitable aluminum container (60 x 30 x 1 mm³) with a minimized 2mm Al total wall thickness which allows a filling gas pressure of 3 bars. Three hydrides were investigated: $LaNi_5H_{5.8}$ in the range temperature 4.2-300K, $LaNi_4MnH_{5.7}$ and $LaNi_4AlH_{4.3}$ at 300K and various Q-values. The experimental backgrounds were respectively determined with the same dehydrided samples and substracted from the measured spectra in order to obtain spectra from the pure quasi-elastic scattering of the hydrogen.

Results

Using the best energy resolution (10.05 Å incident wavelength) we could not detect any difference with the instrumental resolution peak (20 μeV FHWM) for neutrons incoherently scattered at room temperature by hydrogen, this up to momentum transfer Q=1.1 Å⁻¹. Owing to the highly resolvable triangular shape of the resolution peak, the minimum measurable enlargement is 1/10 of the 20 μeV FHWM that is 2 μeV. Assuming a standard Fick's law, generally valid in that range of Q, we can estimate a hydrogen diffusion coefficient at room temperature D [300 K] < 2x10⁻⁷ cm²s⁻¹. This is in agreement with high resolution backscattering neutron results obtained recently by Lebsanft et al. (7). Lowering the temperature down to 4.2 K

Table 2. Atomic Parameters of Deuterium Atoms In $LaNi_{5-x}M_x$ Deuterides Refined in a 5 Sites Model (Space Group $P6/mmm$) – $\Delta V/V$ is the Lattice Expansion.

	Nominal Composition		
	$LaNi_{4.5}Al_{0.5}D_{5.4}$	$LaNi_{4.5}Si_{0.5}D_{4.3}$	$LaNi_4CuD_{5.07}$
$a(\overset{\circ}{A})$	5.353(1)	5.328(1)	5.374(1)
$c(\overset{\circ}{A})$	4.263(1)	4.077(1)	4.190(1)
$\Delta v/v$ (%)	19.5	15.6	19.3
Site 4h			
x	1/3	1/3	1/3
y	2/3	2/3	2/3
z	0.398(8)	0.365(14)	0.397(6)
n	0.23(2)	0.25(3)	0.55(2)
Site 6m			
x	0.138(1)	0.155(1)	0.150(1)
y	0.276(1)	0.310(2)	0.301(2)
z	1/2	1/2	1/2
n	1.91(2)	1.93(3)	2.24(2)
Site 12n			
x	0.470(2)	0.471(4)	0.467(3)
y	0	0	0
z	0.103(1)	0.076(3)	0.085(2)
n	2.62(3)	2.11(3)	2.28(3)
Site 12 o			
x	0.225(3)		
y	0.450(5)		
z	0.319(7)		
n	0.64(3)		
$B(\overset{\circ}{A}{}^2)$			
B_{La}	1.35(7)	1.54(14)	1.31(9)
B_I	1.65(5)	3.12(15)	1.92(8)
B_{II}	1.49(3)	1.71(7)	1.53(4)
B_D	2.15(10)	3.15(20)	3.53(15)
Refined Composition			
La	0.99	0.97	0.98
Ni	4.60	4.56	4.06
M	0.42	0.50	0.98
D	5.40(5)	4.30(5)	5.07(4)
Normalization	$LaNi_{4.64}Al_{0.42}D_{5.45}$	$LaNi_{4.69}Si_{0.51}D_{4.43}$	$LaNi_{4.14}Cu_{1.00}D_{5.17}$
R-factor (%)	4.94	7.26	5.91

with the same good energy resolution we could not see any evidence for tunneling effects.

An interesting effect shows up when using a different energy resolution that is 5.14 Å incident wavelength and 127 µeV FWHM. We compare in Fig. 1 the spectra obtained from a vanadium sample and from hydrogen scattering in $LaNi_5H_{5.8}$. Although they are very similar a marked difference can be seen at the base of the peaks (Fig. 1b and 1d) when they are enlarged 50 times. It has been possible, by using least-squares deconvolution programs available at ILL, to show that the hydrogen peak can be decomposed in a big elastic peak plus a small broadened Lorentzian component.

In the range of momentum transfer Q explored (0.5–2.1 $\overset{\circ}{A}{}^{-1}$) the refined HWHM (half width half maximum) of the small Lorentzian contribution appears to be ~80±30 µeV and fairly constant. Moreover

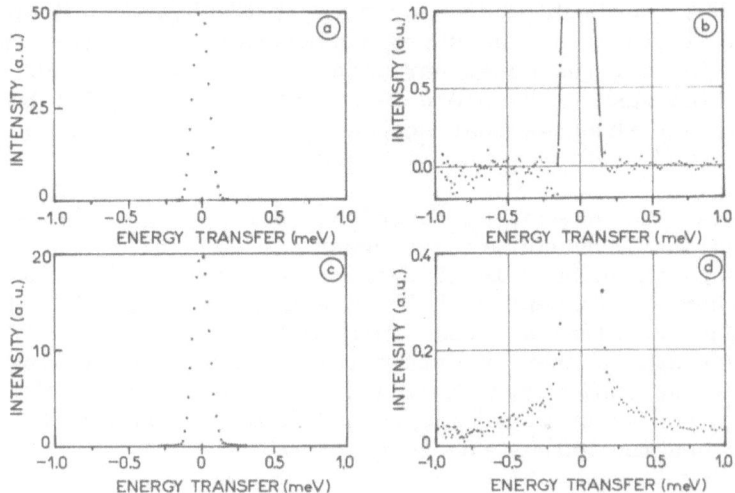

Fig. 1: Neutron TOF Spectra and Details (Intensity x 50) of the Quasielastic Lines from: a,b) Vanadium; c,d) Hydrogen in $LaNi_5H_{5.8}$ (λ_0 = 5.14 Å, T = 300 K, Q = 1.9 Å^{-1}).

the relative intensity of this quasielastic contribution increases substantially when raising Q, it goes from 2 to 5%. These facts reveal a fairly localized (~0.6 Å) hydrogen diffusion mode. This effect disappears at low temperature; it remains the same in $LaNi_4MnH_{5.7}$ but is very much reduced in $LaNi_4AlH_{4.3}$.

CONCLUSIONS

Al substitutions for Ni in $LaNi_5$ result in very important microscopic changes in the behavior of hydrogen stored in these materials:

- a substantial decrease in the hydrogen mobility has been measured with proton NMR technique (8);
- the disappearance of 3 (3f, 4h, 12o) out of the 5 different hydrogen sites has been seen by neutron diffraction (2);
- the quasi elastic neutron scattering (QNS) seen for the first time in $LaNi_5H_{5.8}$ disappears [this work].

Both 12o and 3f sites become empty in $LaNi_4AlH_{4.8}$ and so they can be invoked in the QNS drastic reduction with Al. The distance between 12o and next neighbor 6m site being 0.8 Å we think that the 3f-12n shorter distance (0.6 Å) is more likely to be the cause of a "rapid localized" hydrogen jumping mode required by the general features of the QNS observed. Holding that assumption we have built

a simple but realistic model for this 3f-12n hydrogen jumping mode
which contains only one adjustable parameter (residence time τ on
the 12n site) all the other parameters being fixed by our structural
previous findings. A complete report of this work will be made
later (9), but it shows that such a model is in good agreement with
our QNS data. We find that $\tau = 1.2\pm0.3 \; 10^{-11}$ sec.

Furthermore it must be noted that our previous structural
results for LaNi$_5$ β-deuteride suggest the possibility of a continuous long range path for hydrogen along 12o-6m-12o-12n-3f-12n-12o
in qualitative agreement with the tunnel idea (10). The disappearance of the 12o site, an "easy path way" between 6m and 12n sites
certainly plays another important role in the reduction by Al of
the long range mobility of hydrogen. More systematic NMR measurements on other substituted compounds should be made as the comparison with neutron data seems to be of value.

ACKNOWLEDGMENTS

We thank Dr. F. Volino for helpful discussions on the nature
of the QNS in our samples.

REFERENCES

1.   C. Lartigue, A. Percheron-Guegan, J.C. Achard, F. Tasset,
     J. Less Com. Met, 75:23 (1980).
2.   A. Percheron-Guegan, C. Lartigue, J.C. Achard, P. Germi,
     F. Tasset, J. Less Com. Met., 74:1 (1980).
3.   I. Sheft, D.M. Gruen, G.J. Lamich, J. Less Com. Met., 74:401
     (1980).
4.   J.C. Achard, C. Lartigue, J.C. Mathieu, A. Pasturel,
     A. Percheron-Guegan, F. Tasset, J. Less Com. Met., 79:167 (1981).
5.   H. Diaz, A. Percheron-Guegan, J.C. Achard, C. Chatillon,
     J.C. Mathieu, Int. J. Hydrogen Energy, 4:445 (1979).
6.   H.M. Rietveld, Acta Cryst., 22:151 (1967).
7.   E. Lebsanft, D. Richter, J.M. Topler, Proc. Hydrogen in Metals
     Conf., Münster (1979).
8.   R.C. Bowman, D.M. Gruen, M.H. Mendelsohn, Sol. State Com.,
     32:501 (1979).
9.   To be published.
10. T.K. Halstead, J. of Sol. State Chem., 11:114 (1974).

THERMODYNAMIC AND MAGNETIC PROPERTIES OF LaNi$_{5-x}$Fe$_x$ COMPOUNDS AND THEIR HYDRIDES

J. Lamlouni, C. Lartique, A. Percheron-Guegan and
J.C. Achard, Chimie Metallurgique des Terres Rares,
CNRS, 1 Place A. Briand, 92190 Meudon Bellevue;
G. Jehanno, Physique des Solides, CEN Saclay, B.P. 2,
91190 Gif Yvette, France

The variation of thermodynamic and magnetic properties with partial replacement of Ni by Fe in LaNi$_5$ has been investigated. For compositions of LaNi$_{5-x}$ with x≤1, the unit cell remains hexagonal, P6/mmm, and its volume increases with increasing Fe concentration according to the relation V=86.9 + 1.87X$_{Fe}$ Å^3. The equilibrium pressure of the hydride decreases slightly from 1.7 bars in LaNi$_5$ to 0.6 bar for LaNi$_4$Fe at 25°C (1) and a hydrogen content from 6 to 5H/mole. Figure 1 shows ^{57}Fe Mossbauer spectra for LaNi$_4$Fe. The 298 K spectrum can be fitted to a single quadrupole effect (e^{2}qQ/2= 1.08 mm/s) suggesting that all Fe atoms are on a single site. At 190 K, a weak hyperfine field of 5kOe is observed due to the onset of magnetic order. This field increases to 222 kOe at 4.2 K. Susceptibility as a function of temperature (Fig. 2), measured at 35 Oe, indicates paramagnetic behavior to 190 K. Magnetization isotherms as a function of fields up to 150 kOe (Fig. 3) show that in low fields, the value of the initial slope is large and temperature dependent. For H>30kOe, magnetization increases linearly with field for T<160 K. Here M$_s$ = 20.5 x 10^3 emu/Fe atom corresponding to a superimposed susceptibility of 4.45 x 10^{-3} emu/Fe atom, indicating that partial replacement of Fe for Ni in the exchange enhanced Pauli paramagnet LaNi$_5$ leads to weak ferromagnetic behavior. For the hydride, a large decrease in the magnetization is observed (7.42 x 10^3 emu/Fe atom) compared to the intermetallic (21.17 x 10^3 emu/Fe atom) and no magnetic order to 4.2 K was observed.

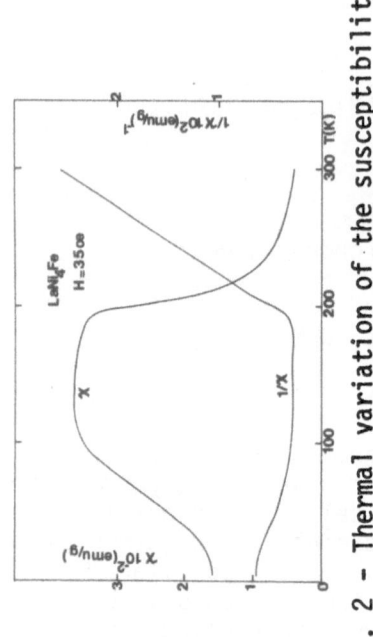

Fig. 2 – Thermal variation of the susceptibility and the reciprocal of LaNi$_4$Fe

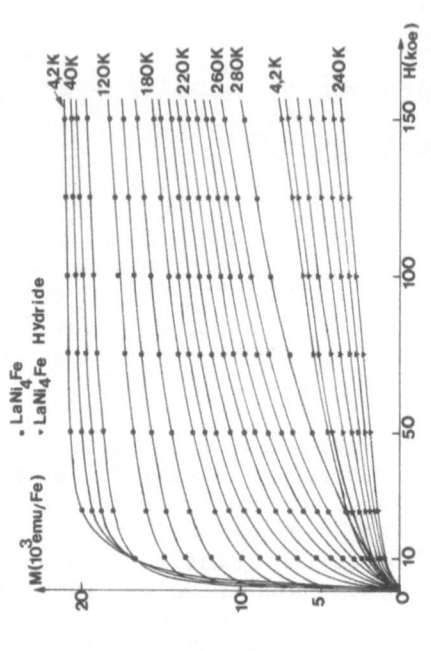

Fig. 3 – Magnetization M = f(H,T) isotherms for LaNi$_4$Fe and its hydride.

Fig. 1 – LaNi$_4$ ^{57}Fe Mössbauer spectra as a function of temperature.

THERMODYNAMIC PROPERTIES OF LaNi$_4$M COMPOUNDS AND THEIR HYDRIDES

A. Pasturel and C. Chatillon. E.N.S.E.E.G. Domaine
Universitaire B.P. 44 - 38401 St. Martin d'Hères;
A. Percheron-Guegan and J.C. Achard, Chimie Métallur-
gique des Terres Rares, C.N.R.S. 1, Place Aristide
Briand 92190 Meudon, France

INTRODUCTION

As part of a systematic study of hydrides related to LaNi$_5$,
we have been attempting to correlate properties of the hydrides to
those of their parent intermetallics. Miedema's model for predict-
ing enthalpy of hydrogenation from estimated values of enthalpy of
formation of the intermetallic compound and of its constituent
binary hydrides is applicable qualitatively to a large number of
hydrides (1,2). In the present work, we have measured the enthalpies of formations of LaNi$_4$M compounds and their hydrides where
M = Al, Cu, Fe, Ni, Mn. For these substituted compounds, Miedema's
model is not suitable. We propose here a relationship that takes
into account the M for Ni substitution and compares the resulting
predicted enthalpies with experimental values.

EXPERIMENTAL

Specimens were prepared by induction melting of the high purity
elements as described previously (5). Phase characterization is
summarized in Table 1. Enthalpies of formation were calculated from
dissolution enthalpies in molten aluminum according to:

$$\Delta H_f^{TO} < LaNi_4M > \ = \ Q_{La}^{\infty} + Q_M^{\infty} + 4\,Q_{Ni}^{\infty} - Q_{LaNi_4M}^{\infty}$$

where Q_X^{∞} is the heat of dissolution at infinite dilution of the
element. The dissolution enthalpies measurements were carried out
in an isoperibolic calorimeter as described previously (3). Enthalpies of hydrogenation of the hydrides were deduced from the

variation of desorption pressure measured over the range 25-60°C
for M= Fe, Ni, Ca and 60-200°C for M = Al and Mn.

RESULTS

The experimental data from the calorimetric measurements are
summarized in Table 2. To correct for temperature differences among
the various experiments, we used enthalpy data for pure elements (8)
and assumed that the partial limit enthalpies of mixing do not
change over an interval of 30°C.

Enthalpies of Hydrogenation

From these experimental values of the enthalpies of formation
of the intermetallic compounds, we have calculated their hydrogenation enthalpies. The hydrogenation reaction is given by the
following relation:

$$AB_n + mH_2 \rightarrow AB_nH_{2m} \qquad\qquad \Delta H \text{ (reaction)}.$$

Knowing ΔH_f of AB_n and ΔH_f of AB_nH_{2m} one calculates ΔH (reaction) =
ΔH_f (AB_nH_{2m}) $- \Delta H_f$ (AB_n) where $\Delta H_f(AB_n)$ are experimental values
(see Table 2). For the estimate of ΔH_f (AB_nH_{2m}), we assumed that
the binary hydrides were AH_3 and B_nH_3 because the maximum value of
2m is close to 6. Moreover, we assumed that these hydrides form an
ideal solution in which the following is true: ΔH_f (AB_nH_6) $= \Delta H_f$
(AH_3) $+ \Delta H_f$ (B_nH_3). For 2m equal to 6, we assumed that all
metallic bonds A-B are destroyed in the ternary hydride AB_nH_6. For
2m less than 6, we assumed that some A-B bonds coexist with the
bonds A-H and B-H in the ternary hydride, for which the following
relation applies:

$$\Delta H_f(AB_nH_{2m}) = \frac{m}{3} \Delta H_f(AH_3) + \frac{m}{3} \Delta H_f(B_nH_3) + \frac{3-m}{3} \Delta H_f(AB_n).$$

If we replace a part of metal B by metal C, the following relations
are obtained:

$$AB_{n-x}C_x + mH_2 \rightarrow AB_{n-x}C_xH_{2m} \qquad\qquad \Delta H_r (AB_{n-x}C_xH_{2m})$$

$$AB_{n-x}C_x \rightarrow A + (n-x) B + x C (- \Delta H_f AB_{n-x}C_x)$$

$$A + (n-x) B + x C + mH_2 \rightarrow AB_{n-x}C_xH_{2m} (\Delta H_f AB_{n-x}C_xH_{2m})$$

where $\Delta H_f AB_{n-x}C_xH_{2m}$ is estimated as done previously:

$$\Delta H_f AB_{n-x}C_xH_{2m} = \frac{m}{3} \Delta H_f(AH_3) + \frac{m}{3} (\frac{n-x}{n}) \Delta H_f(B_nH_3) + \frac{m}{3}\cdot\frac{x}{n}\Delta H_f(C_nH_3)$$

$$+ \frac{3-m}{3} \Delta H_f(AB_{n-x}C_x) + \Delta H_1$$

Table 1.   Characterization of Hexagonal LaNi$_4$M Compounds

COMPOUND	ANNEALING CONDITIONS		MICROPROBE ANALYSIS	a(Å)	c(Å)
LaNi$_5$	quenched		LaNi$_5$ $\pm$ 0.02	5.017	3.986
LaNi$_4$Al	3 hours	1180°C	LaNi$_{4.14}$Al$_{.98}$	5.061	4.07
LaNi$_4$Cu	3 days	800°C	LaNi$_{3.98}$Cu$_{.96}$	5.033	4.007
LaNi$_4$Fe	1 week	1100°C	LaNi$_{3.95}$Fe$_{.99}$	5.049	4.015
LaNi$_4$Mn	3 days	900°C	LaNi$_{3.95}$Mn$_{.95}$	5.089	4.082

Table 2.   Calorimetric Data

COMPOUND	T (K)	ΔH_x^∞ (Kcal)	Q_{298}^∞ (Kcal)	σ	ΔH_f^{298} (Kcal mole^{-1})	δ	Ref.
La	951	-44.06	-39.5	0.46			(6)
Ni	948	-33.44	-28.6	0.48			(6)
Mn	1007	-19.40[a]	-13.4	0.15			(*)
Fe	983	-27.16	-21.55	0.40			(7)
Cu	983	-5.35	-0.87	0.10			(*)
LaNi$_4$Mn	969		-133.4	0.85	-32.0	2.1	(*)
LaNi$_4$Cu	983		-119.7	1.0	-34.2	2.1	(*)
LaNi$_4$Fe	983		-144.3	1.1	-31.0	2.2	(*)

[a]Referenced to β-Mn   *This study

ΔH_1 represent the mixing effect due to the substitution of B atoms by C atoms:

$$\Delta H_1 = \Delta H_f(AB_{n-x}C_x) - \left(\frac{n-x}{n}\right)\Delta H_f(AB_n) - \frac{x}{n}\Delta H_f(AC_n).$$

This relation can be simplified for a low rate of substitution ($x \leq 1$) assuming that the variation of the formation enthalpy of $AB_{n-x}C_x$ with respect to the AB_n formation enthalpy results only from the mixing enthalpy of B and C. Thus, the hydrogenation enthalpy can be expressed by the following reaction

$$\Delta H_{react} = \left(\frac{1-m}{3}\right)\Delta H_f(AB_{n-x}C_x) - \Delta H_f(AB_n) + \frac{m}{3}\Delta H_f(AH_3) +$$

$$\frac{m}{3}\left(\frac{n-x}{n}\right)\Delta H_f(B_nH_3) + \frac{m}{3}\cdot\frac{x}{n}\Delta H_f(C_nH_3). \qquad (1)$$

This leads to the values summarized in Table 3. Here the formation enthalpies of the binary hydrides AH_3, BH_3, CH_3 are either experi-

Table 3. Hydrogenation Enthalpies for $LaNi_4M$ Compounds

INTERMETALLIC COMPOUNDS		HYDRIDES ΔH^b reaction	
	$\Delta H_f{}^a$	calculated	experimental
$LaNi_5$	-31.5 (5)	$LaNi_5H_6$ - 7.1	- 7.3 (5)
$LaNi_4Al$	-58.6 (5)	$LaNi_4AlH_{4.5}$ -11.9	-11.4 (5)
$LaNi_4Mn$	-32 *	$LaNi_4MnH_6$ -10.8	-11.6 (6)
$LaNi_4Fe$	-31 *	$LaNi_4FeH_5$ - 8.3	- 8.2 *
$LaNi_4Cu$	-34.2 *	$LaNi_4CuH_5$ - 8.5	- 8.1 *

*This study ^{a}Kcal mole^{-1} ^{b}Kcal mole^{-1} H_2

mental values (4) or values estimated by Miedema (2). These calculated values of the hydrogenation enthalpies are in agreement with the experimental values (Table 3). Note that ΔH_f values for $LaNi_4M$ with M=Cu, Fe, Ni and Mn are similar, hydrogenation enthalpies differ, especially in the case of Mn. $LaNi_4Al$ is the most stable intermetallic and also forms a stable hydride.

REFERENCES

1. H.H. Van Mal, K.H.J. Buschow and A.R. Miedema, J. Less. Com. Met., 35:65 (1974).
2. P.C.P. Bouten and A.R. Miedema, J. Less Com. Met. 71:147 (1980).
3. J.C. Mathieu, F. Durand and E. Bonnier, Thermodynamics I.A.E.A., Vienna, Vol. 1 (1966).
4. W.M. Mueller, J.P. Blackledge and G.G. Libowitz, Metal Hydrides, Academic Press, New York (1968).
5. H. Diaz, A. Percheron-Guegan, J.C. Achard, C. Chatillon and J.C. Mathieu - Int. J. of Hydrogen Energy 4:445 (1979).
6. C. Lartigue, A. Percheron-Guegan, J.C. Achard and F. Tasset, J. Less Com. Met., 75:23 (1980).
7. J.C. Mathieu, B. Jounel, P. Desré, and E. Bonnier, Thermodynam. of Nuclear Materials, I.A.E.A., Vienna, p. 767 (1968).
8. R. Hultgren, P.D. Desai, D.T. Hawkins and K.K. Kelley, Selected Values of the Thermodynamic Data of the Elements, Am. Soc. Metals, Metals Park, Ohio (1973).

HYDROGEN DESORPTION RATES IN LaNi$_{5-x}$Al$_x$-H

Arthur Tauber and Robert D. Finnegan
U.S. Army Electronics Technology and Devices Laboratory
(ERADCOM)
Fort Monmouth, NJ 07703

The desorption of hydrogen in intermetallic compounds of the system LaNi$_{5-x}$Al$_x$ has been investigated because these compounds exhibit potential as hydrogen reservoirs for high power cold cathode/reservoir thyratrons. Hot cathode/reservoir thyratrons have been used for 25 years to switch high power pulsed radar operation. Cold cathode/reservoir thyratron tubes would be capable of switching high power without the reservoir heater, power supply and power source presently required in hot tubes using Ti as a reservoir. An ideal reservoir material is one that exhibits a plateau pressure between 400 and 600 millitorr in its room temperature P vs C equilibrium diagram, is not easily poisoned by the ambients of a cold cathode (O_2, H_2O, CO_2, CO), and exhibits recovery of pressure in seconds.

The system LaNi$_{5-x}$Al$_x$ has been the subject of several recent studies (1,2,3). A preliminary analysis of ZrVFe in our Laboratory demonstrated the low pressure capability plus very fast kinetics of this compound; however, even in a good vacuum system this compound poisoned after 24 hours. For this reason, further investigation was abandoned in favor of the system LaNi$_{5-x}$Al$_x$. Mendelson and Gruen (1) suggested that a plateau may be present at room temperature for LaNi$_{3.5}$Al$_{1.5}$ in the range 400-600 millitorr. The high tolerance of LaNi$_5$ to such poisoning prompted this investigation (4,5).

EXPERIMENTAL

Compounds were synthesized from 3 nines La, 4 nines Ni and 4 nines purity Al by arc melting on a water cooled Cu hearth in 6 nines pure Ar. They were homogenized at 1020°C for 16 hours in Ta tubes sealed in evacuated quartz tubes. While x-ray diffractometer traces revealed a well-crystallized single phase, metallography revealed microsegregation as previously observed by Mendelsohn et al (6) in $LaNi_{5-x}Al_x$. The second phase was found to occur rarely in $LaNi_{3.75}Al_{1.25}$ (nominal) and more prevalently in $LaNi_{3.45}Al_{1.46}$ (nominal).

Five to ten gram samples were exercised in a Tempress Rene' metal vessel at room temperature by alternately absorbing at 1000 PSI and desorbing in a vacuum (20 millitorr) 6 nines pure H_2 through 20 cycles. Desorption measurements were carried out in a gravimetric apparatus in the pressure range 1 to 10^{-4} atm and temperatures of 120 to 23°C. Samples transferred from the high pressure apparatus were ground under kerosene and sieved through 100 mesh screens. These powders were loaded into Al buckets and carefully weighed to 0.01mg. If $LaNi_5$ was employed as an additive, it too was sieved through 400 mesh screens (37uM). Activation of samples was carried out by first baking the sample and system at about 100°C for 12 to 24 hours at a vacuum of 10^{-5} to 10^{-6} atmospheres. Six nines pure H_2 was introduced until 1 atmosphere was achieved. This was held for 24 hours. Activation could be achieved more rapidly by raising the temperature of the sample with a resistance furnace to 100°C.

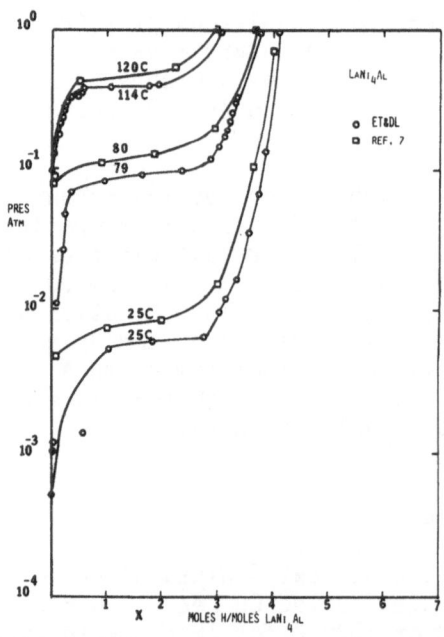

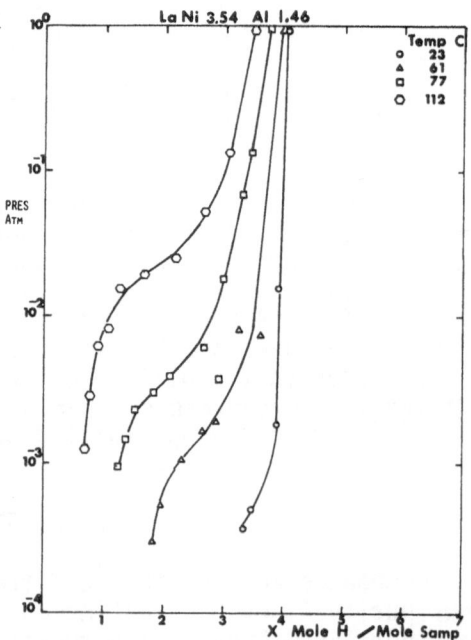

Figure 1. Desorption isotherms $LaNi_{3.45}Al_{1.46}Hx$.

Figure 2. Desorption isotherms for LaNiAl.

RESULTS

P-C-T diagrams are shown in Figs. 1 and 2. Data in this investigation is compared with that of H. Diaz et al (7) and LaNi₄Al in
Fig. 1. Deviations arise probably from small differences in composition. Equilibrium is easily established, although at room
temperature and low pressure, times can be as long as hours.
LaNi₄Al is not very useful as a reservoir because the 400-600 millitorr range cannot be reached. In Fig. 2, the absence of distinct
plateaus is noted even at elevated temperatures. The data points,
for isotherms below 100°C, are not fully equilibrated even after a
week. Data points at room temperature and pressures below 10^{-2}
atmospheres are not equilibrated after two weeks. True equilibrium
may require months. In spite of the apparent absence of plateaus,
sufficient slope is present in the curves to provide a useful
reservoir if desorption rates are fast.

Desorption is plotted as a function of time Fig. 3, 4 and 5.
It is quite clear that desorption rates are very slow (see Fig. 3).
In general, the rate decreases with increasing hydride stability
(increasing Al content), and with decreasing equilibrium pressure
(see Figs. 4 and 5). To determine if slow thermal conductivity was
responsible even in part for the slow desorption, two samples, each
once measured, were mixed with four times their volume of fine
aluminum powder and pressed into a pellet at 10,000 PSI. The pellet
had to be activated by heating to 100°C in 1 atmosphere of H_2. Once
activated, the samples were remeasured, and gave results identical
to those obtained without Al.

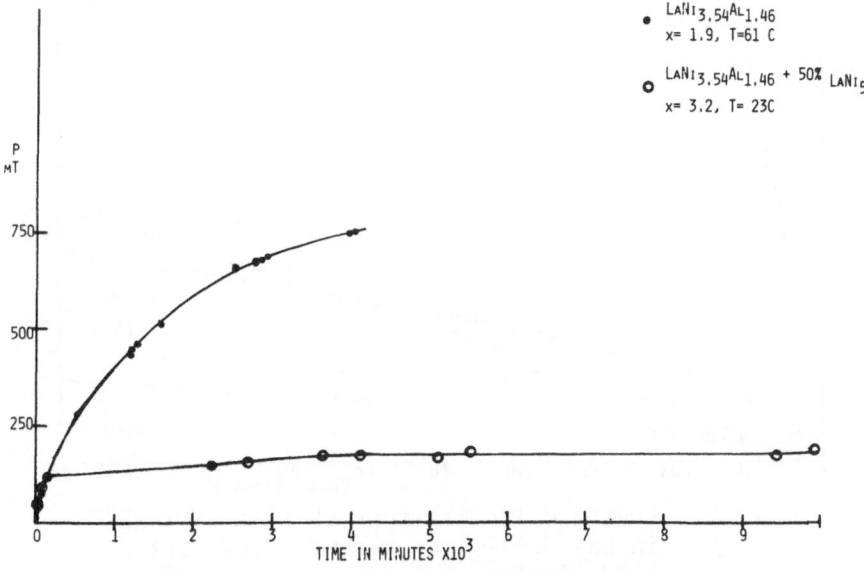

Figure 3. Pressure in millitorr vs time.

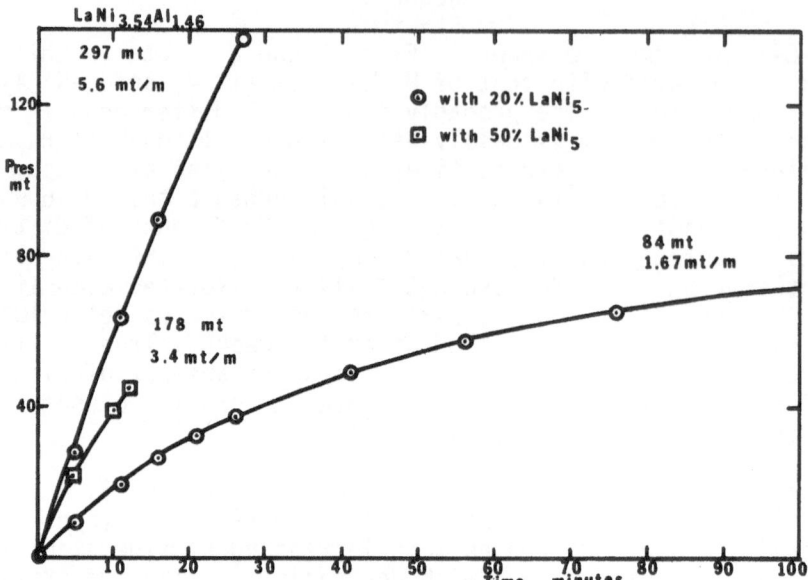

Figure 4.   Pressure in millitorr vs time at t=23°C
            in $LaNi_{3.45}Al_{1.46}H_{3.25}$; mt values after
            5000 minutes.

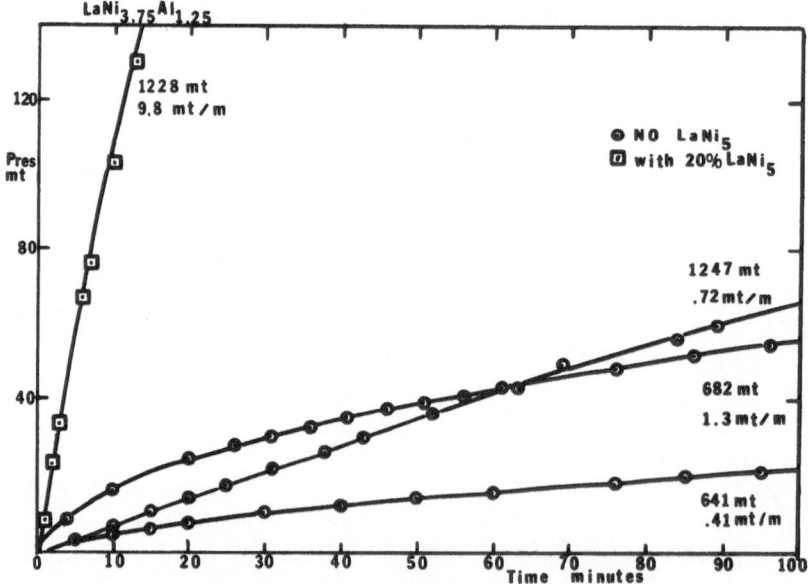

Figure 5.   Pressure in millitorr vs time at t=23°C
            in $La_{3.75}Al_{1.25}H_{0.49}$; mt values after
            5000 minutes.

To increase the desorption rate, LaNi$_5$ was used as a catalytic additive with the samples. Desorption rate increases of more than an order of magnitude were noted. For practical applications, an increase of almost two orders of magnitude is necessary. A catalytic effect of this type was first observed by Reilly and Wiswall (8), and employed extensively by Tanguay et al (9).

DISCUSSION

The desorption results of this study, combined with NMR investigations of Bowman (10,11), suggest that the rate-determining step in desorption of hydrogen from LaNi$_{5-x}$Al$_x$Hy may be different from that in LaNi$_5$Hy. Wallace et al (12) have concluded that release from the interfacial region is the rate-determining step for desorption in LaNi$_5$Hy. Bowman has shown that hydrogen diffusion activation energy increases and the room temperature diffusion constant is reduced by over 2 orders of magnitude with increasing Al content. Hydrogenated LaNi$_5$ releases about 90% of its absorbed hydrogen in five minutes (13) while hydrogenated LaNi$_{3.54}$Al$_{1.46}$ releases hydrogen at a rate orders of magnitude slower. The importance of the surface reaction in the Al-substituted compound is born out by the increase in desorption by more than an order of magnitude in the presence of a catalytic substance. Even with enhanced surface reaction, the desorption rate is still 2 orders of magnitude slower in Al-substituted LaNi$_5$ than in the unsubstituted compound.

REFERENCES

1. M.H. Mendelsohn and D.M. Gruen, Nature, 269:45 (1977).
2. J.C. Achard, A. Percheron-Guegan, H. Diaz, F. Briaucourt, F. Demany, Second International Congress on Hydrogen in Metals, Paper 1E12 (1977).
3. T. Takeshita, S.K. Malik and W.E. Wallace, J. Sol. State Chem., 23:271 (1978).
4. J.J. Reilly and R.H. Wiswall Jr., Brookhaven National Laboratory Report No. 17136 on Proj. Order 71-96083 USA E COMD.
5. G.D. Sandrock and P.D. Goodell, J. Less-Common Metals, 73:161 (1980).
6. M.H. Mendelsohn, D.M. Gruen and A.E. Dwight, J. Less-Common Metals, 63:193 (1979).
7. H. Diaz, A. Percheron-Guegan, J.C. Achard, 2nd World Hydrogen Energy Conference, Aug. 21, 1978.
8. J.J. Reilly and R.H. Wiswall Jr., Inorg. Chem., 6:2220 (1976).
9. B. Tanguay, J.L. Soubeyroux, M. Pezat, J. Portier and P. Hagenmuller, Mat. Res. Bull., 11:1441 (1976).
10. R.C. Bowman Jr., D.M. Gruen and M.H. Mendelsohn, Solid State Commun., 32:501 (1979).

11. R.C. Bowman Jr., B.D. Croft and A. Attalla, J. Less-Common
     Metals, 73:227 (1980).
12. W.E. Wallace, R.F. Korlicek Jr. and H. Imamura, The Rare Earths
     in Modern Science and Technology, Vol. 2, Plenum Press, NY
     583 (1980).
13. J.H.N. Van Vucht, F.A. Kuijpers and H.C.A.M. Bruning, Philips
     Res. Repts. 25:133 (1970).

STRUCTURE-REACTIVITY STUDIES ON THE EXTRACTION OF LANTHANIDES BY DIALKYL ISOPROPYLPHOSPHONATES

Chengye Yuan, Weizhen Ye, Enxin Ma, Fubing Wu, and
Xiaomin Yan
Shanghai Institute of Organic Chemistry
Academia Sinica, Shanghai 200032, China

Extraction of lanthanides by neutral organophosphorous ligands has been studied extensively only with tributylphosphate (TBP). The wide applications of TBP are limited by its insufficiently large separation factor and comparatively higher acidity in aqueous phase required for extraction. Various efforts have been made for the development of more powerful extractants. Investigations of the relationship between the chemical structure and extraction properties of ligands are considered to be one of the effectual ways to design such new reagents (1). Steric effects play an important role in the selectivity of reagents for solvent extraction, especially for the separation of elements with monotonously varying atomic radius like the lanthanides (1,2). This is well demonstrated by the higher selectivity during the solvent extraction of lanthanides by di(1-methylheptyl) methylphosphonate (DSOMP) (3). The influence of chemical structure of dialkyl alkylphosphonates on their extraction behaviors of 3d elements has been reported elsewhere (4) but the data for the rare earth elements are lacking.

In this paper the structure and reactivity studies on the extraction of lanthanides by dialkyl isopropylphosphonates are reported. The alkoxyl alkyl groups in the molecules of these phosphorous-based ligands are n-octyl (abbreviated as DNOPP), 2-ethylhexyl (DIOPP), 1-methyl-heptyl (DSOPP), n-butyl (DNBPP), iso-butyl (DIBPP), sec-butyl (DSBPP), tert-butyl (DTBPP), and cyclo-hexyl (DCHPP). As representatives of light, medium and heavy rare earth the nitrate salts of lanthanum, praseodymium, neodymium, samarium, and yttrium are used.

1. STRUCTURAL EFFECTS

The influence of ester alkyl groups on the distribution ratio (D) and the extraction separation factor (β) are shown in Table 1, compared with that of TBP and DSOMP.

It is evident that the D values in the extraction by dialkyl isopropylphosphonates with the same carbon number in alkoxy group decrease with the strengthening of isomerization of the alkyl radical, i.e. as the steric hinderance toward the phosphorous atom is enhanced. The following order has been observed: DNOPP>DIOPP>DSOPP; DNBPP> DIBPP>DSBPP>DTBPP. The lower extracting ability of DCHPP is attributed to the greater steric hinderance resulting from the geometric conformation of the cyclohexyl group.

As anticipated, the extractive ability of dialkyl isopropylphosphonates is superior to TBP due to the higher basicity of phosphonyl oxygen. The fact that the D values of extraction by dialkyl isopropylphosphonates is less than that by DSOMP can be explained either by hyperconjugation of methyl group (Baker-Nathan effect) in the latter molecule, or by steric hinderance of isopropyl group which is directly linked to the phosphorous atom in the former.

With the proper degree of isomerization in alkoxy alkyl group, both DIOPP and DSOPP possess excellent separation factors for Pr/La, 3.06 and 2.55, respectively. Both values are higher than that for TBP (1.94) and DSOMP (2.27) under the same conditions. The β value of Pr/La for DIBPP is 2.19 which lies between those for TBP and DSOMP. Higher separation factors of Pr/Nd for these extractants in comparison with that of TBP have been obtained. Analogous to the extraction of thorium (5), an intense influence of steric hinderance has also been observed during the lanthanide extraction. This is reasonable as the ionic radii of the lanthanides (La^{3+} = 1.016Å) are close to that of thorium (1.02Å). In contrast to the results reported by Eriemen (6) and Shviedov (7), the present study clearly shows that the steric effect of extractants plays an important role in their separation behavior.

2. TEMPERATURE EFFECT AND THERMODYNAMIC FUNCTIONS

The distribution ratios of lanthanum and praseodymium nitrate extraction by DNOPP, DIOPP, DNBPP, DIBPP, and TBP at various temperatures were estimated. All of D values decrease with the increase of temperature. The plot of log D versus 1/T was found to be a straight line for all extraction systems under investigation. ΔH° was estimated. The relative free energy change of $Pr(NO_3)_3$ to $La(NO_3)_3$ extraction can be expressed as $\Delta G_r^\circ = \Delta(\Delta G_r^\circ) \equiv -RT\ln K_{Pr}/K_{La} = -RT\ln\beta_{Pr}/La$, where $\beta = K_{Pr}/K_{La} \simeq D_{Pr}/D_{La}$ under the same condition. Relating ΔG_r° to ΔH_r° ($\Delta H_r^\circ = \Delta H_{Pr}^\circ - \Delta H_{La}^\circ$), the relative entropy change can also be calculated (Table 2). It follows that the separation between Pr and La resulted from the favorable extraction of the former and it is caused neither by the difference of enthalpy, nor by the difference of

Table 1. Extraction of lanthanides by dialkyl isopropylphosphonates Ln(NO$_3$)$_3$, 0.5M, HNO$_3$, 1, 3M/Extractant 2M, dodecane.

Extractant	DCHPP		DNOPP		DIOPP		DSOPP		DSOMP	
[HNO$_3$] or g.	1	3	1	3	1	3	1	3	1	3
D$_{La}$	0.263	0.123	0.331	0.157	0.220	0.076	0.096	0.026	0.469	0.233
D$_{Pr}$	0.461	0.275	0.528	0.344	0.417	0.233	0.164	0.067	0.782	0.529
D$_{Nd}$	0.509	0.331	0.585	0.434	0.497	0.292	0.190	0.075	0.873	0.646
D$_{Sm}$	0.542	0.357	0.729	0.537	0.609	0.375	0.206	0.084	0.967	0.770
D$_{Y}$	0.253	0.087	0.369	0.231	0.308	0.162	0.082	0.027	0.634	0.426
[HNO$_3$] equil	0.547	2.03	0.634	2.08	0.587	2.07	0.467	1.88	0.603	2.05
β Pr/La	1.75	2.24	1.60	2.19	1.90	3.06	1.70	2.55	1.58	2.27
β Nd/Pr	1.10	1.20	1.11	1.26	1.19	1.25	1.16	1.12	1.12	1.22
β Sm/Nd	1.06	1.08	1.25	1.24	1.23	1.28	1.08	1.12	1.11	1.19

Extractant	DNBPP		DIBPP		DSBPP		DTBPP		TBP	
[HNO$_3$] or g.	1	3	1	3	1	3	1	3	1	3
D$_{La}$	0.291	0.183	0.273	0.149	0.097	0.043	<0.01	<0.01	0.116	0.043
D$_{Pr}$	0.413	0.344	0.379	0.326	0.128	0.075	<0.01	<0.01	0.127	0.082
D$_{Nd}$	0.473	0.401	0.451	0.371	0.136	0.080	<0.01	<0.01	0.151	0.082
D$_{Sm}$	0.546	0.496	0.554	0.461	0.138	0.079	<0.01	<0.01	0.191	0.142
D$_{Y}$									0.114	0.072
[HNO$_3$] equil	0.760	2.10	0.717	2.01	0.639	1.83	0.602	1.94	0.591	2.14
β Pr/La	1.42	1.88	1.39	2.19	1.33	1.74			1.09	1.94
β Nd/Pr	1.15	1.17	1.19	1.14	1.06	1.07			1.19	1.001
β Sm/Nd	1.15	1.24	1.23	1.23	1.01	~ 1			1.27	1.72

Table 2.   The thermodynamic functions for $La(NO_3)_3$ and $Pr(NO_3)_3$ extraction by dialkyl isopropylphosphonates.

Thermodynamic functions	$-\Delta H^{o}_{La}$	$-\Delta H^{o}_{Pr}$	ΔG^{o}_{r}	ΔH^{o}_{r}	$T\Delta S_{r}$
DNOPP	8.60	6.56	-0.279	2.04	2.32
DIOPP	7.13	6.31	-0.531	0.82	1.35
DNBPP	5.34	4.04	-0.256	1.30	1.56
DIBPP	6.60	6.02	-0.232	0.58	0.813
TBP	6.87	4.95	-0.0511	1.92	1.97

relative bond energy between the complex formed and the hydrated lan-thanide ion.   The calculated values of the relative entropy change are all positive.   Considering the change of the relative thermodynamic functions for praseodymium and lanthanum extraction by dialkyl iso-propylphosphonates with various alkoxy group it can be said that the hindered ester of isopropylphosphonic acid reveals both a smaller relative enthalpy change and less reverse contribution to the extrac-tive separation.

3.   COMPOSITION AND CHARACTERISTICS OF THE COMPLEX

The composition of the complexes of $La(NO_3)_3$ with DNOPP, DIOPP, DSOPP and DCHPP were deduced by the method of continuous variation to be $La(NO_3)_3 \cdot 3L$ in agreement with the elemental analyses.   In the IR spectra of the complexes of $La(NO_3)_3$ or $Pr(NO_3)_3$ with dialkyl isopro-pylphosphonates, the characteristic frequency of νPO at 1234-1238 cm^{-1} of the parent alkylphosphonates was replaced by a strong band at 1175-1182 cm^{-1}, designated as the coordinated PO band with $\Delta \nu PO$ of 55-58 cm^{-1}.   The stretching frequency $\nu p-o-c$ of the complex shifts a little toward higher wave number due to the inductive effect of coordinated PO bond.   These IR spectra further indicated that the coordinated NO_3 groups are of low symmetry as is evident by the absorption frequen-cies.   There is no strong absorption of the ionic type in the region of 1390-1350 cm^{-1}.   It follows that there are three monodentate alkyl-phosphates and three ambient nitrate groups coordinated to the lan-thanide ion in the complex.

The proton NMR spectra of the complex of $La(NO_3)_3$ with DIOPP and DNOPP indicate a small shift toward lower field of the α-carbon atom of the alkoxy group resulting from the coordination of the lanthanum ion to the phosphonyl oxygen atom.   The $\Delta \delta - CH_2O-$was observed as 0.18 and 0.12 ppm for the lanthanum complex of DNOPP and DIOPP respectively. Analogous data were obtained for TBP complex of lanthanum. In mixing an equivalent of alkyl phosphonate with the above complex, only one set of resonance peaks can be observed, indicating fast ligand exchange (8).

The molar magnetic susceptibility of $La(NO_3)_3$ and $Pr(NO_3)_3$ was estimated as -7.72×10^{-1} and 4.94×10^{-8} respectively. After correction for the diamagnetism of the parent ligand and nitrate group, the effective magnetic moment of the lanthanum and praseodymium ions were estimated. The estimated value of praseodymium ion in nitrate is 2.98 which is lower than the theoretical value $3.58\mu_\beta$, calculated by the equation $\mu = g[J(J+1)]^{\frac{1}{2}}$. It may be considered as a consequence of the influence of some covalent coordination of nitrate group on the f electron of the lanthanides. The magnetic moment of the praseodymium complexes with DIOPP and TBP were estimated as 3.18 and $3.00\mu_\beta$ respectively. These values approach the value of nitrate.

REFERENCES

1. Chengye Yuan, in Proceedings of International Solvent Extraction Conference, 1980. Liege, Belgium. Paper No. 80–81.
2. Chengye Yuan, Nucl. Sci. and Technol., 908 (1962).
3. Chengye Yuan, Rongyu Zhang, Jifa Xie, and Lilan Shi, ibid., 977 (1964).
4. N.M. Karayannis, C.M. Mikulski, L.S. Gelfand, and L.L Pytlewski, J. Inorg. Nucl. Chem., 40:1513 (1978).
5. Chengye Yuan, Kexue Tongbao, 22:1465 (1977).
6. G.K. Eriemen and A.I. Kamenov, Zh. Neorg. Khim., 6:1487 (1961).
7. V.P. Shviedov and Yu. F. Orlov, Zh. Prikl. Khim., 38:1605 (1965); Zh. Neorg. Khim., 10:693 (1965).
8. T.H. Siddall, III, W.E. Stewart, and D.G. Karraker, Inorg. Nucl. Chem. Letters, 3:279 (1967).

SEPARATION OF THE RARE EARTHS AND SOME POLYVALENT CATIONS ON INORGANIC ION EXCHANGERS

Jai-kai Cheng, Ru-xiu Cai, Xin-quan Zhao, Si-fu Shi, and
Yung-ao Tseng
Laboratory of Analytical Chemistry, Department of Chemistry
Wuhan University, Wuhan, China

Inorganic ion exchangers have been found to have excellent selectivity for the alkali metals and alkaline earths (1). We have investigated inorganic ion exchangers for the separation of the rare earths and other elements. Thallium tungstophosphate, ammonium molybdophosphate and zirconium tungstate have been studied for the absorption and separation of the rare earths and some polyvalent cations.

Thallium tungstophosphate was prepared according to the method proposed by Caron (2). We have studied the ion exchange equilibrium of strontium (II), cobolt (II), cerium (III) and cesium (I) on thallium tungstophosphate in nitric acid and sodium nitrate solution and measured the distribution coefficients (Fig. 1,2). The separation of ^{89}Sr, ^{60}Co, ^{144}Ce(^{152,154}Eu) and ^{137}Cs on the column was obtained (Fig. 3). The percentage recovery was up to 96% for Sr, 95% for Co, 97% for Ce, and 98% for Cs.

Chromatography on paper impregnated with ammonium molybdophosphate (3,4) was used for the separation of thorium, uranium, scandium, and the lanthanons (La, Ce, Eu). Ln (III)/U (VI), Th (IV)/U (VI) may be separated by developing with nitric acid; Th(IV)/Ln (III)/U (VI) and Th (IV)/Ln (III)/Sc (III) may be separated by developing with nitric acid and alcohol. The R_f values were Th 0.21; Ln 0.52; U 0.81; and Sc 0.79 when alcohol to 0.5M nitric acid was in the ratio of 5:1.

The rare earths can be separated from large amounts of uranium, thorium, and titanium, using a solution of HCl and NH$_4$Cl as the eluant on the ion exchange column of zirconium tungstate. This was applied to the separation of trace amounts of rare earths from high-purity compounds of uranium (Fig. 4). The trace amounts of rare earths were determined by extraction and spectrophotometry using arsenazo III.

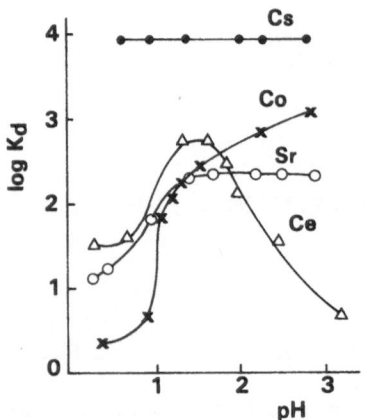

Figure 1.   Effect of pH on the absorption.

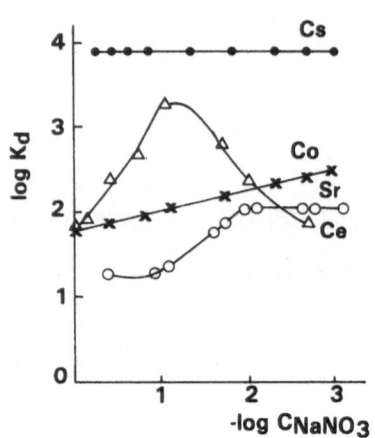

Figure 2.   Effect of sodium ion on the absorption.

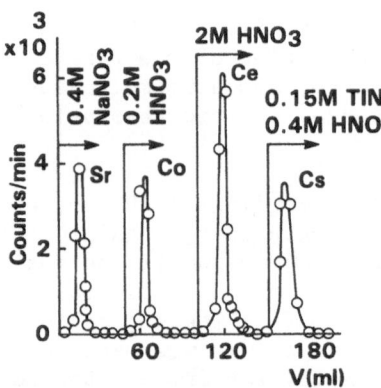

Figure 3.   Separation of Sr(II), Co(II), Ce(III) and Cs(I).

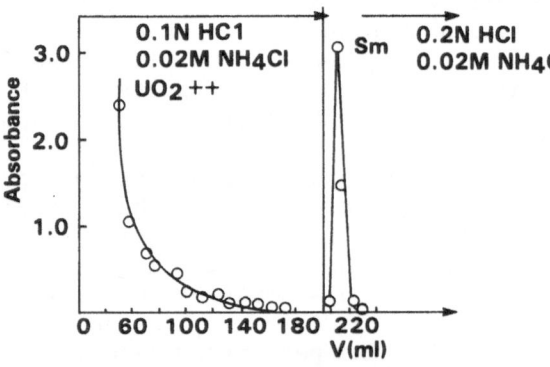

Figure 4.   Separation of rare earths from uranium.

The percentage recovery of rare earths (0.2 to 5.0 µg) determined by extraction-spectrophotometry and radioactivity was 96%.

REFERENCES

1.   Anil K. DE and K. Sen, Separation Science and Technology, 13:517 (1978).
2.   H.L. Caron and T.T. Sugihara, Anal. Chem., 34:1082 (1962).
3.   G. Alberti and G. Grassini, J. Chromatogr., 4:423 (1960).
4.   J. Van Smit, J.J. Jacobs, and W. Robb, J. Inorg. Nucl. Chem., 12:95 (1959).

SOLVENT EXTRACTION BEHAVIOR OF LANTHANIDES WITH DI(2-ETHYLHEXYL) ISOPROPYLPHOSPHONATE

Enxin Ma, Sanyi Wang, Fubing Wu, and Chengye Yuan

Shanghai Institute of Organic Chemistry
Academia Sinica, Shanghai 200032, China

Investigations of the behavior of lanthanides in solvent extraction can contribute both to design an effective extraction systems for metal separation and particularly to the development of the lanthanide theory. Fidelis et al. (1,2) has studied the regularity of variation of thermodynamic functions in the process of lanthanide extraction. Recently, Siekierski (3) examined the position of Y, in respect to $\Delta G°$, $\Delta H°$ and $\Delta S°$ of extraction and complex formation in the lanthanide series. As shown by our studies (4,5) the regularity of variation of the equilibrium constant, free energy and entropy change reveals a tetrad effect.

The regularity in lanthanides extraction (except Pm) and yttrium by di(2-ethylhexyl) isopropylphosphonate (abbreviated as DIOPP and denoted by L) from 3N and 14N HNO_3 in the aqueous phase is described and the thermodynamic properties are calculated between 10-60°C. The observed regularities are discussed.

I. The Variation of the Extraction Equilibrium Parameters and Thermodynamic Functions with the Atomic Number of the Lanthanides

Extraction of lanthanides by neutral organophosphorous-based ligands from 3N HNO_3 solution proceeds by a solvation mechanism. In extraction with DIOPP, the composition of the solvated species has been deduced as $Ln(NO_3)_3 \cdot 3L$ (6). The extraction equilibrium constant (K_{ex}) of lanthanides (except Pm) and yttrium at 10, 25, 40 and 60°C were estimated. The plot of $\log K_{ex}$ versus $1/T$ was found to be a straight line for all lanthanides and the thermodynamic functions of the extraction reaction were determined as in a previous report (4).

507

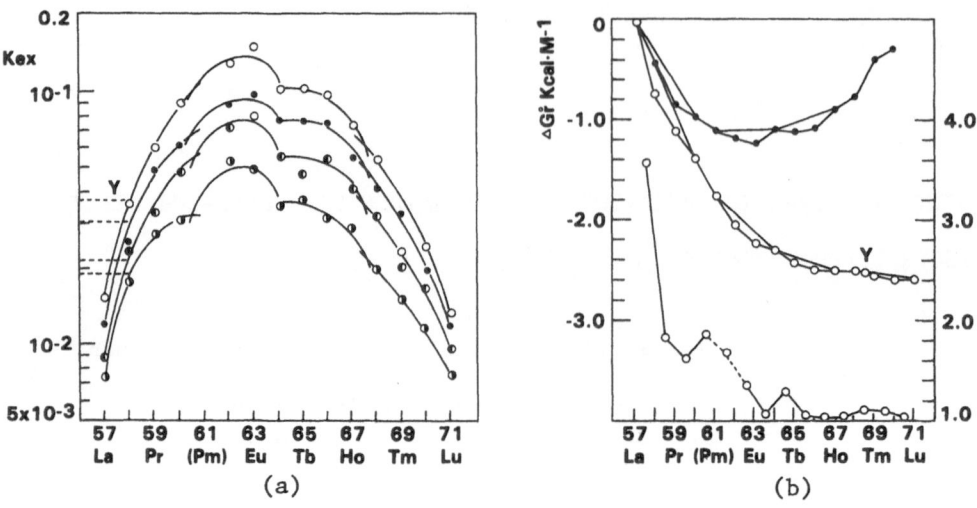

Figure 1.   Extraction of lanthanides by 2M DIOPP in dodecane, (a) Plot
of Kex versus Z number [Ln] 0.1M, [HNO$_3$] 3N, o 10°C,
● 25°C, ◐ 40°C, ◔ 60°C.   (b) Dependence of separation factor (β) and ΔG_r^o on Z number, [HNO$_3$] 3N (●), 14N (o).

    The plots of estimated Kex as well as ΔG_r^o versus atomic number
(Z) of lanthanides are shown in Fig. 1.   These figures show that both
Kex and $\left|-\Delta G_r^o\right|$ values rise gradually with increasing Z from La to Sm
or Eu, and then decrease with higher Z, and a tetrad effect.   Gadolinium lies both in the second and third subgroup.   Yttrium is located
between either Ce and Pr or Er and Tm.   Plots of the logarithms of
ΔH_r^o and ΔS_r^o versus Z were also found to be related to the tetrad
grouping of lanthanides.

    Extraction of lanthanides by alkylphosphonates from 14N HNO$_3$ has
been recognized as a process of the formation of oxonium salt together
with solvation species.   Therefore, D was used instead of Kex in such
systems.   As indicated in Fig. 2, an apparently tetrad effect is also
observed.

    There is no obvious change in the D value in the extraction of
La, Ce and Pr at various temperatures.   However, ΔH^o of the remaining
lanthanides were estimated and plotted against the Z number.   Some
regularity with concave type curve, resembling the latter three sub-
groups in tetrad classification was observed.   The relationship between ΔS_r^o and Z is very similar to that of ΔH^o and Z but a convex
type curve resulted.   Analogous observations were reported in lanthanide extraction by TBP from higher acidic solution (1).   The extraction behavior of Y occurs between Ho and Er.   The average separation
factor for adjacent lanthanides is calculated to be 1.37 in this
extraction system.

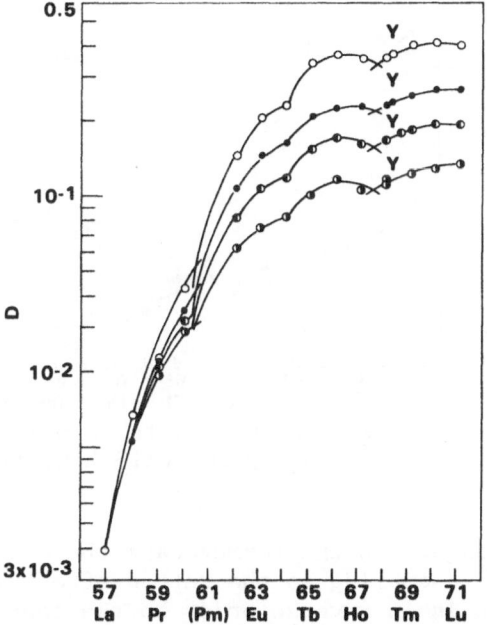

Figure 2.   Plot of D versus Z number, [DIOPP] 2M in
             dodecane, [Ln] 0.05M, [HNO₃] 14N. o 10°C,
             ● 25°C, ◐ 40°C, ◒ 60°C.

II.   The Influence of the Degree of Nitrate Complexing and
      Hydration of the Lanthanide Ions on the Regularity of
      Variation in Extraction Process.

      The behavior of DIOPP in lanthanide extraction is closely re-
lated to the acidity of the aqueous phase. A maximum D value was
found in the mid-lanthanide series in lower acidity, while the extraction ability was enhanced gradually as increasing Z at higher acidity.

      The lanthanide extraction operated by cation exchange mechanism
appears to be a process of conversion of hydrated lanthanide ions to
the chelated molecules soluble in organic diluents (4).   In case of
solvation extraction, the complexing reaction of nitrate ion toward
lanthanides should be considered.   It is interesting to note that a
similar maximum stability constant of $Ln(NO_3)^{2+}$ in water is located
at the mid-lanthanide series (8).   The coordination of lanthanides
with DIOPP is enhanced as Z increases at higher acidity, due to the
lanthanide contraction.

      The water contents in the organic phases under extraction equi-
librium have been determined by Karl-Fischer method as about 0.6-
0.75M(25°C), 0.8-0.9M(10°C) at lower acidity and 0.36-0.46M(25°C) in

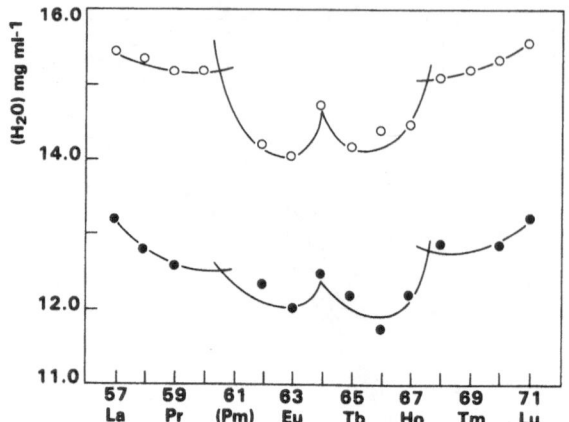

Figure 3.   Variation of water contents in
            organic phase with Z number.

higher acidity range. In the former case the water content varies
with Z regularly resembling to the tetrad effect (Fig. 3). However,
the water content is in reverse order to the concentration of
lanthanides in the organic phase. The observed phenomenon may be
illustrated by the following situation. The lanthanide ions having a
strong tendency toward hydration are complexed with the weaker nitrate
ligand. It is reasonable to consider that the lanthanide ions enter
the organic phase being accompanied with the hydration sphere and
nitrate ion in the extraction process. The metal ion with more
stronger hydration ability will bring more water molecules and is
characterized by lower D value in extraction. The associated hydroxyl
groups were shown in the IR spectra of the organic phases, but it is
difficult to distinguish whether the water molecules are bound to the
lanthanide ions or coexist with the phosphorous ligand. This explanation is, however, unable to exclude the possibility of competitive
coordination of water molecule between the lanthanide ion and the
phosphorous-based ligand.

                              REFERENCES

1.   I. Fidelis and S. Siekierski, in Proceedings of the Tenth Rare
     Earth Research Conference, Vol. II, p 919 (1973).
2.   I. Fidelis, J. Inorg. Nucl. Chem., 32:997 (1970).
3.   S. Siekierski, in Proceedings of International Solvent Extraction
     Conference, 1980, Liege, Belgium. Paper No. 80-49.
4.   Enxin Ma, Xiaomin Yan, Sanyi Wang, Haiyan Long, and Chengye Yuan,
     ibid., Paper No. 80-147.
5.   Enxin Ma, Xiaomin Yan, Guoliang Wang, and Chengye Yuan, Kexue
     Tongbao, 25:911 (1980).

6. Chengye Yuan, Weizhen Ye, Enxin Ma, Fubing Wu, and Xiaomin Yan, Scientia Sinica (to be published).

7. D.F. Peppard, V.J. Driscoll, R.J. Sironen, and S. McCarty, J. Inorg. Nucl. Chem., 4:326 (1957).

8. D.F. Peppard, G.W. Mason, and I. Hucher, J. Inorg. Nucl. Chem., 24:881 (1962).

SPECTROPHOTOMETRIC DETERMINATION OF RARE EARTHS IN LIGAND BUFFER

MASKING SYSTEMS

Jai-kai Cheng, Qing-yao Luo, Ru-xiu Cai, Xin-xiang Li,
and Yung-ao Tseng
Laboratory of Analytical Chemistry, Department of
Chemistry, Wuhan University, Wuhan, China

INTRODUCTION

Metal ions such as Fe(III), Al(III), Cr(III), Ti(IV) and Zr(IV) interfere strongly with the spectrophotometric determination of rare earths with organic reagents. We have utilized "ligand buffers" (1) for the spectrophotometric determination of rare earths with arsenazo III, dicarboxyarsenazo III, chlorophosphonazo III, carboxynitrazo and semixylenolorange cetylpyridiniumbromide (CPB).

Zn-EDTA, Zn-EDDD(Ethylenediaminediaceticacid-diethyl acetate), Zn-HEDTA, Zn-CyDTA and Ca-CyDTA were studied as ligand buffer masking systems. The application of ligand buffer masking of metal chelates to the spectrophotometric determination of rare earths greatly increases the selectivity of masking and the tolerance amounts of various interfering elements from μg to mg.

EXPERIMENTAL

EDTA, HEDTA, CyDTA, arsenazo III, dicarboxyarsenazo III, chlorophosphonazo III, carboxynitrazo and semixylenolorange-CPB were "pro analyst" grade. EDDD was synthesized by esterification of EDTA. All other reagents were analytical grade, including the lanthanide oxides (Specpure, Johnson Matthey Chemical Ltd., England). A Shimadzu UV-300 spectrophotometer, a 721 spectrophotometer (China) were used, and a Leizi 25 pH-meter was used. To the standard solution of rare earth in a standard flask, a ligand buffer masking agent was added, the pH was adjusted with a buffer solution and color-forming reagent. The absorbance was then measured.

RESULTS AND DISCUSSION

Masking agents. Fe(III) seriously interferes with the spectrophotometric determination of rare earths with arsenazo III. Up to
20 mg of Fe(III) can be masked by Zn-EDTA complex because the order
of formation constants of metal-EDTA complexes is Fe(III)>Zn>RE(La,
Ce, Pr). The complexes of EDTA with heavier rare earths are fairly
stable, thus the color development is affected by EDTA. The buffer
effect of Zn(II) in the Zn-EDDD complex is better than that in Zn-EDTA
complex, so the influence of Zn-EDDD complex on the color development
of rare earths is small. Since the rate of the Fe-EDDD reaction is
rather slow, it is necessary to accelerate the reaction by heating.

Using Zn-HEDTA as the masking agent, it has no need of heating.
Since the formation constants of RE-HEDTA complexes are small, the
effect of HEDTA on the determination of rare earth is negligible.
About 100 mg of Fe(III); 10 mg of Al(III); 6 mg of V(V), W(VI); 3 mg
of Mo(VI) and 2 mg of Ni(II), Ti(IV), Co(II) can be masked by Zn-HEDTA
complex and do not interfere with the determination of rare earths in
the spectrophotometric determination of rare earths with dicarboxyarsenazo III.

The interference of heavier rare earths is present when EDTA is
used as the masking agent in the spectrophotometric determination of
lighter rare earths with carboxynitrazo. In this case not only the
amounts of EDTA must be controlled, but also the absorbance of Nd is
affected apparently. When Zn-EDTA complex is used as the masking
agent in the same determination, the interference of heavier rare
earths are effectively suppressed, but also the color development of
lighter rare earths is unchanged.

Buffer metal ion. The formation constants of complexes of buffer
metal ions with the masking agents should be between the elements
being masked and the rare earths. Zn(II), Cd(II), Co(II) and Ni(II)
all can be used as buffer metal ions, among which Zn(II) is the best
one.

The ratio of buffer metal ion and ligand. In a ligand buffer,
the concentration of free ligand is determined by the ratio of the
concentration of buffer metal ion (C_M) and that of the ligand (C_A).
A maximum of buffer capacity of a ligand buffer is obtained when
$C_M/C_A = 2$; that is, a solution of such composition shows a minimum
change in concentration of free ligand when a small amount of metal
ion or ligand is introduced. The ratio of the concentration of buffer
metal ion and that of the ligand is generally between 1:1 to 1:3.

Application. The ligand buffer masking method increases the
selectivity of masking. The tolerance limit of many interfering elements is increased distinctly in the spectrophotometric determination
of rare earths, so the rare earths in various steel and low alloy

Table 1. Analysis results for rare earths in various
standard steel samples.

Sample No.	RE, %				
	Standard contents	Zn–EDTA[*]	Zn–EDDD[**]	Zn–HEDTA[**]	Ca–CyDTA[***]
Castirons D15	0.30	0.29			
" D26	0.14	0.15			
Carbon steels 01	0.007	0.007			
" 04	0.039	0.040			
Steels 16Mn	0.040	0.040			
44MnSi-2	0.046	0.045			
High Mn	0.099	0.11			
20K-1	0.0061	0.0060			
44MnTiRE	0.017		0.018		
GSiMnMoVRE-2	0.035		0.034		
GSiMnMoVRE-3	0.013		0.013		
PCuWMnRE-1	0.029		0.029		
PCuWMnRE-3	0.11		0.11		
MnAlRE	0.045		0.046		
35MoVAlTiRE	0.14		0.14		
20K-2	0.0062		0.0064		
Alloy steels-34	0.0070		0.0070		
" -21	0.0045		0.0047		
Steels 08	0.0030			0.0030	
GSiMnMoVRE-4	0.0075			0.0076	
Alloy steels-31	0.0080			0.0079	
" -36	0.0014			0.0015	
" -37	0.0040			0.0040	
" -38	0.0016			0.0015	
" -39	0.0050			0.0053	
Castirons D52	0.042				0.042
Steels 05	0.040				0.041
" 06	0.032				0.030
Alloy steels-24	0.050				0.052
" -30	0.050				0.052
" -32	0.019				0.018
MnMoVTiB	0.050				0.047
CrMoCu alloy	0.050				0.052

*arsenazo III   **dicarboxyarsenazo III   ***semixylenolorange-CPB

Table 2.   Analysis results for rare earths in standard
high alloy steel and control ore samples.

Standard high alloy steel samples			Control ore samples		
Sample No.	RE, %		Sample No.	RE, %	
	Standard contents	Zn–EDTA*		Control contents	Zn–EDDD**
Ni-base alloy					
GH–159	0.0013	0.0014	1	0.30	0.28
GH–001	0.032	0.032	2	0.088	0.10
GH–002	0.013	0.012	3	0.12	0.14
GH–003	0.0044	0.0045	4	0.18	0.18
GH–004	0.0016	0.0013	5	0.78	0.79
K–264	0.015	0.015	6	1.08	1.07
			7	0.27	0.27
Fe-base alloy			8	0.21	0.20
GH–029	0.021	0.022	9	0.87	0.92
GH–031	0.0026	0.0026			
GH–032	0.0072	0.0075			
GH–033	0.0083	0.0087			

*arsenazo III   **dicarboxyarsenazo III

steel (>0.001% of RE) can be determined directly without separation procedure. The interfering elements of co-precipitation can be masked by Zn–EDTA in the spectrophotometric determination of rare earths separated by precipitation in the Fe–Ni base high temperature alloy steels and ores. The method for the spectrophotometric determination of lighter rare earths in ores with carboxynitrazo, using Zn–EDTA for masking heavier rare earths, have been applied to routine analysis for several years (Table 1, 2). Zn–EDTA can be used as masking agent in the determination of lighter, heavier rare earths and yttrium in castirons with chlorophosphonazo III (2).

REFERENCES

1. Motharu Tanaka, Anal. Chim. Acta, 29:193 (1963).
2. Yung-ao Tseng, et al., Chemical Journal of Chinese Universities, 1:60 (1980).

SPECTROPHOTOMETRIC DETERMINATION OF TRACE AMOUNTS OF COPPER IN HIGH-PURITY RARE EARTH WITH $\alpha,\beta,\gamma,\delta$-TETRA-(4-TRIMETHYLAMMONIUM-PHENYL) PORPHINE

Jai-kai Cheng, Kai-rang Yang and Quan-yu Chang

Laboratory of Analytical Chemistry, Department of Chemistry

Wuhan University, Wuhan, China

High-purity rare earth such as Y_2O_3 is a common material for the preparation of various solid-state luminescent materials. Since the luminescence is affected by the presence of trace copper impurities, attention has recently been given to the spectrophotometric determination of trace amounts of copper with porphyrin as spectrophotometric reagents, because of their high sensitivity and stability (1-3). In this paper, the complex reaction of water-soluble $\alpha,\beta,\gamma,\delta$-tetra-(4-trimethylammoniumphenyl)porphine T(4-TAP)P with copper(II) has been studied. This compound was synthesized according to the method proposed by Krishnamuthy (4). It was found to be a highly sensitive and selective reagent for the direct spectrophotometric determination of copper at ppb level. The soret band of the reagent was laid at 431 nm, which was well separated from that of the complex (411 nm). Apparent molar absorption coefficient of the reagent and sensitivity for 0.001 absorbance were 4.7×10^5 1 mol^{-1} cm^{-1} and 0.13 ng Cu cm^{-2}, respectively, and that of the complex were 4.0×10^5 1 mol^{-1} cm^{-1} and 0.16 ng Cu cm^{-2}, respectively (Fig. 1). The interference of many ions and salts has been examined and found to be minimal. Large amounts of rare earths do not interfere with the determination.

Analysis of high-purity rare earth samples: Take 0.05-0.7g samples in a 50-ml beaker. Dissolve it with hydrochloric acid and evaporate to remove the excess of hydrochloric acid. Transfer to a 25-ml volumetric flask, add 0.5 ml of hexamine buffer solution (pH 4.5) and 1 ml of 6.3×10^{-5} M T(4-TAP)P solution. Heat for 5 min. in a boiling water bath and allow to cool to room temperature in running water. Add 1 ml of 1.8 M sulphuric acid and dilute to the mark with water. Measure the absorbance at 431 nm, in 3-cm cells, against water blank or at 411 nm against reagent.

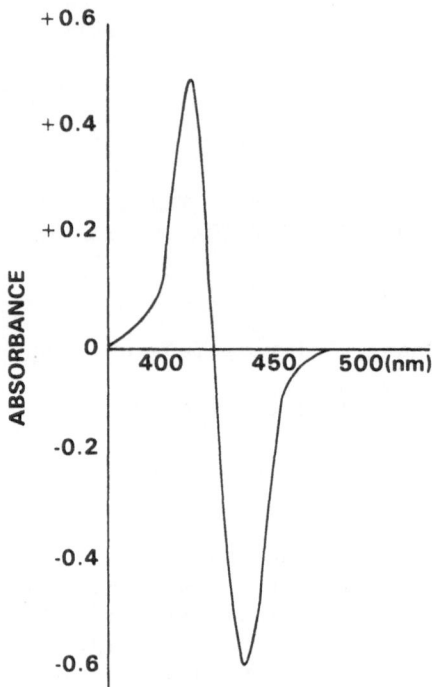

Figure 1.   Absorption spectra of Cu(II)-T(4-TAP)P complex.
In the presence of excess of T(4-TAP)P.
Reference: reagent blank.

The results are summarized in Table 1.

The influence of temperature and pH values on the rate of the
complex formation reaction of T(4-TAP)P with copper(II), the stability
and the composition of the complex have been studied.   The absorbance
of the copper(II)-T(4-TAP)P complex remained unchanged for at least
24h.   Continuous-variation and mole-ratio methods were applied to
determine the composition of the complex.   The results revealed that
the ratio is 1:1.

Twenty-seven standard solutions containing $0.016\mu g\ ml^{-1}$ of copper
were analyzed to determine the precision of the method.   The results
yield a relative standard deviation of ±1.8%.

The sensitivity and selectivity of the method using T(4-TAP)P as
reagent are better than that using other water-soluble porphyrines as
reagents in the determination of copper.

Table 1.   Analysis results for copper in high-purity
rare earth samples.

Sample	Sample taken (mg)	Cu(μg)		Cu content		Mean value (%)
		Added	Found	(μg)	(%)	
Y_2O_3	200	0	0.075	0.075	3.8×10^{-5}	
	200	0	0.070	0.070	3.5×10^{-5}	
(99.995%)	200	0.100	0.170	0.070	3.5×10^{-5}	
	200	0.400	0.475	0.075	3.8×10^{-5}	3.7×10^{-5}
	226	0	0.084	0.084	3.7×10^{-5}	
	226	0.100	0.190	0.090	4.0×10^{-5}	
	253	0.100	0.200	0.100	3.9×10^{-5}	
Y_2O_3	50	0	0.026	0.026	5.2×10^{-5}	
	100	0	0.047	0.047	4.7×10^{-5}	
(99.99%)	200	0	0.100	0.100	5.0×10^{-5}	5.0×10^{-5}
	500	0.400	0.650	0.250	5.0×10^{-5}	
Y_2O_3	700	0	0.020	0.020	2.9×10^{-6}	
	700	0.100	0.123	0.023	3.3×10^{-6}	3.1×10^{-6}
Y_2O_3 (Specpure, Johnson, Lodon)	150	0	0.039	0.039	2.6×10^{-5}	2.6×10^{-5}
	150	0.100	0.137	0.037	2.5×10^{-5}	
Y_2O_3 (Specpure, Shanghai)	100	0	0.072	0.072	7.2×10^{-5}	6.8×10^{-5}
	100	0.400	0.463	0.063	6.3×10^{-5}	
Nd_2O_3	50	0	0.050	0.050	1.0×10^{-4}	
	100	0	0.100	0.100	1.0×10^{-4}	
(99.99%)	184	0	0.164	0.164	8.9×10^{-5}	9.7×10^{-5}
	50	0.100	0.152	0.052	1.0×10^{-4}	

ACKNOWLEDGMENTS

The authors wish to express their gratitude to Associate Professor X. J. Wu and Mr. J. S. Kao (Department of Chemistry, Wuhan University) who synthesized the reagent T(4-TAP)P.

REFERENCES

1. J. Itoh, T. Yotsuyanagi, and K. Aomura, <u>Anal. Chim. Acta</u>, 74:53
    (1975).
2. H. Ishii and H. Koh, <u>Talanta</u>, 24:417 (1977).
3. H. Ishii, K. Hidemasa, and K. Kazulumi, <u>Nippon Kagaku Kaishi</u>,
    75:686 (1978).
4. M. Krishnamurthy, <u>Indian J. Chem.</u>, 15B:964 (1977).

RARE EARTH ION SELECTIVE ELECTRODES: II.   EUROPIUM AND PRASEODYMIUM

COMPOUND MEMBRANES

Yasuo Suzuki, Hisako Itoh, and Taisuke Nakano
Department of Industrial Chemistry, Faculty of
Engineering, Meiji University
Higashi-mita, Tama-ku, Kawasaki 214, Japan

INTRODUCTION

It has been proved that cerium(IV) oxide could be used as a
sensor of the rare earth(III) ions with a response slope of 60mV,
as described in the authors' previous paper (1). This slope even-
tually coincides with those observed on univalent ions, presumably
utilizing the difference of the oxidation states between the sensor
in membrane and the ion in solution. Subsequent studies on the
rare earth(III) ion selective electrodes have been conducted in
using sensing materials containing rare earth elements of the oxidation states except three. The present work includes the results on
11 europium(II) compounds and on praseodymium oxide.

EXPERIMENTAL

Ten europium(II) compounds, EuB_2O_4, $Eu_2B_2O_5$, EuB_4O_7, $Eu_3B_2O_6$,
$EuNbO_3$, $Eu_{0.95}NbO_3$, $Eu_{0.90}NbO_3$, $Eu_{0.88}NbO_{2.31}$, $Eu_{0.73}NbO_{2.81}$, and
$Eu_{0.60}NbO_{3.03}$ were supplied from the Department of Applied Chemistry,
Osaka University. Europium boride, EuB_6, was supplied from the
National Institute for Researches in Inorganic Materials. Rare
earth oxides were obtained from Shin-etsu Chemical Industries, Co.
Among the epoxy adhesives commercially available, Araldite Standard
of Ciba Geigy, Scotch of 3M, and Elmer's of Borden Chemical Co.
were used. Other adhesives employed include Silicone of Shin-etsu
Chemical Industries Co., Soni-Bond of Sony Chemical Co., and Bond
G17 of Konishi Co. All other chemicals were of reagent grade.

The appearance of the electrodes prepared in this work was the
same as described in the previous paper (1). The electromotive

force measurements were made by Corning model M130 and Beckman
model 3500 pH meters, together with Orion model 90-02 double
junction reference electrodes at 25.0 °C.

RESULTS AND DISCUSSION

Europium(II)-Membrane Electrodes

The response behaviors of the membranes of 11 Eu(II) compounds
were studied in the lanthanoid(III) solutions of 10^{-1} - 10^{-7} mol
dm^{-3}. Most of the compounds exhibited either non-linear response
or vigorous drift, while EuB_2O_4, $Eu_{0.90}NbO_3$, and EuB_6 showed
apparently better overall behaviors, particularly with the membranes
containing 35% of the compounds. Figure 1 indicates the responses
plotted against Eu(III) activity for the three membranes, where the
values of mean activity coefficients of the lanthanoid(III) ions
(2-4) were used to calculate the activities. The slopes of the
straight lines in Fig. 1 were as much as 68, 63, and 57 mV for
EuB_2O_4, $Eu_{0.90}NbO_3$, and EuB_6, respectively, whereas those on concentration scale were 52, 47, and 43 mV, respectively.

The responses of the $Eu_{0.90}NbO_3$ electrode against lanthanoid(III)
ions except europium are shown in Fig. 2. Similar responses were
observed on any of the lanthanoid ions, within a pH region of 4 -
6.5. The selectivity coefficients obtained by the separate solution
method (5,6) for eight cations are given in Table 1. Bivalent ions
such as Mg^{2+} and Cu^{2+} gave comparable responses to those of the
lanthanoids with a slope as much as 50 mV. Both the EuB_2O_4 and EuB_6
electrodes showed analogous behaviors. These results may not favor
the Eu(II) membranes to be used in lanthanoid solutions containing
uni- or bivalent cations.

Table 1. Selectivity coefficients of the Eu(II)-membrane
electrodes.

Cation	Sensing material		
	EuB_2O_4	$Eu_{0.90}NbO_3$	EuB_6
Al(III)	6.4×10^{-3}	3.8×10^{-3}	0.77
Fe(III)	6.1×10^{-2}	4.6×10^{-2}	585
Na(I)	3.4×10^{-2}	2.8×10^{-2}	10.1
K(I)	1.9×10^{-2}	2.1×10^{-2}	5.3
Ce(III)	1.9	1.3	2.6
Sm(III)	5.1	9.1	1.1
Gd(III)	0.40	4.3	1.7
Lu(III)	0.23	25	0.17

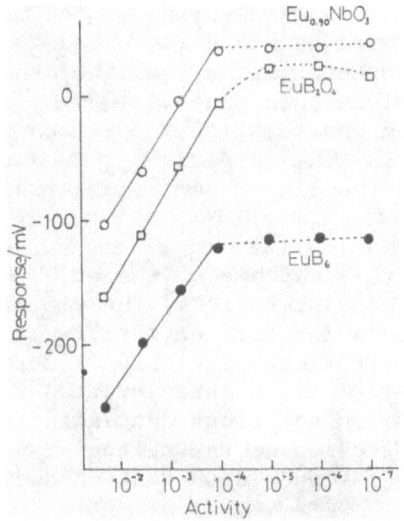

Fig. 1.   Responses of three Eu(II) membranes against Eu^{3+}.

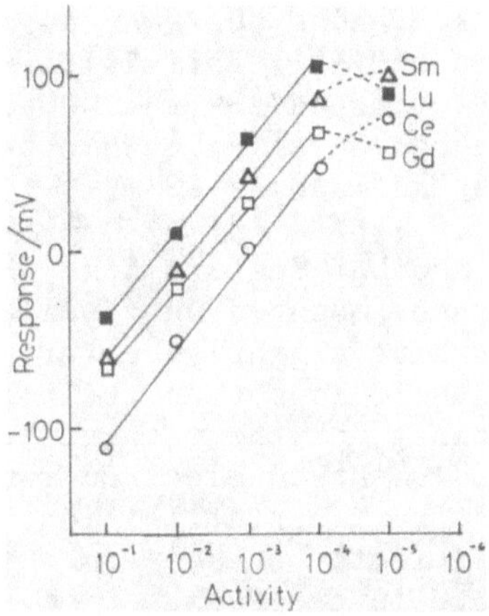

Fig. 2.   Responses of $Eu_{0.90}NbO_3$ membrane electrode against four lanthanoid ions.

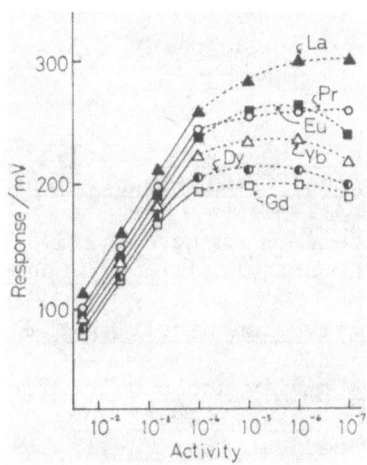

Fig. 3.   Responses of Pr_6O_{11} electrode against six lanthanoid ions.

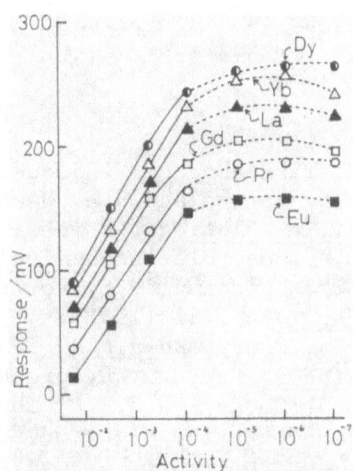

Fig. 4.   Responses of Pd-based Pr_6O_{11} electrodes.

Praseodymium Oxide-Membrane Electrodes

The praseodymium atoms in Pr_6O_{11} are stoichiometrically not quadrivalent. The Pr_6O_{11} membrane electrodes were prepared by mixing the oxide with one of the adhesives as described. The membranes containing 55% oxide with Araldite Standard exhibited apparently better overall response characteristics than the others. Figure 3 shows the responses of the electrodes plotted against six lanthanoid(III) activities, indicating shorter linear portions than those observed for the CeO_2 electrodes. The response curves were close and parallel each other, within the linear portions having slopes as much as 60 mV.

Another feature Fig. 4 shows is the response of the Pr_6O_{11} membranes mounted on a palladium-coated copper disk. The electrodes would have slightly better response behavior for the individual lanthanoid ions than the former. Due to the appreciably different responses for the respective lanthanoid ions, however, the palladium-based electrodes may be used in solutions containing a single rare earth(III) ion. Further studies on the selectivity coefficients, as well as the effects of copper-disk coating and of pH, are in progress.

ACKNOWLEDGMENTS

The authors wish to express their gratitude to Drs. J. Shiokawa and G. Adachi of Osaka University and to Dr. T. Tanaka of the National Institute for Researches in Inorganic Materials for supplying the Eu(II) compounds. The authors are also indebted to Messrs. Y. Sugisaki and M. Itoh for the help in a part of the experiments.

REFERENCES

1. Y. Takasaka and Y. Suzuki, Bull. Chem. Soc. Jpn., 52:3455 (1979).
2. H.S. Harned and B.B. Owen, The Physical Chemistry of Electrolytic Solutions, Reinhold, New York (1957).
3. D. Dobos, Electrochemical Data, Elsevier, Amsterdam (1975).
4. F.H. Spedding, P.E. Porter, and J.M. Wright, J. Amer. Chem. Soc., 74:2781 (1952).
5. G.J. Moody and J.D.R. Thomas, Selective Ion Sensitive Electrodes, Merrow, Watford Hertz (1971).
6. G.J. Moody and J.D.R. Thomas, "Selectivity and Sensitivity of Ion-Selective Electrodes," in Ion-Selective Electrodes, E. Pungor (ed.), Akadémiai Kiadó, Budapest, pp. 97-113 (1973).

ANALYSIS OF RARE EARTH ELEMENTS* IN ORE CONCENTRATE SAMPLES USING DIRECT CURRENT PLASMA SPECTROMETRY

G. W. Johnson and T. E. Sisneros

Molycorp, Inc.

P.O. Box 607, Louviers, CO 80131

INTRODUCTION

Because of the increasing interest in the rare earth (RE) elements for many diverse applications, the need to analyze for all of the RE's in various types of mixtures is becoming more important. Most methods for RE analysis are not suitable for determining all of the individual elements; therefore, it is often necessary to use a combination of methods when doing a complete RE distribution analysis. Methods which have been used for determining all of the RE's in a mixture include x-ray fluorescence, mass spectrometry, neutron activation analysis and optical emission. Of these, x-ray fluorescence and optical emission (using arc or spark excitation) are normally available to most analysts. However, both of these methods suffer from significant drawbacks; x-ray fluorescence is generally limited to concentrations above 100 ppm and is subject to severe matrix effects, while arc- and spark-excited optical spectroscopy are slow because of the need to record and analyze very complex spectra and, without proper precautions, can have poor precision due to instabilities in the excitation processes. A good review is included in a recent four-volume publication covering the RE field (1). This same volume also includes chapters on other analytical methods.

A variation of atomic emission spectroscopy which has been commercially developed in recent years utilizes an argon plasma as the excitation source. Two versions of this source are available; an inductively coupled plasma (ICP) and a direct current plasma jet (DCP). Early reports on the use of plasma emission spectrometers for RE samples have been very encouraging; advantages claimed include

---

*Rare earth elements refer to lanthanides (except Pm), Y, and Th.

detection limits which are two or three orders of magnitude lower than
those obtained by other methods, linear response over a concentration
range of four to five orders of magnitude, reduced matrix effects,
good precision and capability for providing rapid analyses. Several
papers have included data on the analysis of synthetic solutions or
on the determination of some of the RE elements in various types of
samples (2,3,4). The determination of all RE elements in standard
mineral samples has been investigated by Broekaert, et al. (5). It
was noted (5) that in order to realize the inherent power of detection of the plasma emission source used (ICP), a spectrometer with
resolving power greater than about 170,000 ($R_P=\lambda/\Delta\lambda$)* is required.
Results reported therein were obtained using an ICP source coupled
to a 3.4m Ebert spectrograph providing resolution (R_T=460,000 in the
second order)** greater than that offered by most commercial ICP
spectrometer systems.

The purpose of the present work was to evaluate a DCP-echelle
spectrometer system for its ability to determine all RE elements in
four RE ore concentrate samples. Even though R_T>500,000 for these
spectrometers, ubiquitous emission spectra produced by RE elements
(>20,000 emission lines between 200 and 800 nm) warranted an extensive search to locate useful analysis lines. Once one analysis line
was selected for each RE the sixteen line set was used to measure
concentrations of all the RE elements in all four materials. After
compensating these measurements for spectral interferences, excellent
agreement was found between DCP results and those obtained by other
methods.

EXPERIMENTAL

The atomic emission systems utilized in this investigation were
the SpectraSpan IV and SpectraSpan IIIB (SpectraMetrics, Inc.) DCP-
echelle spectrometers. Both were operated in a sequential analysis
mode. Experimental facilities and operating conditions, if not
discussed below, have been previously reported (6).

Single element stock solutions for all analytes except ThO_2
were prepared from 5-9's or better oxides (Spex Industries), while
the ThO_2 stock was prepared from $Th(NO_3)_4 \cdot 4H_2O$ (Baker RG). These
solutions were made up to be 10 g/l of the appropriate REO and 1%
(v/v) HNO_3, and served as reservoirs from which all subsequent standard and test solutions were prepared. Three single element calibra-
tion standard solutions were prepared for each analyte for its
determination in the IGS 36, monazite and bastnasite concentrates.
A second three-standard set was prepared for each analyte for its
determination in the xenotime concentrate. Concentration levels of

---

*This is the definition of practical resolution = R_p.
**R_T is theoretical resolution and may be >R_p in practice.

the standards were determined after assessing previously collected
data (e.g. ref. 7) and assuming a 0.1% (w/v) concentration of total
REO in sample solutions would be used. All standard solutions were
prepared to contain 2000 mg/l of lithium as an ionization buffer and
were 1% HNO_3. In addition, three single element test solutions were
prepared for each RE covering a ten-fold concentration range. The
average concentration in this range was prepared to approximate that
expected in a 1.0% REO solution derived from the IGS 36 sample. All
test solutions contained 2000 mg/l lithium and 1% HNO_3.

Analysis lines were selected on the basis of minimal spectral
interference. With the aid of standard wavelength tables (8) and
previous work (1), several prospective analysis lines were identified
for each RE. Using the standards and test solutions prepared above,
several apparent analyte concentration measurements due to the
presence of other RE's in solution were made on each of these lines.
The line exhibiting the minimum sum in units of apparent analyte concentration was used as the analysis wavelength for ensuing determinations. Prior to these, test solutions which resulted in significant
apparent analyte concentrations on any chosen analyte line were
investigated for analyte contamination. This was done using a wavelength profile method (6). Blanks were subtracted from apparent
analyte concentrations to give the spectral interference equivalent
concentration $(SIEC)_{i,j}$ of the spectrally offensive element i on the
jth analyte.

Four RE ore concentrates ($\approx$60% w/w REO) were selected for these
determinations; an Institute of Geological Sciences monazite (IGS 36),
an Australian monazite, a Malaysian xenotime, and a Mountain Pass (CA)
bastnasite. Inter-laboratory RE distribution results were available
on the IGS 36 sample (7), while comparative data using other analysis
techniques had been previously obtained in analytical laboratories
at Molycorp, Inc. Sample preparation involved the use of standard
fusion and/or acid decomposition procedures. The REO fractions were
separated from other sample constituents using two or three selective
precipitations, the final one in each case was effected using oxalic
acid. The final RE oxalates were ignited at 1000°C and the resulting
REO mixtures dissolved in nitric acid. All sample solutions were
prepared at 0.1% REO (except bastnasite, for which a 1.0% REO solution
was also prepared), 1% HNO_3, and 2000 mg/l of lithium as an ionization
buffer.

RESULTS AND DISCUSSION

Results of this investigation are summarized in Table I. All
measurements by the DCP-echelle spectrometer method reported therein
were corrected for spectral interference by the method outlined in
reference 6. This method amounts to successive subtraction of pro-
portioned $(SIEC)_{i,j}$'s from the measured analyte concentrations.

Table I. Summary of Results by DCP

% or p – µg/g OF TOTAL RECOVERED REO

ANALYSIS FOR:	LINE (nm)	DCP MEASUREMENTS, IGS 36			IGS DATA (7)			MALAYSIAN XENOTIME			AUSTRALIAN MONAZITE			MT. PASS BASTNASITE		
		H₂SO₄ BAKE	Na₂CO₃ FUSION	Na₂O₂ FUSION	IGS VALUE	STANDARD ERROR	STATUS OF IGS VALUE	DCP VALUE	A.M.** VALUE	A.M.# USED	DCP* VALUE	A.M. VALUE	A.M. USED	DCP VALUE	A.M. VALUE	A.M. USED
La₂O₃	433.37	19.1	19.3	18.6	19.74	1.378	Probable	1.24	1.18	1	20.0	20.1	1	33.2	32.1	1
CeO₂	394.28	42.1	44.4	42.0	41.23	1.53	Probable	3.13	2.92	2	43.1	42.8	2	49.1	48.2	2
Pr₆O₁₁	414.31	4.81	4.82	4.79	4.624	0.263	Probable	0.493	0.45	3	4.68	4.54	3	4.34	4.44	3
Nd₂O₃	406.11	17.7	17.9	17.8	17.381	1.148	Probable	1.59	1.65	3	16.3	16.3	3	12.0	12.0	3
Sm₂O₃	443.43	2.70	2.68	2.66	2.55	0.316	Probable	1.14	1.06	1	2.53	2.59	1	0.789	0.830	1
Eu₂O₃	272.78	558p	552p	565p	577p	65p	Probable	0.012	0.020	1	926p	910p	1	0.118	0.118	1
Gd₂O₃	344.00	1.55	1.55	1.52	1.27	0.058	Probable	3.47	4.82	1	1.52	1.29	1	0.166	0.179	1
Tb₄O₇	367.64	0.154	0.154	0.156	0.18	0.088	Probable	0.906	0.68	1	0.163	0.16	1	159p	837p	1
Dy₂O₃	353.17	0.524	0.532	0.530	0.51	0.099	Probable	8.32	7.70	1	0.566	0.51	1	312p	247p	1
Ho₂O₃	345.60	747p	733p	748p	657p	40p	Possible	1.98	1.89	1	743p	750p	1	50.5p	53p	1
Er₂O₃	369.27	0.125	0.128	0.128	0.101	0.003	Possible	6.43	6.27	1	0.106	0.11	1	34.9p	78p	1
Tm₂O₃	313.13	135p	137p	137p	108p	26p	Possible	1.12	1.00	1	84.5p	130p	1	8.5p	17p	1
Yb₂O₃	328.94	646p	661p	687p	488p	65p	Probable	6.77	6.49	1	335p	330p	1	6.26p	13p	1
Lu₂O₃	261.54	79.5p	80p	85p	96p	12p	Possible	0.988	##		31.4p	##		1.02p	##	
Y₂O₃	377.43	1.99	2.03	1.93	1.85	0.309	Probable	61.0	61.3	1	1.91	1.52	1	913p	510p	1
ThO₂	283.71	8.76	8.33	9.86	10.13	0.832	Recommended	1.49	1.27	3	9.60	8.86	4	790p	886p	3
TOTALS:		99.73%	102.04%	100.20%	99.76%			100.08%	98.70%		100.69%	98.99%		99.94%	98.13%	

*Average of measurements on three replicate samples of the Australian monazite.

**A.M. – Alternate Method

#1 = Atomic absorption, 2 = titration, 3 = spectrophotometric, 4 = gravimetric.

##No determination by alternate method.

Satisfactory agreement between DCP results and those made by other methods is shown in Table I. Thus, only one analysis line is necessary for accurate determination of each RE in any of these four complex sample materials. Moreover, accurate determination of RE distributions in ore concentrate samples was demonstrated using the DCP-echelle spectrometer technique.

ACKNOWLEDGMENTS

The authors gratefully acknowledge the preparation of ore concentrate sample solutions by R. D. Witham.

REFERENCES

1. E.L. De Kalb and V.A. Fassel, "Optical Emission and Absorption Methods," in Handbook on the Physics and Chemistry of Rare Earths, Vol. 4, K.A. Gschneidner, Jr. and L. Eyring (eds.), North-Holland Publishing Co., Amsterdam, pp. 405-440 (1979).
2. S. Nikdel, A. Massoumi and J.D. Winefordner, Microchem. Jour., 24:1 (1979).
3. M.A. Floyd, V.A. Fassel and A.P. D'Silva, Anal. Chem., 52:2168 (1980).
4. M.A. McMahon, "Direct Determination of Rare Earths from Mixed Aqueous Solutions by Plasma Emission Spectroscopy," in The Rare Earths in Modern Science and Technology, G.J. McCarthy, J.J. Rhyne and H.B. Silber (eds.), Plenum Press, New York, pp. 593-598 (1978).
5. J.A.C. Broekaert, F. Leis and K. Laqua, Spectrochim Acta, 34B:73 (1979).
6. G.W. Johnson, H.E. Taylor and R.K. Skogerboe, Spectrochim Acta, 34B:197 (1979).
7. Brian Lister, Geostandards Newsletter, 5:75 (1981).
8. W.F. Meggers, C.H. Corliss and B.F. Scribner, Tables of Spectral Line Intensities Part I-Arranged by Elements, 2nd ed., NBS (1975).

BEHAVIOR OF REE IN GEOLOGICAL AND BIOLOGICAL SYSTEMS[*]

J.C. Laul and W.C. Weimer
Physical Sciences Department; Battelle
Pacific Northwest Laboratory
Richland, Washington 99352

ABSTRACT

The REE abundances when normalized to primordial (chondritic) abundances behave as a smooth function of the REE ionic radii in both the geological and biological systems. The REE are hardly fractionated chemically through various stages of their transformation from soil-soil extract-plant-geological systems.

INTRODUCTION

Rare earth elements (REE) are very informative in revealing various chemical fractionation processes in geological and biological systems. The REE's behavior in geological materials is characteristic of different minerals (primary and secondary) which comprise a rock. Their behavior in biological materials is hardly known at all, perhaps mainly because of their inherently low concentrations. Their study in biological materials is particularly important because, in addition to revealing the REE's characteristics, they may also provide information on the long-term behavior of their transuranic analogs in the natural environment (1-3). The REE concentrations in biological samples are at or below the parts-per-billion level. To detect the REE at such low levels, we have used radiochemical neutron activation analysis (RNAA) with a REE group separation scheme (4-6). To maximize the detection sensitivity of the individual REE at this level, we have made use of selective

---

[*] Prepared for the U.S. Department of Energy under Contract DE-AC06-76RLO 1830.

γ-ray/x-ray energies measured by Ge(Li), intrinsic Ge, and coincidence-noncoincidence Ge(Li)-NaI(Tl) counting systems. The chemical yields of the REE are determined by reactivation.

EXPERIMENTAL

The details of the REE group separation scheme are outlined elsewhere (4,6). In brief, the samples and a liquid aliquot of the REE mixed standard are irradiated at a thermal flux of $\sim$1 x 10^{13} n/cm^2/sec for 6 to 8 hours in a reactor. After about a day's decay, the sample is transferred into a Ni crucible in which the REE mixed carriers are already dried therein. The sample is fused with a Na_2O_2-NaOH (10:1) mixture. The fused cake is decomposed with H_2O and neutralized with HCl and the REE are precipitated as a group hydroxide with excess NH_4OH. The REE(OH)$_3$ is then dissolved in 10 N HCl, and loaded onto a 1 cm x 8 cm anion exchange column (AG1-X10, 100-200 mesh). The REE are collected in the first 30-40 ml and precipitated as the group hydroxide with NH_4OH. The REE(OH)$_3$ is dissolved in HCl and the REE are precipitated as fluorides at pH $\sim$4 with 1 M NH_4HF_2 and HF. The REE fluoride is dissolved in HNO_3 and H_3BO_3, and reprecipitated as REE(OH)$_3$ with NH_4OH. The fluoride and hydroxide cycle is repeated at least two times prior to their counting.

The REE group aliquots are counted on a Ge(Li) detector (25% efficiency, FWHM 1.8 keV for ^{60}Co), intrinsic Ge, and coincidence-noncoincidence Ge(Li)-NaI(Tl) counting systems. Two counting decay intervals (2-5 days and 20-30 days) are used for optimum detection sensitivities of the various REE. Based on the x-ray and selected γ-ray energies, Ge(Li) γ-ray counting is favorable for [140]La, [141]Ce, [142]Pr, [153]Sm, [171]Er and ^{177}Lu, whereas intrinsic Ge γ-ray counting is favorable of [143]Ce, [147]Nd, [160]Tb, and [166]Ho, and intrinsic Ge x-ray counting is favorable for [152]Eu and [175]Yb. Gamma-ray counting of [153]Gd and [170]Tm is equally sensitive with Ge(Li) or intrinsic Ge detectors. Greatest sensitivities are achieved for [166]Ho, [147]Nd, [141]Ce, [142]Pr, [153]Sm, [170]Tm and [153]Gd by counting on coincidence-noncoincidence systems. Thus, depending on the levels of REE present in a sample, suitable counting combinations can be used to obtain the trace REE data with high precision.

DISCUSSION

The REE concentrations have been measured in a wide variety of geological and biological samples, and the REE content varies by about six orders of magnitude. We have normalized the REE concentrations in these samples to chondritic (primordial) abundances, and plotted them versus the REE ionic radii. The chondritic values (ppm) used for normalization are La 0.34; Ce 0.91; Pr 0.12;

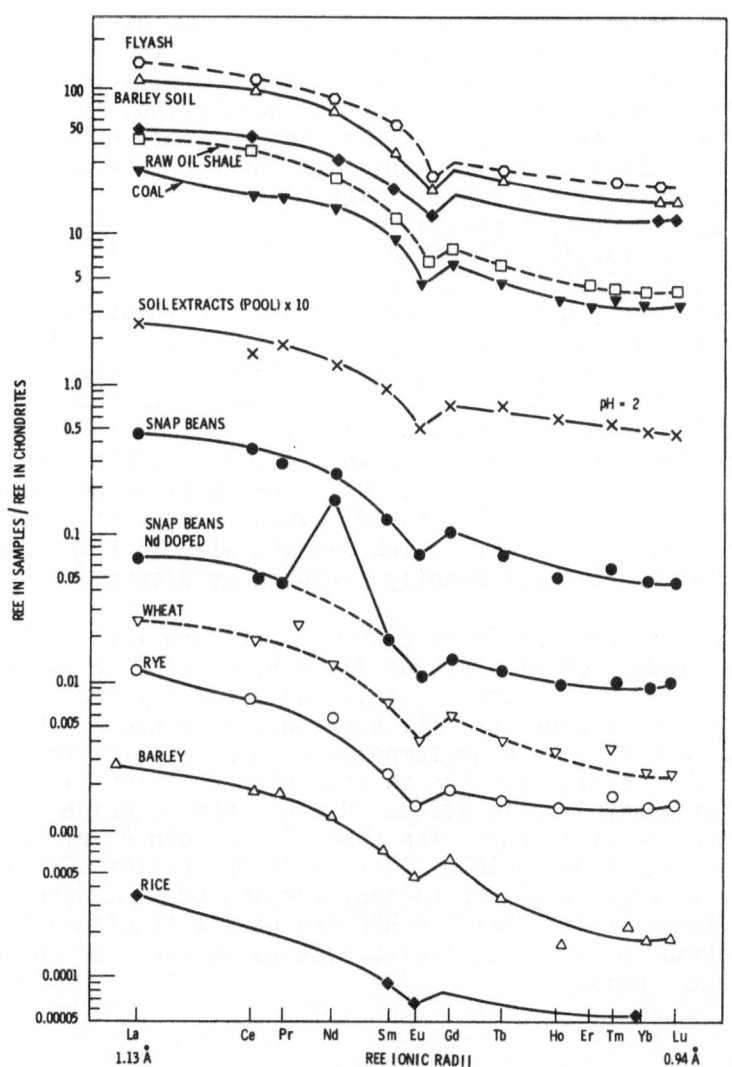

Figure 1. REE patterns in geological and biological samples.

Nd 0.64; Sm 0.195; Eu 0.073; Gd 0.26; Tb 0.047; Ho 0.078; Er 0.20;
Tm 0.032; Yb 0.22 and Lu 0.034 (5). The chondritic normalized REE
patterns are displayed in Figure 1. These patterns in both geological and biological samples behave as a smooth function of the
REE ionic radii. In a geological sample, the primary minerals
(olivine, plagioclase, orthopyroxene, clinopyroxene, etc.) and
secondary minerals (apatite, biotite, hornblend, etc.) have their
own respective characteristic REE patterns, and the secondary
minerals usually have very high distribution coefficients of REE
relative to primary minerals (3,5). On weathering, the secondary
minerals from a rock are the first to become loose or decompose
releasing the REE in a soil. Thus, the observed REE pattern in a
soil, which reflects the overall mineral composites, is usually
dominated by secondary minerals in a soil. The mineral pattern
dominating the overall soil patterns is that of apatite. Interestingly, the biological materials (corn, snap beans, wheat, rye,
barley, rice, etc.) have REE patterns that are identical to their
host soils (Figure 1). This comparison strongly suggests that the
REE are not fractionated during their uptake by plants from their
host soils. This evidence is further supported from soil extracts
(pH 2-4) which represent the biologically available REE pool and
show REE patterns again identical to the soil and biological
materials. Relative to the host soil, the absolute REE concentrations are depleted by 10^{-2} to 10^{-3} in soil extracts and 10^{-4} to
10^{-5} in biological material, which demonstrates that the nonessen-
tial REE elements are not readily taken up by plants.

Another interesting feature shown in Figure 1 is that coal,
which is a product of plants, and fly ash, which is a late stage
residue of coal in coal power plants, have exactly the same REE
characteristics, except that fly ash contains a higher REE content
than coal. The REE are high-temperature non-volatile refractory
elements and thus are expected to concentrate in the fly ash without undergoing any fractionation. Raw oil shale, which is also an
old biogenic material, shows the same REE pattern as plants and
coal. Thus, the remarkable similarity in REE patterns in a wide
variety of matrices with REE content varying over six orders of
magnitude demonstrates that the REE are hardly fractionated via
transformation in the geological-biological-geological chain over
geologic time scales.

REFERENCES

1. J.C. Laul, W.C. Weimer and L.A. Rancitelli, Proc. of 2nd
    Symposium on the Origin and Distribution of the Elements,
    UNESCO, pp. 819-827 (1978).
2. W.C. Weimer, J.C. Laul, J.C. Kutt and E.A. Bondietti, Proc.
    of the 3rd Int. Conf. on Nuclear Methods in Environ. and Energy
    Res., pp 472-481 (1978).

3. W.C. Weimer, J.C. Laul and J.C. Kutt, "Prediction of the
    Ultimate Biological Availability of Transuranium Elements in
    the Environment", in <u>Contaminants and Sediments</u>, <u>2</u>, R.A. Baker
    (ed.), Ann Arbor Science Publisher, pp. 465-484 (1980).
4. J.C. Laul, K.K. Nielson and W.A. Wogman, <u>Proc. of the 3rd Int.
    Conf. on Nuclear Methods in Environ. and Energy Res.</u>, pp 188-
    209 (1978).
5. J.C. Laul, <u>Atomic Energy Review</u>, 17:603-697 (1979).
6. J.C. Laul, E.A. Lepel, W.C. Weimer and N.A. Wogman, "Precise
    Trace Rare Earth Analysis by Radiochemical Neutron Activation",
    <u>J. of Radioanal. Chem.</u> (submitted 1981).

TRENDS IN RARE EARTH METAL CONSUMPTION FOR STEEL APPLICATIONS IN THE 1980'S

Joseph R. Jackman and William H. Trethewey

Reactive Metals & Alloys Corporation

P.O. Box 366, Route 168, West Pittsburg, PA 16160

In the early 1920's and 1930's, the rare earth metals were virtually unknown to most carbon and alloy steel producers. Rare earths were being used at that time to improve hot and cold formability of carbon, alloy, and stainless steel (1-3). Problems encountered with rare earth additions, inconsistent recovery and poor cleanliness, prevented a much broader application of these elements.

The rare earth metals (REM) form extremely stable sulfides and oxysulfides (4). Therefore, their use in aluminum killed steels in the improvement of toughness and formability is accomplished by controlling the sulfide morphology. Maintaining a RE/S ratio of 3 to 1 in the final product results in globular rare earth sulfides instead of the elongated type II manganese sulfides. This was very dramatically demonstrated with the development of an 80,000 psi high strength low alloy steel during the late 1960's (5).

During the early 1950's, there were significant contributions in rare earth metal additions in stainless, tool steel, alloy and electrical steel grades (6). REM (mischmetal) was used to promote improved rollability of stainless, to improve Charpy impact at low temperature in armor plate and in wear and abrasion resistant steels. Again, use and acceptance of REM was limited because of the recovery and cleanliness problems. The development of the HSLA steels in the 1960's provided the necessary incentive for the rare earth metal manufacturer and the steelmaker to solve some of these basic problems. This was done with development of new addition techniques and a more thorough understanding of the various mechanisims and side effects involved in the use of REM.

Since 1968, the free world annual consumption of rare earth metals for metallurgical applications has grown dramatically from 200,000 pounds (100 net tons (NT) to 14,500,000 pounds (7,243 NT) in 1978, Fig. 1 (7-8). As metallurgical technology advanced through the 1970's, several dramatic changes had taken place within the steel industry in rare earth metal consumption. Not only will we review the history of this rare earth metal consumption in the steel industry during the 1970's, but evaluate several trends that will affect consumption for the early 1980's.

CHANGES IN RARE EARTH METAL CONSUMPTION DURING THE 1970'S

During the 1970's there were counteracting trends that affected rare earth metal consumption in the steel making community. Several had a negative impact on actual REM consumption while others provided growth through a broader application of REM in the development of more critical grades of steel. The broader application of REM used in steelmaking the past 10 years has more than offset the decrease in actual pounds of REM used per ton of steel treated.

As can be seen in Fig. 1, for 1968 and 1970, the initial rare earth metal consumption growth came from the United States and Canada. This segment has continued to grow throughout the 1970's. During the 1972-1974 period, REM consumption in the rest of the free world, primarily Europe and Japan, increased substantially. Over 90% of the foreign consumption for that period was in large diameter line pipe production from ingot cast steel.

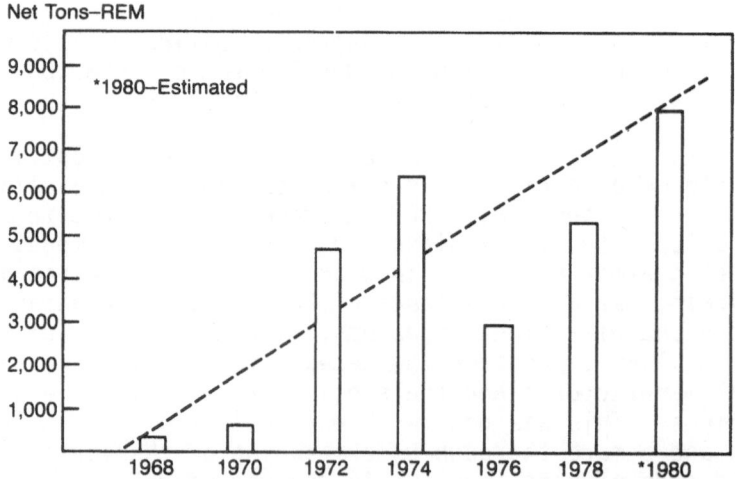

Fig. 1:   Free World Consumption of Rare Earth Metals in Iron and Steel[7]

As the European and Japanese steelmaking communities moved towards continuous casting facilities, rare earth metal consumption dropped due to technical difficulties encountered when attempting to produce rare earth treated continuous cast steels (9). Simultaneously, the world-wide steel community was rapidly developing the technology to obtain lower sulfur level steels. For example, desulfurizing to .010% to .005% sulfur in some cases reduced or eliminated the need for sulfide shape control.

The move toward continuous slab casting and the development of desulfurization technology had a somewhat negative impact on total rare earth metal consumption. The official or unofficial opinion among many metallurgists during the early 1970's, was that REM in steel was trending toward limited usage. Beyond the previously mentioned trends, alleged welding problems with carbon dioxide gas metal-arc techniques, particularly in root passes of field girth welding of gas transmission lines, was creating additional doubt about REM value in steel. Based upon this performance feedback, some metallurgists became very skeptical about the use of REM in many steel products.

About 1973-74, REM market development and steel process development efforts were accelerated by a United States producer of mischmetal and rare earth silicide. The market development approach focused upon the mechanisms of rare earth metal action on steel structures and rare earth metal's subsequent positive impact and influence on steel product properties (9). Moreover, the steel process development approach focused upon REM addition practices during steelmaking (10). Primary emphasis was placed on proper deoxidation and desulfurization to levels of sulfur where the resultant sulfides and oxysulfides created would not cause secondary problems to the steelmaker. These secondary problems were sometimes more serious than what the initial REM treatment was designed to eliminate.

Hence, the metallurgist was beginning to understand many other facets of rare earth metal metallurgy in the microalloying of REM in steelmaking. New areas of rare earth technology were being developed because of the better understanding of liquid steel alloying, liquid steel reactions, solidification, hot rolling and cooling and mechanical working and welding of steel containing REM (9).

REM MARKET DEVELOPMENT

With the knowledge that rare earth metal could have metallurgical benefits in the area of sulfide shape control and hydrogen induced cracking control, and with the following trends in REM consumption patterns:

Negative Trends:

- European and Japanese Trend Ingot Cast to Continuous Cast
- Desulfurization Technology Development

Positive Trends:

- Metallurgical Realization of REM Benefits
  -Sulfide Shape Morphology Control
  -Hydrogen Induced Cracking Control

- Steel Process Technology Leading to More Cost Effective
  Use of REM

steel product groups were then isolated where both were known to be prevalent problems.

Sulfide Shape Control

As is now history, REM is added to high carbon and high strength low alloy steels for resistance improvement to spalling. Examples of steel product applications are sawblades, plow blades and ordinance applications of armor plate.

REM is now being added to improve the weld quality of high strength ERW tubes to the point where they successfully compete with seamless tubes in oil company goods applications.

The addition of REM to large structural beam steels, non semi-killed, has minimized lamellar tearing around the welds as well as increasing low temperature toughness for new rigid welded structure designs.

Small additions of REM after alloying and desulfurizing of pipe and plate steels to .003/.005% sulfur were found to maximize impact properties. Japanese literature shows improvements were available by adding REM to .003% sulfur steels (11). However, steel desulfurization by injecting with calcium compounds did not lead to 100% sulfide shape control. Manganese sulfide inclusions were observed at .005% sulfur levels after calcium treatment. These MnS inclusions were modified with additions of REM which resulted in improved properties.

Hydrogen Induced Cracking Control

During the 1970's, it was found that well calibrated additions of REM could have significant delaying effects on hydrogen induced cracking, both in welds and the base metal without effecting other properties (12). Rare earth metals are added routinely for hydrogen sulfide stress corrosion cracking resistance in high strength deep oil and gas well casing as well as drilling for "sour gas" environments. Considerable progress has been made in reversing the negative thoughts about REM use in gas transmission lines. Again, the benefit of a delaying effect on hydrogen induced cracking, where critical pressures and hydrogen sulfide concentrations are known to exist, is being re-evaluated.

This market development and acceptance has led to a much broader application of rare earth metal in steel products. In the 1960's, there were essentially six broad steel product groupings using REM. They are as follows:

- Super Alloys
- Stainless Steels
- Special Bar Quality
- HSLA Automotive
- ERW Line Pipe
- Armor Plate

Today there are approximately eleven product groupings:

- Super Alloys
- Stainless Steels
- Special Bar Quality
- HSLA Automotive
- ERW Line Pipe
- High Carbon Steel
- ERW Welded Tubes
- Seamless Tubes
- Rails
- Large Structural
- Armor Plate

PROCESS DEVELOPMENT

During the 1960's and early 1970's, a rare earth silicide (approximately 33% rare earth, 33% silicon and 33% iron) alloy was the most commonly accepted form of rare earth metal for REM additions in the ladle and ingot mold. Ladle addition levels generally ran between 10-24 lbs. of alloy per net ton of steel treated, with very low and sometimes erratic recoveries. This often resulted in inconsistent steel property results. The raw material treatment cost of this ladle addition practice was between $12.00 and $25.00 (in 1980 dollars) per net ton of steel produced.

Mold additions of rare earth silicide were also used during this period. Although treatment cost was much lower per ton of steel treated because of higher recoveries, it often caused surface and subsurface problems. Some plants found mischmetal a more attractive mold addition and results usually were much better than a comparable mold practice with rare earth silicide (13-14).

Based on the high cost R.E. silicide ladle practice, along with its very erratic recoveries and combined with the steelmakers desire to eliminate the mold addition technique, ladle plunging of mischmetal canisters won acceptance during the early 1970's (15-16). Addition rates were in the area of 1.5 to 2.5 lbs. per ton of steel

treated. The use of this practice brought the REM treatment down
to $7.00 to $12.00 per net ton. The actual cost per ton depended
on grade and particular application.

As the 1970's progressed, economic balances between reasonably
low sulfurs and residual sulfide control reopened interest in the
use of mischmetal in certain steelmaking applications. Some of
these applications were high strength low alloy (HSLA) automotive
steels where cold formability, surfaces and cost were criteria for
steel product acceptance (13). Steelmaking practices were developed
to add mischmetal balls using standard mold addition and delayed
mold addition techniques. The cost of these practices further
reduced the addition of REM to steel to a level of 1-1 1/2 lbs. per
net ton for a treatment cost of $5.00 to $7.00 per net ton of steel
treated.

Table 1 summarizes this advancing metallurgical technology
from the 1950's through the 1970's. The reduction in both the consumption of REM per net ton and the cost per net ton have been
quite impressive.

After an initial examination of Fig. 1 and the review of the
negative trends, namely development of desulfurization technology
and continuous casting, it would have been easy to conclude that
rare earth metal consumption was static or perhaps declining during
the mid-1970's with little potential growth available for the balance of the 1970's. However, after the review of the market
development and process development changes that impacted rare earth
metal consumption during the late 1970's, rare earth metals are now
positioned for a continued growth cycle during the early 1980's.

If we take all of the positive and negative impact trends and
combine the data, some interesting conclusions can be drawn. Fig. 2
shows that the net tons of steel treated with rare earth metals have
multiplied by a factor of five between 1972 and 1978. The key
factor which was not apparent in Fig. 1, was the dramatic reduction
in consumption of REM per net ton of steel treated during the 1970's.
This increase in net tons of steel treated with rare earth metals
should continue through the early 1980's because of the broader
acceptance and use of rare earth metals in many applications
requiring sulfide shape control and hydrogen induced cracking
control.

TRENDS DURING THE 1980's

As we look forward into the 1980's, sulfur removal and control
in steelmaking continues to be the focus of considerable attention.

Since steelmaking began, the need for lower and lower sulfur
steel has been recognized. The need for improved surface and inter-

Table 1. Technological Development of REM Additions to Steel

Period	Addition Method	REM Product	REM Addition Level (Lbs/NT)	Cost/NT (1980 Dollars)
1950's	1. Mold	Mischmetal 4 oz. Ingots	3/4-6	$ 4-30
1960's	1. Mold	Mischmetal 4 oz. Ingots	2-3	$10-15
	2. Ladle	R.E. Silicide 2" × Down	3-8	$12-25
	3. Mold	R.E. Silicide 2" × Down	2-3	$ 6-12
1970's	1. Mold	Mischmetal 4 oz. Ingots	1-2	$ 5-10
	2. Ladle	R.E. Silicide 2" × Down	3-7	$12-23
	3. Mold	R.E. Silicide 2" × Down	2-3	$ 5-10
	4. Ladle Plunging	Mischmetal Canisters	1½-2½	$ 7-12
	5. Delayed Mold	Mischmetal 4 oz. Ingots	½-1½	$ 3- 7

Note: Mischmetal is 96% R.E.
R.E. Silicide is 33% R.E.

nal soundness varies considerably and is dependent upon product application. Moreover, not only does the incentive for producing low sulfur steels vary, but the approaches to sulfur control vary as well (17).

Today and as we look into the 1980's the need for lower sulfur steels to meet the improved property requirements at reduced costs have forced the steel producer toward intensifying individualized plant searches for the "best" approach to sulfur control. These trends which became apparent during the mid-1970's are being magnified as we move into the 1980's. At one end of the spectrum, steel producers are faced with a trend toward supplying higher performance steels with minimal sulfur contents and sulfide inclusion shape treatment. However, at the other end of the spectrum, only

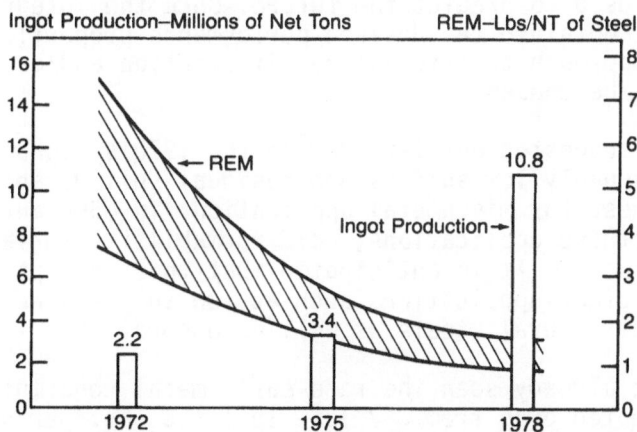

Fig. 2. North America Ingot Tons Steel Treated with REM vs. Pounds REM/NT

higher sulfur raw materials are available, such as higher sulfur
coke, iron ore and scrap.

Due to the capital intensity of the steel industry, investments
in fixed assets such as blast furnaces, electric furnaces and BOF
shops are made with the intent to maximize production. These fixed
assets generally have to operate at higher sulfur levels to maximize
production.

As an example, blast furnaces operating to hot metal specifi-
cations of .045% sulfur will use less coke, work at higher wind
rates and generally increase the net tons of hot metal per square
foot of hearth area compared to similar operations casting hot
metal at .025% sulfur levels. In open hearth, electric and basic
oxygen steelmaking furnaces, lower tap sulfur requirements generally
tend to increase the furnace refining time.

Considering the above factors and in an attempt to reverse
the lower quality/higher sulfur raw material trend which enables
the finished steel product to meet the higher performance specifi-
cations, the steel producer has proceeded toward external desulfurization and sulfide shape control. Currently, the most widely
accepted techniques for desulfurization and sulfide shape control
are as follows:

  · Hot Metal Desulfurization - Injection
  · Steel Ladle Desulfurization - Ladle Additives
  · Steel Ladle Desulfurization - Injection
  · Sulfide Shape Control - Rare Earth Metals

The steelmaker will use the most economical approach to external
desulfurization and sulfide shape control to obtain the desired
performance properties in the steels of the future. If historical
fact can be used to predict the future, once the integrated steel
producer has obtained the desired performance properties, the most
economical approach to external desulfurization and/or sulfide shape
control will be chosen.

As was suggested earlier, during the 1970's economic balances
between reasonably low sulfurs and residual sulfide shape control
opened interest for mischmetal applications in HSLA automotive
steels. In these applications, cold formability, surface and cost
must be balanced. It is anticipated that this concept will be
extended to other applications such as low sulfur line pipe, plate
steels and structural steels as well as others.

We have already seen the rare earth metal consumption per ton
of steel treated drop from 3-7 lbs. to 1-1.5 lbs. per net ton for
the same level of improvements in properties. We are forecasting
that rare earth metal consumption could be hovering around the 1 lb.

per net ton level by 1985 when added to lower deoxidized and desul-
furized steels. We are entering an era where the steelmaker will
be deoxidizing and desulfurizing to much lower levels than ever
before in order to produce the high performance steels required.
This of course reduces the amount of REM required per net ton of
steel produced and also the cost of REM per net ton. However, it
is anticipated that increasing diversification of steel appplica-
tions for rare earth metals will continue in the early 1980's.

To summarize, it is conceivable that REM consumption in the
United States steel industry will reach 10-12,500 net tons by
1985 (18). That tonnage converts to the treatment of approximately
20 to 25 million net tons of steel in the United States.

Major new uses for REM in the 1980's are expected to be in
continuously cast steels, ferritic stainless steels, free-machining
steels, electroslag melted steels, vermicular and nodular graphite
cast irons, high energy magnets, and energy storing devices (19).
The future growth of the rare earth industry looks promising as we
enter a new decade.

REFERENCES

1. H.W. Gillett and E.L. Mack, Bureau of Mines Bulletin No. 199,
   pp. 57-74.
2. L. Eyring, "Progress in the Science and Technology of the
   Rare Earths", Vol. I and II, Pergamon Press - The Macmillan
   Co., New York, 1964.
3. E. Anderson and J. Spreadborough, Rev. Met., 64:177 (1967).
4. R. Kiessling and N. Lange, "Non-Metallic Inclusions in
   Steels", Part II, ISI No. 100, 1966.
5. L. Luyckx, J.R. Bell, A. McLean and M. Korchynsky, "Sulfide
   Shape Control in HSLA Steels", Metallurgical Transactions 1,
   12:3341-3350 (1970).
6. W.E. Knapp and W.T. Bolkcom, "Rare Earths Improve Properties
   of Many Ferrous Metals", Iron Age, April 24 and May 1, 1952.
7. J.G. Cannon, "Rare Earths, Monazite Supplies Tighten Following
   Record Year in 1978", E/MJ, March 1979 and "Rare Earths - '76
   was Slow; Pickup seen for '77", E/MJ, March, 1977.
8. "Rare Earths - Industrial Profile and Market Review",
   Industrial Minerals, March, 1979, pp. 21-59.
9. L. Luyckx, "Mechanisms of Rare Earth Action of Steel Structures", ASM Annual Materials Science Symposium, Cincinnati,
   November, 1975.
10. J.R. Jackman and L. Luyckx, "Mischmetal in Steelmaking
    Today", 13th Annual CIMM Conference, Toronto, Canada, August
    25-28, 1974.

11. H. Matsubara, et al. "New Manufacturing Process and Properties of Arctic Grade Line-Pipe", Proceedings International Conference Materials Engineering in the Arctic ASM book, May 1977, pp. 190-200.

12. T. Ohki, et al. "Effect of Inclusions on Sulfide Stress Cracking", Technical Research Center Nippon Kokan Kabushiki Kaisha, January, 1977.

13. L. Luyckx and J.R. Jackman, "Current Trends in the Use of Rare Earths in Steelmaking", Electric Furnace Conference Proceedings, Cincinnati, 1973.

14. C.J. Bingel and L.V. Scott, "Rare Earth Useage in Steel-making", Electric Furnace Conference, December, 1973.

15. J.R. Jackman, Patents 4,022,444 and 4,060,407, "Methods and Apparatus for Adding Mischmetal to Molten Steel", May 10, 1977 and November 29, 1977.

16. J.J. Bosley and J.J. Oravec, "Steel Ladle Practices for Desulfurization and Sulfide Morphology Control at U.S. Steel", NOH and BOS Conference in Chicago, April, 1978.

17. John P. Orton, "The Importance of Low Sulfur on Steel Processing and Metallurgical Properties", Symposium on External Desulfurization of Hot Metal, McMaster University, May, 1975, pp. 11-120.

18. C.M. Moore, "Rare Earths" Bureau of Mines - United States Department of the Interior, May, 1979.

19. I.S. Hirschhorn, "Trends in the Industrial Uses for Mischmetal", Presented at the Rare Earth Conference in Fargo, North Dakota, June, 1979.

THE USE OF RARE EARTHS IN PHOTOVOLTAICS

P. Munz and E. Bucher

Fakultät für Physik, Universität Konstanz, PF 5560

D-7750 Konstanz (FRG)

ABSTRACT

The metal-insulator semiconductor (MIS) junction used as an alternative solar cell is reviewed. The properties of the new solar cell barrier metals Sc, Y, Lu and Yb are discussed and compared with other barrier metals such as Be, Hf, Cr, etc. It is shown that some, in particular Sc and Lu, are favorable candidates because of excellent barrier forming ability as well as high optical transmittance. These metals actually seem to induce a "cold" p-n junction. Open circuit voltages obtained (550mV with 6Ωcm p-Si) reach the theoretical limit given by diffusion theory.

INTRODUCTION

Photovoltaic energy conversion is the only physical process that gives a direct conversion of the radiant energy of light into electricity. Broad terrestrial use of solar cell as an energy source is hampered today by the fact that the production costs of the type of cells produced commercially are too high by a factor of at least twenty, depending on the type of application considered. Many alternative solar cell materials and cell configurations are being investigated today (1). For a working solar cell, a junction on the base material is needed; it induces the internal electric field for the separation of the electron hole pairs generated in the base material by solar irradiation. Since the discovery a few years ago by several workers that the insertion of an extremely thin insulating layer between the semiconductor and the metal of a Schottky contact can greatly improve the photovoltaic properties, this simple type of junction (an alternative to the classical p-n junction) has become

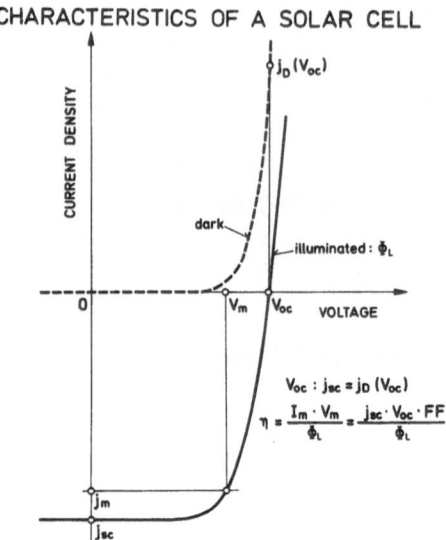

Fig. 1.   Dark and Illuminated I-V curve of a MIS Solar Cell

of considerable interest for photovoltaic energy conversion. High
temperature processes are not necessarily involved in its fabrication.
Ultimately, simple procedures for the deposition of the thin metal
films, such as plating, might possibly be used. It is with this alternative MIS solar cell where rare earth metals appear to be promising materials as a component.

    A p-n solar cell is essentially a current source with a diode in
parallel; this is true also for the MIS solar cell. The typical I-V
curves of such a cell in dark and when illuminated are shown in
Fig. 1. The important physical quantities that characterize the cell
are the open circuit voltage V_{OC}, the short circuit current j_{SC}, the
efficiency η and the fillfactor FF. They are defined by this figure.
The maximum power is delivered by the cell at the voltage V_m which is
somewhat lower than V_{OC}. There are essentially two configurations for
this MIS cell that are investigated today: the grating structure and
the cell with a continuous semitransparent metal layer. They are
sketched in Fig. 2. Outstanding results have been obtained by Green
et al. (2) ($\eta \sim 18\%$) and by others (3,4) with the grating approach
where the discontinuous barrier-forming metal layer is produced by
photolithography. The continuous layer approach appears at least as
attractive owing to its inherent simplicity and it would be desirable
to obtain efficiencies which approach those of the grating configuration even though their maximum efficiencies probably never will be
completely reached. At this time the efficiencies of the continuous
layer approach are considerably lower (5-7,19). The complete structure of a continuous layer cell -- the type that is dealt with in this

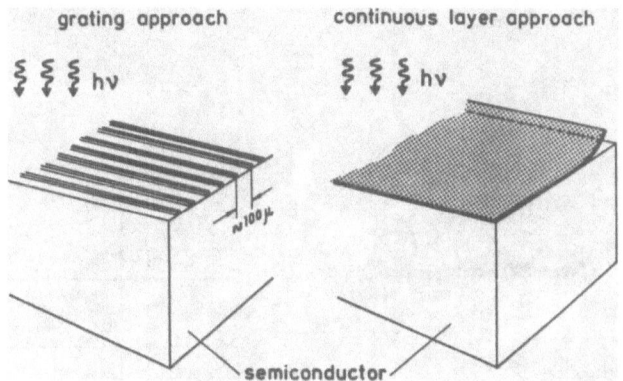

Fig. 2.   The Two Configurations for MIS Solar Cells

paper -- is shown in Fig. 3.   Here only a large spaced current-collecting finger grid as is used in all the other types of solar cells has to be incorporated.   In this sort of junction the properties of the semitransparent barrier layer is of crucial importance and will be the main subject of this paper.   P-type crystalline Si material so far has yielded higher efficiencies than n-type base material and is mainly considered here.   But the favorable properties of the alternative barrier metals described here should also apply to other non-silicon p-type semiconductors.   Some results with amorphous p-silicon will be reported too.

THE CHOICE OF THE BARRIER FORMING METAL

The important requirements for the barrier forming, semi-transparent layers can be described as follows: (1) It should result in a high barrier height and strong inversion of the Si surface. The metal should have a positive influence on the interface state density; (2) It should be highly transparent for a practical thickness; (3) It should have an acceptable low sheet resistance at this thickness; (4) It should lead to stable MIS junction performance. Preferable properties would be: (1) The preparation of the SiO_x/Si surface should not be too critical; (2) A simple layer can be used (no multiple layer required); (3) A large-spaced current collecting grid can be applied.   In a Schottky contact with p-type material only small barrier energies around 0.5eV can be obtained and the simple barrier model $\Phi_{Bp} = \chi + Eg - \Phi_m$ (physical symbols defined in Fig. 4) suggested implicitly earlier by Mott (8) could not be verified, as explained later by Bardeen (9), because of the existence of interface states, as is well known today.

In the MIS structure with an intentionally prepared insulating

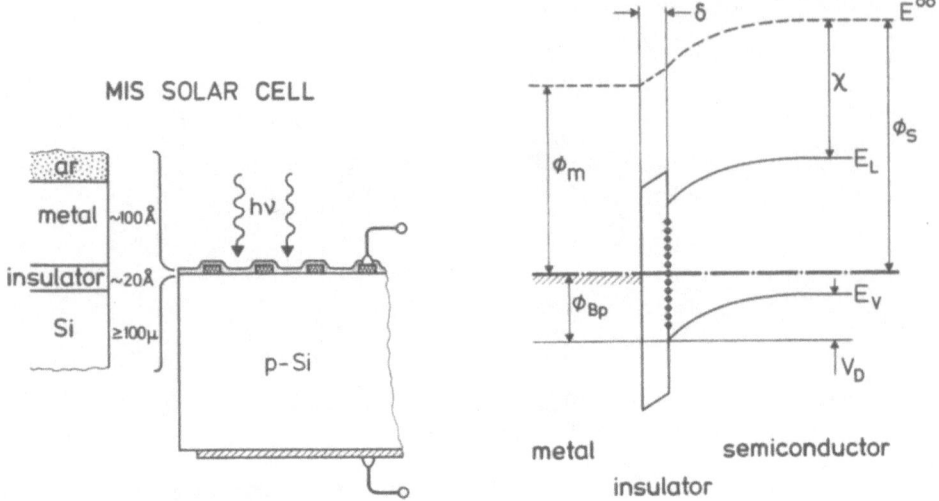

Fig. 3: MIS Cell Cross Section   Fig. 4:   Simple MIS Band Diagram

ultrathin tunnel layer the situation is different (decoupling of
metal and semiconductor) and the particular type of metal may exert
more influence on the barrier formed, e.g. lower work function metals
may prove favorable, as is suggested by the simple band diagram of a
MIS junction (Fig. 4). A crucial condition for a realistic system
will be the stability of the final solar cell. Ponpon and Siffert
(10,11) showed that the degradation of the Schottky or MIS junctions
investigated by them was due to the diffusion of oxygen and water
vapor through the semi-transparent metal layer. They also found a
strong decrease in the diffusion coefficient with increasing heat
of formation of the metal oxides ΔH_O. Therefore reactive barrier
metals might possibly lead to stable MIS junctions. The thin film
properties, the barrier forming abilities and the photovoltaic
performance of MIS junctions formed with Be, Hf, Sc and Y have been
investigated by us in previous papers (12-14). The properties of
the rare earth metals Lu and Yb that have even lower work functions
are presented and compared with the other barrier metals in the
following sections. The preparation of the devices has been performed as much as possible under the same conditions, in an oil free
high vacuum system as described in (12-14).

THIN FILM PROPERTIES OF THE ALTERNATIVE BARRIER METALS

     The necessary information about the properties of films of
these metals prepared under appropriate conditions, measured on
films exposed to atmosphere (an inhomogenious system due to the

Table 1. Optical Properties of Barrier Metal Films

METAL	THICKNESS $\mathring{A}$	AREA MASS DENSITY mg/m^2	DEPOSITION RATE $\mathring{A}/s$	TRANSMITTANCE at 0.62μ	REFLECTANCE
Cr	100	72	1.0	0.22	0.43
Cr/Cu/Cr	40/30/20	65	0.4	0.37	
Al	60	16	0.26	0.38	0.28
Be	80	15	0.25	0.43	0.26
Hf	100	130	1.0	0.46	0.23
Sc	130	39	1.4	0.70	0.14
Y	300	134	1.2	0.50	0.23
Lu	100	98	0.8	0.50	0.21
Yb	130	91	1.2	0.58	0.13

reaction with air), needed for the realization of the MIS junctions could not be found in the literature. Therefore these properties had to be investigated separately (14). The layers were evaporated simultaneously with the preparation of the MIS junctions onto fused quartz substrates in order to check the transmittance and reflectance and the electrical resistance of the films. It should be noticed that the substrate properties of SiO_2 glass and the ultra-thin SiO_x layer on Si need not be fully equivalent.

The relevant spectral range for semi-transparent films used in Si MIS solar cells stretches from about 0.5 μm to 0.85 μm (14). Most metals investigated show a reasonably flat transmittance as a function of wavelength for the film thicknesses required (14). The transmittance and reflectance of semi-transparent films that gave acceptable fill factors (FF) in the corresponding MIS cells are listed in Table 1. The thicknesses indicated are situated in a range where the sheet resistance is still a steeply decreasing function of the thickness (14). The optical transmittance is less critically dependent on thickness as is shown for Yb as an example in Fig. 5. The properties of Lu films were found to be highly dependent on the base pressure in the recipient where the thin film deposition took place, more than with Yb. The values of the area mass densities as given in Table 1 reveal that the consumption of barrier metal as is actually used in the cells is extremely low. It is in the range of 100 mg/m^2 or lower.

Table 1 shows that alternative barrier metals exhibit superior transmittances while retaining an acceptable sheet resistance as compared with the barrier metals commonly used, Cr and Al (5). The best transparency is found with the rare earth metal Yb:0.58 and Sc:0.70. The low Drude absorption in rare earth metals may be one reason for that. This high transmittance should lead to high short circuit current densities in MIS cells.

Sc shows some striking features as described in (14). A
layer of 130 Å leads to a working solar cell with an acceptable
fill factor, whereas such a film deposited on a fused quartz witness
is of very high resistance. One explanation would be that the one
Sc barrier metal leads to such a strong inversion with charge Q_i
as is shown in the band diagram of Fig. 6, that the induced inversion
layer in the silicon surface contributes appreciably to the lateral
conductance in the MIS surface. This effect would be of greatest
interest for the development of MIS cells with better performance,
because it would allow the use of even thinner barrier forming
metal layers with higher transmittances. Experiments with IR
absorption are planned to answer this question.

JUNCTION AND PHOTOVOLTAIC PROPERTIES OF ALTERNATIVE BARRIER METALS

The MIS junctions were prepared with a thermally grown SiO_x
layer (13) of a thickness of about 16Å as measured by a specially
adapted ellipsometric method (15). The details of the method of
preparation are described in (12,13). The dark I-V curves in the
current interval of interest used in obtaining an estimate of the
open circuit voltages V_{oc} are shown in Fig. 7. The nature of the
metals does have a pronounced influence on the magnitude of the
currents. The current is not determined only by the properties of
the substrate as is predicted by the model of a nonequilibrium
minority carrier MIS diode in the semiconductor limited region (16).
A more sophisticated model is necessitated to explain the dependence
of the I-V curves on the type of metal as measured with our diodes.
The I-V curve for Lu is for a relatively thick layer. A thinner Lu
layer is actually used in the MIS junction and gives a slightly
flatter I-V curve.

The maximum open circuit voltage V_0 expected from the measured
dark I-V relation, assuming a short circuit current density
$j_{sc}=25mA/cm^2$ (indicated by a dashed line in the Fig. 7), is plotted
in Fig. 8. Here the usual assumption is made that the illuminated
I-V curve is the dark I-V curve shiftet by j_{sc} (see Fig. 1). This
is only a crude approximation but should give the proper trend. In
Fig. 8 the vacuum work function of the clean metal as given by
Eastman (17) and others is used as an ordering parameter. An
increase of V_0 with lowered work function can be noticed, but this
trend stops at Sc. In order to obtain information about the barriers
formed the barrier energies Φ_{Bp} induced by the different metals
have been measured by several methods (12). Whether these methods
provide proper values of Φ_{Bp} depends on the model applied to
describe the junctions. It is safer to consider them as quasi
barriers. Typical values are indicated in Fig. 8. The Φ_{Bp}'s
follow the trend of V_0 but there is no strict correlation.

The measured V_{oc} values at an irradiation of $92mW/cm^2$ are also

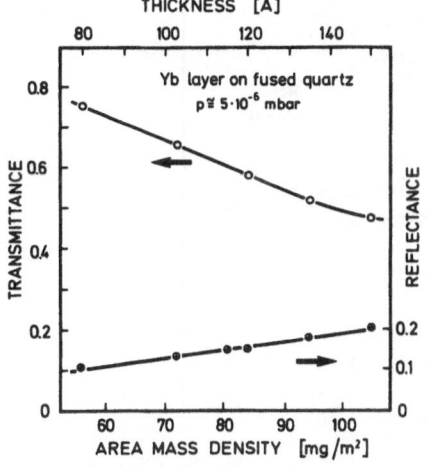

Fig. 5: Optical Properties of
         Yb Films at 0.62 µ.

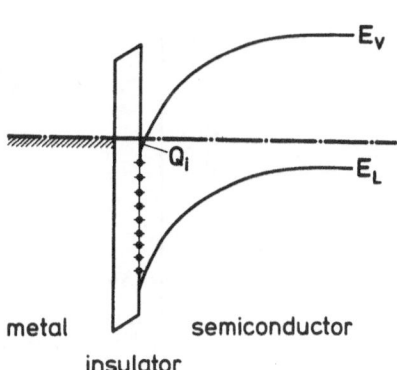

Fig. 6: MIS Junction with
         Inversion Layer.

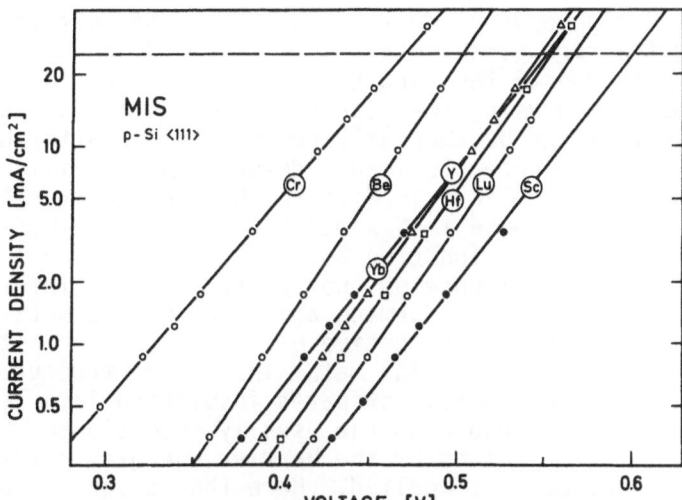

Fig. 7: Dark Current-Voltage Relations. The Current Level of
         25mA/cm^2 is Indicated by a Dashed Line.

indicated in Fig. 8. The highest open circuit voltage is observed
for Sc. It compares favorably with the theoretical maximum value
of V_{OC} for our base material in the simple diffusion model:
$V_{OC} \sim 0.55$ Volt. This corresponds to an ideal p-n junction with
the same base material. Higher open circuit voltages should be
obtained by going from 6Ωcm material as used here to lower resistivity base material. For 0.6Ωcm base material, Sc should yield

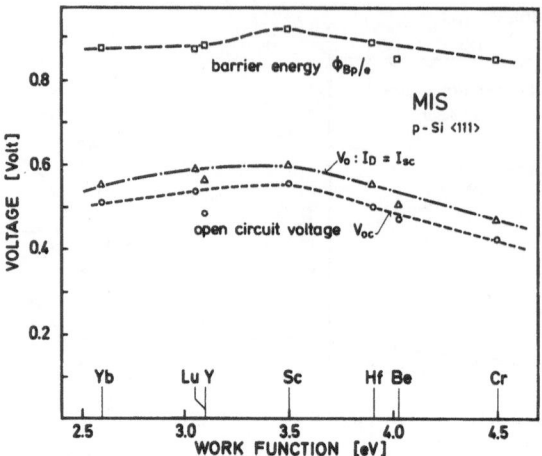

Fig. 8: Barrier energy, expected maximum V_{OC} from dark I-V:V_O
and V_{OC} measured at 92mW/cm^2 as a function of the
vacuum work function

V_{OC}∿620mV. A high open circuit voltage V_{OC}∿0.54 Volt is also observed with the rare earth metal Lu. Y seems to be less favorable,
but this may be due to the fact that this material (as well as Hf)
has to be deposited using the electron beam gun in contrast to the
other metals that can be thermally evaporated. Scattered electrons
of the e-gun are supposed to induce defects in the silicon base
material and the oxide layer. With Yb that has an even lower work
function, only a decrease in V_{OC} was found. This might indicate that
it is not only the work function but also the individual chemical
surface reactions (influence on the spectrum of the interface states)
that determine the barrier forming abilities of the metals. Stuke
et al (18) tested these alternative barrier metals on p-type hydrogenated amorphous silicon. The metal insulator structure as used by
them without an intentionally prepared insulating layer (MS) is shown
in Fig. 9 and is compared with the usually investigated MS structure
with (n)a-Si:H. The values of the barrier energies scatter strongly
for the different methods applied. Here the values Φ_{Bp} as determined
by the dark I-V curve are reproduced in a sequence of increasing
work function of the metal:

Metal	Yb	Sm	Y	Sc	Ag
Φ_{Bp}	0.98	0.98	0.94	0.97	0.81

It can be seen that the alternative barrier metals can induce high
barriers also in amorphous silicon.

     An example of an operational characteristic of a crystalline
Si MIS solar cell with the rare earth barrier metal Yb is shown in

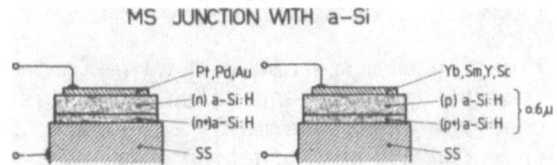

Fig. 9:   Diode structures with hydrogenated amorphous silicon
          on stainless steel

Fig. 10.   Although this cell is not optimized and only a poor anti-
flex coating with SiO_x is used, an efficiency of 9.3% was obtained.
With Sc as a barrier metal efficiencies of about 11% were achieved
with 6Ωcm base material.

      In addition to an acceptable efficiency the stability of the
MIS solar cell will be of crucial importance. There was doubt
about the stability of the open circuit voltage as it is critically
dependent on the junction properties. In Fig. 11 the time
dependence of the normalized open circuit voltage of some of the
earliest MIS junctions prepared are plotted as a function of the
logarithm of time. In some cells with intermediate V_{OC}, this value
increases with time. In the case of Sc it decreases, but when we
extrapolate the curve to a time of twenty years we don't get a
decrease of more than about 10 percent.

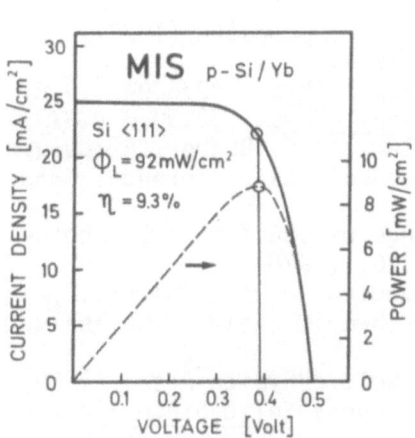

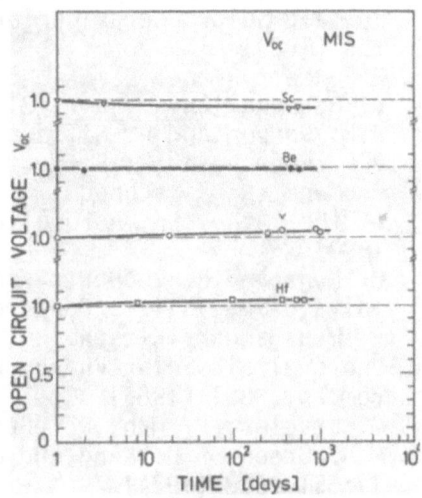

Fig. 10: Illuminated I-V
          characteristics
          of cell with Yb

Fig. 11: Time dependence of
          the normalized open
          circuit voltages of
          MIS cells with Sc,
          Be, Y and Hf

CONCLUSIONS

The investigation of alternative barrier metals: Be, Hf, Sc, Y and the rare earth metals Lu and Yb under realistic conditions (chemically etched surfaces, thermally evaporated barrier metals in an oil free high vacuum) showed that some of these metals are favorable compared with the classical barrier metals Cr and Al. Sc and Lu appear to be superior both in their transparency and in their barrier forming abilities. Open circuit voltages that approximate the theoretical maximum V_{OC} (0.55 Volt with 6Ωcm p-Si) according to the simple diffusion theory can be obtained. The best MIS junctions found appear to be minority carrier dominated devices and can be regarded as metal induced p-n junctions.

REFERENCES

1. Proceedings of the Third E. C. Photovoltaic Solar Energy Conference, D. Reidel Publishing Company, 1980.
2. R.B. Godfrey and M.A. Green, Appl. Phys. Lett., 34:790 (1979).
3. R. Hezel, R. Schörner and T. Meisel, ibid., 1:866 (1980).
4. R.E. Thomas, C.E. Norman, and R.B. North, Proceedings of the Fourteenth IEEE Photovoltaic Specialists Conference, 1350 (1980).
5. D.L. Pulfrey, IEEE Trans. ED-25, 11:1308 (1978).
6. W.A. Anderson, A.E. Delahoye, J.K. Kim, S.H. Hyland, and S.K. Dey, Appl. Phys. Lett., 33:7, 588 (1978).
7. G. Cheek, R. Mertens, Proceedings of the Third E.C. Photovoltaic Solar Energy Conference, D. Reidel Publishing Company, 353 (1980).
8. N.F. Mott, Proc. Cambridge Phil. Soc., 34:568 (1938).
9. J. Bardeen, Phys. Rev., 71:717 (1947).
10. J.P. Ponpon and P. Siffert, J. Appl. Phys., 49:6004 (1978).
11. J.P. Ponpon and P. Siffert, J. Appl. Phys., 50:5050 (1979).
12. P. Munz, K. Kirschbaum, G. Kragler and E. Bucher, Proceedings of the 14th IEEE Photovoltaic Specialists Conference, 1360, 1980).
13. P. Munz and E. Bucher, Proceedings of the 13th IEEE Photovoltaic Specialists Conference, 761 (1978).
14. P. Munz and K. Kirschbaum, Proceedings of the Third E.C. Photovoltaic Solar Energy Conference, D. Reidel Publishing Company, 851 (1980).
15. G. Kragler, P. Munz, E. Bucher, Helv. Physica Acta, (1980).
16. M.A. Green, F.D. King and J. Shewchun, Solid-State Electronics, 17:551, 563 (1974).
17. D.E. Eastman, Phys. Rev., B2:1 (1970).
18. J. Stuke, M. Mell, private communication, Fachbereich Physik Universität Marburg, D-3550 Marburg, FRG.
19. E. Bucher, Appl. Phys., 17:1 (1978).

PREPARING RARE EARTH SILICON IRON ALLOYS

E. Morrice and M. M. Wong

U.S. Department of the Interior, Reno Research Center

Bureau of Mines, Reno, NV 89520

INTRODUCTION

A major use of rare earths produced in the United States is in alloy steels. The rare earths are added as mischmetal or rare earth/iron/silicon alloy for modification of inclusion chemistry and shape (1). Commercial production of rare earth/iron/silicon alloy is by carbothermic reduction of a mixture of rare earth oxides, silica, and iron above 1,930°C (2). Commercial alloys contain from 30 to 50 wt-pct rare earths and rare earth recovery is generally less than 60 pct. To advance mineral technology by maximizing process efficiency, the Bureau of Mines investigated methods to improve preparation of rare earth/iron/silicon alloy (3).

INVESTIGATION OF REDUCTANTS

Tests were made using various reductants to prepare rare earth-silicon and rare earth silicon iron alloys from a rare earth oxide mixture. The mixture (90 wt-pct RE_2O_3) was a commercial product obtained by roasting and leaching bastnasite flotation concentrate. A 90-gram charge of reductant and rare earth oxides, contained in a graphite crucible, was heated in an induction furnace. The rare earth oxides were reduced with calcium silicide, silicon, and ferrosilicon (75 wt-pct Si) according to the following reactions:

$$RE_2O_3 + 3CaSi_2 \rightarrow 2RESi_2 + 2Si + 3CaO, \qquad (1)$$

$$2RE_2O_3 + 11Si \rightarrow 4RESi_2 + 3SiO_2. \qquad (2)$$

When silicon was replaced by ferrosilicon, the reaction proceeded
according to equation (2) and the iron reported to the alloy product.
When 120 pct of the stoichiometric amount of calcium silicide was used
in the reduction, a 50.2 wt-pct RE product was obtained with a 98 pct
rare earth recovery. However, the limited availability and relatively
high cost of calcium silicide would limit its use in an industrial
operation. With silicon and ferrosilicon reductants, rare earth
recoveries were approximately 40 pct and the products contained about
40 wt-pct RE. In an attempt to improve rare earth recovery, aluminum
was added to silicon-mixed rare earth oxides and ferrosilicon-mixed
rare earth oxide charges. The reaction is as follows:

$$3RE_2O_3 + 15Si + 2Al \rightarrow 6RESi_2 + 3SiO_2 + Al_2O_3. \qquad (3)$$

When 660 pct of the stoichiometric amount of aluminum was added
to a silicon-mixed rare earth oxide charge, 76 pct of the rare earths
were recovered in a 47.7 pct RE product. With 660 pct of the stoichiometric amount of aluminum in a ferrosilicon-mixed rare earth
oxide charge, the rare earth recovery was 68 pct in a 36.9 pct RE
product.

Rare earth-silicon-aluminum alloy products disintegrated into
powder after a short-time exposure to air. Rare earth-silicon-iron-
aluminum alloys prepared with ferrosilicon-aluminum reductants remained intact. Rare earth alloys in a powder form are less desirable
as ladle additives. Both alloy products contained approximately equal
amounts of carbon (~ 0.1 wt-pct), so the presence of rare earth carbide is not likely to account for their difference in stability.

FERROSILICON-ALUMINUM REDUCTION

Ferrosilicon-aluminum reduction of rare earth oxide mixtures was
investigated with 600-gram charges. The source of rare earths was a
commercial roasted and leached bastnasite concentrate. The ferrosilicon was a commercial alloy containing approximately 75 wt-pct
silicon and the aluminum was in the form of scrap (99.5 pct Al).
Graphite or densified silicon carbide crucibles were used to contain
the charges.

In the large scale tests, significant amounts of the rare earth
alloy product were entrapped in a viscous slag that made complete
recovery of the product difficult. Various fluxes were added to the
charge in an attempt to minimize this problem by producing a fluid
slag. Fluxes containing CaO, MgO, CaF_2, Al_2O_3, and SiO_2 were cur-
sorily investigated. The amount of RE_2O_3 dissolved in the slags
obtained by using these fluxes decreased with increasing CaO content.
Best results were obtained with a flux composition of 94 wt-pct CaO
and 6 wt-pct MgO. The slag obtained by using this flux had a composition in weight-percent of 52 CaO, 28 Al_2O_3, 8.5 SiO_2, 6.5 RE_2O_3,

Table 1. Results for ferrosilicon-aluminum reduction of mixed rare earth oxides.[1]

Alloy product[2]			Slag[3]	
Element[4]	Wt-pct	Recovery, pct	Oxide[5]	Wt-pct
Rare earth	52.6	80	Re_2O_3	12.7
Silicon	27.6	93	SiO_2	4.7
Iron	10.7	85	FeO	1.2
Aluminum	3.8	16	Al_2O_3	28.9

[1] 1.170-gram charge.
[2] Average weight, 812 grams.
[3] Average weight, 1,100 grams.

[4] Also Ca, 5.0; P, 0.05; S, 0.003; O, 0.28; and C, 0.08.
[5] Also CaO, 49.3; MgO, 2.3; P, 0.15; and S, 0.24.

and 5 MgO. The Al_2O_3 and SiO_2 in the slag were reaction products from aluminum and silicon in the charge.

Tests were made to determine the effect of varying the amount of aluminum in the charge on rare earth recovery and composition of rare earth silicon aluminum iron alloy product. The charges consisted of a 94 wt-pct CaO-6 wt-pct MgO flux, mixed rare earth oxides, 100 pct of the stoichiometric amount of ferrosilicon, and varying amounts of aluminum. The charges were reacted at 1,500°C for 30 minutes. Rare earth recoveries ranged from 36 pct with no aluminum added to 86 pct with 525 pct of the stoichiometric amount of aluminum. Tests were made at holding temperatures of 1,450°, 1,500°, and 1,550°C and at holding durations of 10, 20, and 30 minutes. The charge and flux composition used were those determined best in previous tests. Differences in rare earth recovery and rare earth content of the alloy products were small at temperatures and durations tested.

Larger-scale tests (1,170-gram charges) were also performed in an induction furnace. To minimize reaction between graphite and lime to form calcium carbide at elevated temperature, densitied silicon carbide crucibles were used to contain the melts. Charge composition and operating conditions were those determined to be optimum in the tests using 600-gram charges. The charges consisted of mixed rare earth oxides, 681 grams; ferrosilicon alloy (78 pct Si), 309 grams; aluminum, 182 grams; and a flux containing 681 grams lime and 45 grams magnesia. The charges were held at 1,500°C for 30 minutes. Average results from eight tests are shown in Table 1.

The Bureau of Mines and a major steel company evaluated samples of the alloy products prepared in this investigation (4). The alloy was as effective in improving the properties of steel and cast iron and the rare earth recovery in test heats was as high as those obtained when using commercial rare earth silicon iron alloys.

CONCLUSIONS

Reduction of mixed rare earth oxides by a ferrosilicon-aluminum mixture in combination with a lime-magnesia flux can be utilized to prepare an alloy product with a higher rare earth recovery at a higher rare earth content than the present commercial production method. The alloy product does not disintegrate into powder and is as effective in improving the properties of steel and cast iron as the commercial alloys.

REFERENCES

1. H.W. Bennett and L.F. Sandell, Jr., Rare Earth Additions to Electric Furnace Steels for Sulfide Shape Control, J. Metals, 26:21-24 (1974).
2. J.O. Staggers, Rare Earth Metal Silicide Alloys (assigned to Foote Mineral Co., Exton, PA), U.S. Pat. 4,018,597, Apr. 19, 1977.
3. J.D. Marchant, E. Morrice, B.P. Herve, and M.M. Wong, Preparing Rare Earth-Silicon-Iron-Aluminum Alloys, U.S. Bureau of Mines, RI 8445, 1980.
4. L.A. Neumeier and B.A. Botts, "A New Look at Nodulizing Ductile Iron with Yttrium and Mischmetal Additives," 82d AFS Casting Cong. and Exposition, Detroit, Mich., Apr. 24-28, 1978, Paper 78-60, v. 86, pp. 249-266.

CATALYSIS USING RARE EARTH AND ACTINIDE INTERMETALLICS CONTAINING

Fe, Co, Ni and Cu

W. E. Wallace, J. France and A. Shamsi

Department of Chemistry, University of Pittsburgh

Pittsburgh, PA   15260

I.   INTRODUCTION

Rare earth and actinide elements form numerous intermetallic compounds. The sub-set of substances involving these elements in chemical union with the 3d elements Mn, Fe, Co, Ni and Cu numbers about 300 compounds (1). Many of these compounds absorb hydrogen extensively (2). The absorption is very rapid, dissociative and, in many cases, reversible. In view of the dissociative nature of the absorption, hydrogen must exist on the surface of the metal for at least a fleeting instant as a monatomic species. This suggested that these materials might be effective as hydrogenation catalysts. Numerous studies have been carried out in the last five or six years to test this hypothesis (3-15). These have involved hydrogenation of ethylene, hydrocarbon isomerizations, synthesis of NH_3 from the elements and the production of hydrocarbons and alcohols from synthesis gas. The present account gives a brief summary of the synthesis of NH_3 and syngas conversion studies.

II.   AMMONIA SYNTHESIS

The synthesis of NH_3 from the elements using catalysts formed from intermetallic compounds has been studied by Takeshita, Wallace and Craig (3). The experiments were performed using a stainless steel circulating system. It was operated at 70 atm. and at temperatures up to 600°C. A dry ice-acetone cold trap removed the NH_3 as it was formed. The progress of the reaction was monitored by observing the rate of decline of pressure in the closed system.

561

Thirty-five intermetallic compounds were examined in connection with the NH_3 synthesis. These all involved rare earths or Th in combination with Mn, Fe or Co. $CeRu_2$ and $CeRe_2$ were also examined. Surface areas of the used catalysts were determined by the flow method of Nelsen and Eggertsen (16). These were also examined by x-ray diffraction. The latter measurements revealed that during the reaction the rare earth intermetallic compounds had been transformed into rare earth nitride (RN) and 3d-transition metal. Presumably this is also true of the Th compounds. It therefore appears that the catalytically active material is Fe, Co or Ru supported on rare earth or Th nitride. (The Mn systems and $CeRe_2$ showed little activity.)

The temperature coefficients of the reaction rate were measured and were observed to conform to the Arrhenius expression. Specific reaction rates and the activation energies determined from Arrhenius plots are shown in Table 1. Also included in Table 1 is the sepcific activity of a commercial synthetic NH_3 catalyst, designated 416. Catalyst No. 416 was a doubly promoted Fe system. It is to be noted that decomposition products of many of the intermetallics are more active for NH_3 synthesis than the commercial catalyst, No. 416. The specific activity of Co supported on CeN is approximately 17 times that of No. 416.

Table 1. Rank Ordering for NH_3 Synthesis (at 450°C) and Activation Energies

Precursor Intermetallic	Yield $\left(\dfrac{NH_3 (ml)}{m^2 \text{ of cat. min}} \right)$	E_A (kcal mole^{-1})
$CeCo_3$	17.4	8.7
$CeRu_2$	13.5	10.3
Ce_2Co_7	9	
Ce_2Fe_{17}	6.7	8.0
$PrCo_5$	6	
$CeCo_5$	5.5	9.3
$CeCo_2$	4.3	5.8
$Ce_{24}Co_{11}$	4.3	
$PrCo_3$	3.3	
$ErFe_3$	3.3	12.3
$CeFe_2$	2.3	13.2
416	1.8	20.5
$TbFe_3$	1.7	
$PrCo_2$	1.3	
$ThFe_3$	0.9	
$DyFe_3, HoFe_3$		

III. FORMATION OF METHANE FROM SYNTHESIS GAS

The rationale for these studies is provided in the Introduction.
The first study, carried out by Coon et al. (4), involved the paradigm
hydrogen absorber $LaNi_5$ together with several other intermetallics
with rare earths combined with Fe, Co, Ni and Cu. The case of $LaNi_5$
is instructive and is illustrative of the general behavior of the
entire class of intermetallics. A variety of evidence indicated
that the original $LaNi_5$ was transformed as follows:

$$LaNi_5 \xrightarrow[\text{300--350°C}]{\text{Synthesis Gas}} Ni + La_2O_3. \tag{1}$$

An intimate mixture of Ni and La_2O_3 was formed, indicating <u>oxidation</u>
of $LaNi_5$ by the CO/H_2 mixture. This is rather unusual since syngas
acts normally as a strong reducing agent. The occurrence of reaction (1) was confirmed by x-ray diffraction, magnetic analysis,
SEM-EDAX measurements and Auger spectroscopy. SEM observations
showed nodules of Ni, about 0.5 µM in size, growing out of an La_2O_3
substrate. In the work of Coon et al. the transformation of $LaNi_5$
into Ni/La_2O_3 was partial. In latter studies the conversion was
complete. It has been found that Ni/La_2O_3 formed in this way and
analogous materials formed from other intermetallics are very active
for methanation. This procedure, then, constitutes a novel way to
prepare supported catalysts. In many cases these new catalysts
exhibit exceptional activity. Illustrative results are presented
in the ensuing sections. In all cases the syngas involves H_2 and
CO in a 3/1 ratio.

A. Reactions without a Preoxidizing Step

From the preceding paragraphs it is clear that the rare earth
and actinide intermetallics are such strong reducing agents that
they can be oxidized by syngas alone to produce the new supported
catalysts. The new catalysts can also be produced with other oxi-
dants – O_2, H_2O and NO. Catalysts produced by preoxidation with
O_2 and NO are discussed below. In this section a discussion is
given of the behavior of catalysts formed through contact with syngas alone at elevated temperatures (200–400°C); that is, no preoxidation step is involved.

The early work of Coon et al. (4), Elattar et al. (5) and
Luengo et al. (15) did not take cognizance of the fact that it was
the oxidation products, and not the original intermetallic, which
was responsible for the catalytic activity. Surface areas were
measured only by Ar physisorption. This procedure does not determine the amount of metallic surface, which clearly contains the
site of chemical reactivity. In due course this deficiency was
corrected and in 1979 Elattar, Wallace and Craig (7) published
results on the activity for methanation of Ni/La_2O_3 and Ni/ThO_2

formed by the reaction of syngas with $LaNi_5$ and $ThNi_5$, respectively. Atkinson and Nicks published (14) similar results on a catalyst formed from $MmNi_5$, where Mm represents mischmetal. Results obtained are listed in Table 2 along with corresponding quantities obtained using Ni/SiO_2 in this laboratory and by Vannice (17). The silica-supported catalysts were formed by conventional wet chemistry techniques. The new catalysts are seen to be much more active on a per site basis than those prepared by the conventional wet chemistry method.

To establish that the differences between Ni/SiO_2 and Ni/ThO_2 or Ni/La_2O_3 were not an artifact, several Ni-Si and Co-Si inter-metallic compounds were employed by Imamura and Wallace (19,20) to produce Ni/SiO_2 and Co/SiO_2 by the new technique. Results obtained are given in Table 3. These data make it clear that it is the pre-paration method rather than the nature of the substrate which leads to the high reactivity per site.

The results shown in Tables 2 and 3 indicate that supported nickel catalysts made from the intermetallic compounds are signifi-cantly more active on a per site basis than those formed by conven-tional wet chemical techniques.

B. Reactions Using Catalysts Formed by O_2 Oxidation

Imamura and Wallace observed that intermetallic compounds could be preoxidized with O_2 to form very active methanation catalysts. The first experiments of this nature were performed using a number of intermetallics containing Fe, Co and Ni. Results obtained with Ni_5Si_2, Ni_2Si and Co_2Si are shown in parentheses in column 2 of Table 3. Very active catalysts were produced with turnover numbers exceeding that of a conventional Ni/SiO_2 catalyst by factors of up to 20.

Imamura and Wallace (12) studied a number of Ni-containing intermetallic compounds for use in syngas conversion – RNi_5 with R = La, Ce, Pr, Nd, Ho and Er and ThNi, $ThNi_2$, $ThNi_5$ and Th_7Ni_3. These were oxidized by exposure to O_2 (initially 610 torr) at 350°C in a closed system. The uptake of O_2 was established by following the pressure drop.

Table 2. Turnover Frequencies (N) Measured at 250°C

	$10^3N\ S^{-1}$	Reference
Ni/SiO_2	1.1	18
Ni/SiO_2	0.5 - 1	17
Ni/MmO_x (from $MmNi_5$)	3	14
Ni/La_2O_3 (from $LaNi_5$)	2.7	7
Ni/ThO_2 (from $ThNi_5$)	4.7	7

Table 3. Turnover Frequencies (N) for Ni/SiO$_2$ and Co/SiO$_2$
          Catalysts at 205°C

	10^3N S^{-1}
Ni/SiO$_2$ from Ni$_5$Si$_2$	11[a] (15)[b]
Ni/SiO$_2$ from Ni$_2$Si	3.5 (4.5)
Co/SiO$_2$ from Co$_2$Si	1.0 (1.5)

a. These are turnover numbers of catalysts pro-
    duced by oxidation with syngas.
b. The parenethetical quantities are the turnover
    numbers for catalysts produced by preoxidation
    with O$_2$.

Values obtained for these oxidized materials are given in
Table 4. Again, one notes that the turnover numbers are very high
compared to the results obtained for Ni/SiO$_2$ prepared by conventional wet chemical techniques.

C. Reaction Using Catalysts Formed by NO Oxidation

The intermetallic compounds containing Ni are readily oxidized
by NO at about 350°C to form oxide-supported Ni catalysts. These
also exhibit high activity for methane formation from synthesis gas.
Turnover frequencies for mixtures formed from Ce-Ni intermetallics
are found to be 3 to 6 x 10^{-3}s^{-1} at 205°C, again an order of magnitude larger than that of conventionally formed Ni/SiO$_2$.

IV. FORMATION OF METHANOL FROM SYNTHESIS GAS

The intermetallic compounds represented by the formula RCu$_2$,
with R = La, Ce, Pr, Ho and Th, react with synthesis gas at elevated
temperatures to produce intimate mixtures of the oxides of Cu and R.
These mixtures were found to be quite active in the catalytic formation of CH$_3$OH from synthesis gas. Other oxygenated compounds are
also formed.

The experimental procedure was essentially the same as that
used for NH$_3$ synthesis. The synthesis gas (H$_2$/CO = 2:1) was passed
over the intermetallic compound at about 50 atm. and 300°C. A circulating system was employed. Liquid products were trapped out.
The extent of reaction was established by the pressure drop, which
was monitored by a pressure transducer. Gaseous products were
identified by GC. Results obtained are given in Tables 5 and 6.
A conventional CuO-ZnO catalyst was prepared by normal wet chemi-

Table 4. Activity of Various Catalysts

| | Oxidant: O_2 | | |
Intermetallic Precursor[a]	CO Conversion at 205°C (%)	Activity at 205°C (ml/g.sec)	Turnover Frequency x 10^3 at 205°C (sec^{-1})
ThNi	4.0	4.9×10^{-3}	2.7
ThNi$_2$	3.2	4.0×10^{-3}	2.9
ThNi$_5$[b]	1.5	1.9×10^{-3}	8.7
ThNi$_5$	9.0	1.1×10^{-2}	10.6
ThNi$_5$[c]	7.0(at 190°C)	3.5×10^{-2}	5.6
Th$_7$Ni$_3$	2.6	3.2×10^{-3}	2.8
LaNi$_5$	1.2	1.5×10^{-3}[d]	48[d]
CeNi$_5$	2.0	2.5×10^{-3}[d]	18[d]
" (C-2)[e]	1.0	1.2×10^{-3}	
" (C-3)[e]	2.6	3.7×10^{-3}	1.7
" (C-4)[e]	5.0	6.1×10^{-3}	3.8
PrNi$_5$	1.8	2.2×10^{-3}[d]	26[d]
NdNi$_5$	0.5	6.1×10^{-4}[d]	19[d]
HoNi$_5$	1.6	2.0×10^{-3}[d]	31[d]
ErNi$_5$	4.2	5.2×10^{-3}[d]	59[d]
25%-Ni/ThO$_2$[f]	1.7(at 510°C)		
3.9%-Ni/ThO$_2$[f]	2.0(at 490°C)	2.9×10^{-3}(at 490°C)	

a. In each case, except as noted, the intermetallic compound was treated with sufficient O_2 to produce Ni plus ThO$_2$, CeO$_2$ or R$_2$O$_3$, where R = La, Pr, Nd, Ho or Er.
b. This material was not oxidized completely to Ni or ThO$_2$.
c. This was oxidized to ThO$_2$ + NiO. The turnover frequency listed was obtained by extrapolation to 205°C using the Arrhenius plots.
d. These values correspond to results obtained at 275°C.
e. These catalysts consisted of CeO$_2$ + Ni + NiO.
f. The catalysts were prepared by conventional impregnation techniques as described in the text.

cal techniques (21). Results obtained using this material as the catalyst are also included in Tables 5 and 6.

X-ray diffraction studies of the used catalysts revealed decomposition of the initial intermetallic into R$_2$O$_3$ (or CeO$_2$ or ThO$_2$) and CuO and Cu. From line broadening the particle sizes were estimated as 200-300 Å.

Table 5. Synthesis Gas Conversion over Oxidized RCu_2 Systems

Precursor Intermetallic	% Conversion per pass	% Conversion/m^2 per pass	% H_2O^a	% CH_3OH^a	% C_2,C_3,C_4 Alcohols[a]
$ThCu_2$	0.31	0.022^b	5	94	1
$LaCu_2$	.37	.033	54	25	19
$CeCu_2$	.20	.014	49	37	14
$PrCu_2$	.16	.014	44	37	18
$HoCu_2$	.21	.015	40	50	10
CuO–ZnO	.27	.014	12	87	1

a. These (weight) percentages denote the composition of the liquid phase formed.
b. These are based on the total surface area, measured by Ar adsorption.

Table 6. Gaseous Species Formed in Synthesis Gas Conversion over Oxidized RCu_2 Systems

Precursor Intermetallic	CO	CH_4	CO_2	C_2H_6	$CH_3CH_2CH_3$	CH_3OCH_3
$ThCu_2$	93	2	3	--	---	1
$LaCu_2$	90	6	2	2	---	---
$CeCu_2$	81	11	4	2	2	1
$PrCu_2$	83	10	3	2	2	---
$HoCu_2$	84	10	5	2	---	---
CuO–ZnO	84	2	9	1	1	2

V. CONCLUSION

Rare earth and actinide intermetallics containing Fe, Co and Ni readily oxidize to form mixtures of rare earth or actinide oxides and Fe, Co or Ni. This constitutes a new way to produce supported catalysts. Oxidation can be accomplished by O_2, NO or synthesis gas. These are active for CH_4 formation from synthesis gas. Fe and Co systems are also good synthetic ammonia catalysts. Intermetallics containing Cu catalyze the formation of CH_3OH and other oxygenated species from synthesis gas.

REFERENCES

1. W. E. Wallace, Rare Earth Intermetallics, Academic Press, Inc., New York, chapters 9, 10 and 11 (1973).
2. W. E. Wallace, R. S. Craig and V. U. S. Rao, Advances in Chemistry Series, No. 186:207 (1980).
3. T. Takeshita, W. E. Wallace and R. S. Craig, J. Catal., 44:236 (1976).
4. V. T. Coon, T. Takeshita, W. E. Wallace and R. S. Craig, J. Phys. Chem., 80:1787 (1976).
5. A. Elattar, T. Takeshita, W. E. Wallace and R. S. Craig, Science, 196:1093 (1977).
6. W. E. Wallace, in Hydrides for Energy Storage, A. F. Andresen and A. J. Maeland (eds.), Pergamon Press, Inc., p. 33 (1978).
7. A. Elattar, W. E. Wallace and R. S. Craig, Advances in Chemistry Series, No. 178, E. L. Kugler and F. W. Steffgen (eds.), p. 7 (1979).
8. W. E. Wallace, in The Rare Earths in Modern Science and Technology, Vol. 2, G. J. McCarthy, J. J. Rhyne and H. B. Silber (eds.), Plenum Press, New York, p. 1 (1980).
9. A. A. Elattar and W. E. Wallace, ibid., p. 533 (1980).
10. H. Imamura and W. E. Wallace, J. Catal., 64:238 (1980).
11. H. Imamura, K. Soga, M. Sato and W. E. Wallace, Chem. Lett., 957 (1980).
12. H. Imamura and W. E. Wallace, J. Catal., 65:127 (1980).
13. H. Imamura and W. E. Wallace, J. Phys. Chem., 84:3145 (1980).
14. Gary B. Atkinson and Larry J. Nicks, J. Catal., 46:417 (1977).
15. C. A. Luengo, A. L. Cabrera, H. B. McKay and M. B. Maple, ibid., 47:1 (1977).
16. F. M. Nelsen and F. T. Eggertsen, Anal. Chem., 30:1387 (1958).
17. M. A. Vannice, J. Catal., 44:152 (1976).
18. T. Takeshita and W. E. Wallace, unpublished measurements.
19. H. Imamura and W. E. Wallace, J. Phys. Chem., 83:2009 (1979).
20. H. Imamura and W. E. Wallace, ibid., 83:3261 (1979).
21. R. G. Herman, S. Mehta, G. W. Simmons and K. Klier, J. Catal., 57:339 (1979).

ELECTROCHEMICAL CORROSION OF LANTHANUM CHROMITE AND YTTRIUM CHROMITE IN COAL SLAG

D.D. Marchant and J.L. Bates

Pacific Northwest Laboratory

Richland, WA 99352

INTRODUCTION

Lanthanum (La) chromites have long been considered as electrodes for magnetohydrodynamic (MHD) generator channels (1,2,3,4). These chromites when doped with divalent ions such as Ca, Mg or Sr have adequate electronic and electrical conductivity and have melting points >2500 K (2). In use above ~1850 K, selective vapor loss of chromium (Cr) results in the formation of a hygroscopic La_2O_3 phase which results in a large volume change and loss of mechanical integrity (2).

The analogous yttrium (Y) chromites have thermal and electrical properties similar to those of La chromites (5). Although vapor loss of Cr results in the formation of Y_2O_3, this oxide does not hydrate (5). Corrosion studies (5,6) of Y chromite compositions show that doped $YCrO_3$ may be a viable MHD electrode. This paper describes an electrochemical corrosion study of Mg-doped La and Y chromites in electrolytes of synthetic coal slag. The paper emphasizes possible chemical and electrochemical degradation phenomena, as well as relative rates of corrosion.

EXPERIMENTAL PROCEDURES

The Y and La chromites were nominally doped with 5 mol% MgO to increase the electrical conductivity. The chromites were fabricated into bars by sintering* at high temperature in oxygen partial

---

*$YCrO_3$ from TransTech Inc., Gaithersburg, MD. $LaCrO_3$ from Westinghouse Research Center, Pittsburgh, PA.

pressures ranging from 10^{-10} to 10^{-14} atmospheres. The sintered La chromites were ~91% TD with irregular shaped pores (~3 μm) located at the grain boundaries. The Y chromites were ~95% TD with more uniformly shaped pores (~1 μm) at the grain boundaries. The La chromites contained a uniformly dispersed second phase, of Mg-Cr oxide containing nearly equal amounts of Mg and Cr. A similar phase in the Y chromite was magnesium oxide containing Cr. The matrix grains in both chromites had less than 0.1 wt% Mg.

Two slag compositions were used as electrolytes: (1) Montana Rosebud (MR-1) slag containing (in wt%) 42.4 SiO_2, 18.8 Al_2O_3, 13.4 K_2O, 12.9 CaO, 6.9 Fe_2O_3, 4.1 MgO, 0.7 TiO_2, 0.4 Na_2O, and 0.2 P_2O_5; and (2) Illinois No. 6 (Ill-6-1) slag containing 39.5 SiO_2, 24.3 Fe_2O_3, 18.4 Al_2O_3, 11.7 K_2O, 5.1 CaO, 1.6 MgO, 1.0 P_2O_5, 0.8 TiO_2, and 0.5 Na_2O. Both contained high amounts of potassium (K) to better represent the slags expected in a K-seeded MHD generator.

The electrochemical corrosion tests consisted of partially immersing a chromite anode and cathode into molten coal slag, Fig. 1, and passing a direct electric current between these electrodes through the coal slag. A constant current was maintained by varying the electric potential. An alumina sleeve surrounded each electrode to channel the electric current through the electrode end. A platinum probe was positioned equidistant from the anode and the cathode to measure the electric potentials of the electrodes. Each test was continued for a predetermined time, or until the system resistance prevented the maintenance of a constant current. The chemical corrosion tests were similar to the electrochemical tests, except electric current was not present.

Corrosion rates were determined from geometric measurements of polished cross sections. Loss from the interior of the sample, e.g., grain boundaries, was not included in the corrosion rate. A scanning electron microscope equipped with energy-dispersive x-ray analysis (SEM-EDX) was used to identify the microstructural phases and their distribution. The SEM-EDX was quantified for elements above an atomic number of 10. The relative accuracy is >10% and the detection limit is ±0.1 wt%. The quantitative data in this paper is expressed as the atomic ratio of the major elemental components of each phase.

RESULTS AND DISCUSSION

Overall Corrosion

Results of the electrochemical and chemical corrosion tests are listed in Table 1. The electrochemical corrosion rates are also expressed as g/cm^2-h so they can be compared directly with the chemical corrosion rates of the control sample. The electrochemical corrosion rates of both the anodes and cathodes in all but one test were at

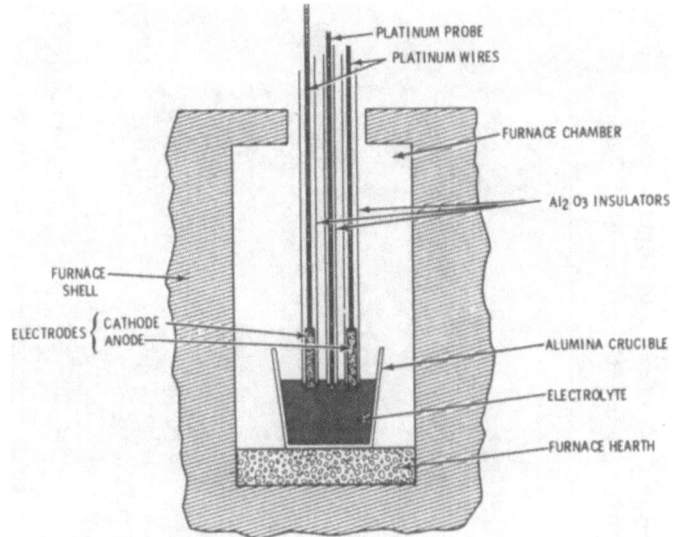

Figure 1. Electrochemical test configuration.

least an order of magnitude greater than the chemical corrosion rates.
The electrochemical corrosion rates in MR-1 slag were greater than in
Ill-6-1 slag. Generally, the cathode corrosion rate was greater than
the anode corrosion rate.

The La chromites had a greater electrochemical and chemical corrosion rate than the Y chromites. This greater corrosion rate is consistent with other experimental data (5). The La chromite degraded
even when stored in air dried by a silica gel dessicant. The degradation results partially from the hydration of La and K oxides (K was
found along grain boundaries) and was greatest for the cathodes. The
Y chromites exhibited no post-test degradation even when exposed to
ambient air for several weeks except in areas where the K was present.

Several processes may occur during the passage of direct electrical current to cause the increased electrochemical corrosion over
chemical corrosion. At the anode/slag interface, gas bubbles containing O_2 are formed. The O_2 may result from the anodic oxidation of
the silicate ions (7). Generation (cavitation) and movement (sweeping
away corrosion products and bringing fresh slag to the reaction interface) of the bubbles can lead to increased corrosion/erosion. More
bubbles were generated in the MR-1 slag than in the Ill-6-1 slag.
The bubbles on the anode surfaces were electrically insulating, causing higher localized current densities, higher voltages and possibly
higher corrosion rates in adjacent surfaces free of bubbles. The formation and migration of bubbles created electric potential instabilities at the anode. The instabilities in the MR-1 slag were as high

Table 1. Summary of chromite electrochemical tests.

	T(K)	Time ks(h)		Coulombs	Anode* µg/coul	Cathode* µg/coul	Anode* g/cm²-h	Cathode* g/cm²-h	Control g/cm²-h
Montana Rosebud (MR-1)									
$YCrO_3$	1733	96.1	(26.7)	8891	13	44	0.04	0.2	--
	1723	86.4	(24.0)	0	0	0	0	0	0.001-0.003
$LaCrO_3$	1723	49.5	(13.8)	3536	131	163	0.6	0.5	--
	1723	86.4	(24.0)	0	0	0	0	0	0.006-0.008
Illinois No. 6 (Ill-6-1)									
$YCrO_3$	1729	173.2	(48.1)	15999	11	1	0.04	0.004	--
	1723	86.4	(24.0)	--	--	--	--	--	0.001-0.003
$LaCrO_3$	1725	199.8	(55.5)	14185	37	42-120	0.1	0.1-0.3	--
	1723	32.4	(9.0)	0	0	0	0	0	0.003-0.005

*Current density was ~1 amp/cm²

as ±50% of the average electric potential, whereas the instabilities
in the Ill-6-1 slag were <±10%. The instabilities were not found at
the cathode. The higher iron concentration in the Ill-6-1 slag resulted in a higher electronic conductivity of the slag reducing the
number of bubbles generated.

Depletion of Ca, K, and Fe cations occurred in the slag adjacent
to the anode surface and concentration of the cations occurred at the
cathode. The higher electronic conducting slag exhibited less cation
depletion and concentration.

In all tests, the average electric potential between the anode
and cathode increased with time and was greatest at the anode and in
MR-1 slag. The formation of reaction products and the depletion of
cations may account for most of the electric potential increase.

Chemical Corrosion

Chemical corrosion in the Ill-6-1 slag occurred primarily at the
immediate chromite/slag interface. An oxide reaction layer containing
Fe, Cr, Al and Mg (atom ratios ~50:17:20:13) was found at this interface. The slag at the interface contained small amounts of La or Y.
The grain boundaries at the reaction interface contained slag and La
or Y. Slag grain boundary penetration and reaction were greater in
the La chromite and contributed to the higher corrosion rate. A slag-
soluble La silicate was formed along the grain boundaries. In Y
chromite, the Y silicate did not form along the grain boundaries in
the interior of the sample, although the boundaries at the immediate
chromite/slag boundary contained up to ~50 at.% Y. Apparently, the Y
was dissolved in the slag and the Cr reacted to form the reaction
product.

Chemical corrosion in the MR-1 slag also occurred at the chro-
mite/slag interface with the formation of an oxide reaction product
containing Fe, Cr, Al and Mg (atom ratios ~12:36:18:34). The lower
Fe content in this reaction product was probably due to the lower Fe
content in the MR-1 slag. Slag penetration along the grain boundaries was significantly greater in the La chromite. Lanthanum silicate
was found along the grain boundaries. In both chromites, neither Ca
or K concentrated along the grain boundaries, in the reaction product
or in the slag. Yttrium silicate was only found along the grain
boundaries near the bulk slag interface.

Electrochemical Corrosion

In the electrochemical corrosion tests, corrosion of the anodes
differed from that of the cathodes and will be discussed separately.

Cathodes. Degradation of the LaCrO$_3$ cathodes was significantly more severe in both coal slags than the YCrO$_3$ cathodes. Although no surface reaction layer was formed, the reaction products at the grain boundaries consisted of La-Al oxides and K-Al oxide phases in the low iron MR-1 slag and Fe, Cr, Al oxides containing some La in the high iron Ill-6-1 slag. No La silicates were formed with the MR-1 slag but an La-Ca-Al oxide was formed at the grain boundaries in the sample interior. In Ill-6-1 slags, La silicates rich in Al, Ca and K were found along the grain boundaries deep inside the sample but not at the slag interface. Yttrium chromite cathodes in the MR-1 slag had a diffuse reaction interface with considerable slag penetration along the grain boundaries. No identifiable reaction product was found. Yttrium silicate was found in the interior of the sample along the grain boundaries.

In the Ill-6-1 slag, the YCrO$_3$ cathodes had a distinct reaction interface. Significantly less grain boundary penetration and interaction with slag occurred. The main degradation appears to be the dissolution of the grains. Two reaction products were formed at the reaction interface: (1) a Y silicate which dissolved in the slag and (2) an oxide containing Fe, Cr, and Al (atom ratios ~29:44:27) which coated the cathode.

Anodes. The LaCrO$_3$ anode formed a distinct reaction layer in the MR-1 slag composed of a Cr-Al oxide (atom ratio ~78:22) and an La-Al silicate (atom ratios ~37:14:49). The slag had penetrated the grain boundaries throughout the sample with the La silicate forming in the slag. A crystalline Cr-Mg oxide (atom ratio ~58:42) formed next to the La chromite grains. The reaction of the LaCrO$_3$ with the Ill-6-1 slag was also extensive and similar to that in the MR-1 slag. La-Al silicate (atom ratios ~16:52:32) and Cr-Al oxide (atom ratio ~64:36) were present. In addition, an Fe-Al oxide (atom ratio ~19:81) phase was present.

The YCrO$_3$ anode in Ill-6-1 slag reacted much less than the La chromite and formed a diffuse reaction layer. Slag had penetrated the grain boundaries and formed a Y silicate and a granular Fe-Cr-Al oxide (atom ratios ~22:64:14). The Y silicate dissolved in the slag leaving the oxide (atom ratios ~33:41:26) on the slag/anode interface. The YCrO$_3$ anode in MR-1 slag formed a distinct reaction layer at the anode surface. This layer was a Cr-Al-Fe-oxide (atom ratios ~69:25:6). The Y dissolved in the slag. Grain boundary penetration and reaction with the slag was greater in the MR-1 slag than in the Ill-6-1 slag.

CONCLUSIONS

· Laboratory tests indicated that Y chromites could be better material for MHD electrodes than La chromites.

- Yttrium chromite with 5 mol% MgO exhibits a greater resistance
  to chemical and electrochemical corrosion than the analogous
  La chromite in molten coal slags.

- Yttrium chromite is less susceptible to hydration than La chromite
  when a rare earth oxide phase is present.

- The chemical and electrochemical corrosion of the chromites is
  less in the high iron Ill-6-1 slag than in the MR-1 slag. The
  reduced corrosion is partially attributed to the higher electrical conductivity and higher electronic transference of the high
  iron slag.

- The electrochemical corrosion rates were 10 to 1000 times greater
  than chemical corrosion rates.

- In general, Y or La in the chromite reacts with slag forming a
  (La or Y) silicate which remains dissolved in the slag. The Cr
  and Mg in the chromite then react with the Al, Fe and Mg in the
  slag, forming a crystalline (Fe, Cr, Al, Mg) oxide with varying
  cation ratios depending on location in the sample and whether
  the sample is anode, cathode or control.

- The reactions of the control are more similar to those at the
  anode than with those at the cathode. This is attributed to the
  oxidizing conditions present at the anode.

ACKNOWLEDGMENT

     Work supported by the U.S. Department of Energy under Contract
No. DE-ACO6-76RLO #1830.

REFERENCES

1. J.B. Heywood and G.J. Womack, Open-Cycle MHD Power Generation,
    Pergamon Press, New York, pp. 568-570, 609-610, 613 (1969).
2. S.J. Schneider, H.P.R. Frederikse, G.P. Telegin and A.I. Romanov,
    "Materials," in Open-Cycle Magnetohydrodynamic Electrical Power
    Generation, M. Petrick and B. Ya. Shumayatsky (eds.), Argonne
    National Laboratory, Argonne, IL, pp. 586-589 (1978).
3. Westinghouse Electric Corporation, "Final Report on the Joint
    U.S.-U.S.S.R. Test of the U.S. MHD Electrode System in the U-O$_2$
    Facility - Phase III," FE-2248-23. Westinghouse Electric,
    Pittsburgh, PA (1919).
4. J. Dong-laing, M. Zhi-giang, J. Wen-hao, W. Da-qian and
    P. Zhen-su, "A Composite Electrode Material Study and its Per-
    formance in a MHD Test Unit," in Seventh International Conference

on MHD Electrical Power Generation, Massachusetts Institute of Technology, Cambridge, MA, June 16-20, pp. 292-299 (1980).

5. D.D. Marchant and J.L. Bates, "Development of Electrodes Based on Yttrium Chromites and Rare Earth Doped Hafnia for MHD Generator Applications," 18th Symposium Engineering Aspects of Magnetohydrodynamics, June 18-20, Butte, MT, pp. D-1.5.1-D-1.5.8 (1979).

6. A. Drozniak and A. Kozlik-kutak, "Yttrium Chromite YCrO₃ as a Material for Electrodes of the MHD Generator," Ibid., pp. 300-305.

7. M.T. Simnod, G. Derge and I. George, "Ionic Nature of Liquid Ion - Silicate Slag," J. of Metals, Dec., pp. 1386-1390 (1954).

AUTHOR INDEX